DATA COMMUNICATIONS NETWORKING DEVICES

SECOND EDITION

DATA COMMUNICATIONS NETWORKING DEVICES

SECOND EDITION

Gilbert Held

4-Degree Consulting
Macon, Georgia
USA

JOHN WILEY & SONS
Chichester · New York · Brisbane · Toronto · Singapore

Library of Congress Cataloging-in-Publication Data

Held, Gilbert, 1943–
Data communications networking devices/Gilbert Held. — 2nd ed.
p. cm.
Bibliography: p.
Includes index.
ISBN 0 471 91869 5 (pbk.)
1. Computer networks. 2. Data transmission systems. I. Title.
TK5105.5.H44 1988
004.6—dc19 88-20702
CIP

British Library Cataloguing in Publication Data

Held, Gilbert, 1943–
Data communications networking devices.—2nd ed.
1. Computer systems. Data transmission equipment
I. Title
621.398'1

ISBN 0 471 91869 5

Typeset by Photo-Graphics, Honiton, Devon
Printed and bound in Great Britain by the Anchor Press, Tiptree, Essex

To Beverly, Jonathan and Jessica
for their patience understanding and support —
I love you all

To Dr Alexander Ioffe and family of Moscow —
congratulations on next year in Jerusalem being this year!

CONTENTS

PREFACE

Data Communications Networking Devices are the building blocks upon which networks are constructed. With this in mind, I have written this book to provide the reader with an intimate awareness of the numerous devices which can be employed in the design, modification or optimization of a data communications network.

The evolution of this book originated from the responses I received to a series of articles I wrote for *Data Communications* magazine. These articles covered the operational characteristics and utilization of a series of specialized communications devices that are usually omitted from most books covering the data communications field. As a result of the publication of these articles, persons responsible for data communications at industrial firms and government agencies, both in the United States and abroad, contacted me for additional information on specific devices as well as requesting additional information on other devices which were briefly mentioned. Using the previous articles as a foundation resulted in the development of my first book, which was published under the title *Data Communications Network Components*. This book was designed to provide the reader with an insight into how communication devices operate, where they can be employed in networks, and the cost and performance parameters which should be considered in selecting such equipment.

Based upon the evolution in data communications technology, the original edition of this book was revised in 1986. In fact, to recognize the emphasis of the book upon the networking of communications devices the title was changed to more accurately reflect the contents and goal of the book. Published under the title *Data Communications Networking Devices*, the revision was adopted by over thirty colleges and universities in North America and Europe as well as by training administrators at many commercial organizations. Due to the continuing evolution in communications technology, a new edition was required to provide readers with up-to-date examples of networking technology and is published here under the title "Data Communications Networking Devices", second edition.

Although many years have passed since my first communications book was published, my goal in providing the reader with a detailed understanding of the operation and utilization of communications devices remains unchanged.

Due to the adoption of the original book by numerous colleges and universities as well as my experience over the last decade in an academic environment, I have structured this book to facilitate its utilization in a one-semester course at a high level undergraduate or first-year graduate course level.

As I rewrote the book, I focused particular attention upon explaining communications concepts and have included a comprehensive introductory section which covers the fundamental concepts common to all phases of data communications. This section should be read first by those new to this field and can be used as a review mechanism for those readers with a background in communications concepts. Thereafter, each section is written to cover a group of components based upon a common function.

Through the use of numerous illustrations and schematic diagrams, I believe the reader will easily be able to see how different devices can be integrated into networks, and some examples should stimulate new ideas for even the most experienced person. At the end of each section I have included a comprehensive series of questions that cover many of the important concepts covered in the section. These questions can be used by the reader as a review mechanism prior to going forward in the book.

For those readers actually involved in the sizing of network devices I have included several appendices in this book that cover this area. Since the mathematics involved in the sizing process can result in a considerable effort to obtain the required data, I have also enclosed two computer programs that readers can use to generate a series of sizing tables. Then, after reading the appendices and executing the computer programs, one can reduce many sizing problems to a table lookup procedure.

GILBERT HELD
Macon, Georgia

ACKNOWLEDGEMENTS

The preparation of a manuscript that gives birth to a book requires the cooperation and assistance of many persons.

First and foremost, I must thank my family for enduring those long nights and missing weekends while I drafted and redrafted the manuscript to correspond to each of the editions of this book. The preparation of the first edition was truly a family affair, since both my wife and son typed significant portions of the manuscript on our mobile Macintosh, with both my family and the Macintosh traveling a considerable distance during the preparation of the manuscript. For the preparation of this edition I am indebted to the fine work of Mrs Carol Ferrell. In addition, I would also like to thank Auerbach Publishers Inc. for permitting me to use portions of articles I previously wrote for their *Data Communications Management* publication. Excerpts from these articles were used for developing the section covering Integrated Services Digital Network (ISDN) presented in Chapter 1, expanding the statistical and T1 multiplexing in Chapter 3 and voice digitization, data compression and fiber optic transmission systems presented in Chapter 6.

Last but not least, one's publishing editor, editorial supervisor and copy editor are the critical link in converting the author's manuscript into the book you are now reading. To Ian Shelley, who enthusiastically backed the first edition of this book, I would like to take the opportunity to thank you for your efforts. To Ian McIntosh who provided me with the opportunity to produce this second edition, I would like to acknowledge your efforts in a multinational way. Cheers!

CHAPTER 1

FUNDAMENTAL CONCEPTS

To transmit information between two locations it is necessary to have a transmitter, a receiver, and a transmission medium which provides a path or link between the transmitter and the receiver. In addition to transmitting signals, a transmitter must be capable of translating information from a form created by humans or machines into a signal suitable for transmission over the transmission medium. The transmission medium provides a path to convey the information to the receiver without introducing a prohibitive amount of signal distortion that could change the meaning of the transmitted signal. The receiver then converts the signal from its transmitted form into a form intelligible to humans or machines.

While the transmission of data may appear to be a simple process, many factors govern the success or failure of a communications session. In addition, the exponential increase in the utilization of personal computers and a corresponding increase in communications between personal computers and other personal computers and large-scale computers has enlarged the number of hardware and software parameters one must consider. Although frequently we will interchangeably use the terms "terminal" and "personal computer" and refer to them collectively as "terminals" in this book, in certain instances we will focus our attention upon personal computers in order to denote certain hardware and software characteristics unique to such devices. In these instances we will use the term "personal computer" to explicitly reference this terminal device.

From and including the transmitter, to and including the receiver, a variety of data communications networking devices can be utilized to perform specialized functions, reduce network costs, increase network reliability, and provide additional levels of transmission redundancy. The utilization of these devices is a function of the characteristics which govern computer-to-computer and computer-to-terminal connections. These characteristics will be examined in this chapter.

1.1 LINE CONNECTIONS

Three basic types of line connections are available to connect terminal devices to computers or to other terminals: dedicated, switched, and leased lines.

Dedicated line

A dedicated line is similar to a leased line in that the terminal is always connected to the device on the distant end, transmission always occurs on the same path, and, if required, the line can be easily tuned to increase transmission performance.

The key difference between a dedicated and a leased line is that a dedicated line refers to a transmission medium internal to a user's facility, where the customer has the right of way for cable laying, whereas a leased line provides an interconnection between separate facilities. The term facility is usually employed to denote a building, office, or industrial plant. Dedicated lines are also referenced as direct connect lines and normally link a terminal or business machine on a direct path through the facility to another terminal or computer located at that facility. The dedicated line can be a wire conductor installed by the employees of a company or by the computer manufacturer's personnel, or it can be a local line installed by the telephone company.

Normally, the only cost associated with a dedicated line in addition to its installation cost is the cost of the cable required to connect the devices that are to communicate with one another.

Leased line

A leased line is commonly called a private line and is obtained from a communications company to provide a transmission medium between two facilities which could be in separate buildings in one city or in distant cities. In addition to a one-time installation charge, the communications carrier will normally bill the user on a monthly basis for the leased line, with the cost of the line usually based upon the distance between the locations to be connected.

Switched line

A switched line, often referred to as a dial-up line, permits contact with all parties having access to the public switched telephone network (PSTN). If the operator of a terminal device wants access to a computer, he or she dials the telephone number of a telephone which is connected to the computer. In using switched or dial-up transmission, telephone company switching centers establish a connection between the dialing party and the dialed party. After the connection is set up, the terminal and the computer conduct their communications. When communications are completed, the switching centers disconnect the path that was established for the connection and restore all paths used so they become available for other connections. The cost of a call on the PSTN is based upon many factors which include the time of day when the call was made, the distance between called and calling parties, the duration of the call and whether or not operator assistance was required in placing the call. Direct dial calls made from a residence or business telephone without operator assistance are billed at a lower rate than calls requiring operator assistance. In addition, most telephone companies have three categories of rates: "weekday", "evening" and "night and weekend". Calls made between 8 a.m. and 5 p.m. Monday through Friday are normally billed at a "weekday" rate, while calls between 5 p.m. and 11 p.m. on weekdays are usually billed at an "evening" rate, which reflects a discount of approximately 25% over the "weekday" rate. The last rate category, "night and weekend", is applicable to calls made between 11 p.m. and 8 a.m. on weekdays as well as anytime on weekends and holidays. Calls during this rate period are usually discounted 50% from the "Weekday" rate.

Table 1.1 contains a sample PSTN rate table which is included for illustrative purposes but which should not be used by readers for determining the actual cost of a PSTN call. This is due to the fact that the cost of intrastate calls by state and interstate calls vary. In addition, the cost of using different communications carriers to place a call between similar locations will typically vary from vendor to vendor and readers should obtain a current State schedule from the vendor they plan to use in order to determine or project the cost of using PSTN facilities.

Factors to consider

Cost, speed of transmission, and degradation of transmission are the primary factors used in the selection process between leased and switched lines.

As an example of the economics associated with comparing the cost of PSTN and leased line usage, assume a personal computer located 50 miles from a mainframe has a requirement to communicate between 8 a.m. and 5 p.m. with the mainframe once each business day for a period of 30 minutes. Using the data in Table 1.1, each call would cost $0.31 \times 1 + 0.19 \times 29$ or \$5.82. Assuming there are 22 working days each month, the monthly PSTN cost for communications between the PC and the mainframe would be \$5.82 × 22 or \$128.04. If the monthly cost of a leased line between the two locations was \$250, it is obviously less expensive to use the PSTN for communications.

Table 1.1 Sample PSTN rate table (cost per minute in cents).

	Rate category					
	Weekend		Evening		Night and weekend	
Mileage between locations	First minute	Each additional minute	First minute	Each additional minute	First minute	Each additional minute
1–100	0.31	0.19	0.23	0.15	0.15	0.10
101–200	0.35	0.23	0.26	0.18	0.17	0.12
201–400	0.48	0.30	0.36	0.23	0.24	0.15

Suppose the communications application lengthened in duration to 2 hours per day. Then, from Table 1.1, the cost per call would become $0.31 \times 1 + 0.19 \times 119$ or $22.92. Again assuming 22 working days per month, the monthly PSTN charge would increase to $504.24, making the leased line more economical.

Thus, if data communications requirements to a computer involve occasional random contact from a number of terminals at different locations, and each call is of short duration, dial-up service is normally employed. If a large amount of transmission occurs between a computer and a few terminals, leased lines are usually installed between the terminal and the computer.

Since a leased line is fixed as to its routing, it can be conditioned to reduce errors in transmission as well as permit ease in determining the location of error conditions since its routing is known. Normally, switched circuits are used for transmission at speeds up to 9600 bits per second (bps); however, in certain situations data rates as high as 19,200 bps are achievable when transmission on the PSTN occurs through telephone company offices equipped with modern electronic switches.

Some of the limiting factors involved in determining the type of line to use for transmission between terminal devices and computers are listed in Table 1.2.

Table 1.2 Line selection guide.

Line type	Distance between transmission points	Speed of transmission	Use for transmission
Dedicated (direct connect)	Local	Limited by conductor	Short or long duration
Switched (dial-up)	Limited by telephone access availability	Normally less than 9600 bps	Short-duration transmission
Leased (private)	Limited by telephone company availability	Limited by type of facility	Long duration or short duration calls

1.2 TYPES OF SERVICE AND TRANSMISSION DEVICES

Digital devices which include terminals, mainframe computers, and personal computers transmit data as unipolar digital signals, as indicated in Figure 1.1(a). When the distance between a terminal device and a computer is relatively short, the transmission of digital information between the two devices may be obtained by cabling the devices together. As the distance between the two devices increases, the pulses of the digital signals become distorted because of the resistance, inductance, and capacitance of the cable used as a transmission medium. At a certain distance between the two devices the pulses of the digital data will distort, such that they are unrecognizable by the receiver, as illustrated

Figure 1.1(a) Digital signaling. Digital devices to include terminals and computers transmit data as unipolar digital signals.

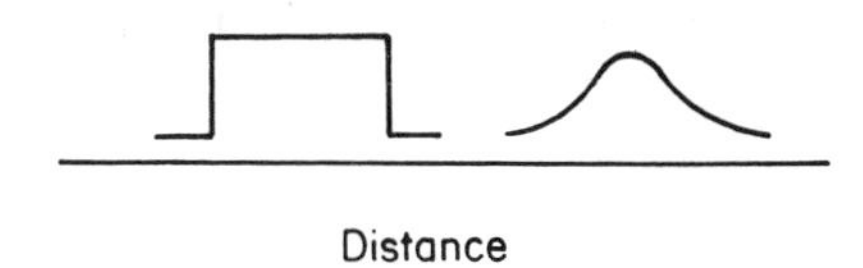

Figure 1.1(b) Digital signal distortion. As the distance between the transmitter and receiver increases digital signals become distorted because of the resistance, inductance, and capacitance of the cable used as a transmission medium.

in Figure 1.1(b). To extend the transmission distance between devices, specialized equipment must be employed, with the type of equipment used dependent upon the type of transmission medium employed.

Digital repeaters

Basically, one can transmit data in a digital or analog form. To transmit data long distances in digital form requires repeaters to be placed on the line at selected intervals to reconstruct the digital signals. The repeater is a device that essentially scans the line looking for the occurrence of a pulse and then regenerates the pulse into its original form. Thus, another name for the repeater is a data regenerator. As illustrated in Figure 1.2, a repeater extends the communications distance between terminal devices to include personal computers and mainframe computers.

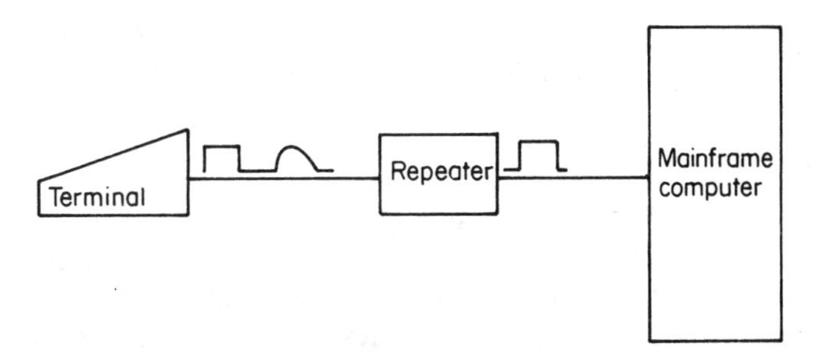

Figure 1.2 Transmitting data in digital format. To transmit data long distances in digital format requires repeaters to be placed on the line to reconstruct the digital signals.

Unipolar and bipolar signaling

Since unipolar signaling results in a dc voltage buildup when transmitting over long distance, digital networks require unipolar signals to be converted into a modified bipolar format for transmission on this type of network. This requires the installation at each end of the circuit of a device known as a digital service unit (DSU) in the United States and a network terminating unit (NTU) in the United Kingdom. The utilization of DSUs for transmission of data on a digital network is illustrated in Figure 1.3. Later in this chapter we will examine digital facilities in more detail.

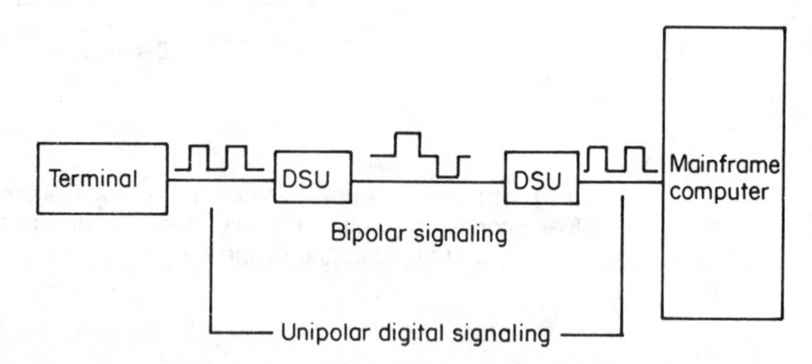

Figure 1.3 Transmitting data on a digital network. To transmit data on a digital network, the unipolar digital signals of terminal devices and computers must be converted into a bipolar signal.

Modems

Since telephone lines were originally designed to carry analog or voice signals, the digital signals transmitted from a terminal to another digital device must be converted into a signal that is acceptable for transmission by the telephone line. To effect transmission between distant points, a data set or modem is used. A modem is a contraction of the compound term modulator–demodulator and is an electronic device used to convert the digital signals generated by computers and terminal devices into analog tones for transmission over telephone network analog facilities. At the receiving end, a similar device accepts the transmitted tones, reconverts them to digital signals, and delivers these signals to the connected device.

Signal conversion

Signal conversion performed by modems is illustrated in Figure 1.4. This illustration shows the interrelationship of terminals, mainframe computers, and transmission lines when an analog transmission service is used. Both leased lines and switched lines employ analog service; therefore, modems can be used for transmission of data over both types of analog line connections. Although an analog transmission medium used to provide a transmission path between

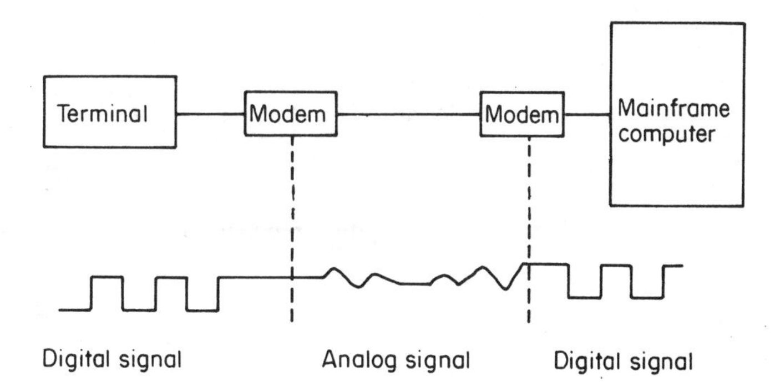

Figure 1.4 Signal conversion performed by modems. A modem converts (modulates) the digital signal produced by a terminal into an analog tone for transmission over an analog facility.

modems can be a direct connect, a leased, or a switched line, modems are connected (hard-wired) to direct connect and leased lines, whereas they are interfaced to a switched facility. Thus, a terminal user can only communicate with the one distant location on a leased line, but he or she can communicate with many devices when there is access to a switched line.

Acoustic couplers

Although popular with data terminal users in the 1970s, today only a small percentage of persons use acoustic couplers for communications. The acoustic coupler is a modem whose connection to the telephone line is obtained by acoustically coupling the telephone headset to the coupler. The primary advantage of the acoustic coupler was the fact that it required no hard-wired connection to the switched telephone network, enabling terminals and personal computers to be portable with respect to their data transmission capability. Owing to the growth in modular telephone jacks, modems that interface the switched telephone network via a plug, in effect, are portable devices. Since many hotels and older office buildings still have hard-wired telephones, the acoustic coupler permits terminal and personal computer users to communicate regardless of the method used to connect a telephone set to the telephone network.

Signal conversion

The acoustic coupler converts the signals generated by a terminal device into a series of audible tones, which are then passed to the mouthpiece or transmitter of the telephone and in turn onto the switched telephone network. Information transmitted from the device at the other end of the data link is converted into audible tones at the earpiece of the telephone connected to the terminal's acoustic coupler. The coupler then converts those tones into the appropriate electrical signals recognized by the attached terminal. The interrelationship of

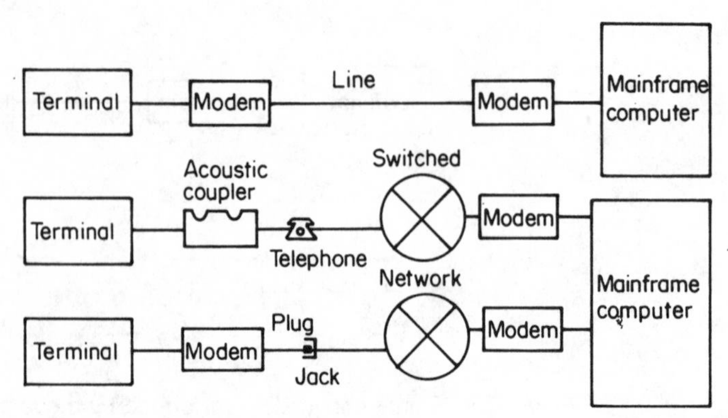

Figure 1.5 Interrelationship of terminals, modems, acoustic couplers, computers and analog transmission mediums. When using modems on an analog transmission medium, the line can be dedicated, leased, or switched facility. Terminal devices can use modems or acoustic couplers to transmit via the switched network.

terminals, acoustic couplers, modems and analog transmission media is illustrated in Figure 1.5.

In examining Figure 1.5, the reader will note that a circle subdivided into four equal parts by two intersecting lines is used as the symbol to denote the public switched telephone network or PSTN. This symbol will be used in the remainder of the book to illustrate communications occurring over this type of line connection.

Analog facilities

Several types of analog switched facilities are offered by communications carriers. Each type of facility has its own set of characteristics and rate structure. Normally, for extensive communications requirements, an analytic study is conducted to determine which type or types of service should be utilized to provide an optimum cost-effective service for the user. The common types of analog switched facilities are direct distance dialing, wide area telephone service, and foreign exchange service.

DDD

Direct distance dialing (DDD) permits the user to dial directly any telephone connected to the public switched telephone network. The dialed telephone may be connected to another terminal device or mainframe computer. The charge for this service, in addition to installation costs, may be a fixed monthly fee if no long-distance calls are made, a message unit rate based upon the number and duration of local calls, or a fixed fee plus any long-distance charges incurred. Depending upon the time of day a long-distance call is initiated and its destination (intrastate or interstate), discounts from normal long-distance tolls are available for selected calls made without operator assistance.

WATS

Introduced by AT&T for interstate use in 1961, wide area telephone service (WATS) is now offered by several communications carriers. Its scope of coverage has been extended from the continental United States to Hawaii, Alaska, Puerto Rico, the US Virgin Islands, and Europe, as well as selected Pacific and Asian countries.

Wide area telephone service (WATS) may be obtained in two different forms, each designed for a particular type of communications requirement. Outward WATS is used when a specific location requires placing a large number of outgoing calls to geographically distributed locations. Inward WATS service provides the reverse capability, permitting a number of geographically distributed locations to communicate with a common facility. Calls on WATS are initiated in the same manner as a call placed on the public switched telephone network. However, instead of being charged on an individual call basis, the user of WATS facilities pays a flat rate per block of communications hours per month occurring during weekday, evening, and night and weekend time periods.

A voice-band trunk called an access line is provided to the WATS users. This line links the facility to a telephone company central office. Other than cost considerations and certain geographical calling restrictions which are a function of the service area of the WATS line, the user may place as many calls as desired on this trunk if the service is outward WATS or receive as many calls as desired if the service is inward. Inward WATS, the well-known "800" area code, permits remotely located personnel to call your facility toll-free from the service area provided by the particular inward WATS-type of service selected. The charge for WATS is a function of the service area. This can be intrastate WATS, a group of states bordering the user's state where the user's main facility is located, a grouping of distant states, or International WATS which extends inbound 800 service to the United States from selected overseas locations. Another service very similar to WATS is AT&T's 800 READYLINE℠ service. This service is essentially similar to WATS, however, calls can originate or be directed to an existing telephone in place of the access line required for WATS service.

Figure 1.6 illustrates the AT&T WATS service area one for the state of Georgia. If this service area is selected and a user in Georgia requires inward WATS service, he or she will pay for toll-free calls originating in the states surrounding Georgia—Florida, Alabama, Mississippi, Tennessee, Kentucky, South Carolina, and North Carolina. Similarly, if outward WATS service is selected for service area one, a person in Georgia connected to the WATS access line will be able to dial all telephones in the states previously mentioned. The states comprising a service area vary based upon the state in which the WATS access line is installed. Thus, the states in service area one when an access line is in New York would obviously differ from the states in a WATS service area one when the access line is in Georgia. Fortunately, AT&T publishes a comprehensive book which includes 50 maps of the United States, illustrating the composition of the service areas for each state. Similarly, a time-of-day rate schedule for each state based upon state service areas is also published by AT&T.

Figure 1.6 AT&T WATS service area one for an access line located in Georgia.

In general, since WATS is a service based upon volume usage its cost per hour is less than the cost associated with the use of the PSTN for long-distance calls. Thus, one common application for the use of WATS facilities is to install one or more inward WATS access lines at a data processing center. Then, terminal and personal computer users distributed over a wide geographical area can use the inward WATS facilities to access the computers at the data processing center.

Since International 800 service enables employees and customers of US companies to call them toll-free from foreign locations, this service may experience a considerable amount of data communications usage. This usage can be expected to include applications requiring access to such databases as hotel and travel reservation information as well as order entry and catalog sales data updating. Persons traveling overseas with portable personal computers as well as office personnel using terminals and personal computers in foreign countries who desire access to computational facilities and information utilities located in the United States represent common International 800 service users. Due to the business advantages of WATS its concept has been implemented in several foreign countries, with inward WATS in the United Kingdom marketed under the term Freephone.

FX

Foreign exchange (FX) service may provide a method of transmission from a group of terminal devices remotely located from a central computer facility at less than the cost of direct distance dialing. An FX line can be viewed as a mixed analog switched and leased line. To use an FX line, a user dials a local number which is answered if the FX line is not in use. From the FX, the information is transmitted via a dedicated voice line to a permanent connection in the switching office of a communications carrier near the facility with which communication is desired. A line from the local switching office which terminates at the user's home office is included in the basic FX service. This is illustrated in Figure 1.7.

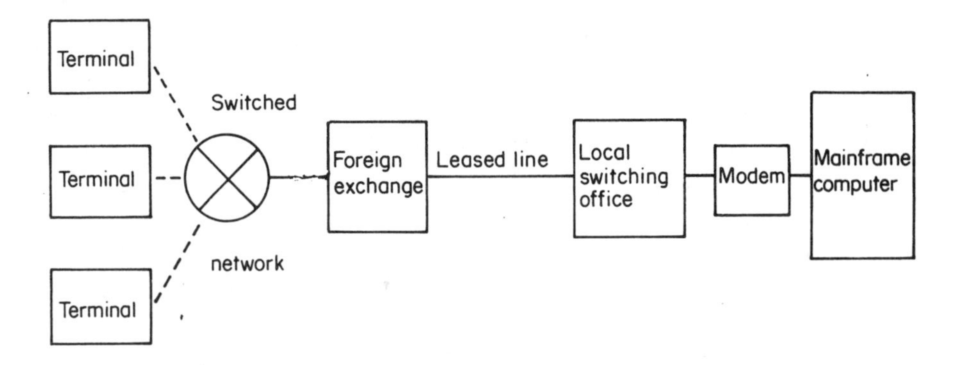

Figure 1.7 Foreign exchange (FX) service. A foreign exchange line permits many terminal devices to use the facility on a scheduled or on a contention basis.

The use of an FX line permits the elimination of long-distance charges that would be incurred by users directly dialing the distant computer facility. Since only one person at a time may use the FX line, normally only groups of users whose usage can be scheduled are suitable for FX utilization. Figure 1.8 illustrates the possible connections between remotely located terminal devices and a central computer where transmission occurs over an analog facility.

The major difference between an FX line and a leased line is that any terminal dialing the FX line provides the second modem required for the transmission of data over the line; whereas a leased line used for data transmission normally has a fixed modem attached at both ends of the circuit.

Digital facilities

In addition to analog service, numerous digital service offerings have been implemented by communications carriers over the last decade. Using a digital service, data is transmitted from source to destination in digital form without the necessity of converting the signal into an analog form for transmission over analog facilities as is the case when modems or acoustic couplers are interfaced to analog facilities.

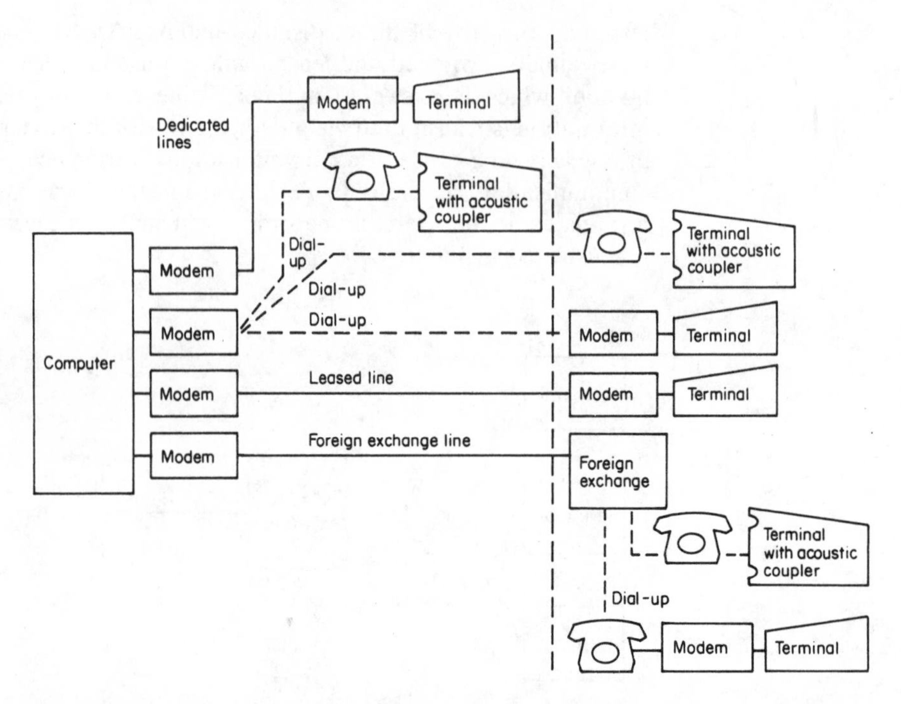

Figure 1.8 Terminal-to-computer connections via analog mediums. A mixture of dedicated, dialup, leased and foreign exchange lines can be employed to connect local and remote terminals to a central computer facility.

AT&T offerings

In the United States, AT&T offers several digital transmission facilities under the ACCUNET℠ Digital Service service mark. Dataphone® Digital Service was the charter member of the ACCUNET family and is deployed in over 100 major metropolitan cities in the United States as well as having an interconnection to Canada's digital network. Dataphone Digital Service operates at synchronous data transfer rates of 2.4, 4.8, 9.6, and 56 kilobits-per-second (kbps), providing users of this service with dedicated, two-way simultaneous transmission capability. At the time this book was being prepared for publication AT&T and several regional communications carriers in the United States were considering providing 19.2 kbps DDS transmission facilities. This data rate would be marketed as an intermediate offering between existing 9.6 kbps and 56 kbps DDS services.

Originally all AT&T digital offerings were leased line services where a digital leased line is similar to a leased analog line in that it is dedicated for full time use to one customer. In the late 1980s, AT&T introduced its Accunet Switched 56 service, a dial-up 56 kbps digital data transmission service. This service enables users to maintain a dial-up backup for previously installed 56 kbps AT&T Dataphone Digital Services leased lines. In addition, this service can be used to supplement existing services during peak transmission periods or for applications that only require a minimal amount of transmission time per day since the service is billed on a per minute basis.

Another offering from AT&T, ACCUNET T1.5 Service is a high capacity 1.544-megabit-per-second (Mbps) terrestrial digital service which permits 24 voice-grade channels or a mixture of voice and data to be transmitted in digital form. This service is only obtainable as a leased line and is more commonly known as a T1 channel or circuit.

European offerings

In Europe, a number of countries have established digital transmission facilities. One example of such offerings is British Telecom's KiloStream service. KiloStream provides synchronous data transmission at 2.4, 4.8, 9.6, 48, and 64 kbps and is very similar to AT&T's Dataphone Digital Service. Each KiloStream circuit is terminated by British Telecom with a network terminating unit (NTU), which is the digital equivalent of the modem required on an analog circuit. In comparison, Dataphone Digital Service users can terminate their digital facilities with either a digital service unit or a channel service unit.

DSUs

A digital service unit (DSU) provides a standard interface to a digital transmission service and handles such functions as signal translation, regeneration, reformatting, and timing. The DSU is designed to operate at one of four speeds: 2.4, 4.8, 9.6, and 56 kbps. The transmitting portion of the DSU processes the customer's signal into bipolar pulses suitable for transmission over the digital facility. The receiving portion of the DSU is used both to extract timing information and to regenerate mark and space information from the received bipolar signal. The second interface arrangement for AT&T's Dataphone Digital Service is called a channel service unit (CSU) and is provided by the communication carrier to those customers who wish to perform the signal processing to and from the bipolar line, as well as to retime and regenerate the incoming line signals through the utilization of their own equipment.

As data is transmitted over digital facilities, the signal is regenerated by the communications carrier numerous times prior to its arrival at its destination. In general, digital service gives data communications users improved performance and reliability when compared to analog service, owing to the nature of digital transmission and the design of digital networks. This improved performance and reliability is due to the fact that digital signals are regenerated whereas, when analog signals are amplified, any distortion to the analog signal is also being amplified.

Although digital service is offered in many locations, for those locations outside the serving area of a digital facility the user will have to employ analog devices as an extension in order to interface to the digital facility. The utilization of digital service via an analog extension is illustrated in Figure 1.9. As depicted in Figure 1.9, if the closest city to the terminal located in city 2 that offers digital service is city 1, then to use digital service to communicate with the

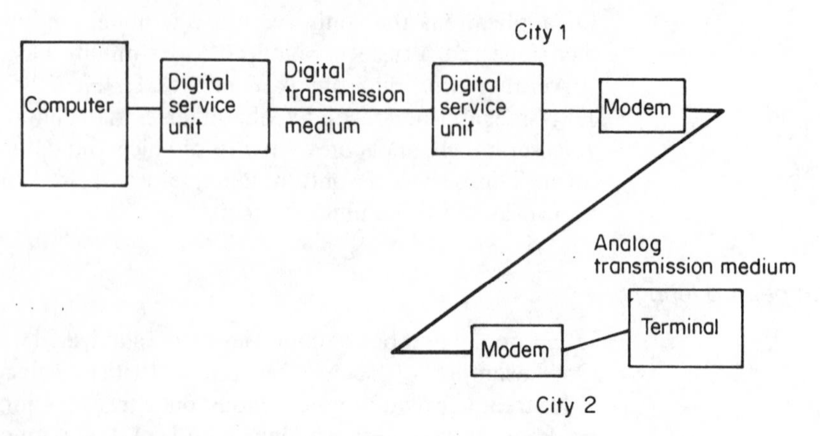

Figure 1.9 Analog extension to digital service. Although data is transmitted in digital form from the computer to city 1, it must be modulated by the modem at that location for transmission over the analog extension.

computer an analog extension must be installed between the terminal location in city 2 and city 1. In such cases, the performance, reliability, and possible cost advantages of using digital service may be completely dissipated.

1.3 TRANSMISSION MODE

One method of characterizing lines, terminal devices, mainframe computers, and modems is by their transmission or communications mode. The three classes of transmission modes are simplex, half-duplex, and full-duplex.

Simplex transmission

Simplex transmission is that transmission which occurs in one direction only, disallowing the receiver of information a means of responding to the transmission. A home AM radio which receives a signal transmitted from a radio station is an example of a simplex communications mode. In a data transmission environment, simplex transmission might be used to turn on or off specific devices at a certain time of the day or when a certain event occurs. -An example of this would be a computer-controlled environmental system where a furnace is turned on or off depending upon the thermostat setting and the current temperature in various parts of a building. Normally, simplex transmission is not utilized where human-machine interaction is required, owing to the inability to turn the transmitter around so that the receiver can reply to the originator.

Half-duplex transmission

Half-duplex transmission permits transmission in either direction; however, transmission can occur in only one direction at a time. Half-duplex transmission is used in citizen band (CB) radio transmission where the operator can either transmit or receive but cannot perform both functions at the same time on the same channel. When the operator has completed a transmission, the other party must be advised that he or she is through transmitting and is ready to receive by saying the term "over". Then the other operator can begin transmission.

When data is transmitted over the telephone network, the transmitter and the receiver of the modem or acoustic coupler must be appropriately turned on and off as the direction of the transmission varies. Both simplex and half-duplex transmission require two wires to complete an electrical circuit. The top of Figure 1.10 illustrates a half-duplex modem interconnection while the lower portion of that illustration shows a typical sequence of events in the terminal's sign-on process to access a computer. In the sign-on process, the user first transmits the word NEWUSER to inform the computer that a new user wishes a connection to the computer. The computer responds by asking for the user's password, which is then furnished by the user.

In the top portion of Figure 1.10, when data is transmitted from a mainframe computer to a terminal, control signals are sent from the mainframe computer to modem A which turns on the modem A transmitter and causes the modem B receiver to respond. When data is transmitted from the terminal to the mainframe computer, the modem B receiver is disabled and its transmitter is turned on while the modem A transmitter is disabled and its receiver becomes active. The time necessary to effect these changes is called a transmission turnaround time, and during this interval transmission is temporarily halted. Half-duplex transmission can occur on either a 2-wire or 4-wire circuit. The

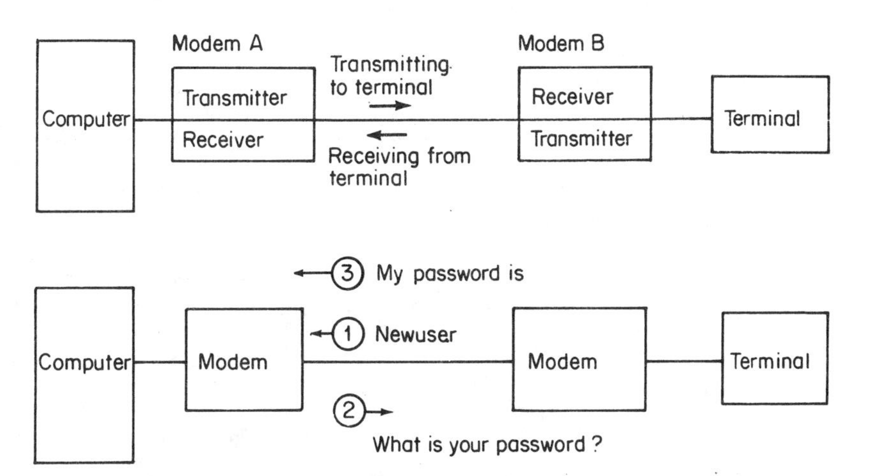

Figure 1.10 Half-duplex transmission. Top: control signals from the mainframe computer and terminal operate the transmitter and receiver sections of the attached modems. When the transmitter of modem A is operating, the receiver of modem B operates; when the transmitter of modem B operates, the receiver of modem A operates. However, only one transmitter operates at any one time in the half-duplex mode of transmission. Bottom: during the sign-on sequence, transmission is turned around several times.

switched network is a 2-wire circuit, whereas leased lines can be obtained as either 2-wire or 4-wire links. A 4-wire circuit is essentially a pair of 2-wire links which can be used for transmission in both directions simultaneously. This type of transmission is called full-duplex.

Full-duplex transmission

Although one would normally expect full-duplex transmission to be accomplished over a 4-wire connection that provides two 2-wire paths, full-duplex transmission can also occur on a 2-wire connection. This is accomplished by the use of modems that subdivide the frequency bandwidth of the 2-wire connection into two distinct channels, permitting simultaneous data flow in both directions on a 2-wire circuit. This technique will be examined and explained in more detail in Chapter 2, when the operating characteristics of modems are examined in detail.

Full-duplex transmission is often used when large amounts of alternate traffic must be transmitted and received within a fixed time period. If two channels were used in our CB example, one for transmission and another for reception, two simultaneous transmissions could be effected. While full-duplex transmission provides more efficient throughput, this efficiency may be negated by the cost of 2-way lines and more complex equipment required by this mode of transmission. In Figure 1.11, the three types of transmission modes are illustrated, while Table 1.3 summarizes the three transmission modes previously discussed.

Readers should note that the column CCITT in Table 1.3 refers to Consultative Committee on International Telephone and Telegraph. The

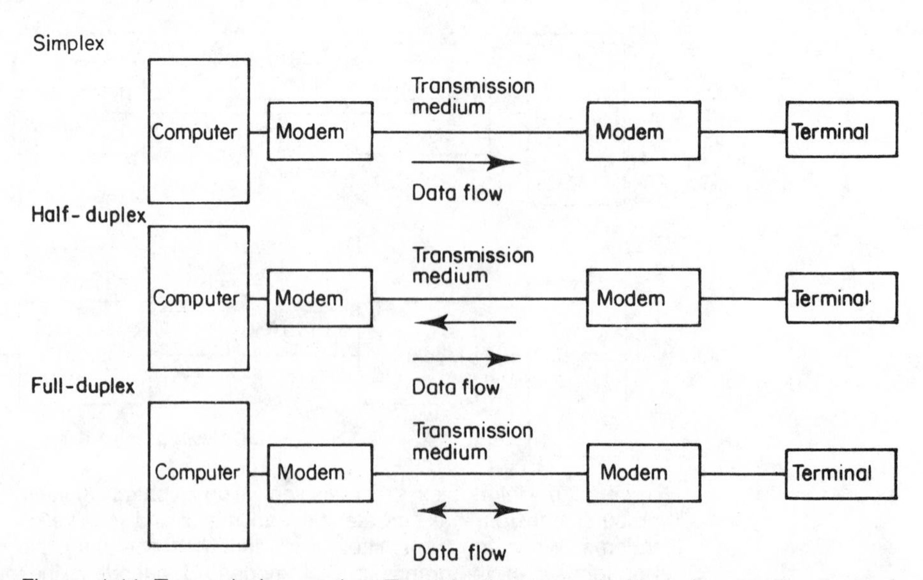

Figure 1.11 Transmission modes. Top: simplex transmission is in one direction only, transmission cannot reverse direction. Center: half-duplex transmission permits transmission in both directions but only one way at a time. Bottom: full-duplex transmission permits transmission in both directions simultaneously.

Table 1.3 Transmission mode comparison.

Symbol	ANSI	US telecommunications industry	CCITT	Historical physical line requirement
←	One-way only	Simplex		2-wire
← →	Two-way alternate	Half-duplex (HDX)	Simplex	2-wire
⇆	Two-way simultaneous	Full-duplex (FDX)	Duplex	4-wire

CCITT operates as part of the International Telecommunications Union (ITU), which is a United Nations agency. Since CCITT modem standards are primarily followed in Europe, these standards may be of particular interest to persons in the USA that have to communicate with overseas locations or ship overseas equipment that was purchased in the USA. In Chapter 2 we will examine some of the more common CCITT modem standards to understand the compatibility problems that can occur between US and European-manufactured modems.

Terminal and mainframe computer operations

When referring solely to terminal operations, the terms half-duplex and full-duplex operation take on meanings different from the communications mode of the transmission medium. Vendors commonly use the term half-duplex to denote that the terminal device is in a local copy mode of operation. This means that each time a character is pressed on the keyboard it is printed or displayed on the local terminal as well as transmitted. Thus, a terminal device operated in a half-duplex mode would have each character printed or displayed on its monitor as it is transmitted.

When one says a terminal is in a full-duplex mode of operation this means that each character pressed on the keyboard is transmitted but not immediately displayed or printed. Here the device on the distant end of the transmission path must "echo" the character back to the originator, which, upon receipt displays or prints the character. Thus, a terminal in a full-duplex mode of operation would only print or display the characters pressed on the keyboard after the character is echoed back by the device at the other end of the line. Figure 1.12 illustrates the terms full- and half-duplex as they apply to terminal devices. The reader should note that although most conventional terminals have a switch to control the duplex setting of the device, personal computer users normally obtain their duplex setting via the software program they are using. Thus, the term "echo on" during the initialization of a communications software program would refer to the process of displaying each character on the user's screen as it is transmitted.

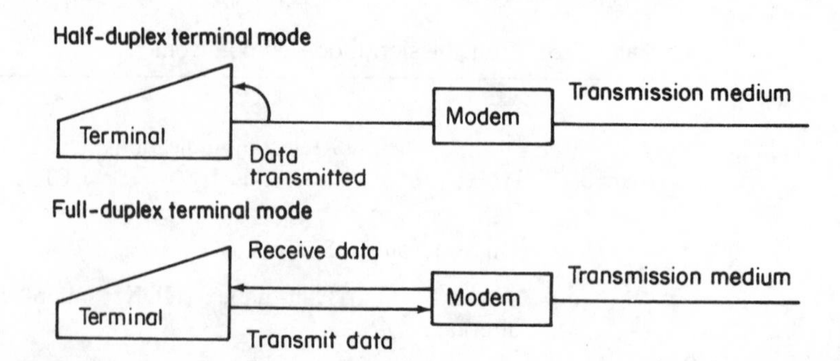

Figure 1.12 Terminal operation modes. Top: the term half-duplex terminal operation implies that data transmitted is also printed on the local terminal. This is known as local copy. Bottom: the term full-duplex terminal operation implies that no local copy is provided.

When we refer to half- and full-duplex with respect to mainframe computer systems we are normally referencing whether or not they echo received characters back to the originator. A half-duplex computer system does not echo characters back, while a full-duplex computer system echoes each character it receives.

Different character displays

When considering the operating mode of the terminal device, the transmission medium, and the operating mode of the mainframe computer on the distant end of the transmission path as an entity, three things could occur in response to each character one presses on a keyboard. Assuming a transmission medium is employed that can be used for either half- or full-duplex communications, our terminal device could print or display no character for each character transmitted, one character for each character transmitted, or two characters for each character transmitted. Here the resulting character printed or displayed would be dependent upon the operating mode of the terminal device and the host computer one is connected to as indicated in Table 1.4.

Table 1.4 Operating mode and character display.

Operating Mode		
Terminal device	Host computer	Character display
Half-duplex	Half-duplex	1 character
Half-duplex	Full-duplex	2 characters
Full-duplex	Half-duplex	No characters
Full-duplex	Full-duplex	1 character

To understand the character display column in Table 1.4, let us examine the two-character display result caused by the terminal device operating in a half-duplex mode while the mainframe computer operates in a full-duplex mode.

When the terminal is in a half-duplex mode it echoes each transmitted character onto its printer or display. At the other end of the communications path, if the mainframe computer is in a full-duplex mode of operation it will echo the received character back to the terminal, causing a second copy of the transmitted character to be printed or displayed. Thus, two characters would appear on one's printer or display for each character transmitted. To alleviate this situation, one would change the transmission mode of one's terminal to full-duplex. This would normally be accomplished by turning "echo" off during the initialization of a communications software program, if using a personal computer; or one would turn a switch to half-duplex if operating a conventional terminal.

1.4 TRANSMISSION TECHNIQUES

Data can be transmitted either synchronously or asynchronously. Asynchronous transmission is commonly referred to as a start–stop transmission where one character at a time is transmitted or received. Start and stop bits are used to separate characters and synchronize the receiver with the transmitter, thus providing a method of reducing the possibility that data becomes garbled.

Most devices designed for human–machine interaction that are teletype compatible transmit data asynchronously. By teletype compatible, we reference terminals and personal computers that operate similarly to the Teletype® terminal manufactured by Western Electric, a subsidiary of AT&T. Various versions of this popular terminal have been manufactured for over 30 years and an installed base of approximately one million such terminals is in operation worldwide. As characters are depressed on the device's keyboard they are transmitted to the computer, with idle time occurring between the transmission of characters. This is illustrated at the bottom of Figure 1.13.

Asynchronous transmission

In asynchronous transmission, each character to be transmitted is encoded into a series of pulses. The transmission of the character is started by a start pulse equal in length to a code pulse. The encoded character (series of pulses) is followed by a stop pulse that may be equal to or longer than the code pulse, depending upon the transmission code used.

The start bit represents a transition from a mark to a space. Since in an idle condition when no data is transmitted the line is held in a marking condition, the start bit serves as an indicator to the receiving device that a character of data follows. Similarly, the stop bit causes the line to be placed back into its previous "marking" condition, signifying to the receiver that the data character is completed.

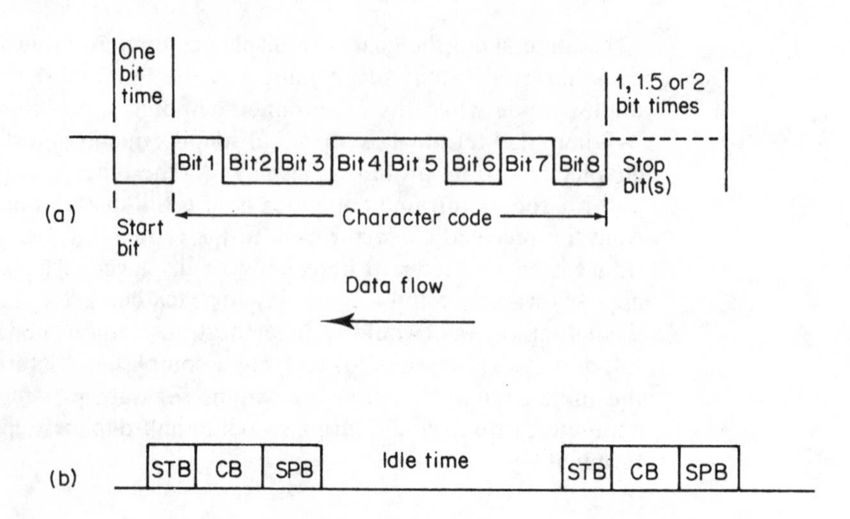

Figure 1.13 Asynchronous (start-stop) transmission. (a) Transmission of one 8-bit character. (b) Transmission of many characters. STB = start bit; CB = character bits; SPB = stop bit(s); idle time is time between character transmission.

As illustrated in the top portion of Figure 1.13, the transmission of an 8-bit character requires either 10 or 11 bits, depending upon the length of the stop bit. In actuality the eighth bit may be used as a parity bit for error detection and correction purposes. The use of the parity bit is described in detail in section 1.10 in this chapter. In the start–stop mode of transmission, transmission starts anew on each character and stops after each character. This is indicated in the lower portion of Figure 1.13. Since synchronization starts anew with each character, any timing discrepancy is cleared at the end of each character, and synchronization is maintained on a character-by-character basis. Asynchronous transmission normally is used for transmission at speeds under 9600 bps over the switched telephone network or on leased lines, while data rates up to 19,200 bps are possible over a direct connect cable whose distance is limited to approximately 50 feet.

The term asynchronous TTY or TTY compatible refers to the asynchronous start–stop protocol employed originally by Teletype® terminals and is the protocol where data is transmitted on a line-by-line basis between a terminal device and a mainframe computer. In comparison, more modern terminals with cathode ray tube (CRT) displays are usually designed to transfer data on a full screen basis.

Personal computer users only require an asynchronous communications adapter and a software program that transmits and receives data on a line-by-line basis to connect to a mainframe that supports asynchronous TTY compatible terminals. Here the software program that transmits and receives data on a line-by-line basis is normally referred to as a TTY emulator program and is the most common type of communications program written for use with personal computers.

Synchronous transmission

A second type of transmission involves sending a grouping of characters in a continuous bit stream. This type of transmission is referred to as synchronous or bit-stream synchronization. In the synchronous mode of transmission, modems located at each end of the transmission medium normally provide a timing signal or clock that is used to establish the data transmission rate and enable the devices attached to the modems to identify the appropriate characters as they are being transmitted or received. In some instances, timing may be provided by the terminal device itself or a communication component, such as a multiplexer or front-end processor channel. No matter what timing source is used, prior to beginning the transmission of data, the transmitting and receiving devices must establish synchronization among themselves. In order to keep the receiving clock in step with the transmitting clock for the duration of a stream of bits that may represent a large number of consecutive characters, the transmission of the data is preceded by the transmission of one or more special characters. These special synchronization or "syn" characters are at the same code level (number of bits per character) as the coded information to be transmitted. However, they have a unique configuration of zero and one bits which are interpreted as the syn character. Once a group of syn characters is transmitted, the receiver recognizes and synchronizes itself onto a stream of those syn characters.

After synchronization is achieved, the actual data transmission can proceed. Synchronous transmission is illustrated in Figure 1.14. In synchronous transmission, characters are grouped or blocked into groups of characters, requiring a buffer or memory area so characters can be grouped together. In addition to having a buffer area, more complex circuitry is required for synchronous transmission since the receiving device must remain in phase with the transmitter for the duration of the transmitted block of information. Synchronous transmission is normally used for data transmission rates in excess of 2000 bps. The major characteristics of asynchronous and synchronous transmission are denoted in Table 1.5.

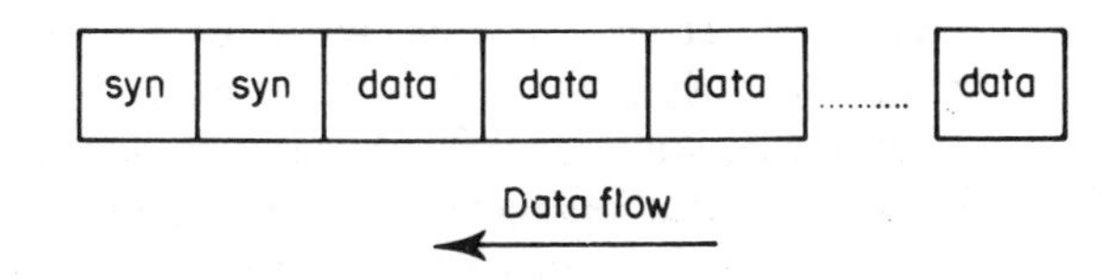

Figure 1.14 Synchronous transmission. In synchronous transmission, one or more syn characters are transmitted to establish clocking prior to the transmission of data.

1.5 TYPES OF TRANSMISSION

The two types of data transmission one can consider are serial and parallel. For serial transmission the bits which comprise a character are transmitted in sequence over one line; whereas, in parallel transmission characters are

Table 1.5 Transmission technique characteristics.

Asynchronous

1. Each character is prefixed by a start bit and followed by one or more stop bits.
2. Idle time (period of inactivity) can exist between transmitted characters.
3. Bits within a character are transmitted at prescribed time intervals.
4. Timing is established independently in the computer and terminal.
5. Transmission speeds normally do not exceed 9600 bps over switched facilities or leased lines and 19,200 bps over dedicated links and leased lines.

Synchronous

1. Syn characters prefix transmitted data.
2. Syn characters are transmitted between blocks of data to maintain line synchronization.
3. No gaps exist between characters.
4. Timing is established and maintained by the transmitting and receiving modems, the terminal, or other devices.
5. Terminals must have buffers.
6. Transmission speeds normally are in excess of 2000 bps.

transmitted serially but the bits that represent the character are transmitted in parallel. If a character consists of eight bits, then parallel transmission would require a minimum of eight lines. Additional lines may be necessary for control signals and for the transmission of a parity bit. Although parallel transmission is used extensively in computer-to-peripheral unit transmission, it is not normally employed other than in dedicated data transmission usage owing to the cost of the extra circuits required.

A typical use of parallel transmission is the in-plant connection of badge readers and similar devices to a computer in that facility. Parallel transmission can reduce the cost of terminal circuitry since the terminal does not have to convert the internal character representation to a serial data stream for transmission. However, the cost of the transmission medium and interface will increase because of the additional number of conductors required. Since the total character can be transmitted at the same moment in time using parallel transmission, higher data transfer rates can be obtained than are possible with serial transmission facilities. For this reason, most local facility communications between computers and their peripheral devices are accomplished using parallel transmission. In comparison, communications between terminal devices and computers normally occur serially, since this requires only one line to interconnect the two devices that need to communicate with one another. Figure 1.15 illustrates serial and parallel transmission.

1.6 LINE STRUCTURE

The geographical distribution of terminal devices and the distance between each device and the device it transmits to are important parameters that must

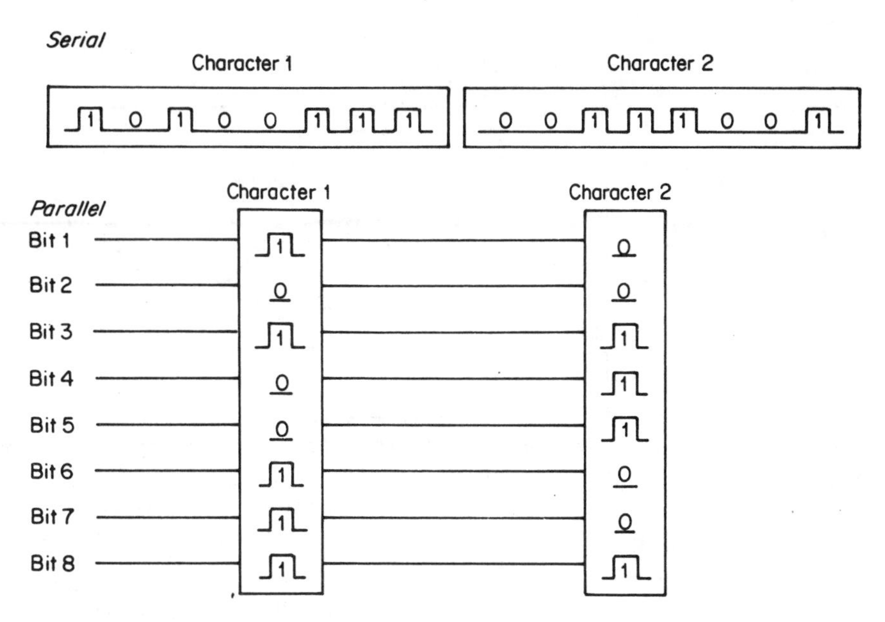

Figure 1.15 Types of data transmission. In serial transmission, the bits that comprise the character to be transmitted are sent in sequence over one line. In parallel transmission, the characters are transmitted serially but the bits that represent the character are transmitted in parallel.

be considered in developing a network configuration. The method used to interconnect personal computers and terminals to mainframe computers or to other devices is known as line structure and results in a computer's network configuration.

Types of line structure

The two types of line structure used in networks are point-to-point and multipoint, the latter also commonly referred to as multidrop lines. Communications lines that only connect two points are point-to-point lines. An example of this line structure is depicted at the top of Figure 1.16. As illustrated, each terminal transmits and receives data to and from a computer via an individual connection that links a specific terminal to the computer. The point-to-point connection can utilize a dedicated circuit or a leased line, or can be obtained via a connection initiated over the switched (dial-up) telephone network.

When two or more terminal locations share portions of a common line, the line is a multipoint or multidrop line. Although no two devices on such a line can transmit data at the same time, two or more devices may receive a message at the same time. The number of devices receiving such a message is dependent upon the addresses assigned to the message recipients. In some systems a "broadcast" address permits all devices connected to the same multidrop line to receive a message at the same time. When multidrop lines are employed, overall line costs may be reduced since common portions of the line are shared

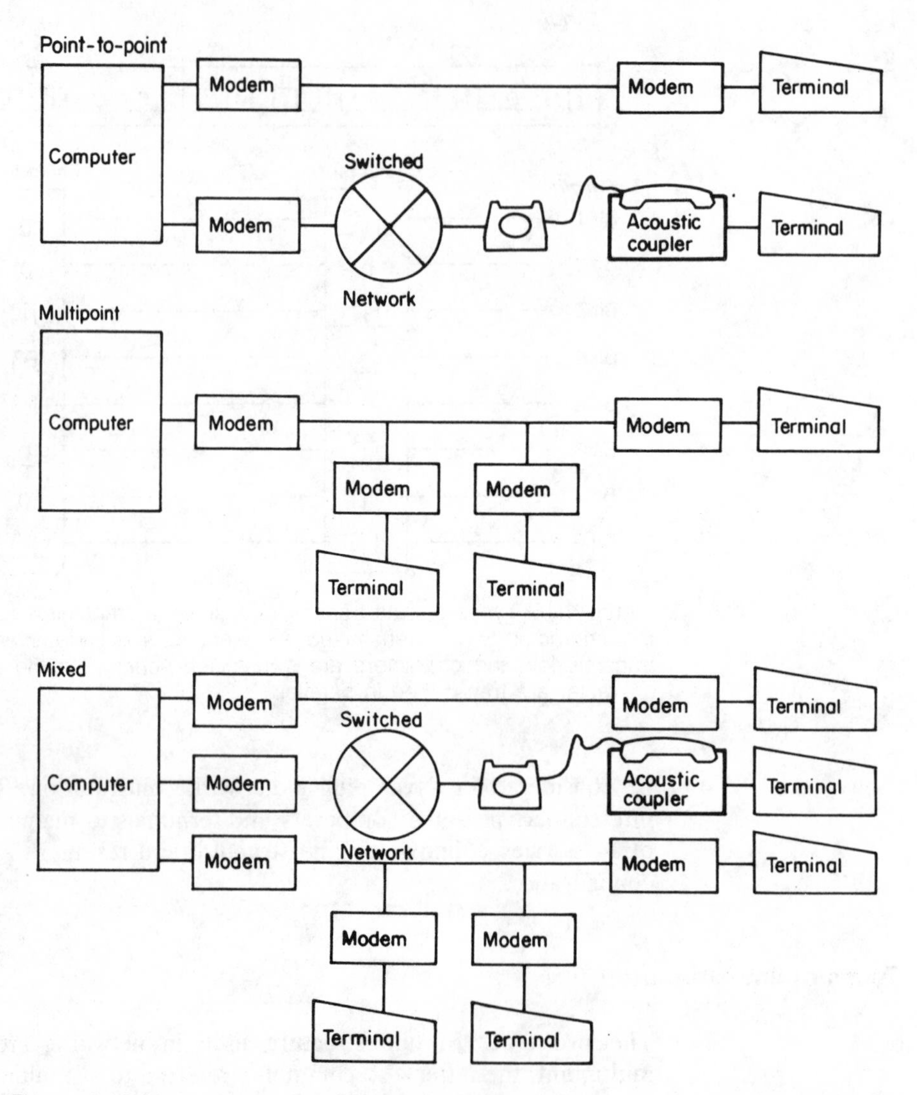

Figure 1.16 Line structures in networks. Top: point-to-point line structure. Center: multipoint (multidrop) line structure. Bottom: mixed network line structure.

for use by all devices connected to that line. To prevent data transmitted from one device from interfering with data transmitted from another device, a line discipline or control must be established for such a link. This discipline controls transmission so no two devices transmit data at the same time. A multidrop line structure is depicted in the second portion of Figure 1.16. For a multidrop line linking n devices to a mainframe computer, $n + 1$ modems are required, one for each device as well as one located at the computer facility.

Both point-to-point and multipoint lines may be intermixed in developing a network, and transmission can be either in the full- or half-duplex mode. This mixed line structure is shown in the lower portion of Figure 1.16.

1.7 LINE DISCIPLINE

When several devices share the use of a common, multipoint communications line, only one device may transmit at any one time, although one or more devices may receive information simultaneously. To prevent two or more devices from transmitting at the same time, a technique known as "poll and select" is utilized as the method of line discipline for multidrop lines. To utilize poll and select, each device on the line must have a unique address of one or more characters as well as circuitry to recognize a message sent from the computer to that address. When the computer polls a line, in effect it asks each device in a predefined sequence if it has data to transmit. If the device has no data to transmit, it informs the computer of this fact and the computer continues its polling sequence until it encounters a device on the line that has data to send. Then the computer acts on that data transfer.

As the computer polls each device, the other devices in the line must wait until they are polled before they can be serviced. Conversely, transmission of data from the computer to each device on a multidrop line is accomplished by the computer selecting the device address to which that data is to be transferred, informing the device that data is to be transferred to it, and then transmitting data to the selected device. Polling and selecting can be used to service both asynchronous or synchronous operating terminal devices connected to independent multidrop lines. Owing to the control overhead of polling and selecting, synchronous high-speed devices are normally serviced in this type of environment. By the use of signals and procedures, polling and selecting line control insures the orderly and efficient utilization of multidrop lines. An example of a computer polling the second terminal on a multipoint line and then receiving data from that device is shown at the top of Figure 1.17. At the bottom of that illustration, the computer first selects the third terminal on the line and then transfers a block of data to that device.

When terminals transmit data on a point-to-point line to a computer or another terminal, the transmission of that data occurs at the discretion of the terminal operator. This method of line control is known as "non-poll-and-select" or "free-wheeling" transmission.

1.8 TRANSMISSION RATE

Many factors can affect the transmission rate at which data is transferred. The types of modems and acoustic couplers used, as well as the line discipline and the type of computer channel to which a terminal is connected via a transmission medium, play governing roles that affect transmission rates; however, the transmission medium itself is a most important factor in determining transmission rates.

Data transmission services offered by communications carriers such as AT&T and Western Union are based on their available plant facilities. Depending upon terminal and computer locations, two types of transmission services may be available. The first type of service, analog transmission, is most readily available and can be employed on switched or leased telephone lines. Digital

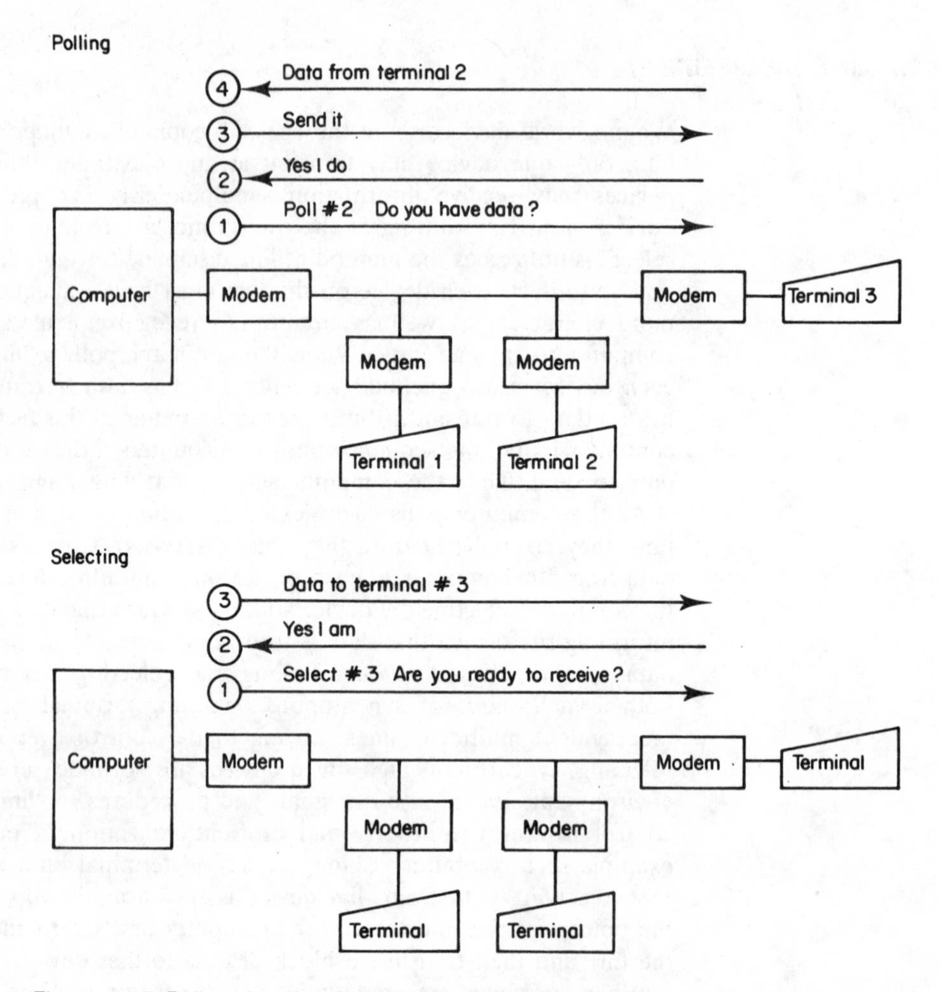

Figure 1.17 Poll and select line discipline. Poll and select is a line discipline which permits several devices to use a common line facility in an orderly manner.

transmission is only available in most large cities, and analog extensions are required to connect to this service from non-digital service locations as previously illustrated in Figure 1.9. Within each type of service several grades of transmission are available for consideration.

Analog service

In general, analog service offers the user three grades of transmission: narrowband, voice-band and wideband. The data transmission rates capable on each of these grades of service is dependent upon the bandwidth and electrical properties of each type of circuit offered within each grade of service. Basically, transmission speed is a function of the bandwidth of the communications line: the greater the bandwidth, the higher the possible speed of transmission.

Narrowband facilities are obtained by the carrier subdividing a voice-band circuit or by grouping a number of transmissions from different users onto a single portion of a circuit by time. Transmission rates obtained on narrowband facilities range between 45 and 300 bps. Teletype® terminals that connect to message switching systems are the primary example of the use of narrowband facilities.

While narrowband facilities have a bandwidth in the range of 200 to 400 Hz, voice-band facilities have a bandwidth in the range of 3000 Hz. Data transmission speeds obtainable on voice-band facilities are differentiated by the type of voice-band facility utilized – switched dial-up transmission or transmission via a leased line. For transmission over the switched telephone network, maximum data transmission is normally between 4800 and 7200 bps, with a 9600-bps data rate obtainable when transmission occurs through modern electronic switches instead of the older, electromechanical switches still used in many telephone offices. Since leased lines can be conditioned, a speed of 9600 to 19,200 bps is employed frequently on such lines.

Although low data speeds can be transmitted on both narrowband and voice-band circuits, one should not confuse the two, since a low data speed on a voice circuit is transmission at a rate far less than the maximum permitted by that type of circuit. Whereas, a low rate on a narrowband facility is at or near the maximum transmission rate permitted by that type of circuit.

Facilities which have a higher bandwidth than voice-band are termed wide-band or group-band facilities since they provide a wider bandwidth through the grouping of a number of voice-band circuits. Wideband facilities are available only on leased lines and permit transmission rates in excess of 19,200 bps. Transmission rates on wideband facilities vary with the offerings of communications carriers. Speeds normally available include 19.2, 40.8, 50, and 230.4 kbps, and 1.544 Mbps.

For direct connect circuits, transmission rates are a function of the distance between the terminal and the computer as well as the gauge of the conductor used.

Digital service

In the area of digital service, several offerings are currently available for user consideration. Digital data service is offered by AT&T as DATAPHONE® digital service (DDS). It provides interstate, full-duplex, point-to-point, and multipoint leased line as well as synchronous digital transmission at speeds of 2400, 4800, 9600, and 56,000 bps. It also provides data transmission at 1.544 Mbps between the servicing areas of many digital cities.

A new high-speed digital switched communications service recently introduced by AT&T offers full-fuplex, synchronous transmission over a common digital network at a transmission rate of 56,000 bps. In addition, several specialized communications carriers now offer or plan to offer digital service in selected areas of the United States and to overseas locations. Table 1.6 lists the main analog and digital facilities, the range of transmission speeds over those facilities, and the general use of such facilities.

Table 1.6 Common transmission facilities.

Facility	Transmission speed	Use
Analog		
Narrowband	45–300 bps	Message switching
Voice-band	less than 4800–19,200 bps	Time sharing; remote job entry; information utility access; file transfer
Switched	up to 19,200 bps	
Leased	up to 19,200 bps	Computer-to-computer; remote job entry; tape-to-tape transmission; high-speed terminal to high-speed terminal
Wideband	Over 19,200 bps	
Digital		
Leased line*	2.4, 4.8, 9.6, 56 kbps and 1.544 Mbps	Remote job entry; computer-to-computer; high-speed facsimile
Switched	56 kbps	Terminal-to-terminal; computer-to-computer; high-speed terminal to computer

* 19.2 kbps service was proposed.

1.9 TRANSMISSION CODES

Data within a computer is structured according to the architecture of the computer. The internal representation of data in a computer is seldom suitable for transmission to devices other than the peripheral units attached to the computer. In most cases, to effect data transmission, internal computer data must be redesigned or translated into a suitable transmission code. This transmission code creates a correspondence between the bit encoding of data for transmission or internal device representation and printed symbols. The end result of the translation is usually dictated by the character code that the remote terminal is built to accept. Frequently available terminal codes include: Baudot, which is a 5-level (5 bits per character code); binary-coded decimal (BCD), which is a 6-level code; American Standard Code for Information Interchange (ASCII), which is normally a 7-level code; and the extended binary-coded decimal interchange code (EBCDIC), which is an 8-level code.

In addition to information being encoded into a certain number of bits based upon the transmission code used, the unique configuration of those bits to represent certain control characters can be considered as a code that can be used to effect line discipline. These control characters may be used to indicate the acknowledgement of the receipt of a block of data without errors (ACK), the start of a message (SOH), or the end of a message (ETX), with the number of permissible control characters standardized according to the code employed.

With the growth of computer-to-computer data transmission, a large amount of processing can be avoided by transferring the data in the format used by the computer for internal processing. Such transmission is known as binary mode transmission, transparent data transfer, code-independent transmission, or native mode transmission.

Morse code

One of the most commonly known codes, the Morse code, is not practical for utilization in a computer communications environment. This code consists of a series of dots and dashes, which, while easy for the human ear to decode, are of unequal length and not practical for data transmission implementation. In addition, since each character in the Morse code is not prefixed with a start bit and terminated with a stop bit, it was initially not possible to construct a machine to automatically translate received Morse transmissions into their appropriate characters.

Baudot code

The Baudot code which is a 5-level (5 bits per character) code was the first code to provide a mechanism for encoding characters by an equal number of bits, in this case, five. The 5-level Baudot code was devised by Emil Baudot to permit teletypewriters to operate faster and more accurately than relays used to transmit information via telegraph.

Since the number of different characters which can be derived from a code having two different (binary) states is 2^m, where m is the number of positions in the code, the 5-level Baudot code permits 32 unique character bit combinations. Although 32 characters could be represented normally with such a code, the necessity of transmitting digits, letters of the alphabet, and punctuation marks made it necessary to devise a mechanism to extend the capacity of the code to include additional character representations. The extension mechanism was accomplished by the use of two "shift" characters: "letters shift" and "figures shift". The transmission of a shift character informs the receiver that the characters which follow the shift character should be interpreted as characters from a symbol and numeric set or from the alphabetic set of characters.

The 5-level Baudot code is illustrated in Table 1.7 for one particular terminal pallet arrangement. A transmission of all ones in bit positions 1 through 5 indicates a letter shift, and the characters following the transmission of that character are interpreted as letters. Similarly, the transmission of ones in bit positions 1, 2, 4, and 5 would indicate a figures shift, and the following characters would be interpreted as numerals or symbols based upon their code structure. Although the Baudot code is quite old in comparison to the age of personal computers, it is the transmission code used by the Telex network which is employed in the business community to send messages throughout the world.

Table 1.7 5-level Baudot code.

Letters	Figures	Bit selection				
		1	2	3	4	5
Characters						
A	-	1	1			
B	?	1			1	1
C	:		1	1	1	
D	$	1			1	
E	3	1				
F	!	1		1	1	
G	&		1		1	1
H				1		1
I	8		1	1		
J	'	1	1		1	
K	(	1	1	1	1	
L	)		1			1
M	.			1	1	1
N	,			1	1	
O	9				1	1
P	Ø		1	1		1
Q	1	1	1	1		1
R	4		1		1	
S		1		1		
T	5				1	
U	7	1	1	1		
V	;		1	1	1	1
W	2	1	1			1
X	/	1		1	1	1
Y	6	1		1		1
Z	"	1				1
Functions						
Carriage return					1	
Line feed			1			
Space				1		
Letters shift	<	1	1	1	1	1
Figures shift	=	1	1		1	1

BCD code

The development of computer systems required the implementation of coding systems to convert alphanumeric characters into binary notation and the binary notation of computers into alphanumeric characters. The BCD system was one of the earliest codes used to convert data to a computer-acceptable form. This coding technique permits decimal numeric information to be represented by 4 binary bits and permits an alphanumeric character set to be represented through the use of 6 bits of information. This code is illustrated in Table 1.8. An advantage of this code is that two-decimal digits can be stored in an 8-bit computer word and manipulated with appropriate computer instructions. Although only 36 characters are shown for illustrative purposes, a BCD code is capable of containing a set of 2^6 or 64 different characters.

Table 1.8 Binary-coded decimal system.

Bit position						
b_6	b_5	b_4	b_3	b_2	b_1	Character
0	0	0	0	0	1	A
0	0	0	0	1	0	B
0	0	0	0	1	1	C
0	0	0	1	0	0	D
0	0	0	1	0	1	E
0	0	0	1	1	0	F
0	0	0	1	1	1	G
0	0	1	0	0	0	H
0	0	1	0	0	1	I
0	1	0	0	0	1	J
0	1	0	0	1	0	K
0	1	0	0	1	1	L
0	1	0	1	0	0	M
0	1	0	1	0	1	N
0	1	0	1	1	0	O
0	1	0	1	1	1	P
0	1	1	0	0	0	Q
0	1	1	0	0	1	R
1	0	0	0	1	0	S
1	0	0	0	1	1	T
1	0	0	1	0	0	U
1	0	0	1	0	1	V
1	0	0	1	1	0	W
1	0	0	1	1	1	X
1	0	1	0	0	0	Y
1	0	1	0	0	1	Z
1	1	0	0	0	0	0
1	1	0	0	0	1	1
1	1	0	0	1	0	2
1	1	0	0	1	1	3
1	1	0	1	0	0	4
1	1	0	1	0	1	5
1	1	0	1	1	0	6
1	1	0	1	1	1	7
1	1	1	0	0	0	8
1	1	1	0	0	1	9

In addition to transmitting letters, numerals, and punctuation marks, a considerable number of control characters may be required to promote line discipline. These control characters may be used to switch on and off devices which are connected to the communications line, control the actual transmission of data, manipulate message formats, and perform additional functions. Thus, an extended character set is usually required for data communications. One such character set is EBCDIC code. The extended binary decimal interchange code (EBCDIC) is an extension of the BCD system and uses 8 bits for character representation. This code permits 2^8 or 256 unique characters to be represented, although currently a lesser number is assigned meanings. This code is primarily used for transmission by byte-oriented computers, where a byte is a grouping of eight consecutive binary digits operated on as a unit by the computer. The use of this code by computers may alleviate the necessity of the computer performing code conversion if the connected terminals operate with the same character set.

Several subsets of EBCDIC exist that have been tailored for use with certain devices. As an example, IBM 3270 type terminal products would not use a paper feed and its character representation is omitted in the EBCDIC character subset used to operate that type of device, as indicated in Table 1.9.

Table 1.9 EBCDIC code implemented for the IBM 3270 information display system.

		00				01				10				11				← Bits 0, 1
	Hex 1 ↓	00	01	10	11	00	01	10	11	00	01	10	11	00	01	10	11	← 2, 3
Bits 4567		0	1	2	3	4	5	6	7	8	9	A	B	C	D	E	F	← Hex 0
0000	0	NUL	DLE			SP	&	-									0	
0001	1	SOH	SBA					/		a	j			A	J		1	
0010	2	STX	EUA		SYN					b	k	s		B	K	S	2	
0011	3	ETX	IC							c	l	t		C	L	T	3	
0100	4									d	m	u		D	M	U	4	
0101	5	PT	NL							e	n	v		E	N	V	5	
0110	6			ETB						f	o	w		F	O	W	6	
0111	7			ESC	EOT					g	p	x		G	P	X	7	
1000	8									h	q	y		H	Q	Y	8	
1001	9		EM							i	r	z		I	R	Z	9	
1010	A					¢	!	¦	:									
1011	B					.	$	,	#									
1100	C		DUP		RA	<	*	%	@									
1101	D		SF	ENQ	NAK	(	)	_	'									
1110	E		FM			+	;	>	=									
1111	F		ITB		SUB	|	¬	?	"									

ASCII code

As a result of the proliferation of data transmission codes, several attempts to develop standardized codes for data transmission have been made. One such code is the American Standard Code for Information Interchange (ASCII). This 7-level code is based upon a 7-bit code developed by the International

Table 1.10 The ASCII character set. This coded character set is to be used for the general interchange of information among information processing systems, communications systems, and associated equipment.

Bits												
b_7					0	0	0	0	1	1	1	1
b_6					0	0	1	1	0	0	1	1
b_5					0	1	0	1	0	1	0	1
b_4	b_3	b_2	b_1	COLUMN / ROW	0	1	2	3	4	5	6	7
0	0	0	0	0	NUL	DLE	SP	0	@	P	\	p
0	0	0	1	1	SOH	DC1	!	1	A	Q	a	q
0	0	1	0	2	STX	DC2	"	2	B	R	b	r
0	0	1	1	3	ETX	DC3			C	S	c	s
0	1	0	0	4	EOT	DC4	$	4	D	T	d	t
0	1	0	1	5	ENQ	NAK	%	5	E	U	e	u
0	1	1	0	6	ACK	SYN	&	6	F	V	f	v
0	1	1	1	7	BEL	ETB	/	7	G	W	g	w
1	0	0	0	8	BS	CAN	(	8	H	X	h	x
1	0	0	1	9	HT	EM	)	9	I	Y	i	y
1	0	1	0	10	LF	SUB	*	:	J	Z	j	z
1	0	1	1	11	VT	ESC	+	;	K	[	k	{
1	1	0	0	12	FF	FS	,	<	L	\	l	|
1	1	0	1	13	CR	GS	–	=	M	]	m	}
1	1	1	0	14	SO	RS	.	>	N	∧	n	~
1	1	1	1	15	SI	US	/	?	O	—	o	DEL

Note that b_7 is the higher order bit and b_1 is the low order bit as indicated by the following example for coding the letter C.

b_7	b_6	b_5	b_4	b_3	b_2	b_1
1	0	0	0	0	1	1

Standards Organization (ISO) and permits 128 possible combinations or character assignments to include 96 graphic characters that are printable or displayable and 32 control characters to include device control and information transfer control characters. Table 1.10 lists the ASCII character set while Table 1.11 lists the ASCII control characters by position and their meaning. A more detailed explanation of these control characters is contained in the section covering protocols in this chapter. The primary difference between the ASCII character set listed in Table 1.10 and other versions of the CCITT International Alphabet Number 5 is the currency symbol. Although the bit sequence 0 1 0 0 0 1 0 is used to generate the dollar ($) currency symbol in the United

Table 1.11 ASCII control characters.

Column/Row	Control character		Mnemonic and meaning
0/0	^@	NUL	Null (CC)
0/1	^A	SOH	Start of Heading (CC)
0/2	^B	STX	Start of Text (CC)
0/3	^C	ETX	End of Text (CC)
0/4	^D	EOT	End of Transmission (CC)
0/5	^E	ENQ	Enquiry (CC)
0/6	^F	ACK	Acknowledgement (CC)
0/7	^G	BEL	Bell
0/8	^H	BS	Backspace (FE)
0/9	^I	HT	Horizontal Tabulation (FE)
0/10	^J	LF	Line Feed (FE)
0/11	^K	VT	Vertical Tabulation (FE)
0/12	^L	FF	Form Feed (FE)
0/13	^M	CR	Carriage Return (FE)
0/14	^N	SO	Shift Out
0/15	^O	SI	Shift In
1/0	^P	DLE	Date Link Escape (CC)
1/1	^Q	DC1	Device Control 1
1/2	^R	DC2	Device Control 2
1/3	^S	DC3	Device Control 3
1/4	^T	DC4	Device Control 4
1/5	^U	NAK	Negative Acknowledge (CC)
1/6	^V	SYN	Synchronous Idle (CC)
1/7	^W	ETB	End of Transmission Block (CC)
1/8	^X	CAN	Cancel
1/9	^Y	EM	End of Medium
1/10	^Z	SUB	Substitute
1/11	^[	ESC	Escape
1/12	^/	FS	File Separator (IS)
1/13	^]	GS	Group Separator (IS)
1/14	-	RS	Record Separator (IS)
1/15	.	US	Unit Separator (IS)
7/15	^-	DEL	Delete

(CC) communications control; (FE) format effector; (IS) information separator.

States, in the United Kingdom that bit sequence results in the generation of the pound sign (£). Similarly, this bit sequence generates other currency symbols when the CCITT International Alphabet Number 5 is used in other countries.

Code conversion

A frequent problem in data communications is that of code conversion. Consider what must be done to enable a computer with an EBCDIC character set to transmit and receive information from a terminal with an ASCII character set. When that terminal transmits a character, that character is encoded according to the ASCII character code. Upon receipt of that character, the computer must convert the bits of information of the ASCII character into an equivalent EBCDIC character. Conversely, when data is to be transmitted to the terminal, it must be converted from EBCDIC to ASCII so the terminal will be able to decode and act according to the information in the character that the terminal is built to interpret.

One of the most frequent applications of code conversion occurs when personal computers are used to communicate with IBM mainframe computers.

Normally, ASCII to EBCDIC code conversion is implemented when an IBM PC or compatible personal computer is required to operate as a 3270 type terminal. This type of terminal is typically connected to an IBM or IBM compatible mainframe computer and the terminal's replacement by an IBM PC requires the PC's ASCII coded data to be translated into EBCDIC. There are many ways to obtain this conversion, including emulation boards that are inserted into the system unit of a PC and protocol converters that are connected between the PC and the mainframe computer. Later in this book, we will explore these and other methods that enable the PC to communicate with mainframe computers that transmit data coded in EBCDIC.

Table 1.12 lists the ASCII and EBCDIC code character values for the ten digits for comparison purposes. In examining the difference between ASCII

Table 1.12 ASCII and EBCDIC digits comparison.

ASCII				
Dec	Oct	Hex	EBCDIC	Digit
048	060	30	F0	0
049	061	31	F1	1
050	062	32	F2	2
051	063	33	F3	3
052	064	34	F4	4
053	065	35	F5	5
054	066	36	F6	6
055	067	37	F7	7
056	070	38	F8	8
057	071	39	F9	9

and EBCDIC coded digits the reader will note that each EBCDIC coded digit has a value precisely Hex C0 (decimal 192) higher than its ASCII equivalent. Although this might appear to make code conversion a simple process of adding or subtracting a fixed quantity depending upon which way the code conversion takes place, in reality many of the same ASCII and EBCDIC coded characters differ by varying quantities. As an example, the slash (/) character is Hex 2F in ASCII and Hex 61 in EBCDIC, a difference of Hex 92 (decimal 146). In comparison, other characters such as the carriage return and form feed have the same coded value in ASCII and EBCDIC, while other characters are displaced by different amounts in these two codes. Due to this, code conversion is typically performed as a table lookup process, with two buffer areas used to convert between codes in each of the two conversion directions. Thus, one buffer area might have the ASCII character set in Hex order in one field of a two-field buffer area, with the equivalent EBCDIC Hex values in a second field in the buffer area. Then, upon receipt of an ASCII character its Hex value is obtained and matched to the equivalent value in the first field of the buffer area, with the value of the second field containing the equivalent EBCDIC Hex value which is then extracted to perform the code conversion.

Modified ASCII

Members of the IBM PC series and compatible computers use a modified ASCII character set which is represented as an 8-level code. The first 128 characters in the character set, ASCII values 0 through 127, correspond to the ASCII character set listed in Table 1.10 while the next 128 characters can be viewed as an extension of that character set since they require an 8-bit representation.

Caution is advised when transferring IBM PC files since characters with ASCII values greater than 127 will be received in error when they are transmitted using 7 data bits. This is because the ASCII values of these characters will be truncated to values in the range 0 to 127 when transmitted with 7 bits from their actual range of 0 to 255. To alleviate this problem from occurring one can initialize one's communications software for 8-bit data transfer, however, the receiving device must also be capable of supporting 8-bit ASCII data.

Although conventional ASCII files can be transmitted in a 7-bit format, many word processing and computer programs contain text graphics represented by ASCII characters whose values exceed 127. In addition, EXE and COM files which are produced by assemblers and compilers contain binary data that must also be transmitted in 8-bit ASCII to be accurately received. While most communications programs can transmit 7 or 8-bit ASCII data, many programs may not be able to transmit binary files accurately. This is due to the fact that communications programs that use the control Z character (ASCII SUB) to identify the end of a file transfer will misinterpret a group of 8 bits in the EXE or COM file being transmitted when they have the same 8-bit format as a control Z, and upon detection prematurely close the file. To avoid this situation one should obtain a communications software program that transfers files by blocks of bits or converts the data into a hexadecimal or octal ASCII equivalent prior to transmission if this type of data transfers will be required.

1.10 ERROR DETECTION AND CORRECTION

As a signal propagates down a transmission medium several factors can cause it to be received in error: the transmission medium employed and impairments caused by nature and machinery.

The transmission medium will have a certain level of resistance to current flow that will cause signals to attenuate. In addition, inductance and capacitance will distort the transmitted signals and there will be a degree of leakage which is the loss in a transmission line due to current flowing across, through insulators, or changes in the magnetic field.

Transmission impairments result from numerous sources. First, Gaussian or white noise is always present as it is the noise level that exists due to the thermal motions of electrons in a circuit. Next, impulse can occur from line hits due to atmospheric static or poor contacts in a telephone system.

Asynchronous transmission

In asynchronous transmission the most common form of error control is the use of a single bit, known as a parity bit, for the detection of errors. Owing to the proliferation of personal computer communications, more sophisticated error detection methods have been developed which resemble the methods employed with synchronous transmission.

Parity checking

Character parity checking which is also known as vertical redundancy checking (VRC) requires an extra bit to be added to each character in order to make the total quantity of 1s in the character either odd or even, depending upon whether one is employing odd parity checking or even parity checking. When odd parity checking is employed, the parity bit is set to 1 if the number of 1s in the character's data bits is even; or it is set at 0 if the number of 1s in the character's data bits is odd. When even parity checking is used, the parity bit is set to 0 if the number of 1s in the character's data bits is even; or it is set to 1 if the number of 1s in the character's data bits is odd.

Two additional terms used to reference parity settings are "mark" and "space". When the parity bit is set to a mark condition the parity bit is always 1 while space parity results in the parity bit always set to 0. Although not actually a parity setting, parity can be set to none, in which case no parity checking will occur. When transmitting binary data asynchronously, such as between personal computers, parity checking must be set to none or off. This enables all 8 bits to be used to represent a character. Table 1.13 summarizes the effect of five types of parity checking upon the eighth data bit in asynchronous transmission.

Table 1.13 Parity effect upon 8th data bit.

Parity type	Parity effect
Odd	Eighth data bit is logical zero if the total number of logical 1s in the first seven data bits is odd.
Even	Eighth bit is logical zero if the total number of logical 1s in the first seven data bits is even.
Mark	Eighth data bit is always logical 1.
Space	Eighth data bit is always logical zero.
None/Off	Eighth data bit is ignored.

For an example of parity checking, let us examine the ASCII character R whose bit composition is 1 0 1 0 0 1 0. Since there are three 1 bits in the character R, a 0 bit would be added if odd parity checking is used or a 1 bit would be added as the parity bit if even parity checking is employed. Thus, the ASCII character R would appear as follows:

```
| data bits |  ┌──────────parity bit
↓          ↓ ↓
10100100                      odd parity check
10100101                      even parity check
10100101                      mark parity check
10100100                      space parity check
```

Since there are three bits set in the character R, a 0 bit is added if odd parity checking is employed while a 1 bit is added if even parity checking is used. Similarly, mark parity results in the parity bit being set to 1 regardless of the composition of the data bits in the character, while space parity results in the parity bit always being set to 0.

Undetected errors

Although parity checking is a simple mechanism to investigate if a single bit error has occurred, it can fail when multiple bit errors occur. This can be visualized by returning to the ASCII R character example and examining the effect of two bits erroneously being transformed as indicated in Table 1.14. Here the ASCII R character has three set bits and a one-bit error could transform the number of set bits to four. If parity checking is employed, the received set parity bit would result in the character containing five set bits, which is obviously an error since even parity checking is employed. Now suppose two bits are transformed in error as indicated in the lower portion of Table 1.14. This would result in the reception of a character containing six set

Table 1.14 Character parity cannot detect an even number of bit errors.

ASCII character R	1 0 1 0 0 1 0
Adding an even parity bit	1 0 1 0 0 1 0 1
	1
1 bit in error	1 Ø 1 0 0 1 0 1
	1 1
2 bits in error	1 Ø 1 Ø 0 1 0 1

bits, which would appear to be correct under even parity checking. Thus, two bit errors in this situation would not be detected by a parity error detection technique.

In addition to the potential of undetected errors, parity checking has several additional limitations. First, the response to parity errors will vary based upon the type of mainframe with which one is communicating. Certain mainframes will issue a "Retransmit" message upon detection of a parity error. Some mainframes will transmit a character that will appear as a "fuzzy box" on one's screen in response to detecting a parity error, while the other mainframes will completely ignore parity errors.

When transmitting data asynchronously on a personal computer, most communications programs permit the user to set parity to odd, even, off, space, or mark. Off or no parity would be used if the system with which one is communicating does not check the parity bit for transmission errors. No parity would be used when one is transmitting 8-bit EBCDIC or an extended 8-bit ASCII coded data, such as that available on the IBM PC and similar personal computers. Mark parity means that the parity bit is set to 1, while space parity means that the parity bit is set to 0.

In the asynchronous communications world, two common sets of parameters are used by most bulletin boards, information utilities and supported by mainframe computers. The first set consists of 7 data bits and 1 stop bit with even parity checking employed, while the second set consists of 8 data bits and 1 stop bit using no parity checking. Table 1.15 compares the communications parameter settings of three popular information utilities.

Table 1.15 Communication parameter settings.

Parameter	Information Utility		
	CompuServe	Dow Jones	The Source
Data rate (bps)	300/1200	300/1200	300/1200
Data bits	7/8	8	8
Parity	even/none	none	none
Stop bits	1	1	1
Duplex	full	full	full

File transfer problems

Although visual identification of parity errors in an interactive environment is possible, what happens when one wishes to transfer a large file over the switched telephone network? For a typical call over the switched telephone network the probability of a random bit error occurring is approximately 1 in 100,000 bits at a data transmission rate of 1200 bps. If one desired to upload or download a 1000-line program containing an average of 40 characters per line, a total of 320,000 data bits would have to be transmitted. During the 4.4 minutes required to transfer this file one can expect 3.2 bit errors to occur, probably resulting in several program lines being received incorrectly if the errors occur randomly. In such situations one would prefer an alternative to visual inspection. Thus, a more efficient error detection and correction method is needed for large data transfers.

Block checking

In this method, data is grouped into blocks for transmission. A checksum character is generated and appended to the transmitted block and the checksum is also calculated at the receiver, using the same algorithm. If the checksums match, the data block is considered to be received correctly. If the checksums do not match, the block is considered to be in error and the receiving station will request the transmitting station to retransmit the block.

One of the most popular asynchronous block checking methods is included in the XMODEM protocol, which is extensively used in personal computer communications. This protocol blocks groups of asynchronous characters together for transmission and computes a checksum which is appended to the end of the block. The checksum is obtained by first summing the ASCII value of each data character in the block and dividing that sum by 255. Then, the quotient is discarded and the remainder is appended to the block as the checksum. Thus, mathematically the XMODEM checksum can be represented as

$$\text{Checksum} = R\left[\frac{\sum_{1}^{128} \text{ASCII value of characters}}{255}\right]$$

where R is the remainder of the division process.

When the data is transmitted using the XMODEM protocol, the receiving device at the other end of the link performs the same operation upon the block being received. This "internally" generated checksum is compared to the transmitted checksum. If the two checksums match, the block is considered to have been received error-free. If the two checksums do not match, the block is considered to be in error and the receiving device will then request the transmitting device to resend the block.

Figure 1.18 illustrates the XMODEM protocol block format. The Start of Header is the ASCII SOH character whose bit composition is 0 0 0 0 0 0 0 1,

Start of header	Block number	One's complement block number	128 data characters	Checksum

Figure 1.18 XMODEM protocol block format. The start of header is the ASCII SOH character whose bit composition is 00000001, while the one's complement of the block number is obtained by subtracting the block number from 255. The checksum is formed by first adding the ASCII values of each of the characters in the 128 character block, dividing the sum by 255 and using the remainder.

while the one's complement of the block number is obtained by subtracting the block number from 255. The block number and its complement are contained at the beginning of each block to reduce the possibility of a line hit at the beginning of the transmission of a block causing the block number to be corrupted but not detected.

The construction of the XMODEM protocol format permits errors to be detected in one of three ways. First, if the Start of Header is damaged, it will be detected by the receiver and the data block will be negatively acknowledged. Next, if either the block number or the one's complement field are damaged, they will not be the one's complement of each other, resulting in the receiver negatively acknowledging the data block. Finally, if the checksum generated by the receiver does not match the transmitted checksum, the receiver will transmit a negative acknowledgement. For all three situations the negative acknowledgement will serve as a request to the transmitting station to retransmit the previously transmitted block.

Data transparency

Since the XMODEM protocol supports an 8-bit, no parity data format it is transparent to the data content of each byte. This enables the protocol to support ASCII, binary and extended ASCII data transmission, where extended ASCII is the additional 128 graphic characters used by the IBM PC and compatible computers through the employment of an 8-bit ASCII code.

Error detection efficiency

While the employment of a checksum reduces the probability of undetected errors in comparison to parity checking, it is still possible for undetected errors to occur under the XMODEM protocol. This can be visualized by examining the construction of the checksum character and the occurence of multiple errors when a data block is transmitted.

Assuming a 128-character data block of all 1s is to be transmitted, each data character has the format 0 1 1 0 0 0 1 0, which is an ASCII 49. When the checksum is computed the ASCII value of each data character is first added, resulting in a sum of 6272(128 × 49). Next, the sum is divided by 255, with the remainder used as the checksum, which in this example is 152.

```
               0                    10
..1001100ø1|00110001|001100øø|...
```

(In the printed figure the last bit of the first character is struck through with "0" written above it, and the last two bits of the third character are struck through with "10" written above them.)

Figure 1.19 Multiple errors on an XMODEM data block.

Suppose two transmission impairments occur during the transmission of a data block under the XMODEM protocol affecting two data characters as illustrated in Figure 1.19.

Here the first transmission impairment converted the ASCII value of the character from 49 to 48, while the second impairment converted the ASCII value of the character from 49 to 50. Assuming no other errors occurred, the receiving device would add the ASCII value of each of the 128 data characters and obtain a sum of 6272. When the receiver divides the sum by 255, it obtains a checksum of 152, which matches the transmitted checksum and the errors remain undetected. Although the preceding illustration was contrived, it illustrates the potential for undetected errors to occur under the XMODEM protocol. To make the protocol more efficient with respect to undetected errors, some bulletin boards have implemented a cyclic redundancy checking (CRC) method into the protocol. The use of CRC error detection reduces the probability of undetected errors to less than one in a million blocks and is the preferred method for insuring data integrity. The concept of CRC error detection is explained later in this section under synchronous transmission, as it was first employed with this type of transmission.

The XMODEM protocol will be examined in more detail in the protocol section of this chapter.

Synchronous transmission

The majority of error-detection schemes employed in synchronous transmission involve geometric codes or cyclic code.

Geometric codes attack the deficiency of parity by extending it to two dimensions. This involves forming a parity bit on each individual character as well as on all the characters in the block. Figure 1.20 illustrates the use of block parity checking for a block of 10 data characters. As indicated, this block parity character is also known as the "longitudinal redundancy check" (LRC) character.

Geometric codes are similar to the XMODEM error-detection technique in the fact that they are also far from foolproof. As an example of this, suppose a 2-bit duration transmission impairment occurred at bit positions 3 and 4 when characters 7 and 9 in Figure 1.20 were transmitted. Here the two "1s" in those bit positions might be replaced by two "0s". In this situation each character parity bit as well as the block parity character would fail to detect the errors.

A transmission system using a geometric code for error detection has a slightly better capability to detect errors than the method used in the XMODEM protocol and is hundreds of times better than simple parity checking. While

			Character parity bit
Character	1	1 0 1 1 0 1 1	0
Character	2	0 1 0 0 1 0 1	0
	3	0 1 1 0 1 0 0	0
	4	1 0 0 1 0 0 1	0
	5	0 1 1 1 1 0 1	0
	6	1 0 1 0 0 0 0	1
	7	0 1 0 1 1 1 0	1
	8	0 1 1 1 0 0 1	1
	9	1 0 0 0 1 1 0	0
	10	0 1 1 0 1 0 1	1
Block parity character (LRC)		1 1 1 0 1 0 1	1

Figure 1.20 VRC/LRC geometric code (odd parity checking).

block parity checking substantially reduces the probability of an undetected error in comparison to simple parity checking on a character by character basis, other techniques can be used to further decrease the possibility of undetected errors. Among these techniques is the use of cyclic or polynomial code.

Cyclic codes

When a cyclic or polynomial code error-detection scheme is employed the message block is treated as a data polynomial $D(x)$, which is divided by a predefined generating polynomial $G(x)$, resulting in a quotient polynomial Q(x) and a remainder polynomial $R(x)$, such that:

$$D(x)/G(x) = Q(x) + R(x)$$

The remainder of the division process is known as the cyclic redundancy check (CRC) and is normally 16 bits in length or two 8-bit bytes. The CRC checking method is used in synchronous transmission similar to the manner in which the checksum is employed in the XMODEM protocol previously discussed. That is, the CRC is appended to the block of data to be transmitted. The receiving device uses the same predefined generating polynomial to generate its own CRC based upon the received message block and then compares the "internally" generated CRC with the transmitted CRC. If the two match, the receiver transmits a positive acknowledgement (ACK) communications control character to the transmitting device which not only informs the distant device that the data was received correctly but also serves to inform the device that if additional blocks of data remain to be transmitted the next block can be sent. If an error has occurred, the internally generated CRC will not match the transmitted CRC and the receiver will transmit a negative acknowledgement (NAK) communications control character which informs the transmitting device to retransmit the block previously sent.

Table 1.16 Common generating polynomials.

Standard	Polynomial
CRC-16 (ANSI)	$X^{16} + X^{15} + X^5 + 1$
CRC (CCITT)	$X^{16} + X^{12} + X^5 + 1$
CRC-12	$X^{12} + X^{11} + X^3 + 1$
CRC-32	$X^{32} + X^{26} + X^{23} + X^{22} + X^{16} + X^{12} + X^{11} + X^{10} +$ $X^8 + X^7 + X^5 + X^4 + X^2 + X + 1$

Table 1.16 lists four generating polynomials in common use today. The CRC-16 is based upon the American National Standards Institute and is commonly used in the United States. The CCITT CRC is commonly used in transmissions in Europe while the CRC-12 is used with 6-level transmission codes and has been basically superseded by the 16-bit polynomials. The 32-bit CRC is defined for use in local networks by the Institute of Electrical and Electronic Engineers (IEEE) and the American National Standards Institute (ANSI). For further information concerning the use of the CRC-32 polynomial the reader is referred to the IEEE/ANSI 802 standards publications.

The column labeled polynomial in Table 1.16 actually indicates the set bits of the 16-bit or 12-bit polynomial. Thus, the CRC-16 polynomial has a bit composition of 1 1 0 0 0 0 0 0 0 0 0 0 1 0 0 0 1.

International transmission

Due to the growth in international communications, one frequently encountered transmission problem is the employment of dissimilar CRC generating polynomials. This typically occurs when an organization in the United States attempts to communicate with a computer system in Europe or a European organization attempts to transmit to a computer system located in the United States. When dissimilar CRC generating polynomials are employed, the two-byte block-check character appended to the transmitted data block will never equal the block-check character computed at the receiver. This will result in each transmitted data block being negatively acknowledged, eventually resulting in a threshold of negative acknowledgements being reached. When this threshold is reached the protocol aborts the transmission session, causing the terminal operator to reinitiate the communications procedure required to access the computer system they wish to connect to. Although the solution to this problem requires either the terminal or a port on the computer system to be changed to use the appropriate generating polynomial, the lack of publication of the fact that there are different generating polynomials has caused many organizations to expend a considerable amount of needless effort. One bank which the author is familiar with monitored transmission attempts for almost 3 weeks. During this period, they observed each block being negatively acknowledged and blamed the communications carrier, insisting that the quality of the circuit was the culprit. Only after a consultant was called and spent approximately a week examining the situation was the problem traced to the utilization of dissimilar generating polynomials.

1.11 PROTOCOLS

Two types of protocol should be considered in a data communications environment: terminal protocols and data link protocols.

The data link protocol defines the control characteristics of the network and is a set of conventions that are followed which govern the transmission of data and control information. A terminal or a personal computer can have a predefined control character or set of control characters which are unique to the terminal and are not interpreted by the line protocol. This internal protocol can include such control characters as the bell, line feed, and carriage return for conventional teletype terminals, blink and cursor positioning characters for a display terminal, and form control characters for a line printer.

For experimenting with members of the IBM PC series and compatible computers readers can execute the one line BASIC program 1Ø PRINT CHR$(X)"DEMO", substituting different ASCII values for the value of X to see the effect of different PC terminal control characters. As an example, using the value 7 for X, the IBM PC will beep prior to displaying the message DEMO, since ASCII 7 is interpreted by the PC as a request to beep the speaker. Using the value 9 for X will cause the message DEMO to be printed commencing in position 9, since ASCII 9 is a tab character which causes the cursor to move on the screen 8 character positions to the right. Another example of a terminal control character is ASCII 11, which is the home character. Using the value of 11 for X will cause the message DEMO to be printed in the upper left-hand corner of the screen since the cursor is first placed at that location by the home character.

Although poll and select is normally thought of as a type of line discipline or control, it is also a data link protocol. In general, the data link protocol enables the exchange of information according to an order or sequence by establishing a series of rules for the interpretation of control signals which will govern the exchange of information. The control signals govern the execution of a number of tasks which are essential in controlling the exchange of information via a communications facility. Some of these information control tasks are listed in Table 1.17.

Table 1.17 Information control tasks.

Connection establishment	Transmission sequence
Connection verification	Data sequence
Connection disengagement	Error control procedures

Although all of the tasks listed in Table 1.17 are important, not all are required for the transmission of data, since the series of tasks required is a function of the total data communications environment. As an example, a single terminal or personal computer connected directly to a mainframe or another terminal device by a leased line normally does not require the establishment and verification of the connection. However, several devices connected to a mainframe computer on a multidrop or multipoint line would require the verification of the identification of each terminal device on the line to insure that data transmitted from the computer would be received by the proper device. Similarly, when a device's session is completed, this fact must be recognized so that the mainframe computer's resources can be made available to other users. Thus, connection disengagement on devices other than those connected on a point-to-point leased line permits a port on the front-end processor to become available to service other users.

Another important task is the transmission sequence which is used to establish the precedence and order of transmission, to include both data and control information. As an example, this task defines the rules for when devices on a multipoint circuit may transmit and receive information. In addition to the transmission of information following a sequence, the data itself may be sequenced. Data sequencing is normally employed in synchronous transmission where a long block is broken into smaller blocks for transmission, with the size of the blocks being a function of the personal computer's or terminal's buffer area and the error control procedure employed. By dividing a block into smaller blocks for transmission, the amount of data that must be retransmitted, in the event that an error in transmission is detected, is reduced.

Although error-checking techniques currently employed are more efficient when short blocks of information are transmitted, the efficiency of transmission correspondingly decreases since an acknowledgement (negative or positive) is returned to the device transmitting after each block is received and checked. For communications between remote job entry terminals and computers, blocks of up to several thousand characters are typically used. However, block lengths from 80 to 512 characters are the most common sizes. Although some protocols specify block length, most protocols permit the user to set the size of the block.

Pertaining to error-control procedures, the most commonly employed method to correct transmitted errors is to inform the transmitting device simply to retransmit a block. This procedure requires coordination between the sending and receiving devices, with the receiving device continuously informing the sending device of the status of each previously transmitted block. If the block previously transmitted contained no detected errors, the receiver will transmit a positive acknowledgement and the sender will transmit the next block. If the receiver detects an error, it will transmit a negative acknowledgement and discard the block containing an error. The transmitting station will then retransmit the previously sent block. Depending upon the protocol employed, a number of retransmissions may be attempted. However, if a default limit is reached owing to a bad circuit or other problems, then the mainframe computer or terminal device acting as the master station may terminate the session, and the terminal operator will have to reestablish the connection.

Communications control characters

Before examining several protocols in more detail, let us first review the communications control characters in the ASCII character set. These characters were previously listed in Table 1.11 with the two-character designator CC following their meaning and will be reviewed in the order of their appearance in the referenced table.

NUL

As its name implies, the Null (NUL) character is a non-printable time delay or filler character. This character is primarily used for communicating with printing devices that require a defined period of time after each carriage return in which to reposition the printhead to the beginning of the next line. Many mainframe computers and bulletin boards operating on personal computers will prompt users to "Enter the numbers of nulls"; this is a mechanism to permit both conventional terminals, personal computers, and personal computers with a variety of printers to use the system without obtaining garbled output.

SOH

The Start of Heading (SOH) is a communications control character used in several character-oriented protocols to define the beginning of a message heading data block. In synchronous transmission on a multipoint or multidrop line structure, the SOH is followed by an address which is checked by all devices on the common line to ascertain if they are the recipient of the data. In asynchronous transmission, the SOH character can be used to signal the beginning of a filename during multiple file transfers, permitting the transfer to occur without treating each file transfer as a separate communications session. Since asynchronous communications typically involves point-to-point communications, no address is required after the SOH character; however, both devices must have the same communications software program that permits multiple file transfers in this manner.

STX

The Start of Text (STX) character signifies the end of heading data and the beginning of the actual information contained within the block. This communications control character is used in the bisynchronous protocol that will be examined later in this chapter.

ETX

The End of Text (ETX) character is used to inform the receiver that all the information within the block has been transmitted. This character is also used to denote the beginning of the block-check characters appended to a transmission block as an error-detection mechanism. This communications control character is primarily used in the bisynchronous protocol.

EOT

The End of Transmission (EOT) character defines the end of transmission of all data associated with a message transmitted to a device. If transmission occurs on a multidrop circuit the EOT also informs other devices on the line to check later transmissions for the occurrence of messages that could be addressed to them. In the XMODEM protocol the EOT is used to indicate the end of a file transfer operation.

ENQ

The Enquiry (ENQ) communications control character is used in the bisynchronous protocol to request a response or status from the other station on a point-to-point line or to a specifically addressed station on a multidrop line. In response to the ENQ character, the receiving station may respond with the number of the last block of data it successfully received. In a multidrop environment, the mainframe computer would poll each device on the line by addressing the ENQ to one particular station at a time. Each station would respond to the poll positively or negatively, depending upon whether or not they had information to send to the mainframe computer at that point in time.

ACK and DLE

The Acknowledgement (ACK) character is used to verify that a block of data was received correctly. After the receiver computes its own "internal" checksum or cyclic code and compares it to the one appended to the transmitted block, it will transmit the ACK character if the two checksums match. In the XMODEM protocol the ACK character is used to inform the transmitter that the next block of data can be transmitted. In the bisynchronous protocol the Data Link Escape (DLE) character is normally used in conjunction with the 0 and 1 characters in place of the ACK character. Alternating DLE0 and DLE1 as positive acknowledgement to each correctly received block of data eliminates the potential of a lost or garbled acknowledgement resulting in the loss of data.

NAK

The Negative Acknowledgement (NAK) communications control character is transmitted by a receiving device to request the transmitting device to retransmit the previously sent data block. This character is transmitted when the receiver's internally generated checksum or cyclic code does not match the one transmitted, indicating that a transmission error has occurred. In the XMODEM protocol this character is used to inform the transmitting device that the receiver is ready to commence a file transfer operation as well as to inform the transmitter of any blocks of data received in error.

SYN

The Synchronous Idle (SYN) character is employed in the bisynchronous protocol to maintain line synchronization between the transmitter and receiver during periods when no data is transmitted on the line. When a series of SYN characters is interrupted, this indicates to the receiver that a block of data is being transmitted.

ETB

The End of Transmission Block (ETB) character is used in the bisynchronous protocol in place of an ETX character when data is transmitted in multiple blocks. This character then indicates the end of a particular block of transmitted data.

Bisynchronous transmission

Among currently used protocols, one of the most frequently used for synchronous transmission is IBM's BISYNC (binary synchronous communications) protocol. This particular protocol is actually a set of very similar protocols that provides a set of rules which effect the synchronous transmission of binary-coded data.

Although there are numerous versions of the bisynchronous protocol in existence, three versions account for the vast majority of devices operating in a bisynchronous environment. These three versions of the bisynchronous protocol are known as 2780, 3780, and 3270. The 2780 and 3780 bisynchronous protocols are used for remote job entry communications into a mainframe computer, with the major difference between these versions the fact that the 3780 version performs space compression while the 2780 version does not incorporate this feature. In comparison to the 2780 and 3780 protocols that are designed for point-to-point communications, the 3270 protocol is designed for operation with devices connected to a mainframe on a multidrop circuit or devices connected to a cluster controller which, in turn, is connected to the mainframe. Thus, 3270 is a poll and select software protocol.

An IBM PC or compatible computer can obtain a bisynchronous communications capability through the installation of a bisynchronous communications adapter card into the PC's system unit. This card is designed to operate in conjunction with a bisynchronous communications software program which with the adapter card enables the PC to operate as an IBM 2780 or 3780 workstation or as an IBM 3270 type of interactive terminal.

The bisynchronous transmission protocol can be used in a variety of transmission codes on a large number of medium- to high-speed equipment. Some of the constraints of this protocol are that it is limited to half-duplex transmission and that it requires the acknowledgement of the receipt of every block of data transmitted. A large number of protocols have been developed owing to the success of the BISYNC protocol. Some of these protocols are bit-oriented, whereas BISYNC is a character-oriented protocol; and some permit full-duplex transmission, whereas BISYNC is limited to half-duplex transmission.

Most bisynchronous protocols support several data codes to include ASCII and EBCDIC. Error control is obtained by using a two-dimensional parity check (LRC/VRC) when transmission is in ASCII. When transmission is in EBCDIC the CRC-16 polynomial is used to generate a block-check character.

Figure 1.21 illustrates the generalized bisynchronous block structure. The start of message control code is normally the STX communications control character. The end of message control code can be either the ETX, ETB, or the EOT character; the actual character, however, depends upon whether the

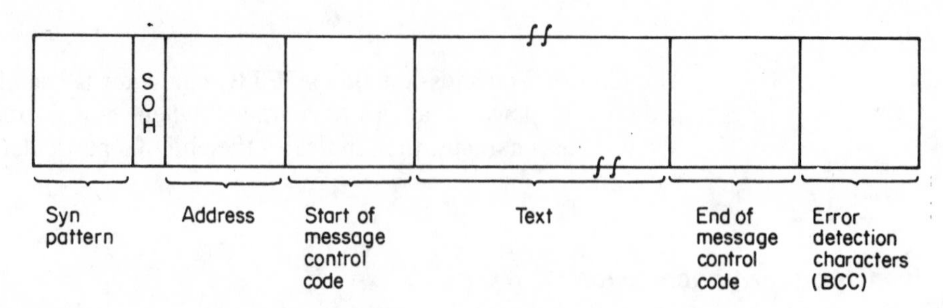

Figure 1.21 Generalized BSC block structure.

block is one of many blocks, the end of the transmission block, or the end of the transmission session.

Figure 1.22 illustrates the error control mechanism employed in a bisynchronous protocol to handle the situation where a line hit occurs during transmission or if an acknowledgement to a previously transmitted data block becomes lost or garbled.

In the example on the left portion of Figure 1.22, a line hit occurs during the transmission of the second block of data from the mainframe computer to a terminal or a personal computer. Note that although Figure 1.22 is an abbreviated illustration of the actual bisynchronous block structure and does not show the actual block-check characters in each block, in actuality they are contained in each block. Thus, the line hit which occurs during the transmission of the second block results in the "internally" generated BCC being different from the BCC that was transmitted with the second block. This causes the terminal device to transmit a NAK to the mainframe, which results in the retransmission of the second block.

In the example on the right-hand part of Figure 1.22, let us assume that the terminal received block 2 and sent an acknowledgement which was lost or garbled. After a predefined timeout period occurs, the master station transmits an ENQ communications control character to check the status of the terminal. Upon receipt of the ENQ, the terminal will transmit the alternating acknowledgement, currently DLE1; however, the mainframe was expecting DLE0. Thus, the mainframe is informed by this that block 2 was never acknowledged and as a result retransmits that block.

XMODEM Protocol

The XMODEM protocol originally developed by Ward Christensen has been implemented into many asynchronous personal computer communications software programs and a large number of bulletin boards. Figure 1.23 illustrates the use of the XMODEM protocol for a file transfer consisting of two blocks of data. As illustrated, under the XMODEM protocol the receiving device transmits a Negative Acknowledgement (NAK) character to signal the transmitter that it is ready to receive data. In response to the NAK the transmitter sends a Start of Header (SOH) communications control character

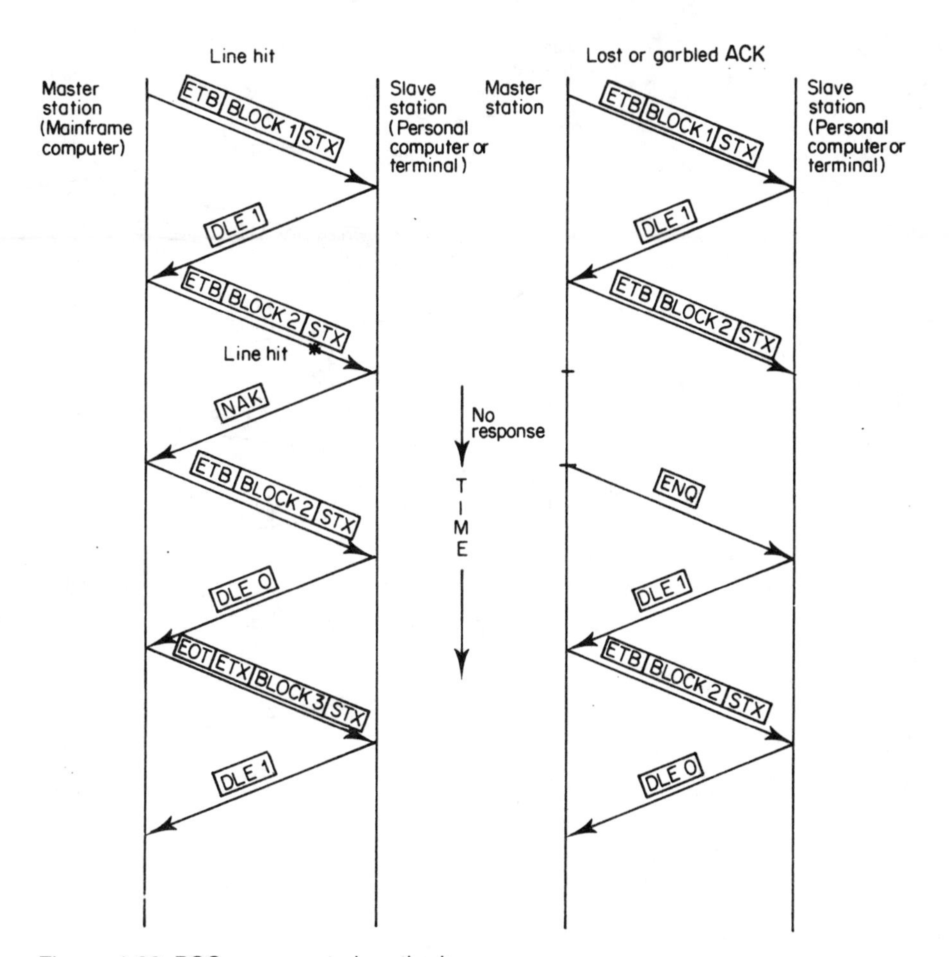

Figure 1.22 BSC error control methods.

followed by two characters that represent the block number and the one's complement of the block number. Here the one's complement is obtained by subtracting the block number from 255. Next a 128-character data block is transmitted which in turn is followed by the checksum character. As previously discussed, the checksum is computed by first adding the ASCII values of each of the characters in the 128-character block and dividing the sum by 255. Next, the quotient is discarded and the remainder is retained as the checksum.

If the data blocks are damaged during transmission, the receiver can detect the occurrence of an error in one of three ways. If the Start of Header is damaged, it will be detected by the receiver and the data block will be negatively acknowledged. If either the block count or the one's complement field are damaged, they will not be the one's complement of each other. Finally, the receiver will compute its own checksum and compare it to the transmitted checksum. If the checksums do not match this is also an indicator that the transmitted block was received in error.

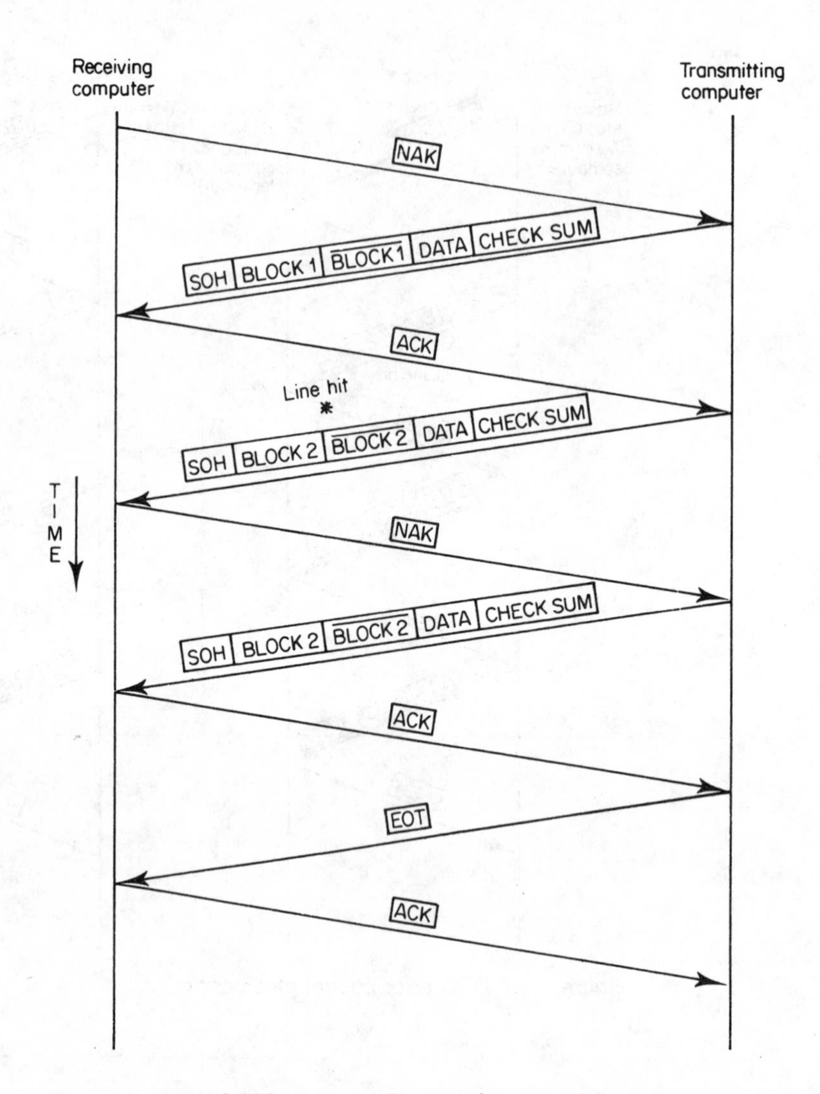

Figure 1.23 XMODEM protocol file transfer operation.

If the two checksums do not match or the SOH was missing or the block count and its complement field are not the one's complement of each other, the block is considered to have been received in error. Then the receiving station will transmit a NAK character which serves as a request to the transmitting station to retransmit the previously transmitted block. As illustrated in Figure 1.23, a line hit occurring during the transmission of the second block resulted in the receiver transmitting a NAK and the transmitting device resending the second block. Suppose more line hits occur which effects the retransmission of the second block. Under the XMODEM protocol the retransmission process will be repeated until the block is correctly received or until nine retransmission attempts occur. If, owing to a thunderstorm or other disturbance, line noise is a problem, after ten attempts to retransmit a block the file transfer process will be aborted. This will require a manual operator intervention to restart the file transfer at the beginning and is one of the major deficiencies of the XMODEM protocol. In comparison to the XMODEM protocol, the Blocked Asynchronous Transmission communications program (BLAST) marketed by Communications Research Group in Baton Rouge, LA, is a more sophisticated commercial protocol which permits files to be retransmitted at the point from which the previous transmission terminated. In addition, this program employs a full-duplex protocol in comparison to the half-duplex XMODEM protocol, permitting transmission of data to be normally accomplished in less time since the acknowledgement to the preceding block can be transmitted at the same time the transmitting station is sending the next data block. BLAST is very similar to a full-duplex synchronous data link protocol since a number of blocks can be outstanding and unacknowledged at any point in time. This obviously adds to one's transmission efficiency in comparison to the XMODEM protocol, since under that protocol block n can only be transmitted after block $n - 1$ is acknowledged.

In spite of the limitations of the XMODEM protocol, it is one of the most popular protocols employed by personal computer users for asynchronous data transfer because of several factors. First, the XMODEM protocol is in the public domain which means it is readily available at no cost for software developers to incorporate into their communications programs. Secondly, the algorithm employed to generate the checksum is easy to implement using a higher level language such as BASIC or Pascal. In comparison, a CRC-16 block-check character is normally generated using assembly language. In addition, the simplistic nature of the protocol is also easy to implement in BASIC or Pascal which enables many personal computer users to write their own routines to transfer files to and from bulletin boards using this protocol. Since the XMODEM protocol only requires a 256-character communications receiver buffer, it can be easily incorporated into communications software that will operate on personal computer systems with limited memory, such as the early systems that were produced with 64K or less RAM.

Several variations of the original XMODEM protocol have been introduced into the public domain. These modified XMODEM protocols incorporate a true CRC block-check character error-detection scheme in place of the checksum character, resulting in a much higher level of error-detection capability.

KERMIT

Kermit was developed at Columbia University in New York City primarily as a mechanism for downloading files from mainframes to microcomputers. Since its original development this protocol has evolved into a comprehensive communications system which can be employed for transferring data between most types of intelligent devices. Although the name might imply some type of acronym, in actuality, this protocol was named after Kermit the Frog, the star of the well-known Muppet television show.

Kermit is a half-duplex communications protocol which transfers data in variable sized packets, with a maximum packet size of 96 characters. Packets are transmitted in alternate directions since each packet must be acknowledged in a manner similar to the XMODEM protocol.

In comparison to the XMODEM protocol which permits 7 and 8-level ASCII as well as binary data transfers in their original data composition, all Kermit transmissions occur in 7-level ASCII. The reason for this restriction is the fact that Kermit was originally designed to support file transfers to 7-level ASCII mainframes. Binary file transfers are supported by the protocol prefixing each byte whose eighth bit is set by the ampersand (&) character. In addition, all characters transmitted to include 7-level ASCII must be printable, resulting in Kermit transforming each ASCII control character with the pound (£) character. This transformation is accomplished through the complementation of the seventh bit of the control character. Thus, 64 modulo 64 is added or subtracted from each control character encountered in the input data stream. When an 8-bit byte is encountered whose low order 7 bits represent a control character, Kermit appends a double prefix to the character. Thus, the byte 100000001 would be transmitted as &£A.

Although character prefixing adds a considerable amount of overhead to the protocol, Kermit includes a run length compression facility which may partially reduce the extra overhead associated with control character and binary data transmission. Here, the tilde (~) character is used as a prefix character to indicate run length compression. The character following the tilde is a repeat count, while the third character in the sequence is the character to be repeated. Thus, the sequence ~XA is used to indicate a series of 88 A's, since the value of X is 1011000 binary or decimal 88. Through the use of run length compression the requirement to transmit printable characters results in an approximate 25% overhead increase in comparison to the XMODEM protocol for users transmitting binary files. If ASCII data is transmitted, Kermit's efficiency can range from more efficient to less efficient in comparison to the XMODEM protocol, with the number of control characters in the file to be transferred and the susceptibility of the data to run length compression the governing factors in comparing the two protocols. Figure 1.24 illustrates the format of a Kermit packet. The Header field is the ASCII Start of Header (SOH) character. The Length field is a single character whose value ranges between 0 and 94. This one-character field defines the packet length in characters less two, since it indicates the number of characters to include the checksum that follow this field.

The Sequence field is another one-character field whose value varies between 0 and 63. The value of this field wraps around to 0 after each group of 64 packets is transmitted.

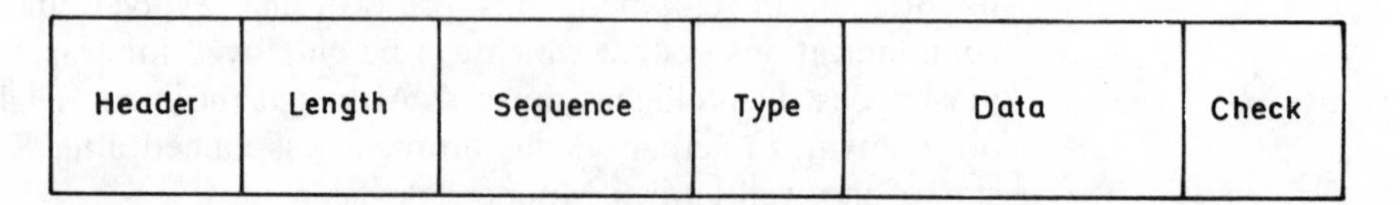

Figure 1.24 The Kermit packet format. The first three fields in the Kermit packet are one character in length and the maximum total packet length is 96 or less characters.

The Type field is a single printable character which defines the activity the packet initiates. Packet types include D (data), Y (acknowledgement), N (negative acknowledgement), B (end of transmission or break), F (file header), Z (end of file) and E (error).

The information contents of the packet are included in the Data field. As previously mentioned, control characters and binary data are prefixed prior to their placement in this field.

The Check field can be one, two or three characters in length depending upon which error-detection method is used since the protocol supports three options. A single character is used when a checksum method is used for error detection. When this occurs, the checksum is formed by the addition of the ASCII values of all characters after the Header character through the last data character and the low order 7 bits are then used as the checksum. The other two error-detection methods supported by Kermit include a two-character checksum and a three-character 16-bit CRC. The two-character checksum is formed similar to the one-character checksum, however, the low order 12 bits of the arithmetic sum are used and broken into two 7-bit printable characters. The 16-bit CRC is formed using the CCITT standard polynomial, with the high order 4 bits going into the first character while the middle 6 and low order 6 bits are placed into the second and third characters, respectively.

By providing the capability to transfer both the filename and contents of files, Kermit provides a more comprehensive capability for file transfers than XMODEM. In addition, Kermit permits multiple files to be transferred in comparison to XMODEM, which requires the user to initiate file transfers on an individual basis.

Bit-oriented line control protocols

A number of bit-oriented line control procedures have been implemented by computer vendors that are based upon the International Standards Organization (ISO) procedure known as High-level Data Link Control (HDLC). Various names for line control procedures similar to HDLC include IBM's Synchronous Data Link Control (SDLC) and Burrough's Data Link Control (BDLC).

The advantages of bit-oriented protocols are three-fold. First, their full-duplex capability supports the simultaneous transmission of data in two directions, resulting in a higher throughput than is obtainable in BISYNC. Secondly, bit-oriented protocols are naturally transparent to data, enabling the transmission of pure binary data without requiring special sequences of control characters to enable and disable a transparency transmission mode of operation as required with BISYNC. Lastly, most bit-oriented protocols permit multiple blocks of data to be transmitted. Then, if an error affects a particular block, only that block has to be retransmitted.

HDLC link structure

Under the HDLC transmission protocol one station on the line is given the primary status to control the data link and supervise the flow of data on the link. All other stations on the link are secondary stations and respond to commands issued by the primary station.

The vehicle for transporting messages on an HDLC link is called a frame and is illustrated in Figure 1.25.

The HDLC frame contains six fields, wherein two fields serve as frame delimiters and are known as the HDLC flag. The HDLC flag has the unique bit combination of 0 1 1 1 1 1 1 0, which defines the beginning and end of the frame. To protect the flag and assure transparency the transmission device will always insert a zero bit after a sequence of five one bits occurs to prevent data from being mistaken as a flag. This technique is known as zero insertion. The receiver will always delete a zero after receiving five 1s to insure data integrity.

The address field is an 8-bit pattern that identifies the secondary station involved in the data transfer while the control field can be either 8 or 16 bits in length. This field identifies the type of frame transmitted as either an information frame or a command/response frame. The information field can be any length and is treated as pure binary information, while the frame check sequence (FCS) contains a 16-bit value generated using a cyclic redundancy check (CRC) algorithm.

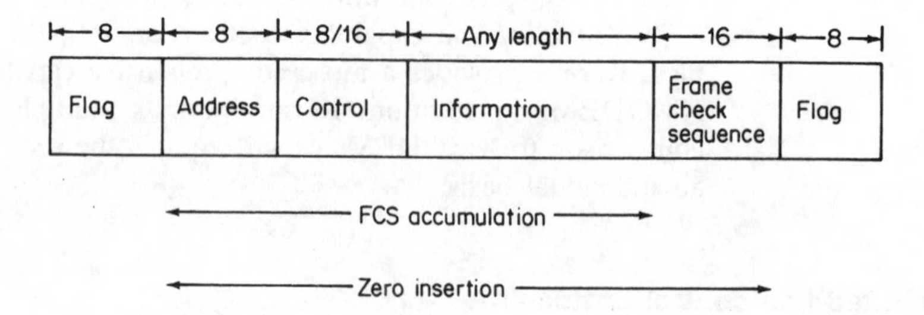

Figure 1.25 HDLC frame format. HDLC flag is 01111110 which is used to delimit an HDLC frame. To protect the flag and assure transparency the transmitter will insert a zero bit after a fifth 1 bit to prevent data from being mistaken as a flag. The receiver always deletes a zero after receiving five 1s.

Control field formats

The 8-bit control field formats are illustrated in Figure 1.26. N(S) and N(R) are the send and receive sequence counts. They are maintained by each station for Information (I-frames) sent and received by that station. Each station increments its N(S) count by one each time it sends a new frame. The N(R) count indicates the expected number of the next frame to be received.

Using an 8-bit control field, the N(S)/N(R) count ranges from 0 to 7. Using a 16-bit control field the count can range from 0 to 127. The P/F bit is a poll/final bit. It is used as a poll by the primary (set to 1) to obtain a response from a secondary station. It is set to 1 as a final bit by a secondary station to indicate the last frame of a sequence of frames.

The supervisory command/response frame is used in HDLC to control the flow of data on the line. Figure 1.27 illustrates the composition of the

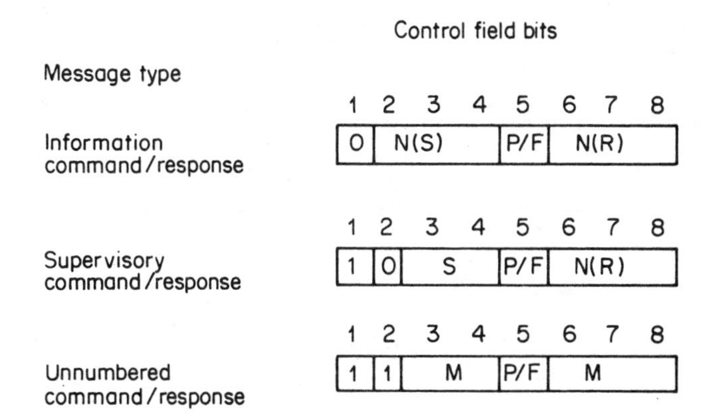

Figure 1.26 HDLC control field formats. N(S) = send sequence count: N(R) = receive sequence count; S = supervisory function bits; M = modifier function bits; P/F = poll/final bit.

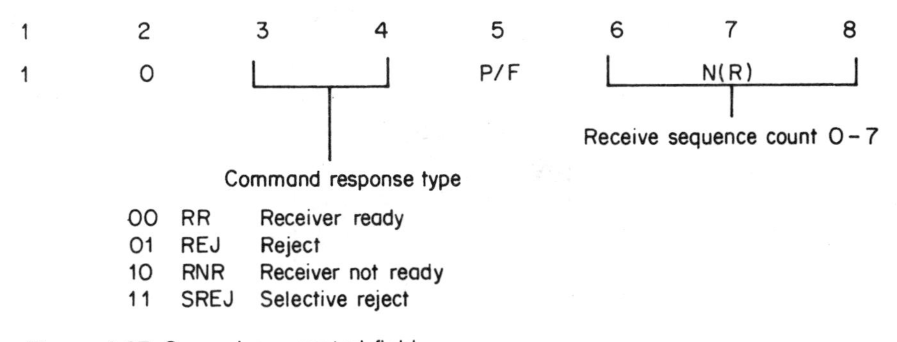

Figure 1.27 Supervisory control field.

supervisory control field: supervisory frames (S-frames) contain an N(R) count and are used to acknowledge I-frames, request retransmission of I-frames, request temporary suspension of I-frames, and perform similar functions.

To illustrate the advantages of HDLC over BISYNC transmission, consider the full-duplex data transfer illustrated in Figure 1.28. For each frame transmitted, this figure shows the type of frame, N(S), N(R), and poll/final (P/F) bit status.

In the transmission sequence illustrated in the left portion of Figure 1.28 the primary station has transmitted five frames, numbered zero through four, when its poll bit is set in frame four. This poll bit is interpreted by the secondary station as a request for it to transmit its status and it responds by transmitting a Receiver Ready (RR) response, indicating that it expects to receive frame five next. This serves as an indicator to the primary station that frames zero through four were received correctly. The secondary station sets its poll/final bit as a final bit to indicate to the primary station that its transmission is completed.

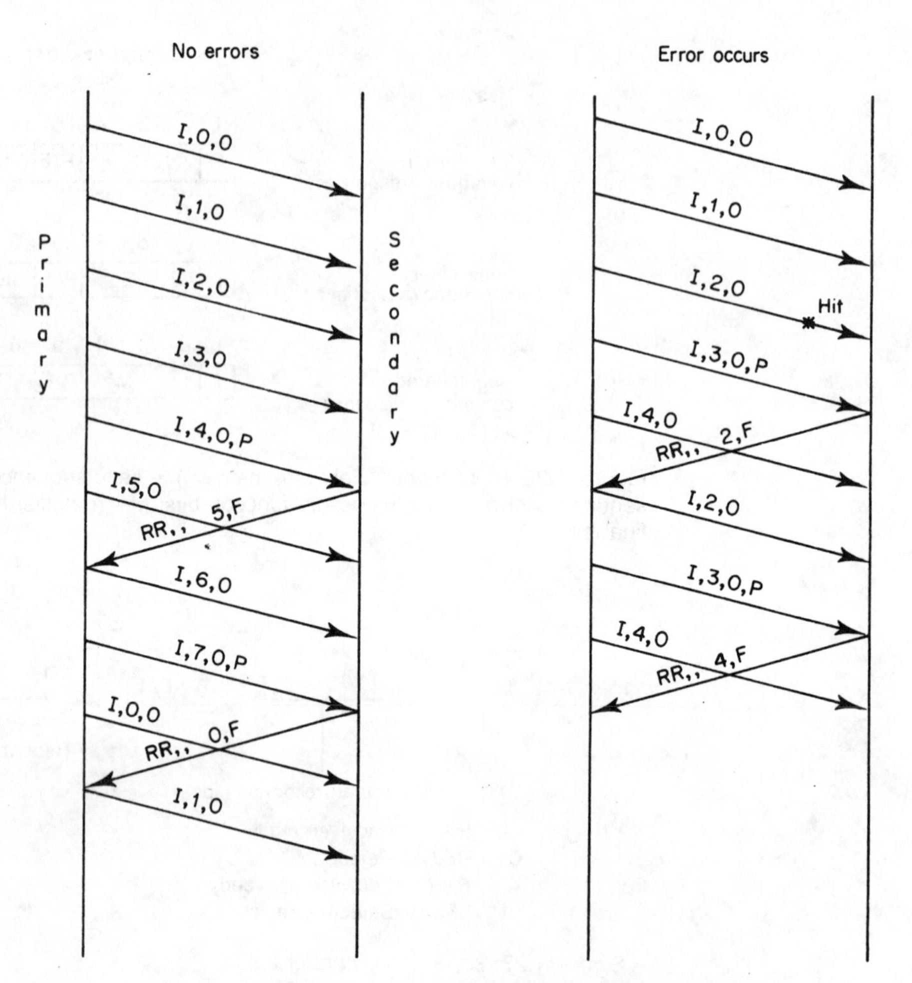

Figure 1.28 HDLC full-duplex data transfer. Format: Type, N(S), N(R), p/f.

Note that since full-duplex transmission is permissible under HDLC, the primary station continues to transmit information (I) frames while the secondary station is responding to the primary's polls. If an 8-bit control field is used, the maximum frame number that can be outstanding is limited to seven since 3 bit positions are used for N(S) frame numbering. Thus, after frame number seven is transmitted, the primary station then begins frame numbering again at N(S) equal to zero. Notice that when the primary station sets its poll bit when transmitting frame seven the secondary station responds, indicating that it expects to receive frame zero. This indicates to the primary station that frames five through seven were received correctly, since the previous secondary response acknowledged frames zero through four.

In the transmission sequence indicated on the right-hand side of Figure 1.28, assume a line hit occurs during the transmission of frame two. Note that in comparison to BISYNC, under HDLC the transmitting station does not have to wait for an acknowledgement of the previously transmitted data block; and it can continue to transmit frames until the maximum number of frames outstanding is reached; or, it can issue a poll to the secondary station to query the status of its previously transmitted frames while it continues to transmit frames up until the maximum number of outstanding frames is reached.

Thus, the primary station polled the secondary in frame three and then sent frame four while it waited for the secondary's response. When the secondary's response was received, it indicated that the next frame the secondary expected to receive N(R) was two. This informed the primary station that all frames after frame one would have to be retransmitted. Thus, after transmitting frame four the primary station then retransmitted frames two and three prior to retransmitting frame four.

It should be noted that if selective rejection is implemented, the secondary could have issued a Selective Reject (SREJ) of frame two. Then, upon its receipt, the primary station would retransmit frame two and have then continued its transmission with frame five. Although selective rejection can considerably increase the throughput of HDLC, even without its use this protocol will provide the user with a considerable throughput increase in comparison to BISYNC.

1.12 THE DTE/DCE INTERFACE

In the world of data communications, equipment that includes terminals and computer ports is referred to as Data Terminal Equipment or DTEs. In comparison, modems and other communications devices are referred to as Data Communications Equipment or DCEs. The physical, electrical, and logical rules for the exchange of data between DTEs and DCEs are specified by an interface standard; the most commonly used is the EIA RS-232-C and RS-232-D standards which are very similar to the CCITT V.24 standard used in Europe and other locations outside of North America. The term EIA refers to the Electronic Industries Association which is a national body that represents a large percentage of the manufacturers in the US electronics industry. The EIA's work in the area of standards has become widely recognized and many of its standards were adopted by other standard bodies. RS-232-C is a recommended standard (RS) published by the EIA in 1969, with the number 232 referencing the identification number of one particular communications standard and the suffix C designating the revision to that standard.

In the late 1970s it was intended that the RS-232-C standard would be gradually replaced by a set of three standards—RS-449, RS-422, and RS-423. These standards were designed to permit higher data rates than obtainable under RS-232-C as well as provide users with added functionality. Although the EIA and several government agencies heavily promoted the RS-449 standard, its adoption by vendors has been limited. Recognizing the fact that the universal adoption of RS-449 and its associated standards was basically impossible the EIA issued RS-232-D (Revision D) in January, 1987 as well as a new standard known as RS-530.

Both RS-232-D and RS-530 formally specify the use of a D-shaped 25-pin interface connector similar to the connector illustrated in Figure 1.29. A cable

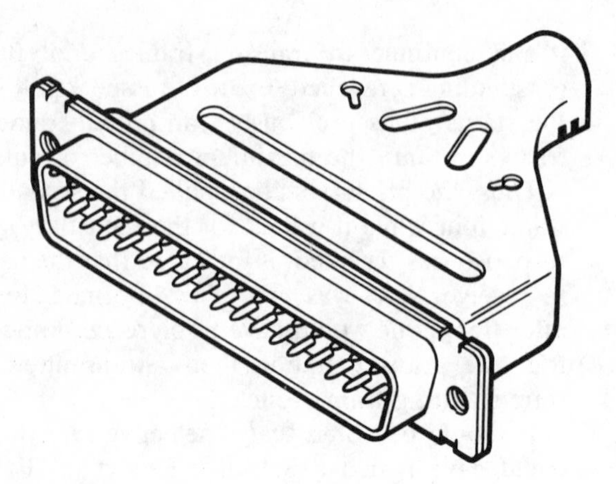

Figure 1.29 The D connector.

containing up to 25 individual conductors is fastened to the narrow part of the connector, while the individual conductors are soldered to predefined pin connections inside the connector.

In comparison to RS-232-D and RS-530, the RS-232-C standard only referenced the connector in an appendix and stated that it was not part of the standard. In spite of this omission, the use of a 25-pin D-shaped connector with RS-232-C is considered as a *de facto* standard.

The major differences between RS-232-D and RS-232-C is that the new revision supports the testing of both local and remote communications equipment by the addition of signals to support this function and modified the use of the protective ground conductor to provide a shielding capability.

Since the use of RS-232-C is basically universal since its publication by the EIA in 1969 we will examine both revisions C and D in this section, denoting the differences between the revisions when appropriate. When both revisions are similar, we will refer to them as RS-232-C/D. In general, devices built to either standard as well as the equivalent International CCITT V.24 recommendation are compatible with one another. However, there are some slight differences that can occur due to the addition of signals to support modem testing under RS-232-D.

Since the RS-232-C/D standards define the interfacing between DTEs and DCEs in the United States, they govern, as an example, the interconnection of terminal devices to stand-alone modems. The RS-232-C/D standards apply to serial data transfers between a DTE and DCE in the range from 0 to 20,000 bits per second. Although the standards also limit the cable length between the DTE and DCE to 50 feet, since the pulse width of digital data is inversely proportional to the data rate, one can normally exceed this 50-foot limitation at lower data rates as wider pulses are less susceptible to distortion than narrower pulses. When a cable length in excess of 50 feet is required, it is highly recommended that low capacitance shielded cable be used and tested prior to going on-line, to ensure that the signal quality is acceptable.

Another part of the RS-232-C/D standards specify the cable heads that serve as connectors to the DTEs and DCEs. Here the connector is known as a DB-25 connector and each end of the cable is equipped with this "male" connector that is designed to be inserted into the DB-25 female connectors normally built into modems. Figure 1.30 illustrates the RS-232-C/D interface between a terminal and a stand-alone modem.

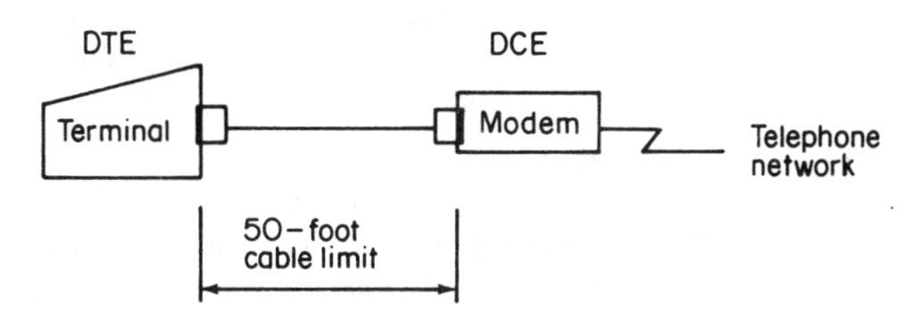

Figure 1.30 The RS-232-C/D physical interface standard cables are typically 6, 10, or 12 feet in length with 'male' connectors on each end.

RS-232-C/D signal characteristics

The RS-232-C/D interface specifies 25 interchange circuits or conductors that govern the data flow between the DTE and DCE. Although one can purchase a 25-conductor cable, normally a smaller number of conductors are required. For asynchronous transmission, normally 9 to 12 conductors are required, while synchronous transmission typically requires 12 to 16 conductors, with the number of conductors required a function of the operational characteristics of the device to be connected. The signal on each of these conductors occurs based upon a predefined voltage transition occurring as illustrated in Figure 1.31. Under RS-232-C a signal is considered to be ON when the voltage (V) on the interchange circuit is between +3 V and +15 V. In comparison, a voltage betwen −3 V and −15 V causes the interchange circuit to be placed in the OFF condition. The voltage range from +3 V to −3 V is a transition region that has no effect upon the condition of the circuit. Under RS-232-D the ON and OFF voltage ranges were extended to +15 V and −15 V, respectively. Table 1.18 provides a comparison between the interchange circuit voltage, its binary state, signal condition and function.

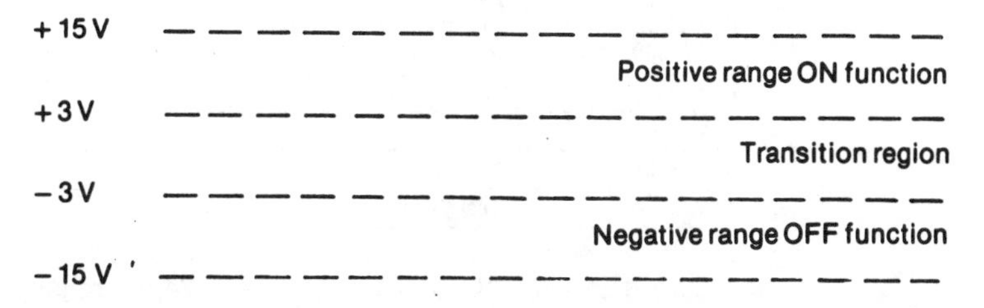

Figure 1.31 Interchange circuit voltage ranges. Under RS-232-D the ON and OFF voltage ranges were extended to +15 V and −15 V, respectively.

Table 1.18 Interchange circuit comparison.

	Interchange circuit voltage	
	Negative	Positive
Binary state	1	0
Signal condition	Mark	Space
Function	OFF	ON

Since the physical implementation of the RS-232-C/D standard is based upon the conductors used to interface a DTE to a DCE, we will examine the functions of each of the interchange circuits. Prior to discussing these circuits, an explanation of RS-232 terminology is warranted since there are three ways one can refer to the circuits in this interface.

PIN Number	Interchange circuit	CCITT equivalent	Description	Gnd	Data		Control		Timing		Testing	
					From DCE	To DCE	From DCE	To DCE	From DCE	To DCE	From DCE	To DCE
1	AA	101	Protective Ground (Shield)	X								
7	AB	102	Signal Ground/Common Return	X								
2	BA	103	Transmitted Data			X						
3	BB	104	Received Data		X							
4	CA	105	Request to Send					X				
5	CB	106	Clear to Send				X					
6	CC	107	Data Set Ready (DCE Ready)				X					
20	CD	108.2	Data Terminal Ready (DTE Ready)					X				
22	CE	125	Ring Indicator				X					
8	CF	109	Received Line Signal Detector				X					
21	(RL)/ CG	110	(Remote Loopback)/Signal Quality Detector				X					
23	CH	111	Data Signal Rate Selector (DTE)					X				
23	CI	112	Data Signal Rate Selector (DCE)				X					
24	DA	113	Transmitter Signal Element Timing (DTE)							X		
15	DB	114	Transmitter Signal Element Timing (DCE)						X			
17	DD	115	Receiver Signal Element Timing (DCE)						X			
14	SBA	118	Secondary Transmitted Data			X						
16	SBB	119	Secondary Received Data		X							
19	SCA	120	Secondary Request to Send					X				
13	SCB	121	Secondary Clear to Send				X					
12	SCF	122	Secondary Received Line Signal Detector				X					
8	—	—	Reserved for Testing									X
9	—	—	Reserved for Testing								X	
18	(LL)		(Local Loopback)									X
25	(TM)		(Test Mode)								X	

Figure 1.32 RS-232-C/D and CCITT V.24 interchange circuits by category. RS 232-D additions/changes to RS-232-C indicated in parentheses.

Circuit/conductor reference

The most commonly used method to refer to the RS-232-C/D circuits is by specifying the number of the pin in the connector which the circuit uses. A second method used to refer to the RS-232-C/D circuits is by the two- or three-letter designation used by the standards to label the circuits. The first letter in the designator is used to group the circuits into one of six circuit categories as indicated by the second column labeled "interchange circuit" in Figure 1.32. As an example of the use of this method, the two ground circuits have the letter A as the first letter in the circuit designator and the signal ground circuit is called "AB", since it is the second circuit in the "A" ground category. Since these designators are rather cryptic, they are not commonly used.

A third method used is to describe the circuits by their functions. Thus, pin 2 which is the transmit data circuit can be easily referenced as transmit data. Many persons have created acronyms for the descriptions which are easier to remember than the RS-232 pin number or interchange circuit designator. For example, transmit data is referred to as "TD", which is easier to remember than any of the RS-232 designators previously discussed.

Although the list of circuits in Figure 1.32 may appear overwhelming at first glance, in most instances only a subset of the 25 conductors are employed. To better understand this interface standard, we will first examine those interchange circuits required to connect an asynchronously operated terminal device to an asynchronous modem. Then we can expand upon our knowledge of these interchange circuits by examining the functions of the remaining circuits, to include those additional circuits that would be used to connect a synchronously operated terminal to a synchronous modem.

Asynchronous operations

Figure 1.33 illustrates the signals that are required to connect an asynchronous terminal device to a particular type of low-speed asynchronous modem known as a 103A type modem. Note that although a 25-conductor cable can be used to cable the terminal to the modem, only ten conductors are actually required. Thus, a 10-conductor cable could be used to connect a 103-type modem to an asynchronously operated terminal device which could result in a significant reduction in cable costs when one is cabling many DTEs to DCEs.

By reading the modem vendor's specification sheet one can easily determine the number of conductors required to cable DTEs to DCEs. Although most cables have straight through conductors, in certain instances the conductor pins at one end of a cable may require reversal or two conductors may be connected onto a common pin. In fact, many times only one conductor will be used for both protective ground and signal ground, with the common conductor cabled to pins 1 and 7 at both ends of the cable. In such instances a 9-conductor cable could be employed to satisfy the cabling requirements illustrated in Figure 1.33. With this in mind, let us review the functions of the ten circuits illustrated in Figure 1.33.

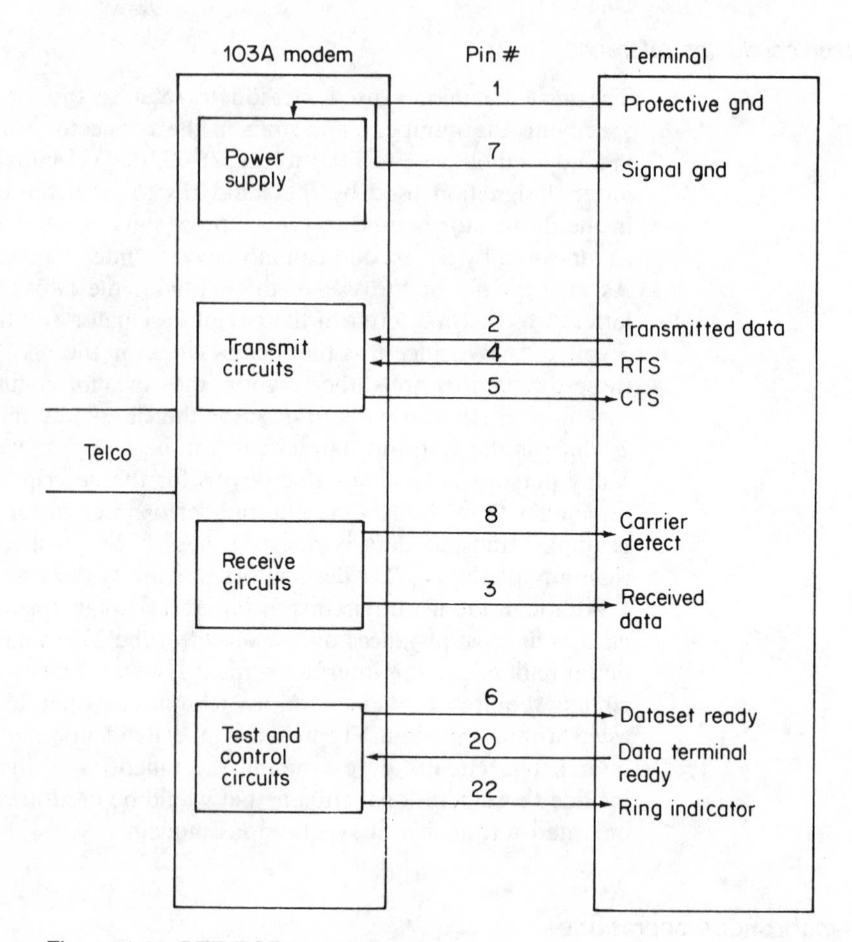

Figure 1.33 DTE-DCE interface example.

Protective Ground (GND, Pin l)

This interchange circuit is normally electrically bonded to the equipment's frame. In some instances, it can be further connected to external grounds as required by applicable regulations. Under RS-232-D this conductor use is modified to provide shielding.

Signal Ground (SG, Pin 7)

This circuit must be included in all RS-232 interfaces as it establishes a ground reference for all other lines. The voltage on this circuit is set to 0 V to provide a reference for all other signals. Although the conductors for pins 1 and 7 can be independent of one another, typical practice is to "strap" pin 7 to pin 1 at the modem. This is known as tying signal ground to frame ground.

Transmitted Data (TD, Pin 2)

The signals on this circuit are transmitted from a terminal device to the modem. When no data is being transmitted the terminal maintains this circuit in a marking or logical 1 condition. This is the circuit over which the actual serial bit stream of data flows from the terminal device to the modem where it is modulated for transmission.

Request to Send (RTS, Pin 4)

The signal on this circuit is sent from the terminal to the modem to prepare the modem for transmission. Prior to actually sending data, the terminal must receive a Clear to Send signal from the modem on pin 5.

Clear to Send (CTS, Pin 5)

This interchange circuit is used by the modem to send a signal to the attached terminal, indicating that the modem is ready to transmit. By turning this circuit OFF, the modem informs the terminal that it is not ready to receive data. The modem raises the CTS signal after the terminal initiates a Request to Send (RTS) signal.

Carrier Detect (CD, Pin 8)

Commonly referred to as Received Line Signal Detector (RLSD), a signal on this circuit is used to indicate to the terminal that the modem is receiving a carrier signal from a remote modem. The presence of this signal is also used to illuminate the carrier detect light-emitting diode (LED) indicator on modems equipped with that display indicator. If this light indicator should go out during a communications session, it indicates that the session has terminated owing to a loss of carrier, and software that samples for this condition will display the message "carrier lost" or a similar message to indicate this condition has occurred.

Receive Data (RD, Pin 3)

After data is demodulated by a modem, it is transferred to the attached terminal over this interchange circuit. When the modem is not sending data to the terminal, this circuit is held in the marking condition.

Data Set Ready (DSR, Pin 6)

Signals on this interchange circuit are used to indicate the status of the data set connected to the terminal. When this circuit is in the ON (logic 0) condition, it serve as a signal to the terminal that the modem is connected to the telephone line and is ready to transmit data. Since the RS-232 standard specifies that the DSR signal is ON when the modem is connected to the communications channel and not in any test condition, a modem using a self-testing feature or automatic dialing capability would pass this signal to the terminal after the self-test is completed or after the telephone number of a remote location was successfully dialed. Under RS-232-D this signal was renamed DCE Ready.

Data Terminal Ready (DTR, Pin 20)

This circuit is used to control the modem's connection to the telephone line. An ON condition on this circuit prepares the modem to be connected to the telephone line, after which the connection can be established by manual or automatic dialing. If the signal on this circuit is placed in an OFF condition, it causes the modem to drop any telephone connection in progress, providing a mechanism for the terminal device to control the line connection. Under RS-232-D this signal was renamed DTE Ready.

Ring Indicator (RI, Pin 22)

This interchange circuit indicates to the terminal device that a ringing signal is being received on the communications channel. This circuit is used by an auto-answer modem to "wake-up" the attached terminal device. Since a telephone rings for one second and then pauses for four seconds prior to ringing again, this line becomes active for one second every five seconds when an incoming call occurs.

Synchronous operations

One major difference between asynchronous and synchronous modems is the timing signals required for synchronous transmission.

Timing signals

When a synchronous modem is used, it puts out a square wave on pin 15 at a frequency equal to the modem's bit rate. This timing signal serves as a clock from which the terminal would synchronize its transmission of data onto pin 2 to the modem. Thus, pin 15 is referred to as transmit clock as well as its formal designator of Transmission Signal Element Timing (DCE), with DCE referencing the fact that the communications device supplies the timing.

Whenever a synchronous modem receives a signal from the telephone line it puts out a square wave on pin 17 to the terminal at a frequency equal to the modem's bit rate, while the actual data is passed to the terminal on pin 3. Since pin 17 provides receiver clocking, it is known as "receive clock" as well as its more formal designator of Receiver Signal Element Timing.

In certain cases a terminal device such as a computer port can provide timing signals to the DCE. In such situations the DTE will provide a clocking signal to the DCE on pin 24 while the formal designator of Transmitter Signal Element Timing (DTE) is used to reference this signal.

Intelligent operations

There are three interchange circuits that can be employed to change the operation of the attached communications device. One circuit can be used to first determine that a deterioration in the quality of a circuit has occurred, while the other two circuits can be employed to change the transmission rate to reflect the circuit quality.

Signal Quality Detector (CG, Pin 21)

Signals on this circuit are transmitted from the modem to the attached terminal whenever there is a high probability of an error in the received data due to the quality of the circuit falling below a predefined level. This circuit is maintained in an ON condition when the signal quality is acceptable and turned to an OFF condition when there is a high probability of an error. Under RS-232-D this circuit can also be used to indicate that a remote loopback is in effect.

Data Signal Rate Selector (CH/CI, Pin 23)

When an intelligent terminal device such as a computer port receives an OFF condition on pin 21 for a predefined period of time, it may be programmed to change the data rate of the attached modem, assuming that the modem is capable of operating at more than one data rate. This can be accomplished by the terminal device providing an ON condition on pin 23 to select the higher data signaling rate or range of rates while an OFF condition would select the lower data signaling rate or range of rates. When the data terminal equipment selects the operating rate the signal on pin 23 flows from the DTE to the DCE and the circuit is known as circuit CH. If the data communications equipment is used to select the data rate of the terminal device, the signal on pin 23 flows from the DCE to the DTE and the circuit is known as circuit CI.

Secondary circuits

In certain instances a synchronous modem will be designed with the capability to transmit data on a secondary channel simultaneously with transmission occurring on the primary channel. In such cases the data rate of the secondary channel is normally a fraction of the data rate of the primary channel.

To control the data flow on the secondary channel the RS-232-C/D standards employ five interchange circuits. Pins 14 and 16 are equivalent to the circuits on pins 2 and 3, except that they are used to transmit and receive data on the secondary channel. Similarly, pins 19, 13, and 12 perform the same functions as pins 4, 5, and 8 used for controlling the flow of information on the primary data channel.

In comparing the interchange circuits previously described to the connector illustrated in Figure 1.29, the reader should note that the location of each interchange circuit is explicitly defined by the pin number assigned to the circuit. In fact, the connector is designed with two rows of pins, with the top row containing 13 while the bottom row contains 12. Each pin has an explicit signal designation that corresponds to a numbering assignment that goes from left to right across the top row and then left to right across the bottom row of the connector. For ease of illustration the assignment of the interchange circuits to each of the pins in the D connector is presented in Figure 1.34 by rotating the connector 90 degrees clockwise. In this illustration, RS-232-D conductor changes from RS-232-C are denoted in parentheses.

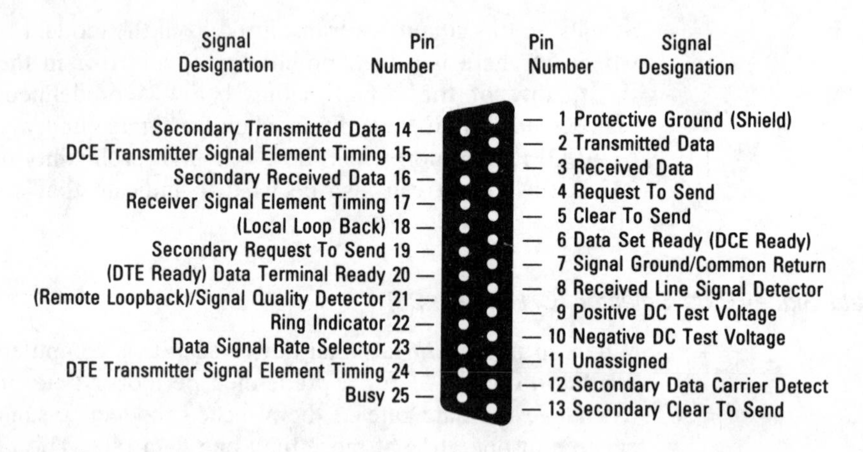

Figure 1.34 RS-232 interface on D connector.

Other interface standards

Three other interface standards that deserve mention are RS-366-A, RS-449, and RS-530.

The RS-366-A interface is employed to connect terminal devices to automatic calling units. This interface standard uses the same type 25-pin connector as RS-232, however, the pin assignments are different. A similar interface to RS-366-A is the CCITT V.25 recommendation, which is also designed for use with automatic calling units. Figures 1.35 and 1.36 illustrate the RS-366-A and CCITT V.25 interfaces. Note that for both interfaces each actual digit to be dialed is transmitted as parallel binary information over circuits 14 through 17. The pulse on pin 14 represents the value 2^0 while the pulses on pins 15 through 17 represent the values 2^1, 2^2, and 2^3 respectively. Thus, to indicate to the automatic calling unit that it should dial the digit 9, circuits 14 and 17 would become active.

Originally, automatic calling units provided the only mechanism to automate communications dialing over the PSTN. Due to the development and wide acceptance of the use of intelligent modems with automatic dialing capability the use of automatic calling units has greatly diminished.

Until recently, only intelligent asynchronous modems had an automatic dialing capability, restricting the use of automatic calling units to mainframe

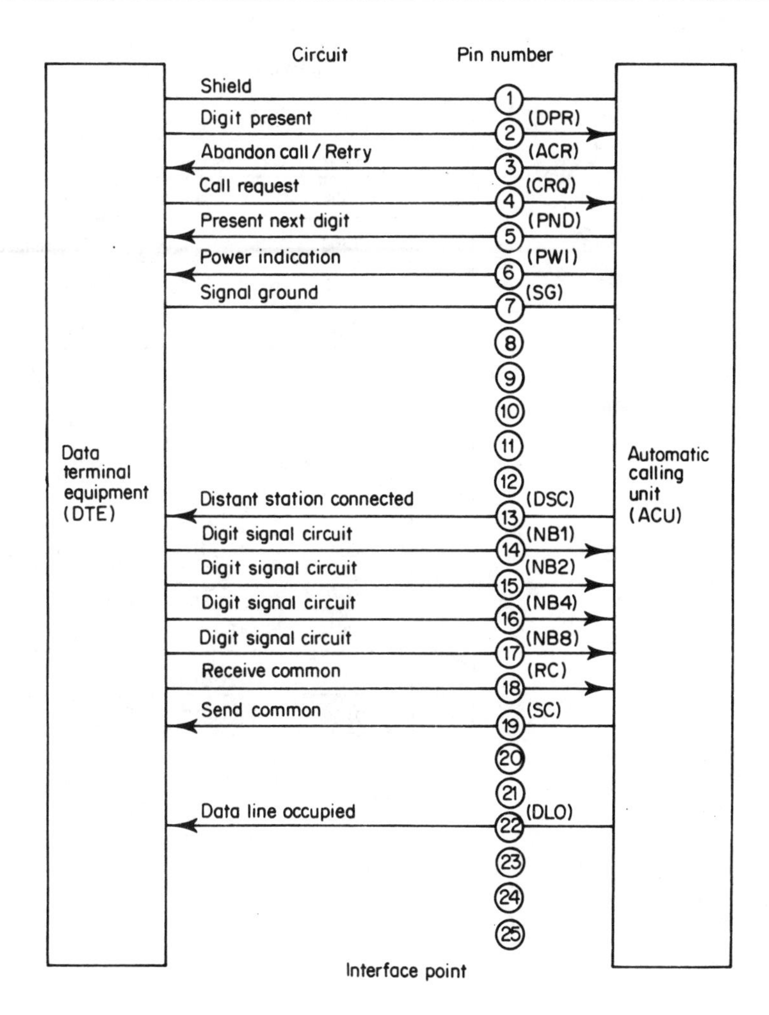

Figure 1.35 RS-366-A interface.

computers that required a method to originate synchronous data transfers over the PSTN. In such situations a special adapter was required to be installed in the communications controller of the mainframe, which controlled the operation of the automatic calling unit. The recent introduction of synchronous modems with automatic dialing capability can be expected to further diminish the requirement for automatic calling units since their use eliminates the requirement to install an expensive adapter in the communications controller as well as the cost of the automatic calling unit. The operation and utilization of intelligent modems is discussed in Chapter 2.

Another standard that requires mentioning is EIA RS-449. RS-449 was introduced in 1977 as an eventual replacement for RS-232-C. This interface specification calls for the use of a 37-pin connector as well as an optional 9-pin connector for devices using a secondary channel. Unlike RS-232-C/D, RS-449 does not specify voltage levels. Two additional specifications known as RS-442-A and RS-423 cover voltage levels for a specific range of data speeds.

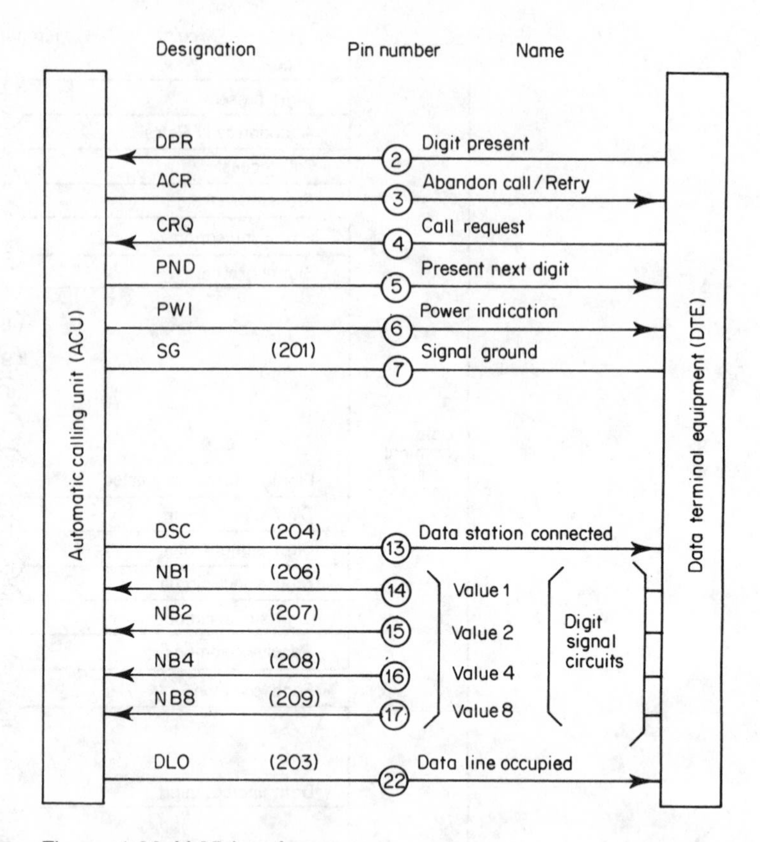

Figure 1.36 V.25 interface.

RS-442-A defines the voltage levels for data rates from 20 kbps to 10 Mbps while RS-423-A defines the voltage levels for data rates between 0 and 20 kbps.

The use of RS-422, RS-423 and RS-449 permits the cable distance between DTEs and DCEs to be extended to 4000 feet in comparison to RS-232-C's 50-foot limitation. Figure 1.37 indicates the RS-449 interchange circuits. If the reader compares RS-449 to RS-232-C/D, he will note the addition of seven circuits which are either new control or status indicators. Although a considerable number of articles have been written describing the use of RS-449, its complexity has served as a constraint in implementing this standard by communications equipment vendors. By mid-1988, less than a few percent of all communications devices were designed to operate with this interface. Due to the failure of RS-449 to obtain commercial acceptance the EIA issued RS-530 in March, 1987. This new standard is intended to gradually replace RS-449.

Circuit mnemonic	Circuit name	Circuit direction	Circuit type	
SG SC RC	Signal Ground Send Common Receive Common	— to DCE from DCE	Common	
IS IC TR DM	Terminal in Service Incoming Call Terminal Ready Data Mode	to DCE from DCE to DCE from DCE	Control	
SD RD	Send Data Receive Data	to DCE from DCE	Data	Primary channel
TT ST RT	Terminal Timing Send Timing Receive Timing	to DCE from DCE from DCE	Timing	Primary channel
RS CS RR SO NS SF SR SI	Request to Send Clear to Send Receiver Ready Signal Quality New Signal Select Frequency Signaling Rate Selector Signaling Rate Indicator	to DCE from DCE from DCE from DCE to DCE to DCE to DCE from DCE	Control	Primary channel
SSD SRD	Secondary Send Data Secondary Receive Data	to DCE from DCE	Data	Secondary channel
SRS SCS SRR	Secondary Request to Send Secondary Clear to Send Secondary Receiver Ready	to DCE from DCE from DCE	Control	Secondary channel
LL RL TM	Local Loopback Remote Loopback Test Mode	to DCE to DCE from DCE	Control	
SS SB	Select Standby Standby Indicator	to DCE from DCE	Control	

Figure 1.37 RS-449 interchange circuits.

RS-530

Like RS-232-C/D, RS-530 uses the near universal 25-pin D-shaped interface connector. Although this standard is intended to replace RS-449, both RS-422 and RS-423 standards specify the electrical characteristics of the interface and will continue in existence. These standards are referenced by the RS-530 standard.

Similar to RS-449, RS-530 provides equipment meeting this specification with the ability to transmit at data rates above the RS-232 limit of 20 kbps. This is accomplished by the standard specifying the utilization of balanced signals in place of several secondary signals and the Ring Indicator signal included in RS-232-C/D. This balanced signaling technique is accomplished by using two wires with opposite polarities for each signal to minimize distortion.

Pin number	Circuit	Description
1	—	Shield
2	BA	Transmitted Data
3	BB	Received Data
4	CA	Request To Send
5	CB	Clear To Send
6	CC	DCE Ready
7	AB	Signal Ground
8	CF	Received Line Signal Detector
9	DD	Receiver Signal Element Timing (DCE Source)
10	CF	Received Line Signal Detector
11	DA	Transmit Signal Element Timing (DTE Source)
12	DB	Transmit Signal Element Timing (DCE Source)
13	CB	Clear To Send
14	BA	Transmitted Data
15	DB	Transmitter Signal Element Timing (DCE Source)
16	BB	Received Data
17	DD	Receiver Signal Element Timing (DCE Source)
18	LL	Local Loopback
19	CA	Request To Send
20	CD	DTE Ready
21	RL	Remote Loopback
22	CC	DCE Ready
23	CD	DTE Ready
24	DA	Transmit Signal Element Timing (DTE Source)
25	TM	Test Mode

Figure 1.38 RS-530 interchange circuits.

Figure 1.38 summarizes the RS-530 interchange circuits based upon their pin assignment. Although no equipment conforming to this standard was being used at the time this book revision was prepared, its elimination of the Ring Indicator signal suggests that its use will be reserved for non-PSTN applications.

1.13 MULTIPLEXING AND DATA CONCENTRATION

As an economy measure multiplexers and data concentrators were developed to combine the serial data stream of two or more digital data sources onto one composite high-speed communications line. Figure 1.39 illustrates the typical employment of multiplexers, which are also referred to as data concentrators, since they "concentrate" data from several data sources onto one line.

The Time Division Multiplexer (TDM), Statistical Time Division Multiplexer (STDM) and Intelligent Time Division Multiplexer (ITDM) can be used on both analog and digital facilities. In comparison, the Frequency Division Multiplexer (FDM) can only be used on analog facilities while the T1 multiplexer can only be used on a digital T1 facility.

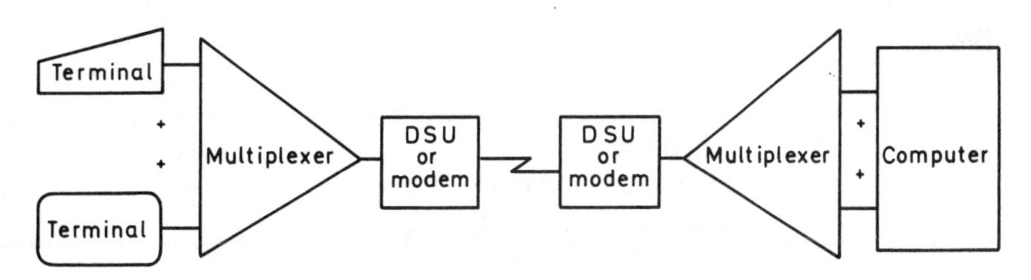

Figure 1.39 Typical multiplexer employment.

Multiplexers can be grouped into five main categories based upon both their method of operation and the type of high-speed line they can be connected to.

FDM

The Frequency Division Multiplexer (FDM) operates by subdividing the 3000 Hz bandwidth of an analog leased line into subchannels or derived bands. As data enters the FDM tones are placed on the line in a predefined subchannel to correspond to the bit composition. That is, a tone is placed on the line in the subchannel at one frequency to correspond to a binary 1 while a second tone at another frequency is placed on the line to correspond to a binary 0.

Figure 1.40 illustrates the operation of a four-channel FDM. Since each digital data stream is converted into a series of tones at predefined frequencies no modem is required when this multiplexer is used. Here a device known as a channel set which converts the digital data stream into tones and received tones into digital data replaces the modem. To prevent tones from one channel drifting into another channel a short segment of frequency known as a guard band is used to separate one channel from another.

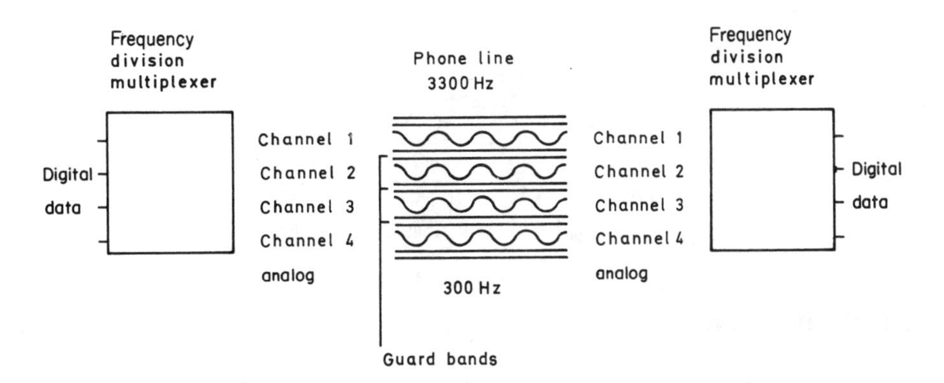

Figure 1.40 FDM operation.

TDM

The Time Division Multiplexer (TDM), as its name implies, separates the high-speed composite line into intervals of time. The TDM assigns a time slot to each low-speed channel whether or not a data source connected to the channel is active. If the data source is active at a particular point in time the TDM places either a bit or a character into the time slot, depending upon whether the TDM performs bit interleaving or character interleaving multiplexing. If the data source is not active the TDM places an idle bit or character into the time slot. At the opposite end of the high-speed data link another multiplexer reconstructs the data for each channel based upon its position in the sequence of time slots. This process is called demultiplexing.

The top of Figure 1.41 illustrates the use of TDMs to multiplex four data sources onto one high-speed line. The lower portion of Figure 1.41 illustrates how the TDMs interleave bits or characters into time slots on the high-speed line at one end of the communications link (multiplexing) and remove the bits or characters from the high-speed line (demultiplexing), sending them to their appropriate channels.

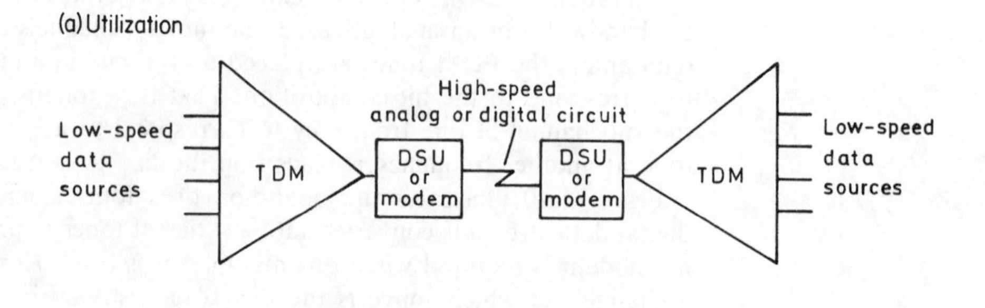

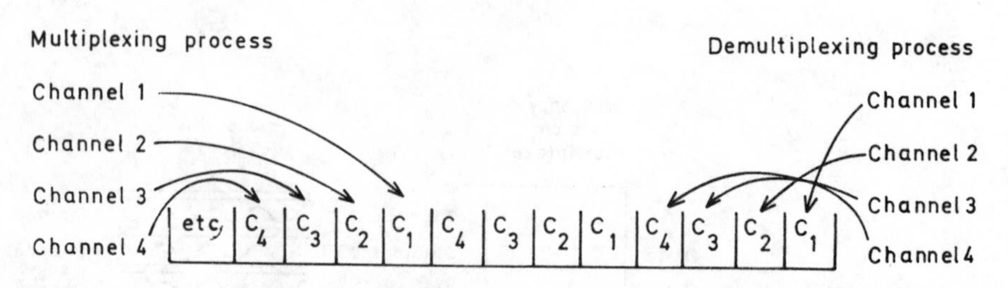

Figure 1.41 Time division multiplexing.

STDM and ITDM

Both statistical and intelligent time division multiplexers (STDM and ITDM) are more efficient than TDMs due to their use of a technique commonly referred to as the dynamic allocation of bandwidth. In this technique a microprocessor in the multiplexer only places data into a time slot when a device connected to the multiplexer is both active and transmitting. Since a time slot can now hold data from any channel, the dynamic allocation of bandwidth requires that an address indicating from what channel the data originated be included in the slot to enable demultiplexing to occur correctly.

The primary advantage of STDMs and ITDMs over conventional TDMs is their greater efficiency in transmitting and receiving data on the high-speed line connecting two multiplexers. Although STDMs and ITDMs both employ the dynamic allocation of bandwidth, ITDMs also perform data compression. Since this reduces the amount of data that has to be transmitted it results in the ITDM being the more efficient device. STDMs and ITDMs service between one and a half and four times as many data sources as a TDM.

If all or a majority of the data sources connected to an STDM or an ITDM became active at the same time the buffer in these multiplexers would overflow, causing data to be lost. This situation would occur since the composite data rate of active channels would exceed the operating rate of the high-speed line, causing the buffers in the multiplexers to fill and eventually overflow. To prevent this situation from occurring, both STDMs and ITDMs incorporate one or more techniques that inhibit data transmission into the multiplexer when the data in its buffer reaches a predefined level. Then, after data is transferred from the buffer onto the composite high-speed line the buffer occupancy is lowered until another predefined level is reached. This lower level then becomes a trigger mechanism for the multiplexer to enable previously inhibited data sources to resume transmission.

The process of inhibiting and enabling data transmission is known as flow control. The most common method of flow control is obtained by lowering and raising the Clear to Send control signal. Another common method of flow control is obtained by having the multiplexer transmit the XOFF and XON characters to devices that recognize those characters as a signal to stop and resume transmission. Figure 1.42 illustrates the statistical multiplexing process.

T1 Multiplexer

The T1 multiplexer is a data concentration device specifically designed to operate at the T1 data rate of 1.544 Mbps in North America and 2.048 Mbps in Europe. Due to the high data rate this device supports it is normally used to multiplex both voice and data as well as video onto a single channel.

Most T1 multiplexers are marketed with several optional voice digitization modules end-users can select. The most common module digitizes voice based upon a technique known as pulse code modulation (PCM) into a 64 kbps data stream. If used only to support digitized voice resulting from PCM, a T1 multiplexer would enable 24 voice conversations to be multiplexed onto one 1.544 Mbps T1 line.

Many T1 multiplexers include modules that permit voice to be digitized at 32 kbps, while some multiplexers support modules that digitize voice at 16 kbps. With these modules users could multiplex 48 or 72 voice conversations

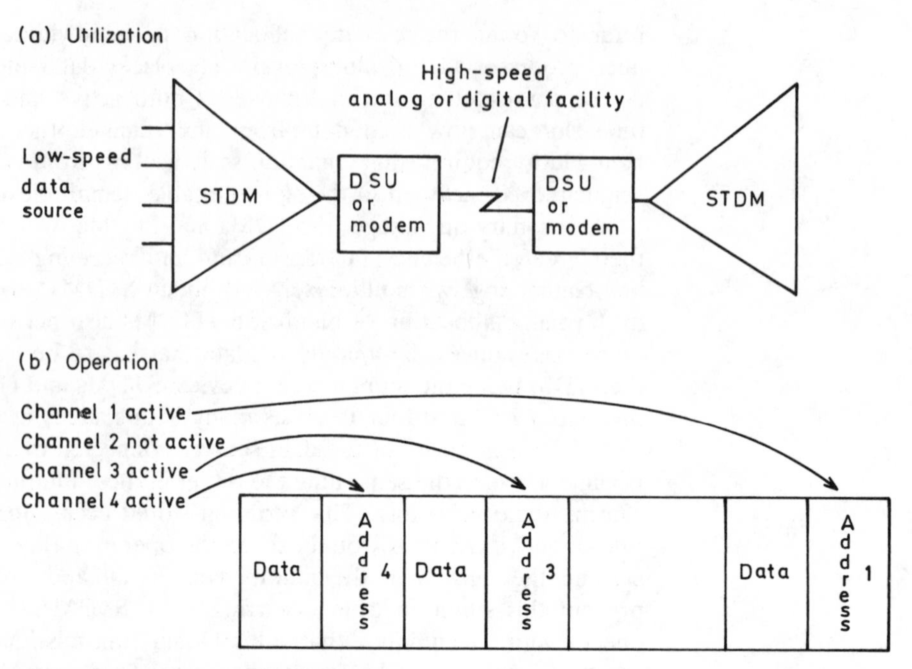

Figure 1.42 Statistical multiplexing.

onto one T1 circuit through the use of T1 multiplexers. The reader is referred to Chapter 3 for additional information covering the operation and utilization of the multiplexers briefly described in this section.

1.14 POWER MEASUREMENTS

In referencing the operations of certain types of communications equipment, the measurement of power gains and losses is frequently expressed in terms of decibels (dB). When used to express a loss, a minus sign is placed before dB, while the absence of a minus sign or the presence of a plus sign is used to indicate a power gain.

Instead of telling one the actual amount of power, the decibel is used to describe the ratio of power in a circuit, describing the power output of a circuit to its input power. If there is less output power than input power, you have a dB loss. If there is more output power than input power, this condition signifies a dB power gain.

Decibel gains and losses are computed using the formula:

$$\text{Number of decibels} = 10 * \log_{10} P_1/P_2$$

where: P_1 is the larger power and P_2 is the smaller power.

To illustrate the use of decibels, assume the input power to a transmission circuit is 1 milliwatt (mW) and its output power is 0.5 mW. The power change on this circuit would become:

$$\text{dB} = 10 * \log_{10} 1/0.5 = 10 * \log_{10} 2$$

From a table of logarithms the log to the base 10 of 2 is .3010, thus the number of decibels is 3.01. Since the input power was greater than the output power, we have a power loss and the dB loss is then −3.01 dB.

From the preceding, it should be apparent that a loss of 3 dB represents a 50% power loss while a gain of 3 dB represents a gain of twice as much power. Table 1.19 lists the relationship between a dB gain or loss and the ratio of power input to output on a circuit.

Table 1.19 Relationship between dB and power.

When you have this dB gain or loss	Larger power divided by smaller power is
1	1.2
2	1.6
3	2.0
4	2.5
5	3.2
6	4.0
7	5.0
8	6.4
9	8.0
10	10.0
20	100.0
30	1,000.0
40	10,000.0

In telephone operations the reference level of power is 0.001 watt (1 mW). This level was selected as it represents the average amount of power generated in the telephone transmitter during a voice conversation. Thus, by using a 1 mW power level, telephone company personnel have a reference level for comparing gains and losses in a circuit. For convenience, 1 mW of power is designated as being equal to 0 dB. To ensure that no one forgets that 1 mW is the reference level, the letter m is attached to the power level. Thus, 0 dB becomes 0 dBm. As a voice circuit is routed by the telephone company through amplifiers, the losses and gains are added algebraically. Thus, a −20 dBm loss on a circuit followed by a gain of +12 dBm due to an amplifier would result in a −8 dBm overall loss.

In addition to using a reference power level, telephone company personnel also use a standard frequency for testing voice circuits. In the United States a frequency of 1004 hertz (1kHz) is used, resulting in a device that supplies 0 dBm of power to a circuit at a frequency of approximately 1 kHz for line-testing purposes.

Transmission Level Point

The transmission level point (TLP in North America and dBr International) is the power in dBm that should be measured when a specific test tone signal

(0 dBm, 1004 Hz North America or 800 Hz International) is transmitted at some point selected as a reference point. As an example, a point where a reading of −12 dBm is expected would be a −12 TLP.

To reference the TLP back to the test tone level a unit of measurement known as the dBm0 is used where:

zero transmission level = actual measurement − test tone level
(dBm0 = dBm − TLP)

Here, dBm0 is a measurement which indicates the departure of a system from its design value.

Figure 1.43 illustrates the relationship between the transmission level point (TLP) and the zero transmission level (dBm0) based upon measured readings at five locations from a telephone company installed line termination block installed at an end-user location. By measuring the actual dBm, the resulting computation of the zero transmission level provides communications carrier personnel with an indication of the adjustment required at each point to bring the system back to its design goal.

BLOCK — 18 dB loss — 20 dB AMPLIFIER — 10 dB loss — 6 dB AMPLIFIER

	*	*	*	*	*
TLP (dBr):	11	−7	+13	+3	+9
Measured dBm:	10	−10	+9	−1	+6
dBm0:	−1	−3	−4	−4	−3

Figure 1.43 Measurement comparison.

Noise Measurement Units

The noise level at any point in a transmission system is the ratio of channel noise at that point to some predefined amount of noise chosen as a reference. This ratio is usually expressed in decibels above reference noise, abbreviated as dBrn.

For noise measurements, the dB scale is shifted so that −90 dBm becomes the "reference noise" level as illustrated in Figure 1.44. Thus, 0 dBrn = −90 dBm.

When a meter is used to measure noise a "C-message" filter is used between the meter and the line. When this filter is used in noise measurements the units are called decibels above reference noise C-message weighted, abbreviated as dBrnc.

Similar to power measurements, a transmission level point is used to provide a reference level for noise where:

C-message weighted zero level = actual noise level − test tone level
(dBrnc0 = dBrnc − dBm)

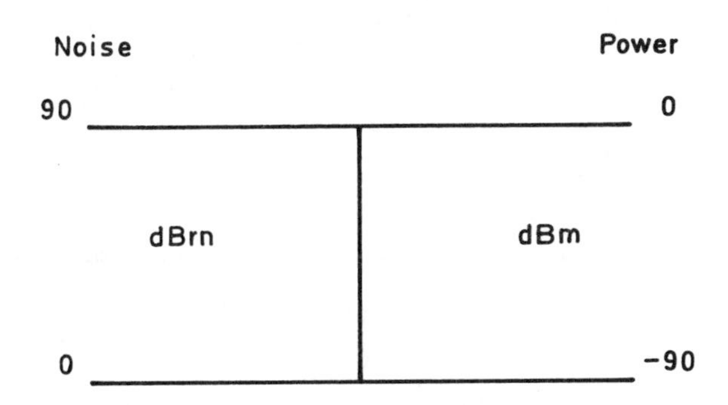

Figure 1.44 Noise and power relationship.

Levels

The level of a circuit references the signal intensity or noise intensity and is usually measured in dBm. If levels are too high, the amplifiers in a circuit path can become overloaded, resulting in an increase in crosstalk or intermodulation. If levels are too low, the signal may not be received correctly. Normally, on leased lines the communications carrier adjusts the receive level to provide a signal strength between -13 and -16 dBm based upon a modem transmit level of 0 dBm. For switched telephone network applications the central office of the communications carrier should not receive a signal greater than -12 dBm.

1.15 CABLES, CONNECTORS, PLUGS AND JACKS

A variety of cables and connectors can be employed in data transmission systems. Cables and connectors are normally employed to connect data terminal equipment (DTE) to data communications equipment (DCE) while plugs and jacks are used to connect communications equipment to telephone company circuits.

Twisted-pair cable

The most commonly employed data communications cable is the twisted-pair cable. This cable can usually be obtained with 4, 7, 9, 12, 16, or 25 conductors, where each conductor is insulated from another by a PVC shield.

For EIA RS-232 and CCITT V.24 applications, those standards specify a maximum cabling distance of 50 feet between DTE and DCE equipment for data rates ranging from 0 to 19,200 bps; and, normal industry practice is to use male connectors at the cable ends which mate with female connectors normally built into such devices as terminals and modems. Figure 1.45 illustrates the typical cabling practice employed to connect a DTE to a DCE.

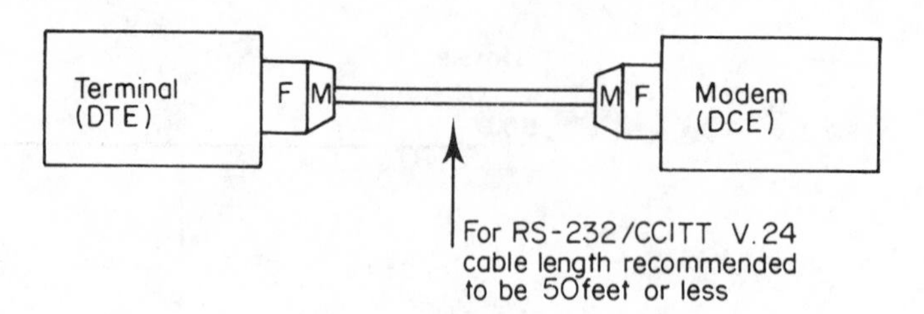

Figure 1.45 DTE to DCE cabling.

Low-capacitance shielded cable

In certain environments where electromagnetic interference and radio frequency emissions could be harmful to data transmission, one should consider the utilization of low-capacitance shielded cable in place of conventional twisted-pair cable. Low-capacitance shielded cable includes a thin wrapper of lead foil that is wrapped around the twisted-pair conductors contained in the cable, thereby providing a degree of immunity to electrical interference that can be caused by machinery, fluorescent ballasts, and other devices.

Ribbon cable

Since an outer layer of PVC houses the individual conductors in a twisted-pair cable, the cable is rigid with respect to its ability to be easily bent. Ribbon or flat cable consists of individually insulated conductors that are insulated and positioned in a precise geometric arrangement that results in a rectangular rather than a round cross-section. Since ribbon cable can be easily bent and folded, it is practical for those situations where one must install a cable that must follow the contour of a particular surface.

The null modem

No discussion of cabling would be complete without a description of a null modem, which is also referred to as a modem eliminator. A null modem is a special cable that is designed to eliminate the requirement for modems when interconnecting two collocated data terminal equipment devices. One example of this would be a requirement to transfer data between two collocated personal computers that do not have modems and use different types of diskettes, such as an IBM PC which uses a $5\frac{1}{4}$-inch diskette and an IBM PS/2 which uses a $3\frac{1}{2}$-inch diskette. In this stiuation, the interconnection of the two computers via a null modem cable would permit programs and data to be transferred between each personal computer in spite of the media incompatibility of the two computers. Since DTEs transmit data on pin 2 and receive data on pin 3, one could never connect two such devices together with a conventional cable as the data transmitted from one device would never be received by the other.

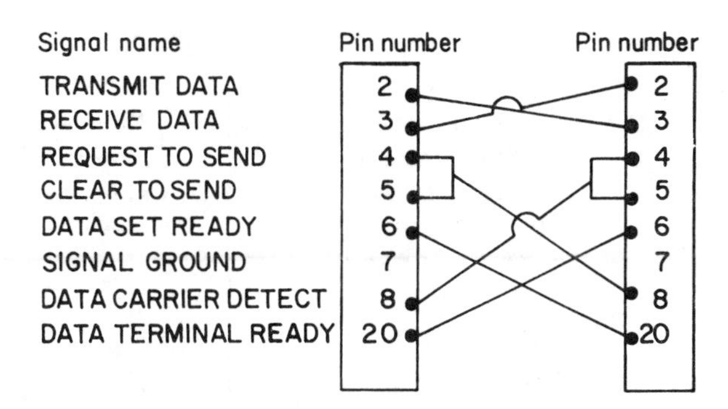

Figure 1.46 Null modem cable.

In order for two DTEs to communicate with one another, a connector on pin 2 of one device must be wired to connector pin 3 on the other device. Figure 1.46 illustrates an example of the wiring diagram of a null modem cable, showing how pins 2 and 3 are cross-connected as well as the configuration of the control circuit pins on this type of cable.

Since a terminal will raise or apply a positive voltage in the 9- to 12-volt range to turn on a control signal, one can safely divide this voltage to provide up to three different signals without going below the signal threshold of 3 volts previously illustrated in Figure 1.31. In examining Figure 1.46, we should note the following control signal interactions are caused by the pin cabling:

(1) Data Terminal Ready (DTR, pin 20) raises Data Set Ready (DSR, pin 6) at the other end of the cable. This makes the remote DTE think a modem is connected to the other end and powered ON.
(2) Request to Send (RTS, pin 4) raises Data Carrier Detect (CD, pin 8) on the other end and signals Clear to Send (CTS, pin 5) at the original end of the cable. This makes the DTE believe that an attached modem received a carrier signal and is ready to modulate data.
(3) Once the handshaking of control signals is completed, we can transmit data onto one end of the cable (TD, pin 2) which becomes receive data (RD, pin 3) at the other end.

The cable configuration illustrated in Figure 1.46 will work for most data terminal equipment interconnections; however, there are a few exceptions. The most common exception is when a terminal device is to be cabled to a port on a mainframe computer that operates as a "ring-start" port. This means that the computer port must obtain a Ring Indicator (RI, pin 22) signal. In this situation, the null modem must be modified so that Data Set Ready (DSR, pin 6) is jumpered to Ring Indicator (RI, pin 22) at the other end of the cable to initiate a connect sequence to a "ring-start" system.

Owing to the omission of transmit and receive clocks, the previously described null modem can only be used for asynchronous transmission. For synchronous transmission one must either drive a clocking device at one end of the cable or employ another technique. Here one would use a modem eliminator which differs from a null modem by providing a clocking signal to the interface. If

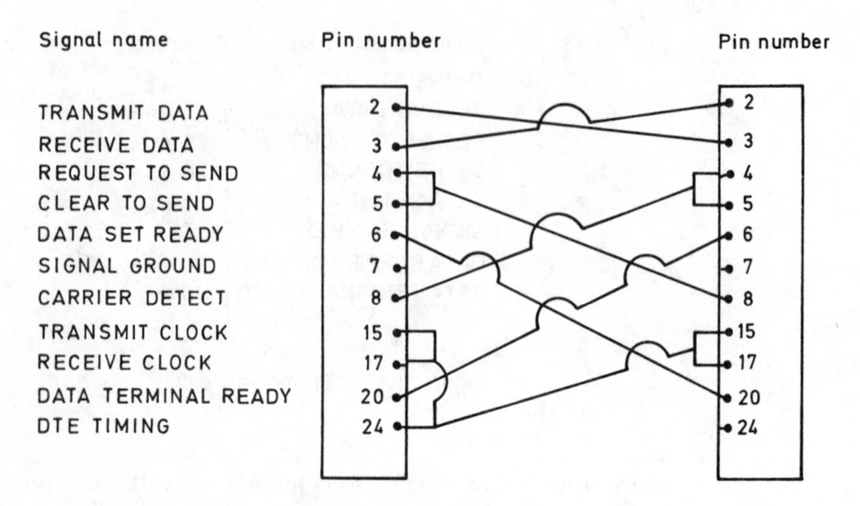

Figure 1.47 Synchronous null modem cable.

a clocking source is to be used, DTE Timing (pin 24) is normally selected to develop a synchronous null modem cable. In developing this cable, pin 24 is strapped to pins 15 and 17 at each end of the cable as illustrated in Figure 1.47. Then, DTE Timing provides transmit and receive clocking signals at both ends of the cable.

Cabling tricks

A general purpose 3-conductor cable can be used when there is no requirement for hardware flow control and a modem will not be controlled. Figure 1.48 illustrates the use of a 3-conductor cable for DTE to DCE and DTE to DTE

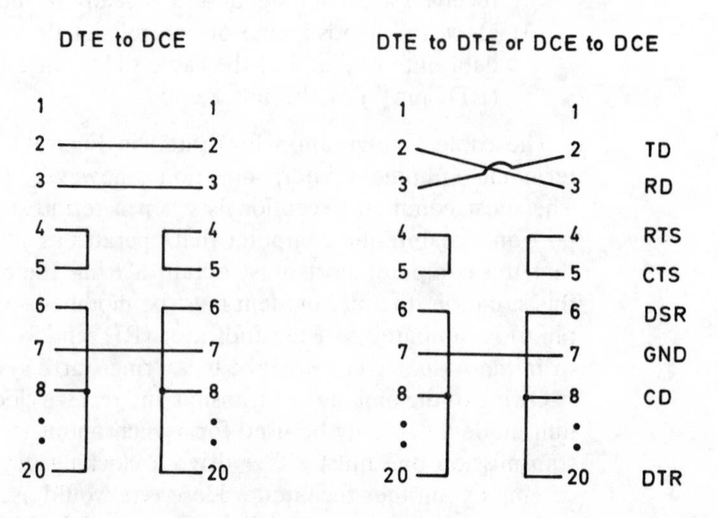

Figure 1.48 General purpose 3-conductor cable.

or DCE to DCE connections. When this situation occurs it becomes possible to use a 9-conductor cable with three D-shaped connectors at each end, with each connected to three conductors on the cable connector. Doing so eliminates the necessity of installing three separate cables.

Figure 1.49 illustrates a 5-conductor cable that can be installed between a DTE and DCE (modem) when asynchronous control signals are required. Similar to the use of 9-conductor cable to derive three 3-conductor connections, standard 12-conductor cable can be used to derive two 5-conductor connections.

Terminal			Modem
	1		1
	2	TD	2
	3	RD	3
RDS	4		4
CTS	5		5
	7	GND	7
	8	DCD	8
	20	DTR	20

Figure 1.49 General purpose 5-conductor cable.

Plugs and jacks

Data communications equipment is connected to telephone company facilities by a plug and jack arrangement as illustrated in Figure 1.50. Although the connection appears to be, and in fact is, simplistic, the number of connection arrangements and differences in the types of jacks offered by telephone companies usually ensures that the specification of an appropriate jack can be a complex task. Fortunately, most modems and other communications devices include explicit instructions covering the type of jack the equipment must be connected to as well as providing the purchaser with information that must be furnished to the telephone company in order to legally connect the device to the telephone company line.

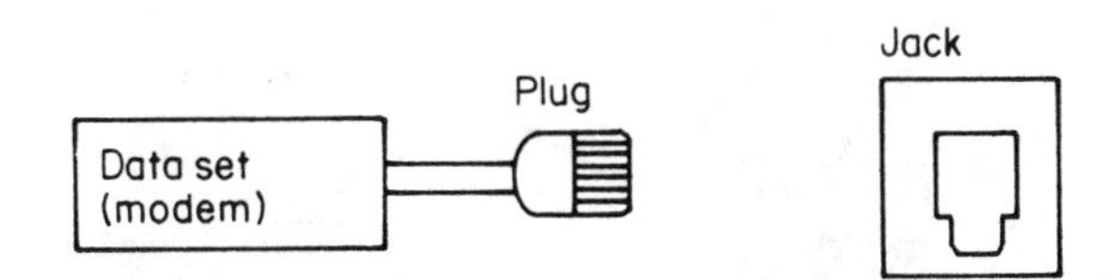

Figure 1.50 Connection to telephone company facilities. Data communications equipment can be connected to telephone company facilities by plugging the device into a telephone company jack.

Most communications devices designed for operation on the PSTN interface the telephone company network via the use of an RJ11C permissive or an RJ45S programmable jack.

Figure 1.51 illustrates the conductors in the RJ11 and RJ45 modular plugs. The RJ11 plug is primarily used on 2-wire dial lines. This plug is used in both the home and in an office for connecting a single instrument telephone to the PSTN. In addition, the RJ11 also serves as an optional connector for 4-wire private lines. Although the RJ11 connector is fastened to a cable containing 4 or 6 stranded-copper conductors, only 2 wires in the cable are used for switched network applications. When connected to a 4-wire leased line, 4 conductors are used.

The development of the RJ11 connector can be traced to the evolution of the switchboard. The plugs used with switchboards had a point known as the "tip" which was colored red, while the adjacent sleeve known as the ring was colored green.

The original color coding used with switchboard plugs was carried over to telephone wiring. If you examine a 4-wire (two-pair) telephone cable, you will note that the wires are colored yellow, green, red and black. The green wire is the tip of the circuit while the red wire is the ring. The yellow and black wires can be used to supply power to the light in a telephone or used to control a secondary telephone using the same 4-wire conductor cable.

The most common types of telephone cable used for telephone installation are 4-wire and 6-wire conductors. Normally, a 4-wire conductor is used in a residence that requires one telephone line. A 6-wire conductor is used in either a residential or business location that requires two telephone lines and can also be used to provide three telephone lines from one jack. Table 1.20 compares the color identification of the conductors in 4-wire and 6-wire telephone cable.

During the late 1970s, telephone companies replaced the use of multiprong plugs by the introduction of modular plugs which in turn are connected to modular jacks.

The RJ11C plug was designed for use with any type of telephone equipment that requires a single telephone line. Thus, regardless of the use of either 4-wire or 6-wire cable only 2 wires in the cable need be connected to an RJ11C

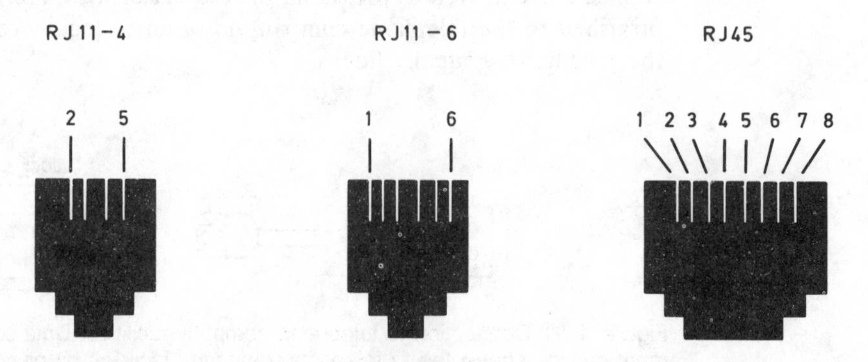

Figure 1.51 RJ11 and RJ45 modular plugs.

Table 1.20 Color identification of telephone cables.

4-Wire		6-Wire	
Pair	Color	Pair	Color
1	YEL GRN	1	BLUE YEL
2	RED BLK	2	GRN RED
		3	BLK WHT

jack. The RJ11 plug can also be used to service an instrument that supports two or three telephone lines, however, RJ14C and RJ25C jacks must then be used to provide that service. These two jacks are only used for voice. For data transmission both 4- and 6-conductor plugs are available for use, with conductors 1, 2, 5, and 6 in the jack normally reserved for use by the telephone company. Then, conductor 4 functions as the ring circuit while conductor 5 functions as the tip to the telephone company network.

The RJ45 plug is also designed to support a single line although it contains 8 positions. In this plug, positions 4 and 5 are used for ring and tip and a programmable resistor on position 8 in the jack is used to control the transmit level of the device connected to the switched network.

The physical size of the plugs used to wire equipment to each jack as well as the size of each of the previously discussed jacks are the same. The only difference between jacks is in the number of wires cabled to the jack and the number of contacts in the jack which are used to pass telephone wire signals.

Connecting arrangements

There are three connecting arrangements that can be used to connect data communications equipment to telephone facilities. The object of these arrangements is to ensure that the signal received at the telephone company central office does not exceed −12 dBm.

Permissive arrangement

The permissive arrangement is used when one desires to connect a modem to an organization's switchboard, such as a private branch exchange (PBX). When a permissive arrangement is employed, the output signal from the modem is fixed at a maximum of −9 dBm and the plug that is attached to the data set cable can be connected to three types of telephone company jacks as illustrated in Figure 1.52. The RJ11 jack can be obtained as a surface mounting (RJ11C) for desk sets or as a wall-mounted (RJ11W) unit, while the RJ41S and RJ45S are available only for surface mounting.

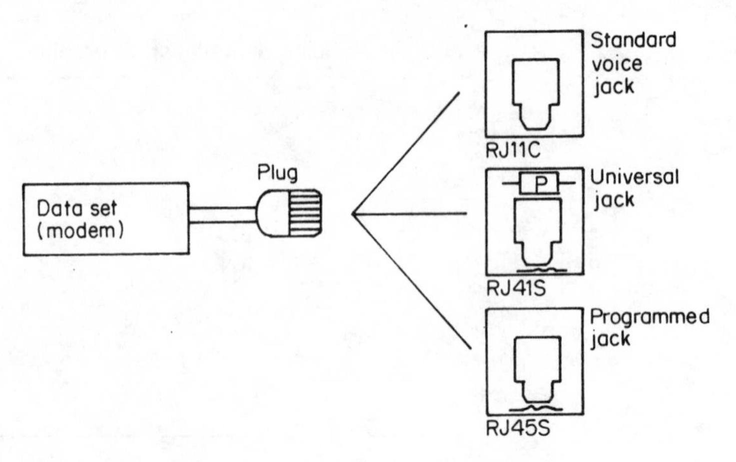

Figure 1.52 Permissive arrangement jack options.

Since permissive jacks use the same 6-pin capacity miniature jack used for standard voice telephone installations, this arrangement provides for good mobility of terminals and modems.

Fixed loss loop arrangement

Under the fixed loss loop arrangement the output signal from the modem is fixed at a maximum of −4 dBm and the line between the subscriber's location and the telephone company central office is set to 8 dBm of attenuation by a pad located within the telephone company provided jack. As illustrated in Figure 1.53, the only jack that can be used under the fixed loss loop arrangement is the RJ41S. This jack has a switch labeled FLL–PROG, which must be placed in the FLL position under this arrangement. Since the modem output is limited to −4 dBm, the 8 dB attenuation of the pad ensures that the transmitted signal reaches the telephone company office at −12 dBm. As the pad in the jack reduces the receiver signal-to-noise ratio by 8 dB, this type of arrangement is more susceptible to impulse noise and should only be used if one cannot use either of the two other arrangements.

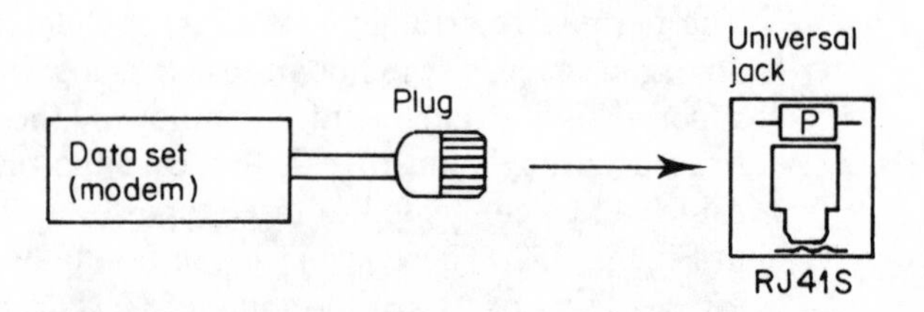

Figure 1.53 Fixed loss loop arrangement.

Programmable arrangement

Under the programmable arrangement configuration a level setting resistor inside the standard jack provided by the telephone company is used to set the transmit level within a range between 0 and −12 dBm. Since the line from the user is directly routed to the local telephone company central office at installation time, the telephone company will measure the loop loss and set the value of the resistor based upon the loss measurement. As the resistor automatically adjusts the transmitted output of the modem so the signal reaches the telephone company office at −12 dBm, the modem will always transmit at its maximum allowable level. As this is a different line interface in comparison to permissive or fixed loss data sets, the data set must be designed to operate with the programmability feature of the jack.

Either the RJ41S universal jack or the RJ45S programmed jack can be used with the programmed arrangement as illustrated in Figure 1.54. The RJ41S jack is installed by the telephone company with both the resistor and pad for programmed and fixed loss loop arrangements. By setting the switch to PROG, the programmed arrangement will be set. Since the RJ45S jack can operate in either the permissive or programmed arrangement without a switch, it is usually preferred as it eliminates the possibility of an inadvertent switch reset.

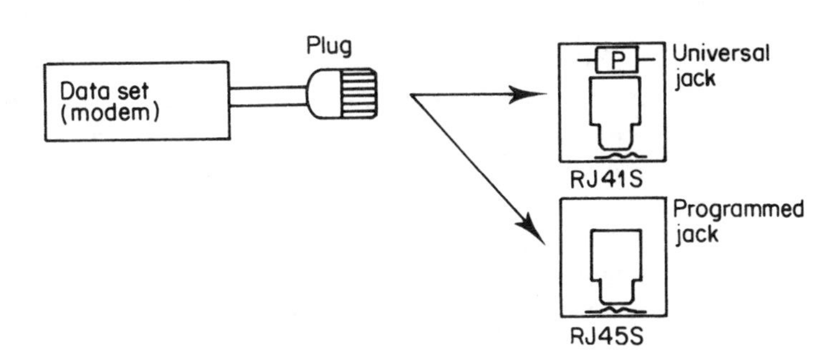

Figure 1.54 Programmed arrangement jack.

Telephone options

A telephone set can be connected to one's data line and used for voice conversations, call origination, and call answering. As part of the ordering procedure you must specify a series of specific options that are listed in Table 1.21.

When the telephone set is optioned for telephone set controls the line, calls are originated or answered with the telephone by lifting the handset off-hook. To enable control of the line to be passed to a modem or data set an "exclusion key" is required.

The exclusion key telephone permits calls to be manually answered and then transferred to the modem using the exclusion key. The exclusion key telephone is wired for either "telephone set controls line" or "data set controls line".

Table 1.21 Telephone ordering options.

Decision		Description
A	1	Telephone set controls line
	2	Data set controls line
B	3	No aural monitoring
	4	Aural monitoring provided
C	5	Touchtone dialing
	6	Rotary dialing
D	7	Switchhook indicator
	8	Mode indicator

Data set control is normally selected if one has an automatic call or automatic answer modem since this permits calls to be originated or answered without taking the telephone handset off-hook. To use the telephone for voice communications the handset must be raised and the exclusion key placed in an upward location.

The telephone set control of the line option is used with manual answer or manual originate modems or automatic answer or originate modems that will be operated manually. To connect the modem to the line the telephone must be off-hook and the exclusion key placed in an upward position. To use the telephone for voice communications the telephone must be off-hook while the exclusion key is placed in the downward position.

When the data set controls the line option is selected, calls can be automatically originated or answered by the data equipment without lifting the telephone handset.

Aural monitoring enables the telephone set to monitor call progress tones as well as voice answer back messages without requiring the user to switch from data to voice.

Users can select option B3 if aural monitoring is not required, while option B4 should be selected if it is required. Option C5 should be selected if touchtone dialing is to be used, while option C6 should be specified for rotary dial telephones. Under option D7, the exclusion key will be bypassed, resulting in the lifting of the telephone handset causing the closure of the switchhook contact in the telephone. In comparison, option D8 results in the exclusion key contacts being wired in series with the switchhook contacts, indicating to the user whether he or she is in a voice or data mode.

Ordering the business line

Ordering a business line to transmit data over the switched telephone network in the United States currently requires one to provide the telephone company with four items of information. First, one must supply the telephone company with the Federal Communications Commission (FCC) registration number of the device to be connected to the switched telephone network. This 14-

character number can be obtained from the vendor who must first register their device for operation on the switched network prior to making it available for use on that network.

Next, one must provide the ringer equivalence number of the data set to be connected to the switched network. This is a three-character number, such as 0.4A, and represents a unitless quotient formed in accordance with certain circuit parameters. Finally, one must provide the jack numbers and arrangement to be used as well as the telephone options if you intend to use a handset.

1.16 LINK TERMINOLOGY

A circuit over which data is transmitted provides a link between those devices used to transmit and receive data. Thus, a circuit used to transmit data between two terminals is commonly referred to as a terminal-to-terminal link.

When data is transmitted from one terminal to a computer, the circuit may be called a computer-to-terminal link. When more than one terminal transmits data over a common circuit (multidrop circuit) to a computer, the line is called a multiterminal-to-computer link. Although terminals can communicate directly to a computer over individual computer-to-terminal links, economics may justify the utilization of a device to combine the data from many low-speed terminals onto one or more high-speed paths for retransmission to a computer. One such device that can be used to combine the data transmitted from many terminals, is called a concentrator. Circuits which connect the terminals to the concentrator are called concentrator-to-terminal links, while the high-speed lines that connect the concentrator to the computer or host processor are known as concentrator-to-host links. When one concentrator transmits data to another concentrator, this type of circuit is known as a concentrator-to-concentrator link. Finally, the transmission path from one computer to another computer is known as a host-to-host link. Such link terminology is illustrated in Figure 1.55.

1.17 THE ISO REFERENCE MODEL

The International Standards Organization (ISO) established a frame-work for standardizing communications systems called the Open System Interconnection (OSI) Reference Model. The OSI architecture defines the communications process as a set of seven layers, with specific functions isolated to and associated with each layer. Each layer, as illustrated in Figure 1.56, covers lower layer processes, effectively isolating them from higher layer functions. In this way, each layer performs a set of functions necessary to provide a set of services to the layer above it. Layer isolation permits the characteristics of a given layer to change without impacting the remainder of the model, provided that the supporting services remain the same. The major advantage of this layered approach is that users can mix and match OSI conforming communications products to tailor their communications system to satisfy a particular networking requirement.

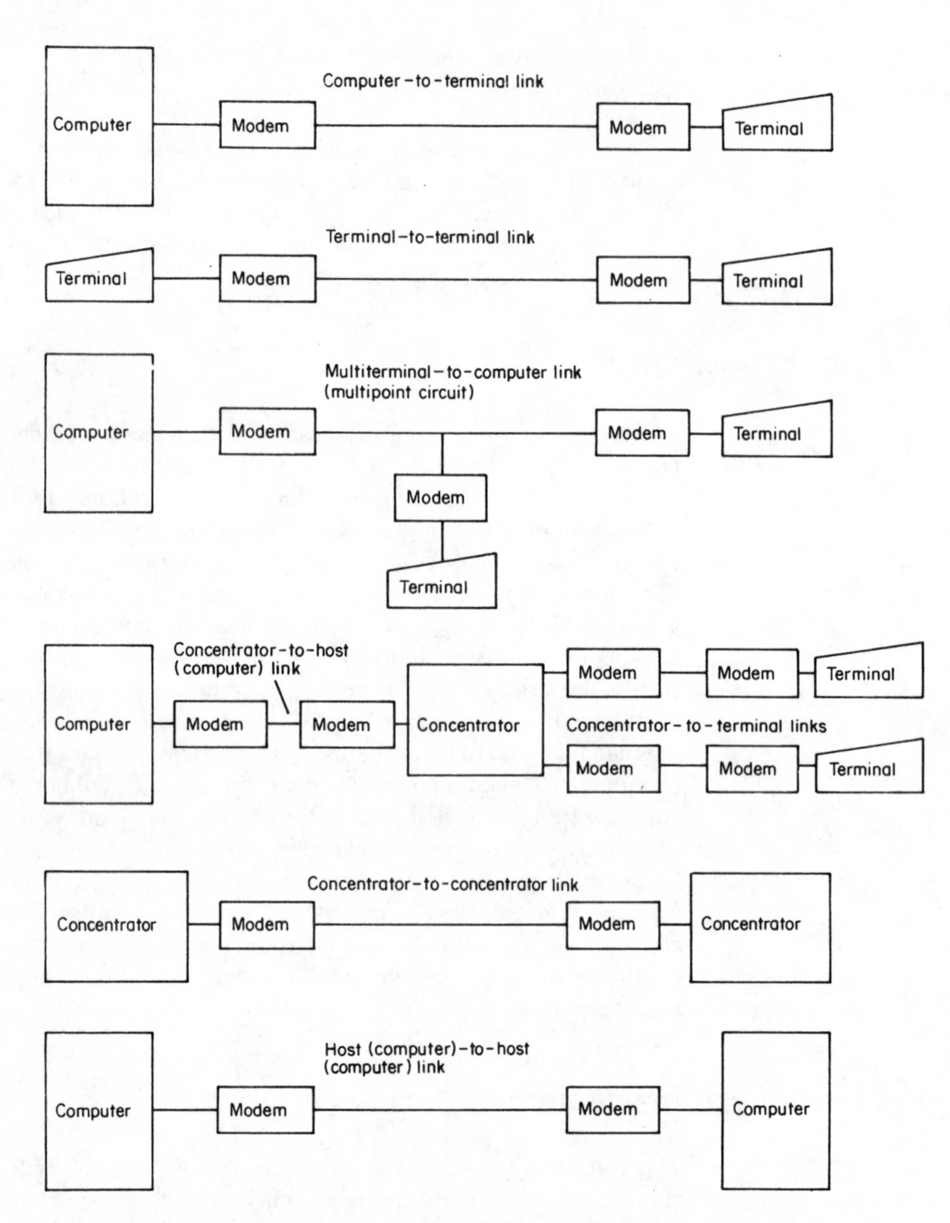

Figure 1.55 Link terminology. Link terminology defines the data path between communications components, computers, and remote terminals.

Layer 1

At the lowest or most basic level, the physical layer (level 1) is a set of rules that specifies the electrical and physical connection between devices. This level specifies the cable connections and the electrical rules necessary to transfer data between devices. Typically, the physical link corresponds to established interface standards such as RS-232-C/D, RS-422-A, or RSA-423-A.

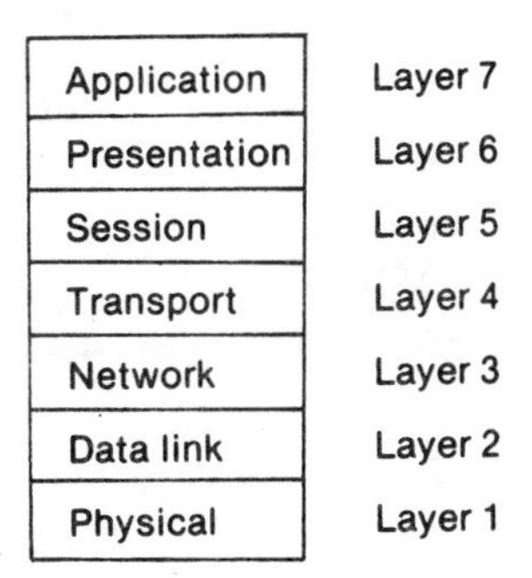

Figure 1.56 ISO reference model.

Layer 2

The next layer, which is known as the data link layer (level 2), denotes how a device gains access to the medium specified in the physical layer; it also defines data formats, to include the framing of data within transmitted messages, error control procedures, and other link control activities. Data link control protocols such as Binary Synchronous Communications (BSC) and High-level Data Link Control (HDLC) reside in this layer.

Layer 3

The network layer (level 3) is responsible for arranging a logical connection between a source and destination on the network. This layer provides services associated with the movement of data through a network, to include addressing, routing, switching, sequencing, and flow control procedures. In a complex network the source and destination may not be directly connected by a single path, but instead require a path to be established that consists of many subpaths. Thus, routing data through the network onto the correct paths is an important feature of this layer.

Layer 4

The transport layer (level 4) is responsible for guaranteeing that data is correctly received by the destination. Thus, the primary function of this layer is to control the communications session between network nodes once a path has been established by the network control layer. Error control, sequence checking, and other end-to-end data reliability factors are the primary concern of this layer.

Layer 5

The session layer (level 5) provides a set of rules for establishing and terminating data streams between nodes in a network. The services that this session layer can provide include establishing and terminating node connections, message flow control, dialogue control and end-to-end data control.

The presentation layer (level 6) services are concerned with data transformation, formatting, and syntax. One of the primary functions performed by the presentation layer is the conversion of transmitted data into a display format appropriate for a receiving device. Data encryption/decryption and data compression and decompression are examples of the data transformation that could be handled by this layer.

Layer 6

Finally, the application layer (level 7) acts as a window through which the application gains access to all of the services provided by the model. Examples of functions performed at this level include file transfers, resource sharing and data base access. While the first four layers are fairly well defined, the top three layers may vary considerably, depending upon the network used. Figure 1.57 illustrates the OSI model in schematic format, showing the various levels of the model with respect to a terminal accessing an application on a host computer system.

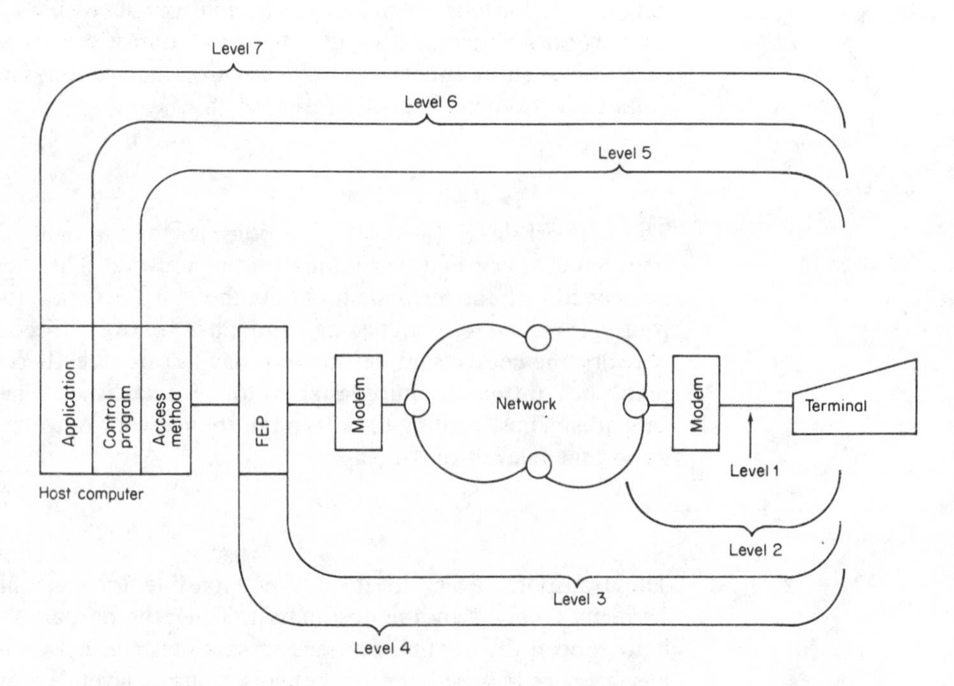

Figure 1.57 OSI model schematic.

Data flow

As data flows within an ISO network each layer appends appropriate heading information to frames of information flowing within the network while removing the heading information added by a lower layer.

Figure 1.58 illustrates the appending and removal of frame header information as data flows through a network constructed according to the ISO Reference Model. Since each higher level removes the header appended by a lower level the frame traversing the network arrives in its original form at its destination.

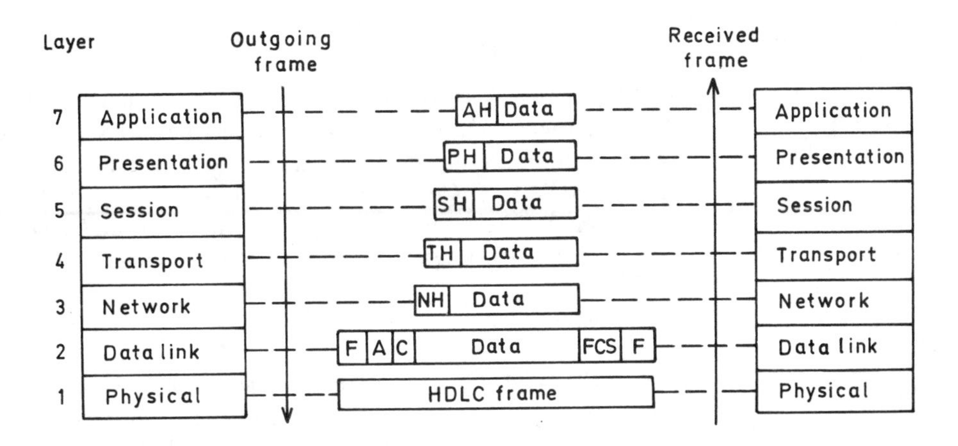

Figure 1.58 Data flow within an ISO reference model network.

As the reader will surmise from the previous illustrations the ISO Reference Model is designed to simplify the construction of data networks. This simplification is due to the eventual standardization of methods and procedures to append appropriate heading information to frames flowing through a network, permitting data to be routed to its appropriate destination following a uniform procedure.

1.18 NETWORK ARCHITECTURE AND SNA

To satisfy the requirements of customers for remote computing capability, mainframe computer manufacturers developed a variety of network architectures. Such architectures define the interrelationship of a particular vendor's hardware and software products necessary to permit communications to flow through a network to the manufacturer's mainframe computer. IBM's System Network Architecture (SNA) is a very complex and sophisticated network architecture which defines the rules, procedures and structure of communications from the input/output statements of an application program to the screen display on a user's personal computer or terminal. SNA consists of protocols, formats and operational sequences which govern the flow of information within a data communications network linking IBM mainframe computers, minicomputers, terminal controllers, communications controllers, personal computers and terminals.

SNA Concepts

An SNA network consists of one or more domains, where a domain refers to all of the logical and physical components that are connected to and controlled by one common point in the network. This common point of control is called the System Services Control Point, which is commonly known by its abbreviation as the SSCP. There are three types of network addressable units in an SNA network—SSCPs, physical units and logical units.

SSCP

The SSCP resides in the communications access method operating in an IBM mainframe computer, such as Virtual Telecommunications Access Method (VTAM), operating in a System/360, System/370, 4300 series of 308X computer, or in the system control program of an IBM minicomputer, such as a System/34, System/36 or System/38. The SSCP contains the network's address tables, routing tables and translation tables which it uses to establish connections between nodes in the network as well as to control the flow of information in an SNA network. Figure 1.59 illustrates single and multiple domain SNA networks.

Network Nodes

Each network domain will include one or more nodes, with an SNA network node consisting of a grouping of networking components which provides it with a unique characteristic. Examples of SNA nodes include cluster controllers, communications controllers and terminal devices, with the address of each device in the network providing its unique characteristic in comparison to a similar device contained in the network.

The Physical Unit

Each node in an SNA network contains a physical unit (PU) which controls the other resources contained in the node. The PU is not a physical device as its name appears to suggest, but rather a set of SNA components which provide services used to control terminals, controllers, processors and data links in the network. In programmable devices, such as mainframe computers and communications controllers, the PU is normally implemented in software. In less intelligent devices, such as cluster controllers and terminals the PU is typically implemented in read-only memory.

The Logical Unit

The third type of network addressable unit in an SNA network is the logical unit, known by its abbreviation as the LU. The LU is the interface or point of access between the end-user and an SNA network. Through the LU an end-user gains access to network resources and transmits and receives data

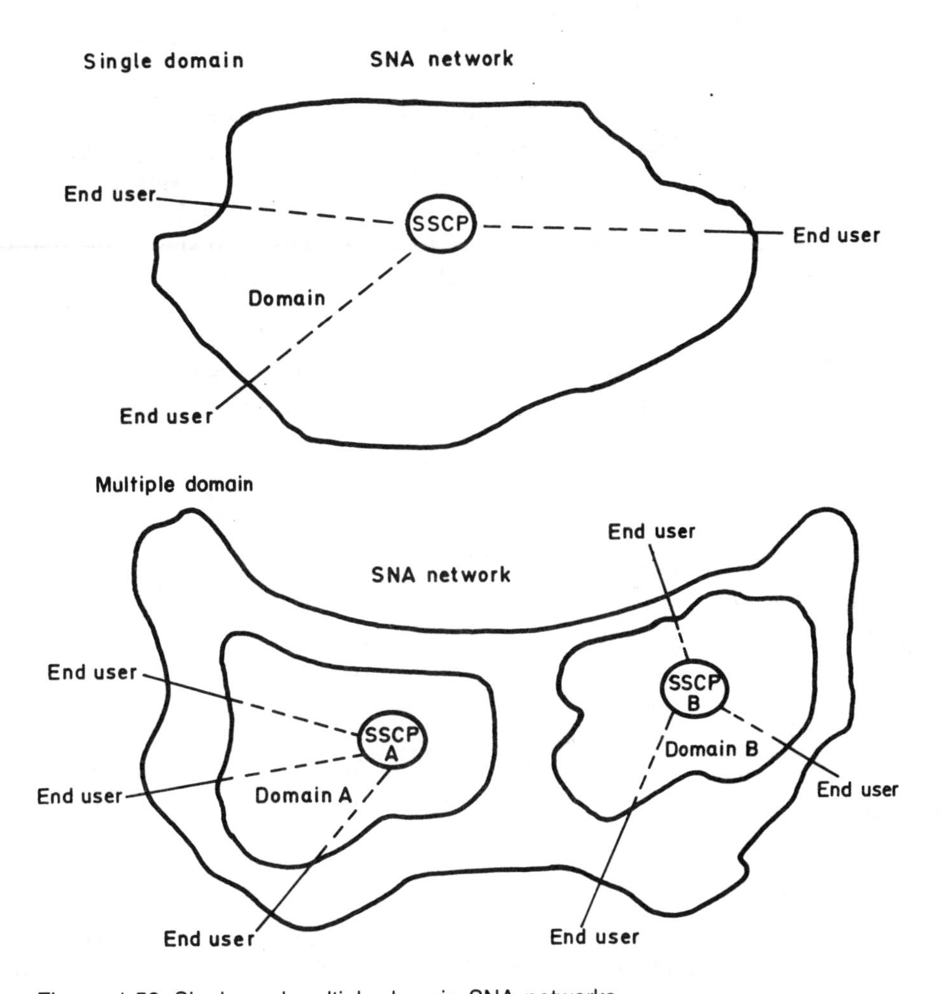

Figure 1.59 Single and multiple domain SNA networks.

over the network. Each PU can have one or more LUs, with each LU having a distinct address.

Multiple Session Capability

As an example of the communications capability of SNA, consider an end-user with an IBM PC and an SDLC communications adapter who establishes a connection to an IBM mainframe computer. The IBM PC is a PU, with its display and printer considered to be LUs. After communications is established the PC user could direct a file to his or her printer by establishing an LU-to-LU session between the mainframe and printer while using the PC as an interactive terminal running an application program as a second LU-to-LU session.

Types of Physical Units

In Table 1.22 the reader will find a list of five types of physical units in an SNA network and their corresponding node type. In addition, this table contains representative examples of hardware devices that can operate as a specific type of PU. As indicated in Table 1.22, the different types of PUs form a hierarchy of hardware classifications. At the lowest level, PU type 1 is a single terminal. PU type 2 is a cluster controller which is used to connect many SNA devices onto a common communications circuit. PU type 4 is a communications controller which is also known as a front-end processor. This device provides communications support for up to several hundred line terminations, where individual lines in turn can be connected to cluster controllers. At the top of the hardware hierarchy, PU type 5 is a mainframe computer.

Table 1.22 SNA PU summary.

PU type	Node	Representative hardware
PU type 5	Mainframe	S/370, 43XX, 308X
PU type 4	Communications controller	3705, 3725, 3720, 3745
PU type 3	Not currently defined	N/A
PU type 2	Cluster controller	3274, 3276, 3174
PU type 1	Terminal	3180, PC with SNA adapter

The communications controller is also commonly referred to as a front-end processor. This device relieves the mainframe of most communications processing functions by performing such activities as sampling attached communications lines for data, buffering the data and passing it to the mainframe as well as performing error detection and correction procedures. The cluster controller functions similar to a multiplexer or data concentrator by enabling a mixture of up to 32 terminals and low-speed printers to share a common communications line routed to a communications controller or directly to the mainframe computer.

Figure 1.60 illustrates a two domain SNA network. By establishing a physical connection between the communications controller in each domain and coding appropriate software for operation on each controller, cross domain data flow becomes possible. When cross domain data flow is established terminal devices connected to one mainframe gain the capability to access applications operating on the other mainframe computer.

SNA was originally implemented as a networking architecture in which users establish sessions with application programs that operate on a mainframe computer within the network. Once a session is established a Network Control Program (NCP) operating on an IBM communications controller, which in turn is connected to the IBM mainframe, would control the information flow between the user and the applications program. With the growth in personal

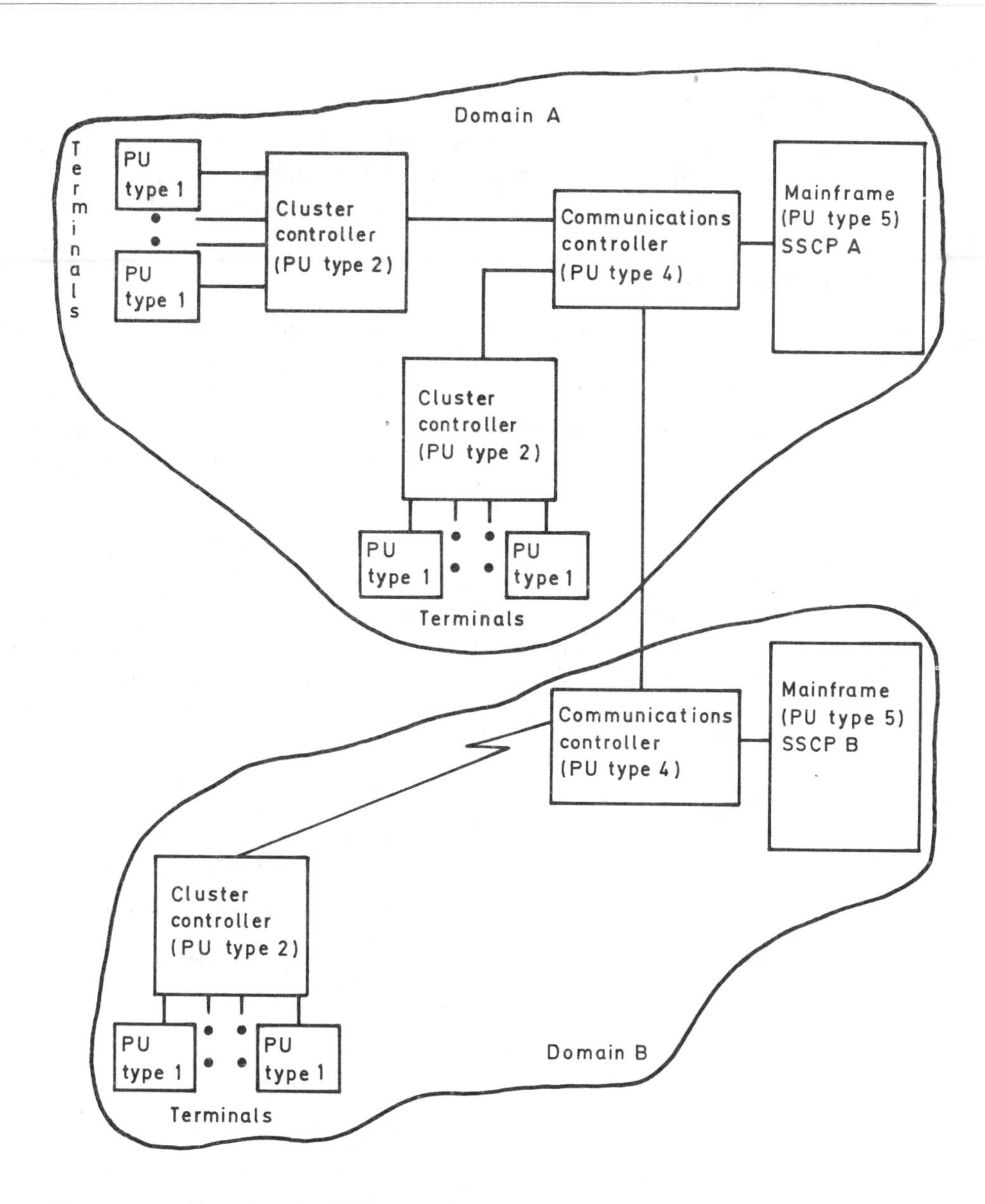

Figure 1.60 Two domain SNA network.

computing many users no longer required access to a mainframe to obtain connectivity to another personal computer connected to the network. Thus, IBM modified SNA to permit peer-to-peer communications capability in which two devices on the network with appropriate hardware and software could communicate with one another without requiring access through a mainframe computer.

SNA Layers

IBM's SNA is a layered protocol which provides six layers of control for every message that flows through the network. Figure 1.61 illustrates the six SNA layers and provides a comparison to the seven-layer ISO Reference Model.

Data Link Layer

In comparison to the ISO model which defines the physical level, in SNA the data link level is the lowest defined layer. This layer formats messages into SDLC frames for transmission across an SNA network and is responsible for the orderly and successful transmission of data. Although SDLC is the only data link protocol defined by SNA, some implementations of this architecture can support bisynchronous and asynchronous transmission.

Path Control Layer

Two of the major functions of the path control layer are routing and flow control. Concerning routing, since there can be many data links connected to a node, path control is responsible for insuring that data is correctly passed through intermediate nodes as it flows from source to destination. At the

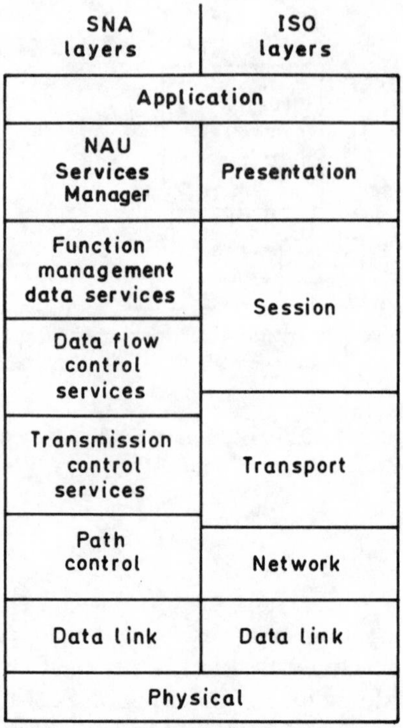

<table>
<tr><th>SNA layers</th><th>ISO layers</th></tr>
<tr><td colspan="2">Application</td></tr>
<tr><td>NAU Services Manager</td><td>Presentation</td></tr>
<tr><td>Function management data services</td><td rowspan="2">Session</td></tr>
<tr><td>Data flow control services</td></tr>
<tr><td>Transmission control services</td><td>Transport</td></tr>
<tr><td>Path control</td><td>Network</td></tr>
<tr><td>Data link</td><td>Data link</td></tr>
<tr><td colspan="2">Physical</td></tr>
</table>

Figure 1.61 Comparing SNA to the ISO reference model.

beginning of an SNA session both sending and receiving nodes as well as all nodes between those nodes cooperate to select the most efficient route for the session. Since this route is only established for the duration of the session it is known as a virtual route. To increase the efficiency of transmission in an SNA network the path control layer at each node through which the virtual route is established has the ability to divide long messages into shorter segments for transmission by the data link layer. Similarly, path control may block short messages into larger data blocks for transmission by the data link layer.

Pacing

Transmission control services layer functions include session level pacing as well as encryption and decryption of data when so requested by a session. Here pacing insures that a transmitting device does not send more data than a receiving device can accept during a given period of time. Thus, pacing can be viewed as similar to the flow control of data in a network.

Data Flow Control Services

The data flow control services layer handles the order of communications within a session for error control and flow control. Here the order of communications is set by the layer controlling the transmission mode. Transmission modes available include full-duplex which permits each device to transmit any time, half-duplex flip-flop where devices can only transmit alternately and half-duplex contention, where one device is considered a master device and the slave cannot transmit until the master completes its transmission.

Function Management

The function management data services layer performs the connection and disconnection of sessions as well as updating the network configuration and performing network management functions. At the highest layer in an SNA network the NAU (Network Addressable Unit) Services Manager is responsible for formatting of data from an application to match the display or printer that is communicating with the application. Other functions performed at this layer include the compression and decompression of data to increase the efficiency of transmission on an SNA network.

SNA Developments

The most significant development to SNA can be considered the addition of new LU and PU subtypes to support what is known as Advanced Peer to Peer Communications (APPC). Previously, LU types used to define an LU-to-LU session were restricted to application-to-device and program-to-program sessions. LU1 through LU4 and LU7 are application-to-device sessions as indicated in Table 1.23, where LU4 and LU6 are program-to-program sessions.

The addition of LU6.2 which operates in conjunction with PU 2.1 to support LU6.2 connections permits devices supporting this new LU to transfer data to

Table 1.23 SNA LU session types.

LU type	Session type
LU1	Host application and a remote batch terminal
LU2	Host application and a 3270 display terminal
LU3	Host application and a 3270 printer
LU4	Host application and SNA word processor or between two terminals via mainframe
LU6	Between applications programs typically residing on different mainframe computers
LU6.2	Peer-to-peer
LU7	Host application and a 5250 terminal

any other device also supporting this LU without first sending the data through a mainframe computer. As new software products are introduced to support LU6.2 a more dynamic flow of data through SNA networks will occur, with many data links to mainframes that were previously heavily utilized or saturated gaining capacity as sessions between devices permit data flow to bypass the mainframe.

1.19 INTEGRATED SERVICES DIGITAL NETWORK

No discussion of fundamental communications concepts would be complete without discussing the future. In this section we will examine the future of both data and voice communications in the form of Integrated Services Digital Network (ISDN), which can be expected to eventually replace most, if not all existing analog networks.

ISDN offers the potential for the development of a universal international digital network, with a series of standard interfaces that will facilitate the connection of a wide variety of telecommunications equipment to the network. Although the transition to ISDN may require several decades and some ISDN functions may never be offered in certain locations, its potential cannot be overlooked. Since many ISDN features offer a radical departure from existing services and current methods of communications, we will review the concept behind ISDN and projected features and new services that may result from its implementation in this section. This information is included to provide the reader with a familiarity of ISDN that will be used to illustrate how to plan to effectively utilize this evolving digital service.

Concept behind ISDN

The original requirement to transmit human speech over long distances resulted in the development of telephone systems designed for the transmission of analog data. Although such systems satisfied the basic requirement to transport human speech, the development of computer systems and the introduction of remote processing required a conversion of digital signals into an analog format. This conversion was required to enable computers and business machines to use existing telephone company facilities for the transmission of digital data. Not only was this conversion awkward and expensive due to the requirement to employ modems, but, in addition, the analog facilities of telephone systems limited the data transmission rate obtainable when such facilities were used.

The evolution of digital processing and the rapid decrease in the cost of semiconductors resulted in the application of digital technology to telephone systems. By the late 1960s, telephone companies began to replace their electromechanical switches in their central offices with digital switches, while by the early 1970s, several communications carriers were offering end-to-end digital transmission services. By the mid-1980s, a significant portion of the transmission facilities of most telephone systems were digital. On such systems, human speech is encoded into digital format for transmission over the backbone network of the telephone system. At the local loop of the network, digitized speech is reconverted into its original analog format and then transmitted to the subscriber's telephone.

Based upon the preceding, ISDN can be viewed as an evolutionary progression in the conversion of analog telephone systems into an eventual all-digital network, with both voice and data to be carried end-to-end in digital form.

ISDN Architecture

Under the evolving ISDN architecture, access to this digital network will result from one of two major connection methods—Basic Access and Primary Access.

Basic Access

Basic Access defines a multiple channel connection derived by multiplexing data on twisted-pair wiring. This multiple channel connection is between an end-user terminal device and a telephone company office or a local Private Automated Branch Exchange (PABX). The ISDN Basic Access channel format is illustrated in Figure 1.62.

As indicated in Figure 1.62, Basic Access consists of framing (F), two bearer (B) channels and a data (D) channel that are multiplexed by time onto a common twisted-pair wiring media. Each bearer channel can carry one pulse

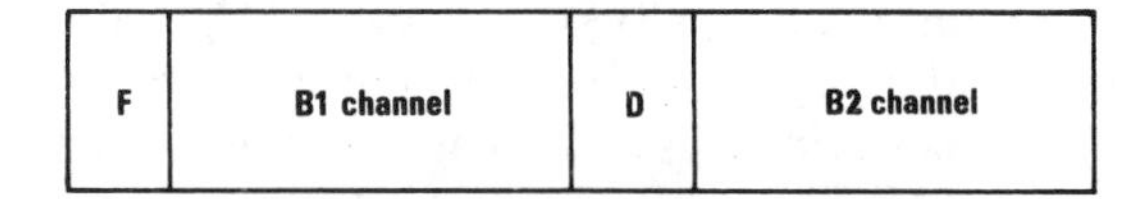

Figure 1.62 ISDN basic access channel format.

code modulation (PCM) voice conversation or data at a transmission rate of 64 kbps. This enables Basic Access to provide the end-user with the capability to simultaneously transmit data and conduct a voice conversation on one telephone line or to be in conversation with one person and receive a second telephone call. In the case of the latter situation, assuming the end-user has an appropriate telephone instrument, he or she could place one person on hold and answer the second call.

The D channel was designed for both controlling the B channels through the sharing of network signaling functions on this channel as well as for the transmission of packet switched data. Concerning the transmission of packet switched data, the D channel provides the capability for a number of new applications to include monitoring home alarm systems and the reading of utility meters upon demand. Since these types of applications have minimum data transmission requirements, the D channel can be expected to be used for a variety of applications in addition to providing the signaling required to set up calls on the B channels.

Primary Access

Primary Access can be considered as a multiplexing arrangement whereby a grouping of Basic Access users shares a common line facility. Typically, Primary Access will be employed to directly connect a Private Automated Branch Exchange (PABX) to the ISDN network. This access method is designed to eliminate the necessity of providing individual Basic Access lines when a group of terminal devices shares a common PABX which could be directly connected to an ISDN network via a single high-speed line. Due to the different types of T1 network facilities in North America and Europe, two Primary Access standards have been developed.

In North America, Primary Access consists of a grouping of 23 B channels and one D channel to produce a 1.544 Mbps composite data rate, which is the standard T1 carrier data rate. In Europe, Primary Access consists of a grouping of 30 B channels plus one D channel to produce a 2.048 Mbps data rate, which is the T1 carrier transmission rate in Europe.

Network Characteristics

Four of the major characteristics of an ISDN network are listed in Table 1.24. These characteristics can also be considered as driving forces for the implementation of the network by communications carriers.

Due to the digital nature of ISDN, voice, data, and video services can be integrated, alleviating the necessity of end-users obtaining separate facilities for each service. Since the network is designed to provide end-to-end digital transmission, pulses can be easily regenerated throughout the network, resulting in the generation of new pulses to replace distorted pulses. In comparison, analog transmission facilities employ amplifiers to boost the strength of transmission signals, which also increases any impairments in the signal. As a

Table 1.24 ISDN characteristics.

Integrates voice, data and video services
Digital end-to-end connection resulting in high transmission quality
Improved and expanded services due to B and D channel data rates
Greater efficiency and productivity resulting from the ability to have several simultaneous calls occur on one line

result of regeneration being superior to amplification, digital transmission has a lower error rate and provides a higher transmission signal quality than an equivalent analog transmission facility.

Due to Basic Access in effect providing three signal paths on a common line, ISDN offers the possibilities of both improvements to existing services and an expansion of services to the end-user. Concerning existing services, current analog telephone line bandwidth limitations normally preclude data transmission rates over 9.6 kbps occurring on the switched telephone network. In comparison, under ISDN each B channel can support a 64 kbps transmission rate while the D channel will operate at 16 kbps. In fact, if both B channels and the D channel were in simultaneous operation a data rate of 144 kbps would be obtainable on a Basic Access ISDN circuit, which would exceed current analog circuit data rates by a factor of between 7 and 15.

Since each Basic Access channel in effect consists of three multiplexed channels, different operations can occur simultaneously without requiring an end-user to acquire separate multiplexing equipment. Thus, the end-user could receive a call from one person, transmit data to a computer and have a utility company read his or her electric meter at a particular point in time. Here, the ability to conduct up to three simultaneous operations on one ISDN line should result in both greater efficiency and productivity. Efficiency should increase since one line can now support several simultaneous operations, while the productivity of the end-user can increase due to the ability to receive telephone calls and then conduct a conversation while transmitting data.

Terminal Equipment and Network Interface

One of the key elements of ISDN is a small set of compatible multipurpose user–network interfaces that were developed to support a wide range of applications. These network interfaces are based upon the concept of a series of reference points for different user terminal arrangements which is then used to define these interfaces. Figure 1.63 illustrates the relationship between ISDN reference points and network interfaces.

The ISDN reference configuration consists of functional groupings and reference points at which physical interfaces may exist. The functional groupings are sets of functions that may be required at an interface, while reference points are employed to divide the functional groups into distinct entities.

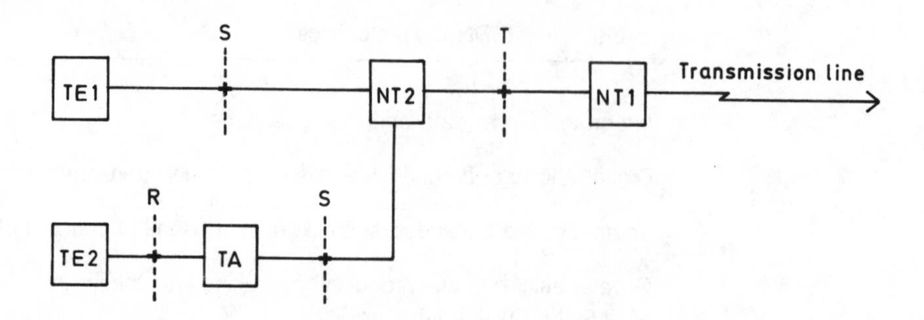

TE1 (terminal equipment 1) type devices comply with the ISDN network interface.

TE2 (terminal equipment 2) type devices do not have an ISDN interface and must be connected through a TA (terminal adapter) functional grouping.

NT2 (network termination 2) includes switching and concentration equipment which perform functions equivalent to layers 1 through 3 of the OSI reference model.

NT1 (network termination 1) includes functions equivalent to layer 1 of the OSI reference model.

Figure 1.63 ISDN reference points and network interfaces. (Reprinted with permission from *Data Communications Management*, © 1987 Auerbach Publishers, New York, NY.)

The TE (terminal equipment) functional grouping is comprised of TE1 and TE2 type equipment. Examples of TE equipment include digital telephones, conventional data terminals, and integrated voice/data workstations. TE1 type equipment complies with the ISDN user–network interface and permits such equipment to be directly connected to an ISDN "S" type interface which supports multiple B and D channels.

TE2 type equipment are devices with non-ISDN interfaces, such as RS-232 or the CCITT X or V-Series interfaces. This type of equipment must be connected through a TA (terminal adapter) functional grouping, which in effect converts a non-ISDN interface (R) into an ISDN Sending interface (S), performing both a physical interface conversion and protocol conversion to permit a TE2 terminal to operate on ISDN.

The NT2 (network termination 2) functional group includes devices that perform switching and data concentration functions equivalent to the first three layers of the OSI Reference Model. Typical NT2 equipment can include PABXs, terminal controllers, concentrators, and multiplexers.

The NT1 (network termination 1) functional group is the ISDN digital interface point and is equivalent to layer 1 of the OSI Reference Model. Functions of NT1 include the physical and electrical termination of the loop, line monitoring, timing, and bit multiplexing. In Europe, where most communications carriers are government owned monopolies, NT1 and NT2 functions may be combined into a common device, such as a PABX. In such situations, the equipment serves as an NT12 functional group. In comparison, in the United States the communications carrier may provide only the NT1, while third-party vendors can provide NT2 equipment. In such situations, the third-party equipment would connect to the communications carrier equipment at the T interface.

Preparing for the Future

Currently, ISDN is several years away from implementation by most communications carriers. Although there are several well-publicized field trials currently being evaluated, most experts predict that it will take between 5 and 10 years prior to ISDN services being available in a significant percent of communications carrier offices. In fact, some experts speculate that even by the year 2020, some analog technology will coexist with ISDN due to the significant capital investment required to implement the technology.

Since it appears that the evolutionary process required for ISDN implementation will span several decades, one logical question readers may have probably concerns how to plan for this future service. This planning process becomes very difficult, since some network locations may obtain ISDN services in the near future while other locations may only obtain such services in the next century.

The best approach for planning for ISDN is to carefully investigate the communications hardware products one must acquire with a view towards their utilization in an ISDN environment. Although the only equipment currently available that supports the ISDN interface are those devices used in field trials, the manufacture of conversion devices known as terminal adapters or TAs can be expected to provide the interface and protocol conversion facilities for currently manufactured equipment to operate in an ISDN environment. Thus, although it is important to monitor vendor and communications carrier plans and follow ISDN standards activities, from a practical perspective, you should focus your attention upon the bandwidth capacity of hardware you anticipate acquiring over the next few years. Commonly used communications devices that should warrant attention with respect to their bandwidth capacity include port selectors, matrix switches, multiplexers, concentrators and PABXs. If these devices support ISDN data rates it will probably be possible to use them in an ISDN networking environment in the future through the use of a terminal adapter. In comparison, if they do not support ISDN data rates a terminal adapter may still enable them to be used in an ISDN networking environment, however, they will serve as a data "bottleneck". This will occur since their below ISDN data rate will have to be converted by the terminal adapter into an ISDN data rate. Then, in the reverse direction the data flow through an ISDN network to the device operating at a subrate of the ISDN data rate will cause the terminal adapter to issue flow control signals to the ISDN network. These flow control signals will inhibit transmission after a block of data is received at one data rate by the terminal adapter and transmitted to the non-ISDN device at a lower data rate, enabling the TA to perform its function without losing data.

REVIEW QUESTIONS

1.1 Discuss the function of each of the major elements of a transmission system.

1.2 What is the relationship between each of the three basic types of line connections and the use of that line connection for short or long duration data transmission sessions?

1.3 Discuss the relationship of modems and digital service units to digital and analog transmission systems. Why are these devices required and what general functions do they perform?

1.4 Name four types of analog facilities offered by communications carriers and discuss the utilization of each facility for the transmission of data between terminal devices and computer systems.

1.5 What is the function of an analog extension?

1.6 What is the difference between simplex, half-duplex and full-duplex transmission?

1.7 Discuss the relationship between the modes of operation of terminals and computers with respect to the printing and display of characters on a terminal in response to pressing a key on the terminal's keyboard.

1.8 In asynchronous transmission how does a receiving device determine the presence of a start bit?

1.9 What is the difference between asynchronous and synchronous transmission with respect to the timing of the data flow?

1.10 What is the difference between serial and parallel transmission? Why do most communications systems use serial transmission?

1.11 Discuss the difference in terminal requirements with respect to point-to-point and multidrop line usage.

1.12 Why is the Morse code basically unsuitable for transmission by terminal devices?

1.13 How does Baudot code which is a 5-level code permit the representation of more than 32 unique characters?

1.14 What is the bit composition of the ASCII characters A and a?

1.15 Assuming even parity checking is employed, what are the parity bits assigned to the ASCII characters A, E, I, O and U? What are the parity bits if odd parity checking is employed?

1.16 What are the major limitations of parity checking?

1.17 Assume a file on your personal computer contains 3000 lines of data, with an average of 60 characters per line. If you transmit the file using 8-bit character transmission and the probability of an error occurring is

1.5 per 100,000 bits, how many characters can be expected to be received in error if the bit errors occur randomly and are singular in occurrence per transmitted character?

1.18 Under the XMODEM protocol, what would be the value of the checksum if the data contained in a block consisted of all ASCII X characters?

1.19 Discuss the relationship between a transmitted cyclic redundancy check character and an internally generated cyclic redundancy check character with respect to the data integrity of the block containing the transmitted cyclic redundancy check character.

1.20 Why are alternating DLE0 and DLE1 characters transmitted as positive acknowledgements in bisynchronous transmission?

1.21 What procedure is used to prevent a stream of binary data from being misinterpreted as an HDLC flag? Explain the operation of this procedure.

1.22 What are the advantages of a bit-oriented protocol in comparison to a character-oriented protocol?

1.23 If a secondary station responds to the poll of a primary station by setting N(R) equal to five in its response, what does this signify to the primary station?

1.24 Discuss the relationship between the rate of data transmission, pulse width and the susceptibility of a pulse to a transmission impairment to explain why the 50-foot cable limit of the RS-232-C and CCITT V.24 interface standards can be violated?

1.25 What are the functions of the Request to Send, Clear to Send, Data Set Ready and Data Terminal Ready control circuits?

1.26 What is the difference between dB and dBm?

1.27 Assume the loss associated with a circuit is −12 dBm prior to its gain of 16 dBm due to the installation of an amplifier. If the cable from the amplifier to its destination has a loss of −8 dBm, what is the overall gain or loss associated with the circuit?

1.28 What function does a null modem cable perform? Why is a null modem cable unsuitable for synchronous transmission?

1.29 What is the operational difference between selecting the telephone set optioned "telephone set controls the line" and "data set controls the line".

1.30 What is the goal of the ISO Open System Interconnection Reference Model?

1.31 Why can you expect transmission quality on ISDN facilities to be superior to existing analog facilities?

1.32 Discuss the data transmission rate differences between a Basic Access ISDN circuit and that obtainable on the switched telephone network.

CHAPTER 2

DATA TRANSMISSION EQUIPMENT

Our method of categorizing data communications components is by the function or group of functions they are designed to perform. In this chapter, the operation and utilization of components designed primarily to effect data transmission via a communications medium will be covered. Specific devices which will be explored in this chapter include a variety of couplers, modems, or devices that can be used to transmit data based upon the interrelationship of lines, terminals, and other data communications components which may be employed on a particular occasion. In addition, a device which permits the extension of parallel transmission from a computer to peripheral units located at a distance from the computational facility will also be covered.

2.1 ACOUSTIC COUPLERS

Unlike conventional modems which may require a permanent or semi-permanent connection to a telephone line, an acoustic coupler is in essence a modem which permits data transmission through the utilization of the handset of an ordinary telephone. Similar in functioning to a modem, an acoustic coupler is a device which will accept a serial asynchronous data stream from terminal devices, modulates that data stream into the audio spectrum, and then transmit the audio tones over a switched or dial-up telephone connection.

Acoustic couplers are equipped with built-in cradles or fittings into which a conventional telephone headset is placed. Through the process of acoustic coupling, the modulated tones produced by the acoustic coupler are directly picked up by the attached telephone headset. Likewise, the audible tones transmitted over a telephone line are picked up by the telephone earpiece and demodulated by the acoustic coupler into a serial data stream which is acceptable to the attached terminal.

Acoustic couplers normally use two distinct frequencies to transmit information, while two other frequencies are employed for data reception. A frequency from each pair is used to create a mark tone which represents an encoded binary one from the digitial data stream, while another pair of frequencies generates a space tone which represents a binary zero. This utilization of two pairs of frequencies permits full-duplex transmission to occur over the 2-wire switched telephone network.

Since acoustic couplers enable any conventional telephone to be used for data transmission purposes, the coupler does not have to be physically wired to the line and thus permits considerable flexibility in choosing a terminal working area which can be anywhere a telephone handset and standard electrical outlet are located. Acoustic couplers are manufactured both as separate units and as built-in units to data terminals, as shown in Figure 2.1.

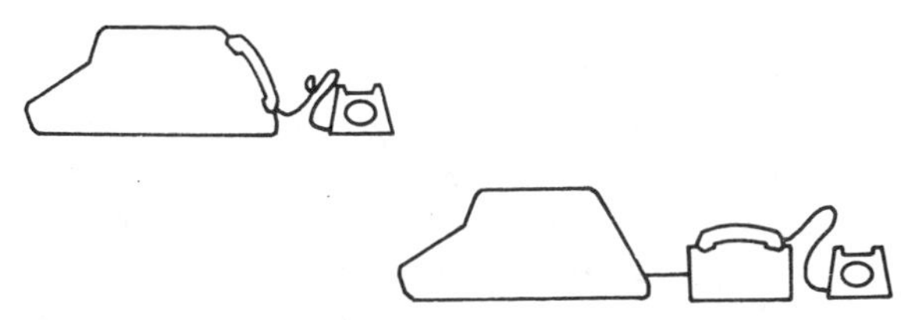

Figure 2.1 Varying coupler connections. Left: terminal with built-in coupler. Right: terminal connected to coupler.

US and European compatibility

Since acoustic couplers are normally employed to permit portable terminals to communicate with data-processing facilities, and since a large portion of low-speed modems at such facilities in the United States were originally furnished by American Telegraph and Telephone Company (AT&T) and its operating companies prior to its break-up into independent organizations, most manufacturers of acoustic couplers have designed them to be compatible with low-speed "Bell System" modems. Here the term "Bell System" refers to the operating characteristics of modems that were manufactured by Western Electric for use by AT&T operating companies prior to those operating companies becoming independent organizations.

In Europe, most acoustic couplers are designed to be compatible with CCITT recommendations that govern the operation of low-speed modems. To understand the differences between low speed Bell System and CCITT modems, we will examine acoustic couplers that operate at data rates between 0 and 450 bps. In the United States, such couplers are compatible with Bell System 103 and 113 type modems while in Europe such couplers are compatible with the CCITT V.21 recommendation. Table 2.1 lists the operating frequencies of acoustic couplers designed to operate with Bell System 103/113 type modems and modems that follow the CCITT V.21 recommendation.

Basically, couplers like low-speed modems must operate in one of two modes – originate or answer. This operational mode should not be confused with a transmission mode of simplex, half-duplex, or full-duplex. What the operational mode refers to is the frequency assignments for transmitting marks and spaces. Thus, from Table 2.1, an acoustic coupler compatible with a Bell System 103/113 type modem would transmit a tone at 1270 Hz to represent a mark and a tone at 1070 Hz to represent a space when it is in the originate mode of operation. To communicate effectively, the device (modem or coupler) at the other end of the line must be in the answer mode of operation. If so, then it would receive a mark at 1270 Hz and a space at 1070 Hz, ensuring that the tones transmitted by the originate mode device would be heard by the receive mode device. This explains why two terminal operators, each with an originate mode coupler, could not communicate with one another. This communications incompatibility results from the fact that one coupler would transmit a mark at 1270 Hz, while the other coupler would be set to receive the mark at 2225 Hz. Thus, the second coupler would never hear the tone originated by the first coupler.

By convention, originate mode couplers and modems are connected to terminals while answer mode devices are connected to computer ports, since terminals originate calls and computers typically answer such calls. Some couplers can be obtained with an originate/answer mode switch. By changing the position of the switch, one changes the coupler's operating frequency assignments. Couplers with an originate/answer mode switch should be obtained when there is a requirement to communicate between terminals or if one believes this requirement could materialize.

In Figure 2.2 the frequency assignments of couplers designed to be compatible with Bell System 103/113 type modems is graphically illustrated. Note that 1170 Hz and 2125 Hz are the channel center frequencies and two independent data channels are derived by frequency, permitting full-duplex transmission to occur over the 2-wire public switched telephone network.

Returning to Table 2.1, note that the operating frequencies of Bell System 103/113 type modems are completely different from modems designed to operate according to the V.21 recommendation. This frequency incompatibility explains why, as an example, a traveling American in Europe more likely than

Table 2.1 Acoustic coupler modem compatibility (operating frequencies in Hz).

	Bell System 103/113 type		CCITT V.21	
	Originate	Answer	Originate	Answer
Transmit				
Mark	1270	2225	980	1650
Space	1070	2025	1180	1850
Receive				
Mark	2225	1270	1650	980
Space	2025	1070	1850	1180

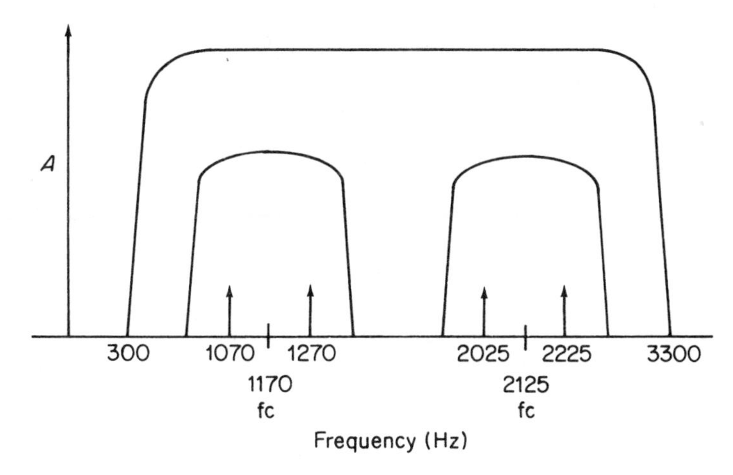

Figure 2.2 Bell System 103/113 frequency spectrum.

not will be unable to use his or her portable personal computer to communicate with either a public packet network or his or her company's mainframe computer located in Europe.

Originally acoustic couplers were developed to transmit and receive data at 300 bps. Today, most vendors market devices that operate at 1200 bps. Such couplers are compatible with either Bell System 202 or 212A or CCITT V.22 modems and their method of modulation will be described in Section 2.2 where modems are covered.

Operation

When a terminal is attached to or has a built-in acoustic coupler and the operator wishes to send data to a computer, he or she merely dials the computer's telephone access number and upon establishing the proper connection by hearing a high-pitched tone, places the telephone headset into the coupler. Although terminal usage varies by their numerous applications, the prevalent utilization of acoustic couplers is in obtaining access to time-sharing networks. In a time-sharing network, a group of dial-in computer telephone access numbers may be interfaced to rotary which enables users to dial the low telephone number of the group and automatically "step" or bypass currently busy numbers. Each telephone line is then connected to a modem on a permanent basis, and the modem in turn is connected to a computer port or channel. An automatic answering device in each modem automatically answers the incoming call and in effect establishes a connection from the user who dialed the number to the computer port, as shown in Figure 2.3.

A disadvantage associated with the use of acoustic couplers is a reduction of transmission rates when compared to rates which can be obtained by using modems. Owing to the properties of carbon microphones in telephone headsets, the frequency band that can be passed is not as wide as the band modems can

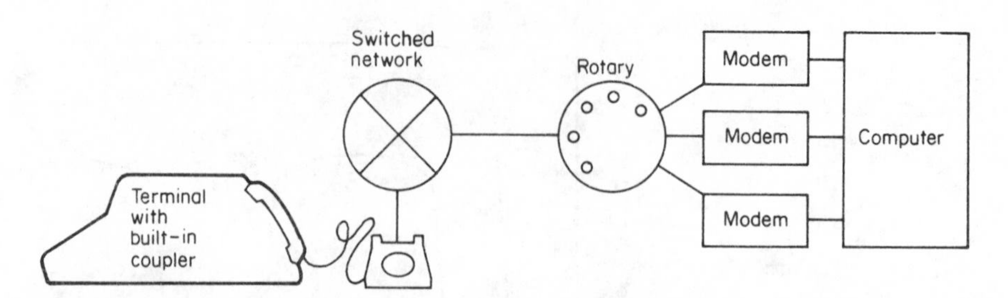

Figure 2.3 Network access in a time-sharing environment. After dialing the computer access number the terminal user places the telephone headset into the cradle of the acoustic coupler.

pass. Although typical data rates of acoustic couplers vary between 110 and 300 bps, some units manufactured permit transmission at 450, 600, and 1200 bps. For usage with slow-speed terminals, the acoustic coupler can be viewed as a low-cost alternative to a modem while increasing user transmission location flexibility.

Problems in usage

A possible cause of errors in the transmission of data can occur from ambient noise leaking into the acoustic coupler. If the coupler is separate from the terminal, one should try to move it as far away from that device as possible to reduce noise levels. Similarly, if the terminal is not in use, one should remove the telephone from the coupler since the continuous placement of the headset in the device can cause crystallization of the speaker and receiver elements of the telephone to occur which will act to reduce the level of signal strength. Another item which may warrant user attention is the placement of a piece of cotton inside the earpiece behind the receiver of the telephone. Although the placement of cotton at this location is normally done by most telephone companies, this should be checked, since the cotton keeps speaker and receiver noise from interfering with each other and acts to prevent transmitted data from interfering with received data.

An easily resolved problem is the placement of the telephone headset into the coupler. On many occasions users have hastily placed the handset only partially into the coupler, and this will act to reduce the level of signal strength necessary for error-free transmission.

2.2 MODEMS

Today, despite the introduction of a number of all-digital transmission facilities by several communications carriers, the analog telephone system remains the primary facility utilized for data communications. Since terminals and computers produce digital pulses, whereas telephone circuits are designed to transmit analog signals which fall within the audio spectrum used in human speech, a device to interface the digital data pulses of terminals and computers with the analog tones carried on telephone circuits becomes necessary when one wishes to transmit data over such circuits. Such a device is called a modem, which derives its meaning from a contraction of the two main functions of such a unit – modulation and demodulation. Although modem is the term most frequently used for such a device that performs modulation and demodulation, "data set" is another common term whose use is synonymous in meaning.

In its most basic form a modem consists of a power supply, a transmitter, and a receiver. The power supply provides the voltage necessary to operate the modem's circuitry. In the transmitter a modulator and amplifier, as well as filtering, waveshaping, and signal control circuitry convert digital direct current pulses; these pulses, originated by a computer or terminal, are converted into an analog, wave-shaped signal which can be transmitted over a telephone line. The receiver contains a demodulator and associated circuitry which reverse the process by converting the analog telephone signal back into a series of digital pulses that is acceptable to the computer or terminal device. This signal conversion is illustrated in Figure 2.4.

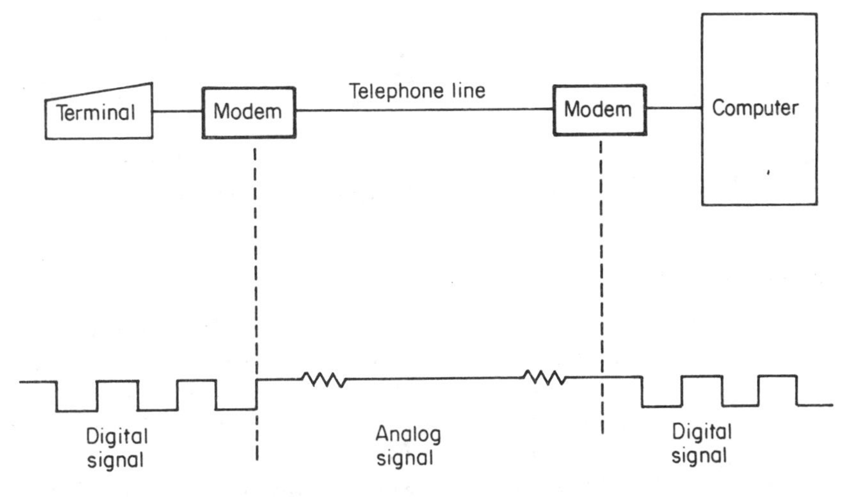

Figure 2.4 Signal conversion performed by modems. A modem converts a digital signal to an analog tone (modulation) and reconverts the analog tone (demodulation) into its original digital signal.

The modulation process

The modulation process alters the characteristics of a carrier signal. By itself, a carrier is a repeating signal that conveys no information. However, when the carrier is changed by the modulation process information is impressed upon the signal. For analog signals, the carrier is a sine wave, represented by:

$$a = A \sin(2\pi ft + \phi)$$

where a = instantaneous value of voltage at time t, A = maximum amplitude, f = frequency, ϕ = phase.

Thus, the carrier's characteristics that can be altered are the carrier's amplitude for amplitude modulation (AM), the carrier's frequency for frequency modulation (FM), and the carrier's phase angle for phase modulation (ϕM).

Amplitude modulation

The simplest method of employing amplitude modulation is to vary the magnitude of the signal from a zero level to represent a binary zero to a fixed peak-to-peak voltage to represent a binary one. Figure 2.5 illustrates the use of amplitude modulation to encode a digital data stream into an appropriate series of analog signals. Although pure amplitude modulation is normally used for very low data rates, it is also employed in conjunction with phase modulation to obtain a method of modulating high-speed digital data sources.

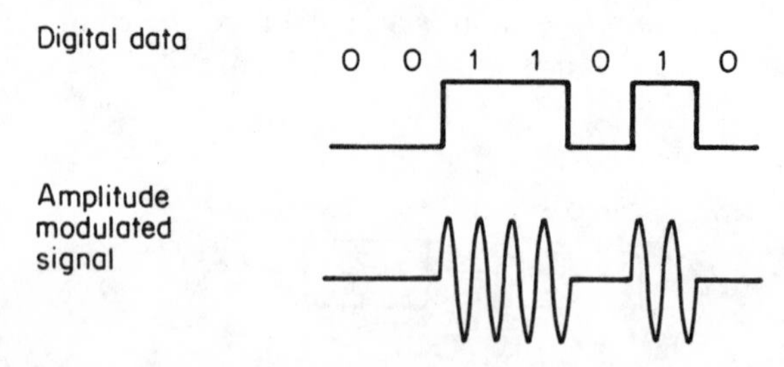

Figure 2.5 Amplitude modulation.

Frequency modulation

Frequency modulation refers to how frequently a signal repeats itself at a given amplitude. One of the earliest uses of frequency modulation was in the design of low-speed acoustic couplers and modems where the transmitter shifted from one frequency to another as the input digital data changed from a binary one to a binary zero or from a zero to a one. This shifting in frequency is known as Frequency Shift Keying (FSK) and is primarily used by modems operating at data rates up to 300 bps in a full-duplex mode of operation and up to 1200 bps in a half-duplex mode of operation. Figure 2.6 illustrates Frequency Shift Keying frequency modulation.

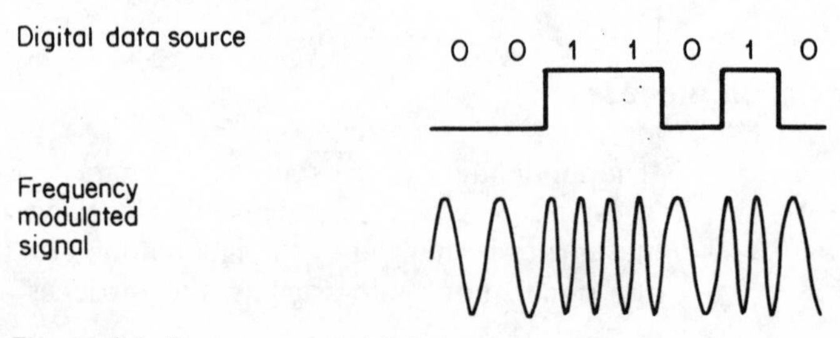

Figure 2.6 Frequency modulation.

Phase modulation

Phase modulation is the process of varying the carrier signal with respect to the origination of its cycle as illustrated in Figure 2.7. Several forms of phase modulation are used in modems to include single and multiple-bit phase-shift keying (PSK) and the combination of amplitude and multiple-bit phase-shift keying.

In single-bit, phase-shift keying, the transmitter simply shifts the phase of the signal to represent each bit entering the modem. Thus, a binary one might be represented by a 90-degree phase change while a zero bit could be represented by a 270-degree phase change. Owing to the variance of phase between two-phase values to represent binary ones and zeros this technique is known as two-phase modulation.

Prior to discussing multiple-bit, phase-shift keying let us examine the basic parameters of a voice circuit and the difference between the data rate and signaling speed. This will enable us to understand the rational for the utilization of multiple-bit, phase-shift keying, where two or more bits are grouped together and represented by one phase shift in a signal.

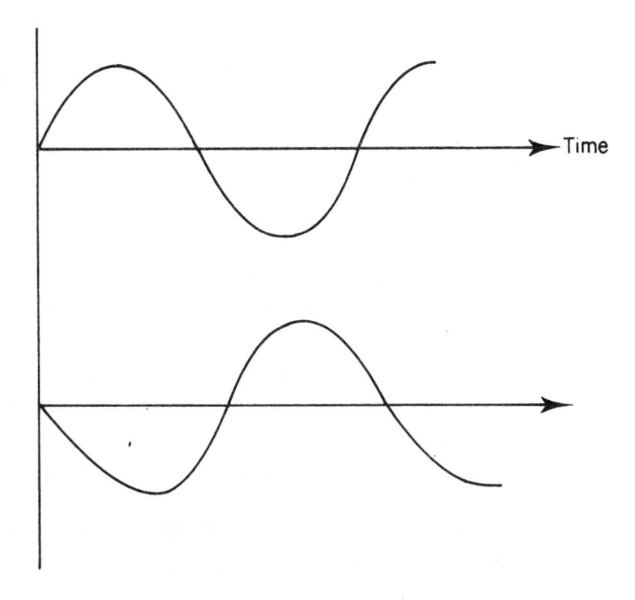

Figure 2.7 Phase modulation. Phase is the position of the wave form of a signal with respect to the origination of the carrier cycle. In this illustration, the bottom wave is 180 degrees out of phase with a normal site wave illustrated at the top.

Bps vs. Baud

Bits per second is the number of binary digits transferred per second and represents the data transmission rate of a device. Baud is the signaling rate of a device such as a modem. If the signal of the modem changes with respect to each bit entering the device, then 1 bps = 1 baud. Suppose a modem is constructed such that one signal change is used to represent two bits. Then the baud rate would be one-half the bps rate.

When one baud is used to represent two bits the encoding technique is known as dibit encoding. Similarly, the process of using one baud to represent three bits is known as tribit encoding and the bit rate is then one-third of the baud rate. Both dibit and tribit encoding are known as multilevel coding techniques and are commonly implemented using phase modulation.

Voice circuit parameters

Bandwidth is a measurement of the width of a range of frequencies. A voice-grade telephone channel has a passband, which defines its slot in the frequency spectrum, which ranges from 300 to 3300 Hz. Thus, the bandwidth of a voice-grade telephone channel is 3300–300 or 3000 Hz.

As data enters a modem it is converted into a series of analog signals, with the signal change rate of the modem known as its baud rate. In 1928, Nyquist developed the relationship between the bandwidth and the baud rate on a circuit as:

$$B = 2W$$

where B = baud rate and W = bandwidth in Hz.

For a voice-grade circuit with a bandwidth of 3000 Hz, this relationship means that data transmission can only be supported at baud rates lower than 6000 symbols or signaling elements per second, prior to one signal interfering with another and causing intersymbol interference.

Since any oscillating modulation technique immediately halves the signaling rate, this means that most modems are limited to operating at one-half of the Nyquist limit. Thus, in a single-bit, phase-shift keying modulation technique, where each bit entering the modem results in a phase shift, the maximum data rate obtainable would be limited to approximately 3000 bps. In such a situation the bit rate would equal the baud rate, since there would be one signal change for each bit.

To overcome the Nyquist limit required engineers to design modems that first grouped a sequence of bits together, examined the composition of the bits, and then implemented a phase shift based upon the value of the grouped bits. This technique is known as multiple bit, phase-shift keying or multilevel, phase-shift keying. Two-bit codes called dibits and three-bit codes known as tribits are formed and transmitted by a single phase shift from a group of four or eight possible phase states.

Most modems operating at 600 to 4800 bps employ multilevel, phase-shift keying modulation. Some of the more commonly used phase patterns employed by modems using dibit and tribit encoding are listed in Table 2.2.

Table 2.2 Common phase-angle values used in multilevel, phase-shift keying.

Bits transmitted	Possible phase-angle values (degrees)		
00	0	45	90
01	90	135	0
10	180	225	270
11	270	315	180
000	0	22.5	45
001	45	67.5	0
010	90	112.5	90
011	135	157.5	135
100	180	202.5	180
101	225	247.5	225
110	270	292.5	270
111	315	337.5	315

Combined modulation techniques

Since the most practical method to overcome the Nyquist limit is obtained by placing additional bits into each signal change, modem designers have combined modulation techniques to obtain very high-speed data transmission over voice-grade circuits. One combined modulation technique commonly used involves both amplitude and phase modulation. This technique is known as quadrature amplitude modulation (QAM) and results in four bits being placed into each signal change, with the signal operating at 2400 (baud), causing the data rate to become 9600 bps.

The first implementation of QAM involved a combination of phase and amplitude modulation, in which 12 values of phase and 3 values of amplitude are employed to produce 16 possible signal states as illustrated in Figure 2.8. One of the earliest modems to use QAM in the United States was the Bell System 209, which modulated a 1650 Hz carrier at a 2400 baud rate to effect data transmission at 9600 bps. Today, most 9600 bps modems manufactured adhere to the CCITT V.29 standard. The V.29 modem uses a carrier of 1700 Hz which is varied in both phase and amplitude, resulting in 16 combinations of 8 phase angles and 4 amplitudes. Under the V.29 standard, fallback data rates of 7200 and 4800 bps are specified.

In addition to combining two modulation techniques, QAM also differs from the previously discussed modulation methods by its use of two carrier signals. Figure 2.9 illustrates a simplified block diagram of a modem's transmitter employing QAM. The encoder operates upon four bits from the serial data stream and causes both an in-phase (IP) cosine carrier and a sine wave that serves as the quadrature component (QC) of the signal to be modulated. The IP and QC signals are then summed and result in the transmitted signal being changed in both amplitude and phase, with each point placed at the *x–y*

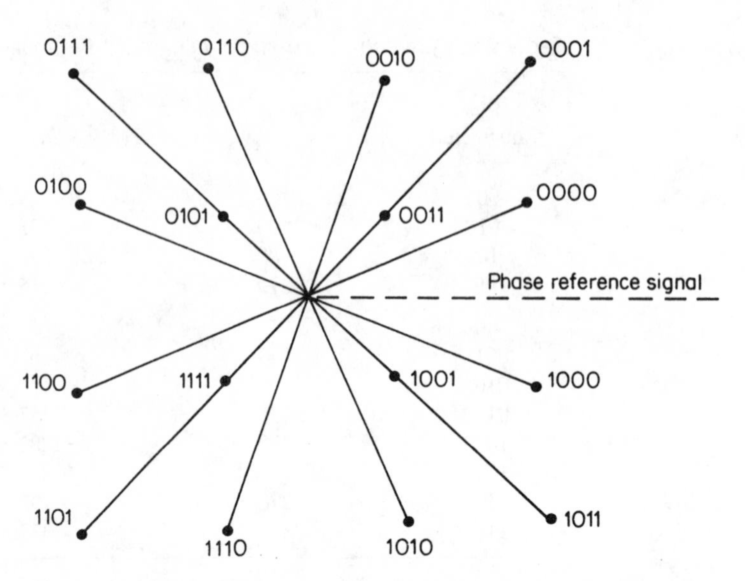

Figure 2.8 Quadrature amplitude modulation produces 16 signal states from a combination of 12 angles and 3 amplitude levels.

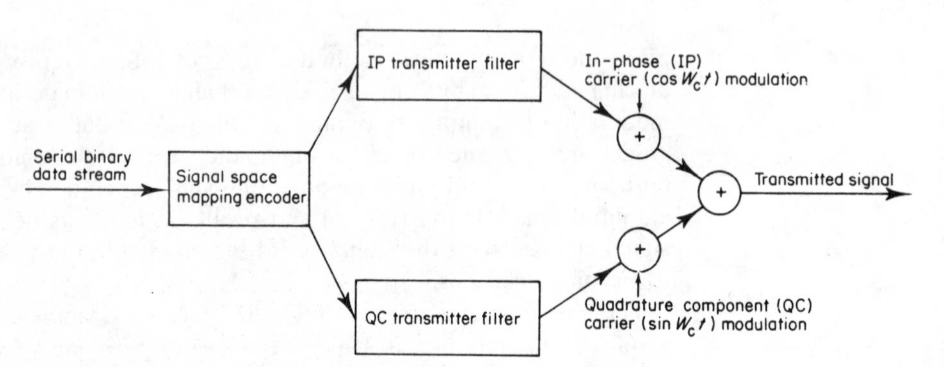

Figure 2.9 QAM modem transmitter.

coordinates representing the modulation levels of the cosine carrier and the sine carrier.

If one plots the signal points previously illustrated in Figure 2.8 which represent all of the data samples possible in that particular method of QAM, the series of points can be considered to be the signal structure of the modulation technique. Another popular term used to describe these points is the constellation pattern. By an examination of the constellation pattern of a modem, it becomes possible to predetermine its susceptibility to certain transmission impairments. As an example, phase jitter which causes signal points to rotate about the origin can result in one signal being misinterpreted for another, which would cause four bits to be received in error. Since there are 12 angles in the QAM method illustrated in Figure 2.8, the minimum rotation angle is 30 degrees, which provides a reasonable immunity to phase jitter.

New techniques

By the late 1980s several vendors were offering modems that operated at data rates up to 19,200 bps over leased voice-grade circuits. Originally, modems that operated at 14,400 bps employed a quadrature amplitude modulation technique, collecting data bits into a 6-bit symbol 2400 times per second, resulting in the transmission of a signal point selected from a 64-point signal constellation. The signal pattern of one vendor's 14,400 bps modem is illustrated in Figure 2.10. Note that this particular signal pattern appears to form a hexagon and according to the vendor was used since it provides a better performance level with respect to signal-to-noise (S/N) ratio and phase jitter than conventional rectangular grid signal structures. However, in spite of hexagonal packed signal structures, it should be obvious that the distance between signal points for a 14,400-bps modem are closer than the resulting points from a 9600-bps modem. This means that a 14,400-bps conventional QAM modem is more susceptible to transmission impairments and the overall data throughput under certain situations can be less than that obtainable with 9600-bps modems. Figure 2.11 illustrates the typical throughput variance of 9600- and 14,400-bps modems with respect to the ratio of noise to the strength of the signal (N/S) on the circuit. From this illustration, it should be apparent that 14,400-bps modems using conventional quadrature amplitude modulation should only be used on high-quality circuits.

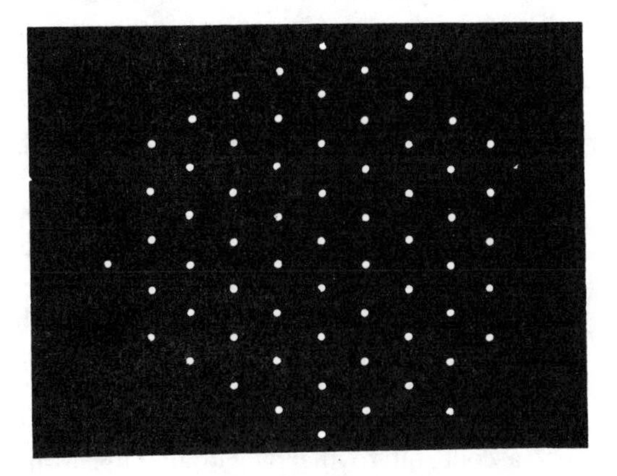

Figure 2.10 14,400-bps hexagonal signal constellation pattern.

Modems that transmit data at 16,000 bps are very similar to 14,400-bps devices, with the major difference being in the baud rate. Thus, most 16,000-bps modems encode data into 6-bit symbols and transmit the signals 2667 times per second. This method also employs a total of 64 signal points; however, the baud rate is increased from 2400 to 2667 to obtain the higher data transfer rate.

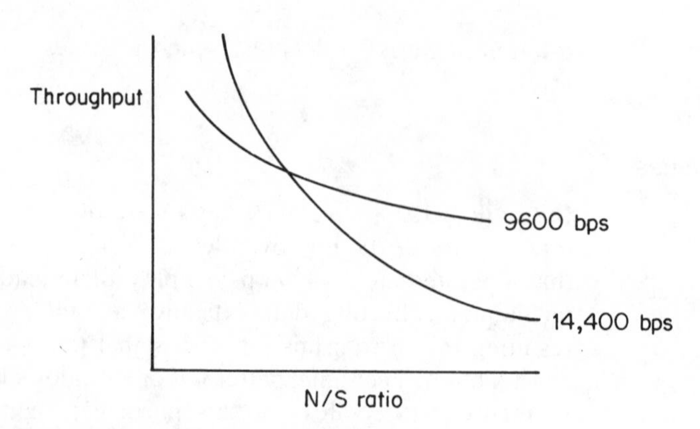

Figure 2.11 Throughput variance. Under certain conditions the throughput obtained by using 9600-bps modems can exceed the throughput obtained when using 14,400-bps devices.

Trellis coded modulation

Owing to the susceptibility of conventional QAM modems to transmission impairments, a new generation of modems based upon Trellis coded modulation (TCM) has been developed. Such modems tolerate more than twice as much noise power as conventional QAM modems, permitting 9600 bps transmission over the switched telephone network and reliable data transmission at speeds ranging from 14,400 to 19,200 bps over good-quality leased lines.

To understand how TCM provides a higher tolerance to noise and other line impairments, to include phase jitter and distortion, let us consider what happens when a line impairment occurs when conventional QAM modems are used. Here the impairment causes the received signal point to be displaced from its appropriate location in the signal constellation. The receiver then selects the signal point in the constellation that is closest to what it received. Obviously, when line impairments are large enough to cause the received point to be closer to a signal point that is different from the one transmitted, an error occurs. To minimize the possibility of such errors, TCM employs an encoder that adds a redundant code bit to each symbol interval.

In actuality, at 14,400 bps the transmitter converts the serial data stream into 6-bit symbols and encodes two of the six bits employing a binary convolutional encoding scheme as illustrated in Figure 2.12. The encoder adds a code bit to the two input bits, forming three encoded bits in each symbol interval. As a result of this encoding operation, three encoded bits and four remaining data bits are then mapped into a signal point which is selected from a 128-point (2^7) signal constellation.

The key to the ability of TCM to minimize errors at high data rates is the employment of forward error correcting (FEC) in the form of convolutional coding. With convolutional coding, each bit in the data stream is compared with one or more bits transmitted prior to that bit. The value of each bit, which can be changed by the convolutional encoder, is therefore dependent upon the value of other bits. In addition, a redundant bit is added for every

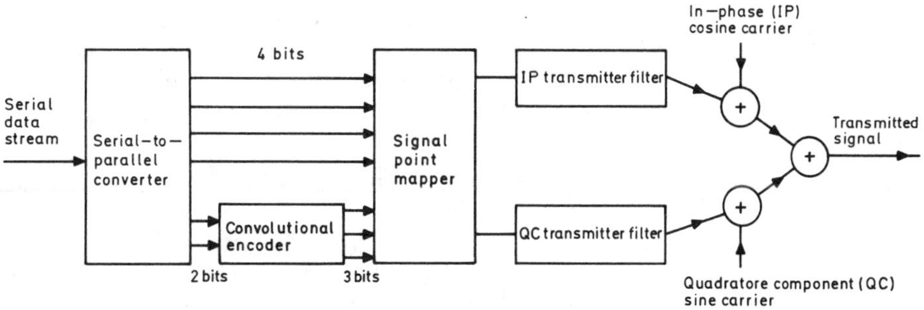

Figure 2.12 Trellis coded modulation.

group of bits compared in this manner. The following examination of the formation of a simple convolutional code clarifies how the convolutional encoder operates.

Convolutional encoder operation

Assuming that a simple convolutional code is formed by the modulo 2 sum of the two most recent data bits, then two output bits will be produced for each data bit — a data bit and a parity bit. If we also assume that the first output bit from the encoder is the current data bit then the second output bit is the modulo 2 sum of the current bit and its immediate predecessor. Figure 2.13 illustrates the generation of this simple convolutional code.

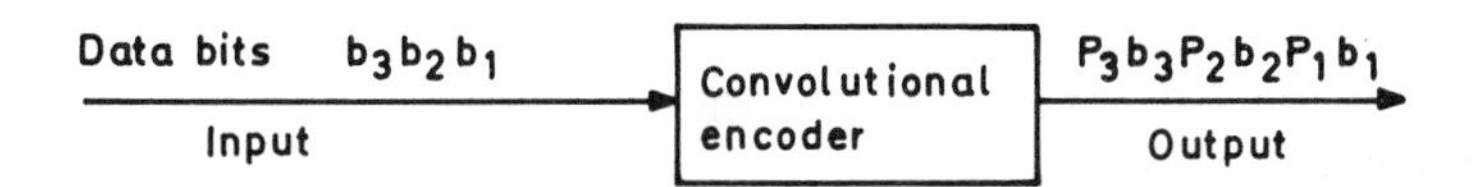

Figure 2.13 Simple convolutional code generation.

Because each parity bit is the modulo 2 sum of the two most recent data bits, the relationship between the parity bits and the data bits becomes:

$$P_i = b_i + b_{i-1} \qquad i = 1, 2, 3 \ldots$$

If the composition of the first four data bits entering the encoder was 1101 ($b_4b_3b_2b_1$), the four parity bits are developed as follows:

$$P_1 = b_1 + b_0 = 1 + 0 = 1$$
$$P_2 = b_2 + b_1 = 0 + 1 = 1$$
$$P_3 = b_3 + b_2 = 1 + 0 = 1$$
$$P_4 = b_4 + b_3 = 1 + 1 = 0$$

Thus, the four-bit sequence 1101 is encoded as 01111011.

The preceding example also illustrates how dependencies can be constructed. In actuality, there are several trade offs in developing a forward error correction scheme based upon convolutional coding. When a bit is only compared with a previously transmitted bit the number of redundant bits required for decoding at the receiver is very high. However, the complexity of the decoding process is minimized. When the bit to be transmitted is compared with a large number of previously transmitted bits the number of redundant bits required is minimized. However, the processing required at both ends increases in complexity.

In a TCM modem, the signal point mapper uses the three encoded bits to select one of eight (2^3) subsets consisting of 16 points developed from the four data bits. This encoding process ensures that only certain points are valid. At the receiving modem, the decoder compares the observed sequence of signal points and selects the valid point closest to the observed sequence. The encoder makes this selection process possible by generating redundant information that establishes dependencies between successive points in the signal constellation. At the receiving modem, the decoder uses an algorithm that compares previously received data with currently received data. The convolutional decoding algorithm then enables the modem to select the optimum signal point. Because of this technique, a TCM modem is twice as immune to noise as a conventional QAM modem. In addition, the probability of an error occurring when a TCM modem is used is substantially lower than when an uncoded QAM modem is used.

TCM Modem Developments The first TCM modems marketed in 1984 operated at 14.4 kbps. Since then, several manufacturers have introduced modems that use more complex TCM techniques. These modems operate at 16.8 kbps and 19.2 kbps. By 1989, the application of TCM to modems enabled network users to double the transmission speed on analog circuits from that achievable only a few years before.

Mode of transmission

If the transmitter or the receiver of the modem is such that the modem can send or receive data in one direction only, the modem will function as a simplex modem. If the operations of the transmitter and receiver are combined so that the modem may transmit and receive data alternately, the modem will function as a half-duplex modem. In the half-duplex mode of operation, the transmitter must be turned off at one location, and the transmitter of the modem at the other end of the line must be turned on before each change in transmission direction. The time interval required for this operation is referred to as turnaround time. If the transmitter and receiver operate simultaneously, the modem will function as a full-duplex modem. This simultaneous transmission in both directions can be accomplished either by splitting the telephone line's bandwidth into two channels on a 2-wire circuit or by the utilization of two 2-wire circuits, such as are obtained on a 4-wire leased line.

Transmission technique

Modems are designed for asynchronous or synchronous data transmission. Asynchronous transmission is also referred to as start–stop transmission and is usually employed by unbuffered terminals where the time between character transmission occurs randomly.

In asynchronous transmission, the character being transmitted is initialized by the character's start bit as a mark-to-space transition on the line and terminated by the character's stop bit which is converted to a "space/marking" signal on the line. The digital pulses between the start and stop bits are the encoded bits which determine the type of character which was transmitted. Between the stop bit of one character and the start bit of the next character, the asynchronous modem places the line in the "marking" condition. Upon receipt of the start bit of the next character the line is switched to a mark-to-space transition, and the modem at the other end of the line starts to sample the data.

The marking and spacing conditions are audio tones produced by the modulator of the modem to denote the binary data levels. These tones are produced at predefined frequencies, and their transition between the two states as each bit of the character is transmitted defines the character. Asynchronous transmission was originally employed with low-speed, teletype-compatible terminals at data rates up to 1800 bps. In the mid-1980s, the manufacture of CCITT compatible V.22 and V.22bis modems raised the data rate commonly obtainable over the PSTN to 2400 bps. By 1988 several vendors were manufacturing V.32 compatible modems which permitted a data rate of 9600 bps to be obtained while other vendors had introduced modems using proprietary modulation schemes that enable data rates up to 19200 bps over the PSTN.

Synchronous transmission permits more efficient line utilization since the bits of one character are immediately followed by the bits of the next character, with no start and stop bits required to delimit individual characters. In synchronous transmission, groups of characters are formed into data blocks, with the length of the block varying from a few characters to a thousand or more. Often, the block length is a function of the terminal's physical characteristics or its buffer size. As an example, for the transmission of data that represents punched card images, it may be convenient to transmit 80 characters of one card as a block, as there are that many characters if one constructed the card image from an 80-column card deck. If punched cards are being read by a computer for transmission, and data is such that every two cards contain information about one employee, the block size could be increased to 160 characters. In synchronous transmission, the individual bits of each of the characters within each block are identified based upon a transmitted timing signal which is usually provided by the modem and which places each bit into a unique time period. This timing or clock signal is transmitted simultaneously with the serial bit stream as shown in Figure 2.14.

Modem classification

Modems can be classified into many categories to include the mode of transmission and transmission technique as well as by the application features they contain and the type of lines they are built to service. Generally, modems can be classified into four line-servicing groups: subvoice or narrowband lines,

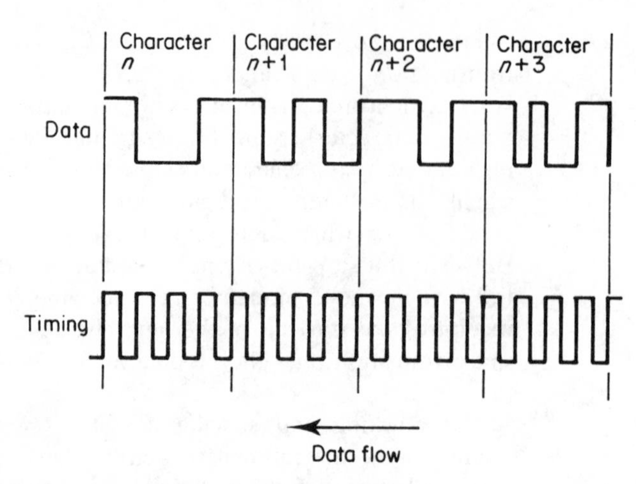

Figure 2.14 Synchronous timing signals. The timing signal is used to place the bits that form each character into a unique time period.

voice-grade lines, wideband lines, and dedicated lines. Subvoice-band modems require only a portion of the voice-grade channel's available bandwidth and are commonly used with equipment operating at speeds up to 300 bps. On narrow-band facilities, modems can operate in the full-duplex mode by using one-half of the available bandwidth for transmission in each direction and use an asynchronous transmission technique.

Modems designed to operate on voice-grade facilities may be asynchronous or synchronous, half-duplex or full-duplex. Asynchronous transmission is normally employed at speeds up to and including 2400 bps, although newer modems operate up to and including 19,200bps asynchronously. Although a leased, 4-wire line will permit full-duplex transmission at high speeds, transmission via the switched telephone network normally occurs in the full-duplex mode at data rates up to 2400 bps.

Voice-grade modems currently transfer data at rates up to 19,200 bps, usually requiring leased facilities for transmission at speeds in excess of 9600 bps. Wideband modems, which are also referred to as group-band modems since a wideband circuit is a grouping of lower-speed lines, permits users to transmit synchronous data at speeds in excess of 19,200 bps. Although wideband modems are primarily used for computer-to-computer transmission applications, they are also used to service multiplexers which combine the transmission of many low- or medium-speed terminals to produce a composite higher transmission speed. The use of group-band modems and multiplexers is explained later in this book. Dedicated or limited-distance modems, which are also known by such names as shorthaul modems and modem bypass units, operate on dedicated solid conductor twisted pair or coaxial cables, permitting data transmission at distances ranging up to 15 to 20 miles, depending upon the modem's operating speed and the resistance of the conductor.

Limited-distance modems

Modems in this category can operate at speeds up to approximately 1.5 million bps and are particularly well suited for in-plant usage where users desire to install their own communications lines between terminals and a computer located in the same facility or complex. Also, in comparison with voice-band and wideband modems, these modems are relatively inexpensive since they are designed to operate only for limited distances. In addition, by using this type of modem and stringing their own in-plant lines, users can eliminate a monthly telephone charge that would occur if the telephone company furnished the facilities. Limited-distance modems are explained in greater detail in Section 2.8. In Table 2.3, the common application features of modems are denoted by the types of lines to which they can be connected.

Table 2.3 Common modem features.

	Line type					
		Voice-grade				
Features	Subvoice (up to 300 bps)	Low, up to 1800 bps	Medium, 2000–4800 bps	High, 7200–19,200 bps	Wideband (19,200 bps and up)	Limited distance (up to 1.5 Mbps)
Asynchronous	√	√				√
Synchronous			√	√	√	√
Switched network	√	√	√	√		
Leased only		√	√	√	√	√
Half-duplex	√	√	√	√		√
Full-duplex	√	√	√	√	√	√
Fast turnaround for dial-up use			√	√		
Reverse/secondary channel	√	√	√			√
Manual equalization			√			
Automatic equalization		√	√	√		
Multiport capability			√	√		√
Voice/data			√	√		√

Line-type operations

Most modems with a rated transmission speed of up to 4800 bps and some modems which transmit data at 9600 to 19,200 bps can operate over the switched, dial-up telephone network. Since a circuit obtained from a dial-up telephone connection is a 2-wire line, when this line is used to carry traffic in both directions, alternately, the line and the modem are said to operate in the half-duplex mode. In this mode, traffic can only flow in one direction at a time, and the line must be turned around to reverse the direction of the flow of traffic. Such turnaround time varies by device and can become a considerable overhead factor if short bursts of data are transmitted, with each burst requiring a short acknowledgement. To visualize some of the overhead problems associated with line turnaround, a short examination of an error-control procedure for synchronous transmission follows.

A common error-control procedure used in synchronous transmission is obtained by the use of an acknowledgement–negative acknowledgement (ACK–NAK) sequence, commonly referred to as PAR (positive acknowledgement retransmit) protocol. When this type of sequence is used, the terminal or computer transmits a block of data to the receiving station. Appended to the end of the block is a block check character which is computed based upon a predefined algorithm. At the receiving device the block of data is examined, and a new block check character is developed, using the same algorithm, which is then compared to the transmitted block check character. If both block check characters are equal, the receiving device sends an ACK signal. If the block check characters do not match, then an error in transmission has occurred, and the receiving device transmits an NAK to signal the device which transmitted the block. This informs the transmitting device that the block should be retransmitted. This procedure is also referred to as automatic request for repeat (ARQ) and requires that the line upon which transmission occurs be turned around twice for each block. Returning to our 80-character punched card image block, transmitting this data as a 960-bit block with control characters appended, at 9600 bps, would take 100 ms (milliseconds); whereas if the modem turnaround time was 150 ms, a total of 300 ms would be necessary to turn the line around. Although recently developed modem features have reduced modem turnaround time, this problem can be avoided or greatly eliminated by using a modem with a reverse channel or by establishing full-duplex transmission over a leased 4-wire circuit, depending upon the type of protocol employed for transmission.

Reverse and secondary channels

To eliminate turnaround time when transmission is over the 2-wire switched network or to relieve the primary channel of the burden of carrying acknowledgement signals on 4-wire dedicated lines, modem manufacturers have developed a reverse channel which is used to provide a path for the acknowledgement of transmitted data at a slower speed than the primary channel. This reverse channel can be used to provide a simultaneous transmission path for the acknowledgement of data blocks transmitted over the higher speed primary channel at up to 150 bps.

A secondary channel is similar to a reverse channel. However, it can be used in a variety of applications which include providing a path for a high-speed terminal and a low-speed terminal. When a secondary channel is used as a reverse channel, it is held at one state until an error is detected in the high-speed data transmission and is then shifted to the other state as a signal for retransmission. Another application where a secondary channel can be utilized is when a location contains a high-speed, synchronous terminal and a slow-speed, asynchronous terminal such as a Teletype. If both devices are required to communicate with a similar distant location, one way to alleviate dual line requirements as well as the cost of extra modems to service both devices is by using a pair of modems that have secondary channel capacity, as shown in Figure 2.15. Although a reverse channel is usable on both 2-wire and 4-wire telephone lines, the secondary channel technique is usable only on a 4-wire circuit. A secondary channel modem derives two channels from the same line, a wide one to carry synchronous data usually at speeds of 2000, 2400, 3600, or 4800 bps and a narrow channel to carry asynchronous teletype-like data. Some modems with the secondary channel option can actually provide two slow-speed channels as well as one high-speed channel, with the two slow-speed channels being capable of transmitting asynchronous data up to a composite speed of 150 bps.

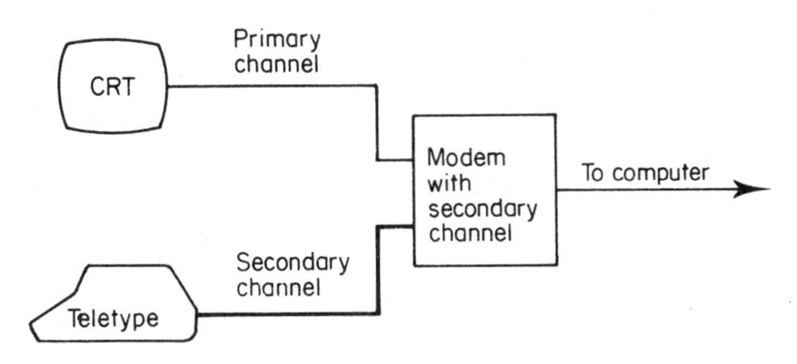

Figure 2.15 Secondary channel operation. Two terminals can communicate with a distant location by sharing a common line through the use of a modem with a secondary channel.

Equalization

Owing to the inconsistencies inherent in a transmission medium that was designed for voice rather than data transmission, modem manufacturers build equalizers into a modem to compensate for those inconsistencies produced by the telephone circuit, amplifiers, switches, and relays, as well as other equipment that data may be transmitted across in establishing a data link between two or more points. An equalizer is basically an inverse filter which is used to correct amplitude and delay distortion which, if uncorrected, could lead to intersymbol interference during transmission. A well-designed equalizer matches line conditions by maintaining certain of the modem's electrical parameters at the widest range of marginal limits in order to take advantage of the data rate capability of the line while eliminating intersymbol interference. The design of the equalizer is critical, since if the modem operates too near or outside these marginal limits, the transmission error rate will increase. There are three basic methods for achieving equalization. The first method, the utilization of fixed equalizers, is typically accomplished by using marginally adjustable high-Q filter sections. Modems with transversal filters use a tapped delay line with manually adjustable variable tap gains, while automatic equalization is usually accomplished by a digital transversal filter with automatic tap gain adjustments. The faster the modem's speed, the greater the need for equalization and the more complex the equalizer. Most modems with rated speeds up to 4800 bps incorporate non-adjustable, fixed equalizers which have been designed to match the average line conditions that have been found to occur on the dial-up network. Thus, most modems with fixed or non-adjustable equalizers are designed for a normal, randomly routed call between two locations over the dial-up network. If the modem is equipped with a signal-quality light which indicates an error rate that is unacceptable, or if the operator encounters difficulty with the connection, the problem can be alleviated by simply disconnecting the call and dialing a new call, which should reroute the connection through different points on the dial-up network.

Manual equalization

Manually adjusted equalization was originally employed on some 4800-bps modems used for transmission over leased lines, with the parameters being tuned or preset at installation time, and re-equalization usually not required unless the lines are reconfigured. One modem manufacturer uses an 8-level (4-phase, 2-amplitude) modulation technique and has eight equalizer controls on the front panel of the modem for transmission and eight additional controls for reception. Using this type of modem, equalization may be performed from a single end of a point-to-point network without assistance from the opposite end. Primarily designed to operate over unconditioned leased telephone lines, manually equalized modems allow the user to eliminate the monthly expense associated with line conditioning. Owing to the incorporation of microprocessors into modems for signal processing they were soon employed to perform automatic equalization. This has resulted in most modems incorporating automatic equalization.

Automatic equalization

Automatic equalization is used on most 4800 bps modems designed for operation over the switched telephone network and on all 7200 bps and above modems which are primarily designed to operate over dedicated lines but which can operate over the switched network in a fallback operational mode. With automatic equalization, a certain initialization time is required to adapt the modem to existing line conditions. This initialization time becomes important during and after line outages, since long initial equalization times can extend otherwise short dropouts unnecessarily. Recent modem developments have shortened the initial equilization time to between 15 and 25 ms, whereas only a few years ago a much longer time was commonly required. After the initial equalization, the modem continuously monitors and compensates for changing line conditions by an adaptive process. This process allows the equalizer to "track" the frequently occurring line variations that occur during data transmission without interrupting the traffic flow. On one 9600-bps modem, this adaptive process occurs 2400 times a second, permitting the recognition of variations as they occur.

Synchronization

For synchronous communications, generally in speeds exceeding 1800 bps, the start–stop bits characteristic of asynchronous communications can be eliminated. Bit synchronization is necessary so that the receiving modem samples the link at the exact moment that a bit occurs. The receiver clock is supplied by the modem in phase coherence with the incoming data bit stream, or more simply stated, tuned to the exact speed of the transmitting clock. The transmitting clock can be supplied by either the modem (internal) or the terminal (external).

The transmission of synchronous data is generally under the control of a master clock which is the fastest clock in the system. Any slower data clock rates required are derived from the master clock by digital division logic, and those clocks are referred to as slave clocks. For instance, a master clock oscillating at a frequency of 96 kHz could be used to derive 9.6 kbps (1/10), 4.8 kbps (1/20), and 2.4 kbps (1/40) clock speeds.

Multiport capability

Modems with a multiport capability offer a function similar to that provided by a multiplexer. In fact, multiport modems contain a limited function time division multiplexer (TDM) which provides the user with the capability of transmitting more than one synchronous data stream over a single transmission line, as shown in Figure 2.16.

In contrast with typical multiplexers, the limited function multiplexer used in a multiport modem combines only a few high-speed synchronous data streams, whereas multiplexers can normally concentrate a mixture of asynchronous and synchronous, high- and low-speed data streams. A further description, as well as application examples, will be found in Section 2.4.

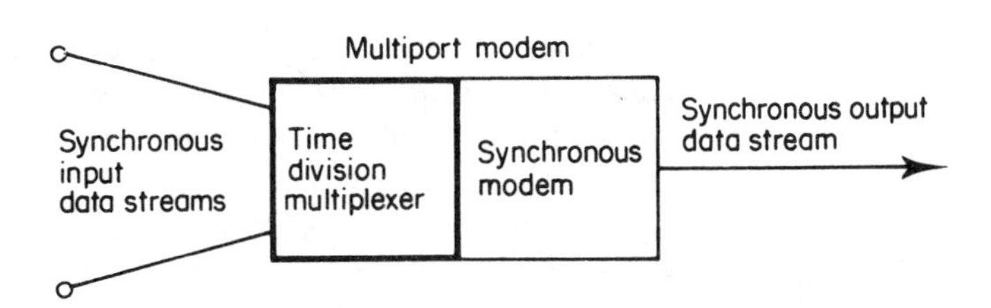

Figure 2.16 Multiport modem. Containing a limited function time division multiplexer, a multiport modem combines the input of a few synchronous input data streams for transmission.

Security capability

To provide an additional level of network protection for calls originated over the PSTN several vendors market "security" modems. In essence, these modems contain a buffer area into which a network administrator enters authorized passwords and associated telephone numbers. When a potential network user dials the telephone number assigned to the security modem that device prompts the person to enter his password. If a valid match between the entered and previously stored password occurs the security modem disconnects the line and then dials the telephone number associated with the password. Thus, a modem with a security capability provides a mechanism to verify the originator of calls over the PSTN by his telephone number. A further description of security modems is contained in Section 2.6.

Multiple speed selection capability

For data communication systems which require the full-time service of dedicated lines but need to access the switched network if the dedicated line should fail or degrade to the point where it cannot be used, dial backup capability for the modems used is necessary. Since transmission over dedicated lines usually occurs at a higher speed than one can obtain over the switched network, one method of facilitating dial backup is through switching down the speed of the modem. Thus, a multiple speed modem which is designed to operate at 9600 bps over dedicated lines may be switched down to 7200 or 4800 bps for operation over the dial-up network until the dedicated lines are restored.

Voice/data capability

Many modems can now be obtained with a voice/data option which permits a specially designed telephone set, commonly called a voice adapter, to be used to provide the user with a voice communication capability over the same line which is used for data transmission. Depending upon the modem, this voice capability can be either alternate voice/data or simultaneous voice/data. Thus, the user may communicate with a distant location at the same time as data transmission is occurring, or the user may transmit data during certain times of the day and use the line for voice communications at other times. Voice/data capability can also be used to minimize normal telephone charges when data transmission sequences require voice coordination. Voice adapters are examined in more detail in Chapter 6, Section 6.4.

Modem operations and compatibility

Many modem manufacturers describe their product offerings in terms of compatibility or equivalency with modems manufactured by Western Electric for the Bell System, prior to its breakup into independent telephone companies, or with Consultative Committee for International Telephone and Telegraph (CCITT) recommendations. The CCITT, which is part of the International Telecommunications Union based in Geneva, has developed a series of modem standards for recommended use. These recommendations are primarily adapted by the Post, Telephone and Telegraph (PTT) organizations that operate the telephone networks of many countries outside the United States; however, owing to the popularity of certain CCITT recommendations, they have also been followed in designing certain modems for operation on communications facilities within the USA. The following examination of the operation and compatibility of the major types of Bell System and CCITT modems is based upon their operating rate.

300 bps

Modems operating at 300 bps use a frequency shift keying (FSK) modulation technique as previously described during the discussion of acoustic couplers in Section 2.1. In this technique the frequency of the carrier is alternated to one of two frequencies, one frequency representing a space or zero bit while the other frequency represents a mark or a one bit. Table 2.4 lists the frequency assignments for Bell System 103/113 and CCITT V.21 modems which represent the two major types of modems that operate at 300 bps.

Bell System 103 and 113 series modems are designed so that one channel is assigned to the 1070- to 1270-Hz frequency band while the second channel is assigned to the 2025- to 2225-Hz frequency band. Modems that transmit in the 1070- to 1270-Hz band but receive in the 2025- to 2225-Hz band are designated as originate modems, while a modem which transmits in the 2025- to 2225-Hz band but receives in the 1070- to 1270-Hz band is designated as an answer modem. When using such modems, their correct pairing is important, since two originate modems cannot communicate with each other. Bell System 113A modems are originate only devices that should normally be used when calls are to be placed in one direction. This type of modem is mainly used to enable teletype-compatible terminals to communicate with time-sharing systems where such terminals only originate calls. Bell System 113B modems are answer only and are primarily used at computer sites where users dial in to establish communications. Since these modems are designed to transmit and receive on a single set of frequencies, their circuitry requirements are less than other modems and their costs are thus more economical.

Modems in the 103 series, which includes the 103A, E, F, G, and J modems, can transmit and receive in either the low or the high band. This ability to switch modes is denoted as "originate and answer", in comparison to the Bell 113A which operates only in the originate mode and the Bell 113B which operates only in the answer mode.

As indicated in Table 2.4, modems operating in accordance with the CCITT V.21 recommendation employ a different set of frequencies for the transmission and reception of marks and spaces. Thus, Bell System 103/113 type modems and CCITT V.21 devices can never communicate with one another. The two pairs of frequencies used by the modems listed in Table 2.4 permit the bandwidth of a communications channel to be split into two subchannels by

Table 2.4 Frequency assignments (Hz) for 300-bps modems.

Major modem types	Originate	Answer
Bell System (103/113 type)	Mark 1270 Space 1070	2225 2025
CCITT V.21	Mark 980 Space 1180	1650 1850

frequency. This technique was illustrated in Figure 2.2 for Bell System 103/113 modems. Since each subchannel can permit data to be transmitted in a direction opposite that transmitted on the other subchannel, this technique permits full-duplex transmission to occur on the switched telephone network which is a 2-wire circuit that normally can only support half-duplex transmission.

300 to 1800 bps

There are several Bell System and CCITT V series modems that operate in the range between 300 to 1800 bps. Some of these modems such as the Bell System 212A and CCITT V.22 devices can operate at either of two speeds; and other modems such as the Bell System 202 and the CCITT V.23 only operate at one data rate. We will examine these modems in pairs, enabling their similarities and differences to be compared.

Bell System 212A and V.22 modems

The Bell System 212A modem permits either asynchronous or synchronous transmission over the public switched telephone network. The 212A contains a 103-type modem for asynchronous transmission at speeds up to 300 bps. At this data rate FSK modulation is employed, using the frequency assignments previously indicated in Table 2.4. At 1200 bps, dibit phase shift keyed (DPSK) modulation is used which permits the modem to operate either asynchronously or synchronously. The phase shift encoding of the 212A type modem is illustrated in Table 2.5.

One advantage in the use of this modem is that it permits the reception of transmission from terminals at two different transmission speeds. Before the

Table 2.5 212A type modem phase shift encoding.

Dibit	Phase shift (degrees)
00	90
01	0
10	180
11	270

operator initiates a call, he or she selects the operating speed at the originating set. The manner in which the operating speed is selected depends upon the type of 212A modem used. If the modem is what is now commonly referred to as a "dumb" modem the operator selects the higher operating speed by pressing a "HS" (high speed) button on the front panel of the modem. If the modem is an intelligent modem built to respond to software commands the operators can either use a communications program or send a series of commands through the serial port of a personal computer or terminal connected to the modem to set its operating speed. Due to the substantial use of intelligent modems with personal computers these modems will be reviewed as a separate entity later in this chapter. When the call is made, the answering 212A modem automatically switches to that operating speed. During data transmission, both modems remain in the same speed mode until the call is terminated, when the answering 212A can be set to the other speed by a new call. The dual-speed 212A permits both terminals connected to Bell System 100 series data sets operating at up to 300 bps or terminals connected to other 212A modems operating at 1200 bps to share the use of one modem at a computer site and thus can reduce central computer site equipment requirements.

The V.22 standard is for modems that operate at 1200 bps on the PSTN or leased circuits and has a fallback data rate of 600 bps. The modulation technique employed is 4-phase PSK at 1200 bps and 2-phase PSK at 600 bps, with five possible operational modes specified for the modem at 1200 bps. Table 2.6 lists the V.22 modulation phase shifts with respect to the bit patterns entering the modem's transmitter. Modes 1 and 2 are for synchronous and asynchronous data transmission at 1200 bps respectively, while mode 3 is for synchronous transmission at 600 bps. Mode 4 is for asynchronous transmission at 600 bps while mode 5 represents an alternate phase change set for 1200 bps asynchronous transmission.

In comparing V.22 modems to the Bell System 212A devices it should be apparent that they are totally incompatible at the lower data rate, since both the operating speed and modulation techniques differ. At 1200 bps the modulation techniques used by a V.22 modem in modes one through four are exactly the same as that used by a Bell System 212A device. Unfortunately, a Bell 212A modem that answers a call sends a tone of 2225 Hz on the line that the originating modem is supposed to recognize. This frequency is used because of the construction of the switched telephone network in the United States and other parts of North America. Under V.22, the answering modem first sends a tone of 2100 Hz since this frequency is more compatible with the

Table 2.6 V.22 modulation phase shift vs. bit patterns.

Dibit values (1200 bps)	Bit values (600 bps)	Phase change modes 1, 2, 3, 4	Phase change mode V
00	0	90	270
01	–	0	180
11	1	270	90
10	–	180	0

design of European switched telephone networks. Then, the V.22 modem sends a 2400 Hz tone that would not be any better except that the V.22 modem also sends a burst of data whose primary frequency is about 2250 Hz, which is close enough to the Bell standard of 2225 Hz that many Bell 212A-type modems will respond. Thus, some Bell 212A modems can communicate with CCITT V.22 modems at 1200 bps while other 212-type modems may not be able to communicate with V.22 devices, with the ability to successfully communicate being based upon the tolerance of the 212 type modem to recognize the V.22 modem's data burst at 2250 Hz.

Bell System 202 series modems

Bell System 202 series modems are designed for speeds up to 1200 or 1800 bps. The 202C modem can operate on either the switched network or on leased lines, in the half-duplex mode on the former and the full-duplex mode on the latter. The 202C modem can operate half-duplex or full-duplex on leased lines. This series of modems uses frequency shift keyed (FSK) modulation, and the frequency assignments are such that a mark is at 1200 Hz and a space at 2200 Hz. When either modem is used for transmission over a leased 4-wire circuit in the full-duplex mode, modem control is identical to the 103 series modem in that both transmitters can be strapped on continuously which alleviates the necessity of line turnarounds.

Since the 202 series modems do not have separate bands, on switched network utilization half-duplex operation is required. This means that both transmitters (one in each modem) must be alternately turned on and off to provide two-way communication.

The Bell 202 series modems have a 5 bps reverse channel for switched network use, which employs amplitude modulation for the transmission of information. The channel assignments used by a Bell System 202 type modem are illustrated in Figure 2.17, where the 387 Hz signal represents the optional 5 bps AM reverse channel. Owing to the slowness of this reverse channel, its use is limited to status and control function transmission. Status information such as "ready to receive data" or "device out of paper" can be transmitted on this channel. Also owing to the slow transmission rate, error detection of received messages and an associated NAK and request for retransmission is

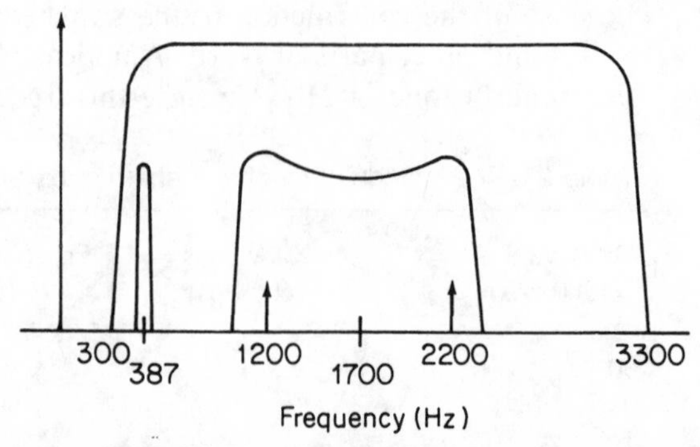

Figure 2.17 Bell System 202-type modem channel assignments.

normally accomplished on the primary channel since even with the turnaround time, it can be completed at almost the same rate one obtains in using the reverse channel for that purpose. Non-Bell 202-equivalent modems produced by many manufacturers provide reverse channels of 75 to 150 bps which can be utilized to enhance overall system performance. Reverse keyboard-entered data as well as error detection information can be practically transmitted over such a channel.

While a data rate of up to 1800 bps can be obtained with the 202D modem, transmission at this speed requires that the leased line be conditioned for transmission by the telephone company. The 202S and 202T modems are additions to the 202 series and are designed for transmission at 1200 and 1800 bps over the switched network and leased lines, respectively. At speeds in excess of 1400 bps, the 202T requires line conditioning when interfaced to either 2- or 4-wire circuits, whereas for a 2-wire circuit, conditioning is required at speeds in excess of 1200 bps when an optional reverse channel is used.

V.23 modems

The V.23 standard is for modems that transmit at 600 or 1200 bps over the PSTN. Both asynchronous and synchronous transmission is supported by using FSK modulation; and, an optional 75 bps backward or reverse channel can be used for error control. Figure 2.18 illustrates the channel assignments for a V.23 modem. In comparing Figure 2.18 with Figure 2.17, it is obvious that Bell System 202 and V.23 modems are incompatible with each other.

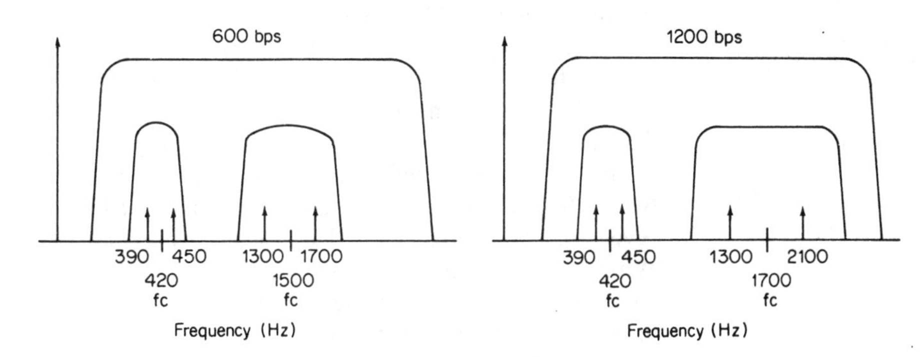

Figure 2.18 V.23 channel assignments.

2400 bps

Examples of modems that operate at 2400 bps include the Bell System 201, CCITT V.26 series, and the V.22 bis modem. The Bell System 201 and CCITT V.26 series modems are designed for synchronous bit serial transmission at a data rate of 2400 bps, while the V.22 bis standard governs 2400 bps asynchronous transmission.

Bell System 201B/C

Current members of the 201 series include the 201B and 201C models. Both of these modems use dibit phase shift keying modulation, with the phase shifts

Table 2.7 Bell System 201 B/C phase shift vs. bit pattern.

Dibit values	Phase shift
00	225°
10	315°
11	45°
10	135°

based upon the dibit values listed in Table 2.7. The 201B modem is designed for half- or full-duplex synchronous transmission at 2400 bps over leased lines. In comparison, the 201C is designed for half-duplex, synchronous transmission over the PSTN. A more modern version of the 201C is AT&T's 2024A modem, which is compatible with the 201C.

V.26 modem

The V.26 standard specifies the characteristics for a 2400 bps synchronous modem for use on a 4-wire leased line. Modems operating according to the V.26 standard employ dibit phase shift keying, using one of two recommended coding schemes. The phase change based upon the dibit values for each of the V.26 coding schemes is listed in Table 2.8.

Two similar CCITT recommendations to V.26 are V.26 bis and V.26 ter. The V.26 bis recommendation defines a dual speed 2400/1200 bps modem for use on the PSTN. At 2400 bps the modulation and coding method is the same as the V.26 recommendation for pattern B listed in Table 2.8. At the reduced data rate of 1200 bps a 2-phase shift modulation scheme is employed, with a binary zero represented by a 90° phase shift while a binary one is represented by a 270° phase shift. The V.26 bis recommendation also includes an optional reverse or backward channel that can be used for data transfer up to 75 bps. When employed, frequency shift keying is used to obtain this channel capacity, with a mark or one bit represented by a 390 Hz signal and a space or zero bit represented by a 450 Hz signal.

The V.26 ter recommendation uses the same phase shift scheme as the V.26 modem, but incorporates an echo-cancelling technique that allows transmitted and received signals to occupy the same bandwidth. Thus, the V.26 ter modem

Table 2.8 V.26 modulation phase shift vs bit pattern.

	Phase change	
Dibit values	Pattern A	Pattern B
00	0	45
01	90	135
11	180	225
10	270	315

is capable of operating in full duplex at 2400 bps on the PSTN. Echo cancelling will be described later in this chapter when the V.32 modem is examined.

V.22 bis

The CCITT V.22 bis recommendation governs modems designed for asynchronous data transmission at 2400 bps over the PSTN, with a fallback rate of 1200 bps. Since V.22 bis defines operations at 1200 bps to follow the V.22 format, communications capability with Bell System 212A type modems at that data rate may not always be possible owing to the answer tone incompatibility usually encountered between modems following Bell System specifications and CCITT recommendations. In addition, V.22 bis modems manufactured in the USA may not be compatible with such modems manufactured in Europe, at fallback data rates. This is because V.22 bis modems manufactured in Europe follow the V.22 format, with fallback data rates of 1200 and 600 bps. At 1200 bps the incompatibility between most European telephone networks, which are designed to accept only 2100 Hz answer tones while the US telephone network usually accepts an answer tone between 2100 and 2225 Hz, may preclude communications between a US and a European manufactured V.22 bis modem at 1200 bps. At a lower fallback speed the European modem will operate at 600 bps while the US V.22 bis modem operates at 300 bps, insuring incompatibility.

In spite of the previously mentioned problems, V.22 bis modems are becoming a *de facto* standard for use with terminals and personal computers communicating over the PSTN. This is due to several factors, to include the manufacture in the United States of V.22 bis modems that are Bell System 212A compatible, permitting persons with such modems to be able to communicate with other persons and mainframe computers connected to either 212A or 103/113 type modems. In addition, at 2400 bps US V.22 bis modems can communicate with European V.22 bis, in effect providing worldwide communications capability over the PSTN.

4800 bps

The Bell System 208 series and CCITT V.27 modems represent the most common types of modems designed for synchronous data transmission at 4800 bps. The Bell System 208 Series modems use a quadrature amplitude modulation technique. The 208A modem is designed for either half-duplex or full-duplex operation at 4800 bps over leased lines. The 208B modem is designed for half-duplex operation at 4800 bps on the switched network.

Newer versions of the 208A are offered by AT&T as the 2048A and 2048C models, which are also designed for 4-wire leased line operation. The 2048C has a start-up time less than one half of the 2048A, which makes it more suitable for operations on multidrop lines.

Both Bell 208 type modems and CCITT V.27 modems pack data three bits at a time, encoding them for transmission as one of eight phase angles. Unfortunately, since each type of modem uses different phase angles to represent a tribit value, they cannot talk to each other. Table 2.9 lists the V.27 modulation phase shifts with respect to each of the eight possible tribit values.

Table 2.9 V.27 modulation phase drift vs bit pattern.

Tribit values	Phase change
001	0
000	45
010	90
011	135
111	180
110	225
100	270
101	315

9600 bps

Three common modems that are representative of devices that operate at 9600 bps are the Bell System 209, and the CCITT V.29 and V.32 modems. Modems equivalent to the Bell System 209 and CCITT V.29 devices are designed to operate in a full-duplex, synchronous mode at 9600 bps over private lines. The Bell System 209A modem operates by employing a quadrature amplitude modulation technique as previously illustrated in Figure 2.8. Included in this modem is a built-in synchronous multiplexer which will combine up to four data rate combinations for transmission at 9600 bps. The multiplexer combinations are shown in Table 2.10. The use of a multiplexer incorporated into a modem is discussed more thoroughly in Section 2.4. A newer version of the 209A offered by AT&T is the 2096A. This modem is noteworthy because it has an EIA RS-449/423 interface with RS-232-C/D compatibility.

Table 2.10 Bell 209A multiplexer combinations.

2400–2400–2400–2400 bps
4800–2400–2400 bps
4800–4800 bps
7200–2400 bps
9600 bps

With the exception of Bell System 209-type modems, a large majority of 9600-bps devices manufactured throughout the world adhere to the CCITT V.29 standard. The V.29 standard governs data transmission at 9600 bps for full- or half-duplex operation on leased lines, with fallback data rates of 7200 and 4800 bps allowed. At 9600 bps the serial data stream is divided into groups of four consecutive bits. The first bit in the group is used to determine the amplitude to be transmitted while the remaining three bits are encoded as a phase change, with the phase changes identical to those of the V.27 recommendation listed in Table 2.9.

Table 2.11 V.29 signal amplitude construction.

Absolute phase	1st bit	Relative signal element amplitude
0,90,180,270	0	3
	1	5
45,135,225,315	0	$\sqrt{2}$
	1	$3\sqrt{2}$

Table 2.11 lists the relative signal element amplitude of V.29 modems, based upon the value of the first bit in the quadbit and the absolute phase which is determined from bits two through four. Thus, a serial data stream composed of the bits 1 1 0 0 would have a phase change of 270° and its signal amplitude would be 5. The resulting signal constellation pattern of V.29 modems is illustrated in Figure 2.19.

A recent CCITT recommendation that warrants attention is the V.32 standard. V.32 is based upon a modified quadrature amplitude modulation technique and is designed to permit full-duplex 9600 bps transmission over the switched telephone network.

A V.32 modem establishes two high-speed channels in the opposite direction of one another as illustrated in Figure 2.20. Each of these channels shares approximately the same bandwidth, with an echo cancelling technique employed to enable transmitted and received signals to occupy the same bandwidth. This is made possible by designing intelligence into the modem's receiver that permits it to cancel out the effects of its own transmitted signal enabling the

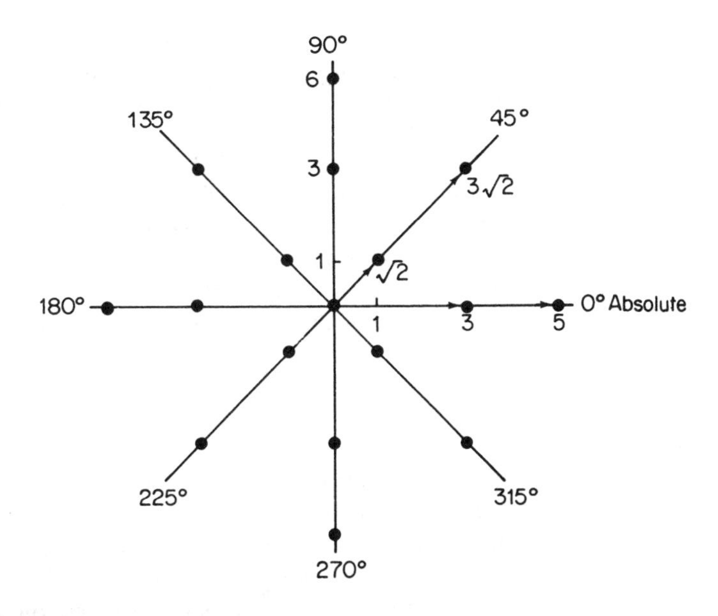

Figure 2.19 V.29 signal constellation pattern.

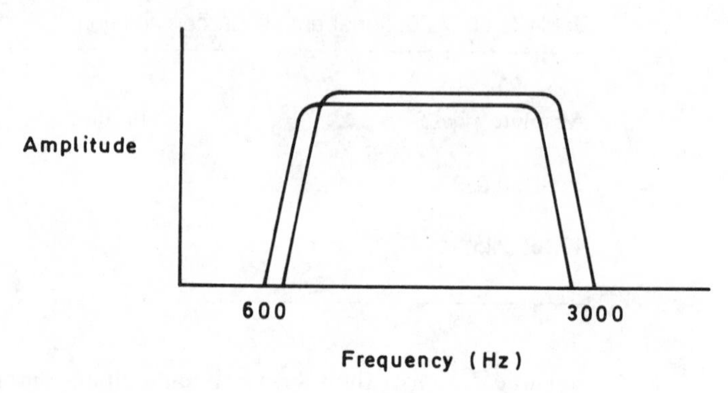

Figure 2.20 V.32 channel derivation.

modem to distinguish its sending signal from the signal being received.

V.32 modems employ Trellis coded modulation (TCM) at 9600 bps and quadrature amplitude modulation (QAM) at 4800 bps, with the lower data rate used when line conditions become seriously degraded.

A V.32 modem obtains high-speed full-duplex transmission by deriving two channels that share approximately the same bandwidth through the use of an echo cancelling technique.

14400 bps

Currently, the only standardized 14,400 bps modem is the CCITT V.33 recommendation.

V.33 modem

The V.33 modem is intended to be used on conditioned leased lines. Trellis coded modulation is employed when the modem operates at its primary data rate of 14,400 bps as well as a fallback data rate of 12,000 bps. Under the V.33 standard an optional multiplexer for combining data rates of 12,000, 9600, 7200, 4800, and 2400 bps is specified. When the modem operates at 14,400 bps eleven multiplex configurations are specified while seven configurations are specified when the modem operates at 12,000 bps. For specific information concerning the operation of a multiplexer in a modem the reader is referred to the section in this chapter covering multiport modems.

A summary of the operational characteristics of Bell System and CCITT V series type modems is listed in Table 2.12.

Non-standard modems

The requirements of personal computer users for higher transmission rates for file transfer and interactive full screen display operations resulted in several vendors designing proprietary operating modems to achieve data rates that would have been beyond the realm of belief several years ago. Some of these modems incorporate data compression and decompression algorithms, permitting data to be compressed prior to transmission and then expanded

Table 2.12 Modem operational characteristics.

Modem type	Maximum data rate	Transmission technique	Modulation technique	Transmission mode	Line use
Bell System					
103A,E	300	asynchronous	FSK	Half, Full	Switched
103F	300	asynchronous	FSK	Half, Full	Leased
201B	2400	synchronous	PSK	Half, Full	Leased
201C	2400	synchronous	PSK	Half, Full	Switched
202C	1200	asynchronous	FSK	Half	Switched
202S	1200	asynchronous	FSK	Half	Switched
202D/R	1800	asynchronous	FSK	Half, Full	Leased
202T	1800	asynchronous	FSK	Half, Full	Leased
208A	4800	synchronous	PSK	Half, Full	Leased
208B	4800	synchronous	PSK	Half	Switched
209A	9600	synchronous	QAM	Full	Leased
212	0–300	asynchronous	FSK	Half, Full	Switched
	1200	asynchronous/ synchronous	PSK	Half, Full	Switched
CCITT					
V.21	300	asynchronous	FSK	Half, Full	Switched
V.22	600	asynchronous	PSK	Half, Full	Switched/ Leased
	1200	asynchronous/ synchronous	PSK	Half, Full	Switched/ Leased
V.22 bis	2400	asynchronous	QAM	Half, Full	Switched
V.23	600	asynchronous/ synchronous	FSK	Half, Full	Switched
	1200	asynchronous/ synchronous	FSK	Half, Full	Switched
V.26	2400	synchronous	PSK	Half, Full	Leased
	1200	synchronous	PSK	Half	Switched
V.26 bis	2400	synchronous	PSK	Half	Switched
V.26 ter	2400	synchronous	PSK	Half, Full	Switched
V.27	4800	synchronous	PSK		
V.29	9600	synchronous	QAM	Half, Full	Leased
V.32	9600	synchronous	TCM/QAM	Half, Full	Switched
V.33	14,400	synchronous	TCM	Half, Full	Leased

back into its original form by a modem at the opposite end of the communications path.

Since compression decreases the amount of data requiring transmission the modem can accept a higher data rate input than it is capable of transmitting. Thus, a V.29 operating modem incorporating data compression that has a 2 to 1 compression ratio is theoretically capable of transmitting data at 19,200 bps, even though the modem operates at 9600 bps. Since the compression efficiency depends upon the susceptibility of the data to the compression algorithms built into the modem, in actuality the modem operates at a variable data rate. When no compression is possible the modem operates at 9600 bps, while the actual throughput of the device will increase as the data input into the modem becomes more susceptible to compression.

Packetized Ensemble Protocol

A second type of non-standard modem reached the market-place in 1986. More formally known as a Packetized Ensemble Protocol Modem, this modem incorporates a revolutionary advance in technology due to the incorporation of a high-speed microprocessor and approximately 70,000 lines of instructions built into read-only memory (ROM) chips on the modem board.

To better understand the operation of a Packetized Ensemble Protocol Modem let us review the operation of a conventional modem, using the Bell System 212 device for illustrative purposes.

When a Bell 212A modem modulates data, two carrier signals are varied to impress information onto the line. One carrier signal operates at a frequency of 1200 Hz while the second carrier frequency operates at 2400 Hz, resulting in two paths, which enables full-duplex transmission to occur on a 2-wire circuit. This is illustrated in Figure 2.21. Thus, at any instant in time data can only flow on two frequencies.

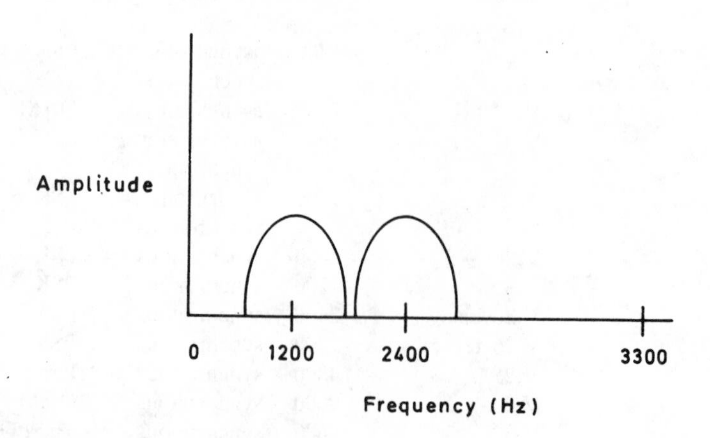

Figure 2.21 Bell 212A modem carrier signals.

Under the Packetized Ensemble Protocol, the originating modem simultaneously transmits 512 tones onto the line. The receiving modem evaluates the tones and the effect of noise on the entire voice bandwidth, reporting back to the originating device the frequencies that are unusable. The originating modem then selects a transmission format most suitable to the useful tones, employing 2-bit, 4-bit or 6-bit quadrature amplitude modulation (QAM) and packetizes the data prior to its transmission. Figure 2.22 illustrates the carrier utilization of a Packetized Ensemble Modem.

As an example of the efficiency of this type of modem, let us assume that 400 tones are available for a 6-bit QAM scheme. This would result in a packet

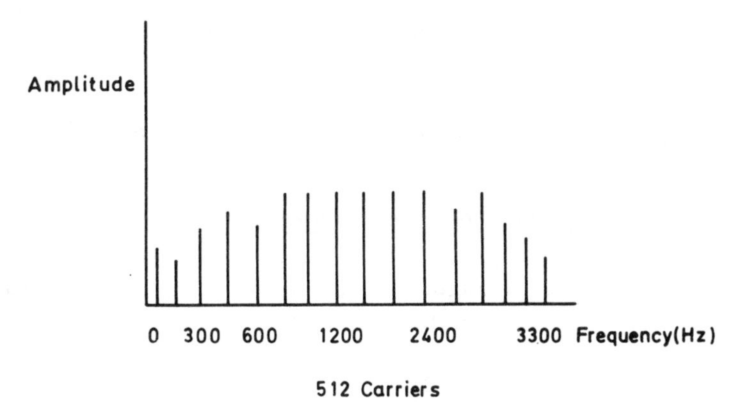

Figure 2.22 Packetized ensemble modem. (Reprinted with permission from *Data Communications Management*, © 1988 Auerbach Publishers, New York.)

size of 400×6 or 2400 bits. If each of the 400 tones is varied four times per second, a data rate of approximately 10,000 bps is obtained. It should be noted that the modem automatically generates a 16-bit cyclic redundancy check (CRC) for error detection, which is added to each transmitted packet. At the receiving modem, a similar CRC check is performed. If the transmitted and locally generated CRC characters do not match, the receiving modem will then request the transmitting modem to retransmit the packet, resulting in error correction by retransmission.

Some of the key advantages of a Packetized Ensemble Protocol Modem are its ability to automatically adjust to usable frequencies which greatly increases the use of the line bandwidth and its ability to lower its fallback rate in small increments. The latter is illustrated in Figure 2.23, which shows how this type of modem loses the ability to transmit on one or a few tones as the noise level on a circuit increases, resulting in a slight decrease in the data rate of the

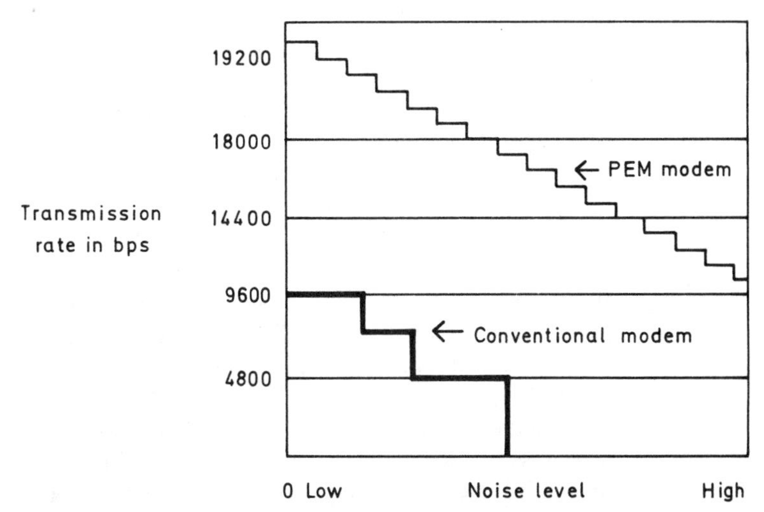

Figure 2.23 Transmission rate versus noise level. (Reprinted with permission from *Data Communications Management*, © 1988 Auerbach Publishers, New York.)

modem. In comparison, a conventional modem, such as a 9600 bps device, is designed to fallback to a predefined fraction of its main data rate, typically 7200 or 4800 bps.

To appreciate the advantages of using high-speed modems, let us examine the transmission of a 250-word document, 20 such documents and the contents of a 360K byte diskette at different data rates. A comparison of the utilization of 300, 1200, 2400, 4800 and 10,000 bps modems is presented in Table 2.13. As indicated, a 10,000 bps modem can substantially reduce the transmission time required to send long files between personal computers or a personal computer and a mainframe computer.

Table 2.13 Comparing throughput efficiencies.

	Transmission rate (bps)				
Item to transmit	300	1200	2400	4800	10,000
1 page 250 words (s)	40	10	5	2.5	1.2
20 pages (s)	800	200	100	50	24
360K diskette (min)	163.84	40.96	20.48	10.24	4.92

The original Packetized Ensemble Modem was designed by Telebit Corporation and is marketed as the Trailblazer. Several vendors now market similar modems under license to include a modem card for insertion into the system unit of an IBM PC or compatible personal computer and a stand-alone unit that can be attached to any computer with a standard RS-232 communications port.

In addition to being compatible with other packetized ensemble protocol modems, these devices are compatible with V.22 bis, V.22, 212A, and 103 type modems. This compatibility permits the personal computer user to use the device for high-speed file transfer operations when connected to another Packetized Ensemble Protocol modem as well as to access information utilities, other personal computers and mainframes that are connected to industry standard modems.

Asymmetrical modems

Borrowing an old modem design concept, several vendors have introduced asymmetrical modems. These modems in essence contain two channels which, in the early days of modem developments, were known as the primary and secondary channel.

Originally, modems with a secondary channel were used for remote batch transmission, where the primary high-speed channel was used to transmit data to a mainframe computer while a lower speed secondary channel was used by the mainframe to acknowledge each transmitted block. Since the acknowledgements were much shorter than the transmited data blocks it was possible to obtain efficient full-duplex transmission even though the secondary channel might have one-tenth of the bandwidth of the primary channel.

In the late 1980s, several modem vendors realized that while high-speed transmission might be required to refresh a terminal's screen when the device was connected to a mainframe or for a file transfer, transmission in the opposite direction is typically limited by the user's typing speed or the shortness of acknowledgements in comparison to data blocks of information. Realizing this, modem vendors developed a new category of devices which use wide and narrow channels to transmit in two directions simultaneously as illustrated in Figure 2.24. The wide bandwidth channel permits a data rate of 9600 bps while the narrow bandwidth channel is used to support a data rate of 300 bps. Where these asymmetrical modems differ from older modems with secondary channels is in the incorporation of logic to monitor the output of attached devices and to then reverse the channels, permitting an attached terminal device to access the higher speed (wider bandwidth) channel when necessary. Although no standards existed for asymmetrical modems when this book was written, several manufacturers were attempting to formulate the use of common frequency assignments for channels.

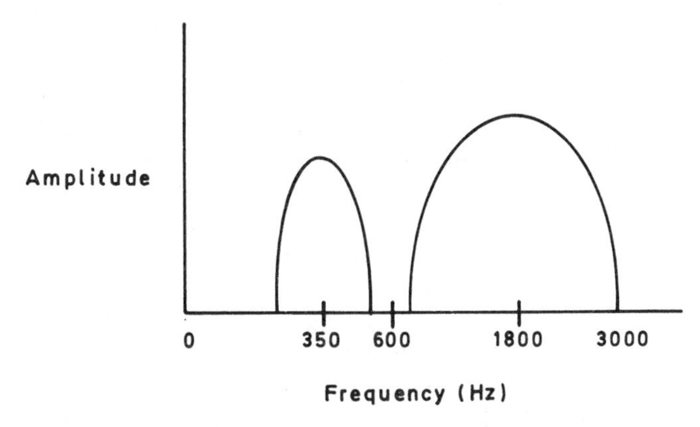

Figure 2.24 Asymmetrical modem channel assignment. (Reprinted with permission from *Data Communications Management*, © 1988 Auerbach Publishers, New York.)

Modem handshaking

Modem handshaking is the exchange of control signals necessary to establish a connection between two data sets. These signals are required to set up and terminate calls, and the type of signalling used is predetermined according to one of three major standards, such as the Electronics Industry Association (EIA) RS-232-C/D or RS-449 standard or the CCITT V.24 recommendation. RS-232-C/D and CCITT V.24 standards are practically identical and are used by over 95 per-cent of all modems currently manufactured. To better understand modem handshaking, let us examine the control signals used by 103-type modems. The handshaking signals of 103-type modems and their functions are

Table 2.14 Modem handshaking signals and their functions.

Control signal	Function
Transmit data	Serial data sent from device to modem
Receive data	Serial data received by device
Request to send	Set by device when user program wishes to transmit
Clear to send	Set by modem when transmission may commence
Data set ready	Set by modem when it is powered on and ready to transfer data; set in response to data terminal ready
Carrier detect	Set by modem when signal present
Data terminal ready	Set by device to enable modem to answer an incoming call on a switched line; reset by adaptor disconnect call
Ring indicator	Set by modem when telephone rings

listed in Table 2.14, while the handshaking sequence is illustrated in Figure 2.25.

The handshaking routine commences when an operator at a remote terminal dials the telephone number of the computer. At the computer site, a ring indicator (RI) signal at the answering modem is set and passed to the computer. The computer then sends a Data Terminal Ready (DTR) signal to its modem, which then transmits a tone signal to the modem connected to the terminal. Upon hearing this tone, the terminal operator presses the data pushbutton on the modem, if it is so equipped for manual operation. Upon depression of the data button for manually operated modems, the originating modem sends a Data Set Ready (DSR) signal to the terminal, and the answering modem sends the same signal to the computer. At this point in time both modems are placed in the data mode of operation.

In a time-sharing environment the computer normally transmits a request for identification to the terminal. To do this the computer sets Request to Send (RTS) which informs its modem that it wishes to transmit data. The modem will respond with the Clear to Send (CTS) signal and will transmit a carrier signal. The computer's port detects the Clear to Send and Carrier ON

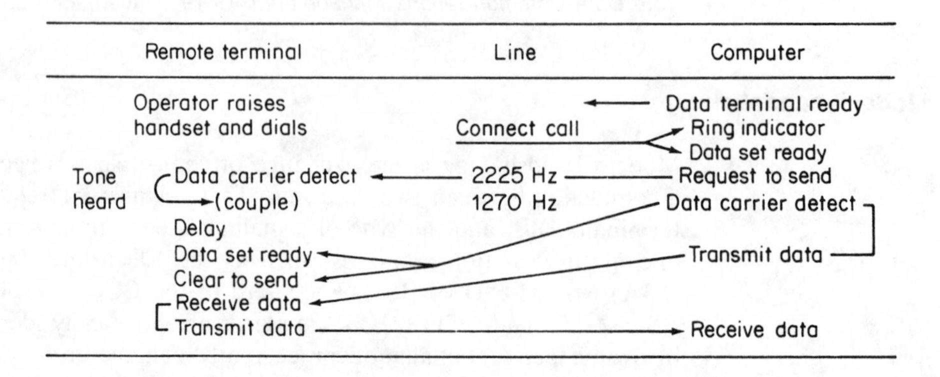

Figure 2.25 Handshake sequence.

signals and begins its data transmission to the terminal. When the computer completes its transmission it drops the DTR signal, and the computer's modem then terminates its carrier signal. Depending upon the type of circuit on which transmission occurs, some of these signals may not be required. For example, on a switched 2-wire telephone line, the RTS signal determines whether a terminal is to send or receive data, whereas on a leased 4-wire circuit RTS can be permanently raised. For further information the reader should refer to specific vendor literature or appropriate technical reference publications.

Applying modems to communications requirements

For a point-to-point data communications system two modems (one at each end of the line) are necessary, as shown in Figure 2.26. Depending upon the modem's date of manufacture and the type of circuit (switched or leased) on which the modem is used, a device known as a data access arrangement (DAA) may be required in the United States. The DAA acts as a protective interface between the modem and the switched telephone network, performing several functions to include ring detection, direct current (dc) isolation, and surge protection, which protects the transmission line from being disturbed by a modem malfunction.

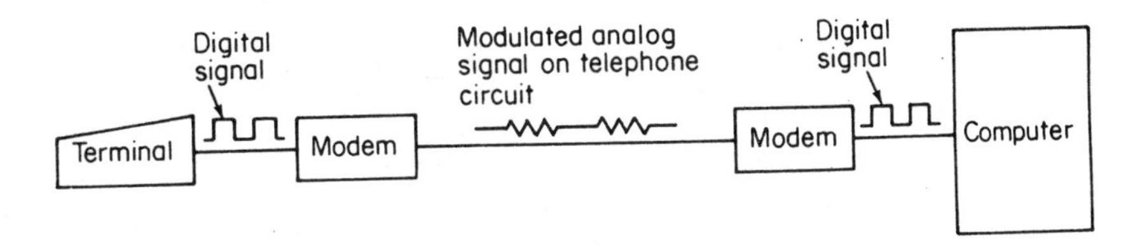

Figure 2.26 Modems used on a point-to-point data communications system. On a point-to-point line, two modems are required to modulate and demodulate the transmission.

In 1969 the Federal Communications Commission (FCC), in a ruling known as the Carterphone decision, permitted the connection of non-Bell modems to the switched network. This connection of non-Bell modems was permitted as long as a network-protecting device known as a DAA was connected between the non-Bell modem and the telephone line. A further revision in FCC regulations now permits all modems to be directly connected to the telephone line in the United States without a DAA under the following conditions.

(1) The modem is certified for use under the FCC Equipment Registration Program. In this case the modem is connected to the telephone line via a telephone company provided jack.
(2) The modem manufacturer includes a built-in line-coupling unit in the modem which is required for a certification under the FCC registration program.

(3) The modem manufacturer provides a standard miniature plug and cable for the connection of the modem to the telephone company jack.

Although DAAs are an endangered species, they are still in use today since there are tens of thousands of modems in operation that were manufactured prior to the FCC's Equipment Registration Program going into effect. There are three types of DAAs and their characteristics and use are examined in detail in Chapter 6, Section 6.3.

Many telephone companies lease DAAs, with the cost dependent upon the type of DAA required and the tariff schedule of the particular telephone company. A number of independent manufacturers also offer these devices for purchase as well as on a rental basis.

The installation of modems manufactured to conform to the FCC Equipment Registration Program is a relatively simple process. Figure 2.27 illustrates the installation of a modem designed for originating and answering calls over the PSTN while Figure 2.28 illustrates the installation of a modem designed only to receive calls over the PSTN. Note that since the answer-only modem will never originate calls, it does not require the installation of an auxiliary telephone set.

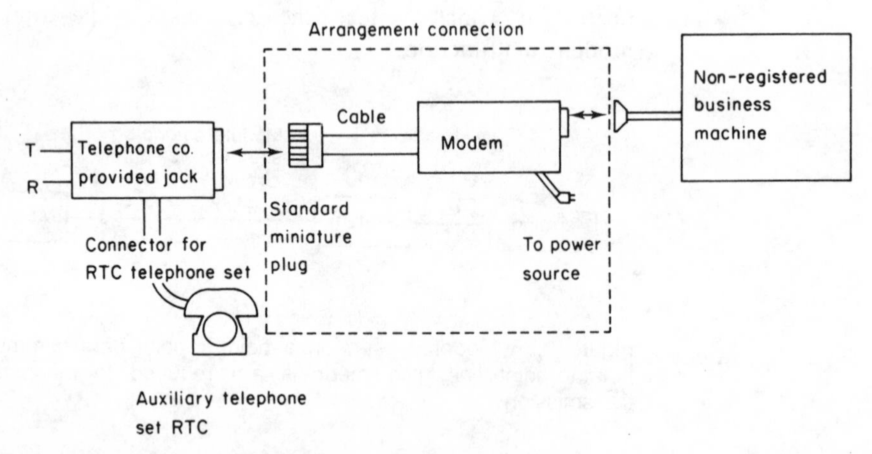

Figure 2.27 Typical originate/answer modem installation. In this arrangement, an exclusion key telephone, supplied by the telephone company (USOC, RTC) is used for call origination.

Self-testing features

Many low-speed and most high-speed modems have a series of pushbutton test switches which may be used for local and remote testing of the data set and line facilities.

In the local or analog test mode, the transmitter output of the modem is connected to the receiver input, disconnecting the customer interface from the modem. A built-in word generator is used to produce a stream of bits which are checked for accuracy by a word comparator circuit, and errors are displayed

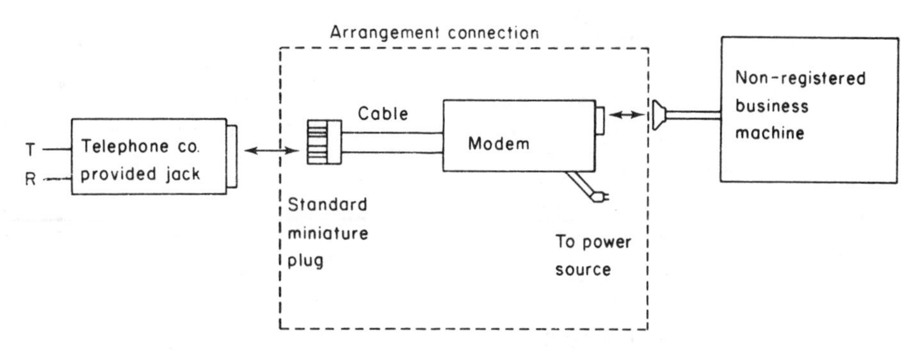

Figure 2.28 Typical answer only modem installation. This arrangement is normally at the computer site, where multiple modem assemblies (rack mounts) are used. No telephones are required.

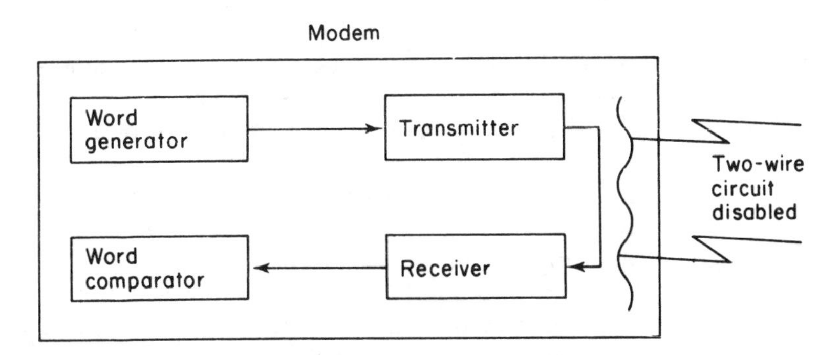

Figure 2.29 Local (analog) testing. In local testing the transmitter is connected to the receiver, and the bit stream produced by the word generator is checked by the word comparator.

on an error lamp as they occur. The local test is illustrated in Figure 2.29.

To check the data sets at both ends as well as the transmission medium, a digital loop-back, self-test may be employed. To conduct this test, personnel may be required to be located at each data set to push the appropriate test buttons, although a number of vendors have introduced modems that can be automatically placed into the test mode at the distant end when the central site modem is switched into an appropriate test mode of operation. In the digital loop-back test, the modem at the distant end has its receiver connected to its transmitter, as shown in Figure 2.30. At the other end, the local modem transmits a test bit stream from its word generator, and this bit stream is looped back from the distant end to the receiver of the central site modem where it is checked by the comparator circuitry. Again, an error lamp indicates abnormal results and indicates that either the modems or the line may be at fault.

The analog loop-back self-test should normally be used to verify the internal operation of the modem, while the digital loop-back test will check both modems and the transmission medium. While analog and digital tests are the

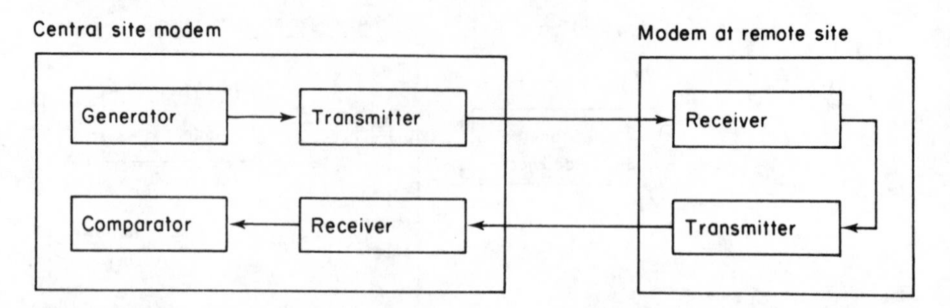

Figure 2.30 Digital loop-back self-test. In the digital loop-back test both the modems and the transmission facility are tested.

main self-tests built into modems, several vendors offer additional diagnostic capabilities that may warrant attention. A few of these diagnostic tests that deserve mention include bit error rate testing and alarm threshold monitoring. Normally these two diagnostic functions are implemented by the user obtaining a network control system (NCS) designed to operate with modems manufactured by the company that produced the NCS. Typically, the NCS is a microcomputer-based system that monitors the status of the modems in an organization's network, generates alarms when certain predefined conditions occur, and in some cases it actually performs connective action prior to a disruption in network operation occurring.

The NCS is normally installed at the central computer site and on a periodic basis queries the status of the remote modems on both point-to-point and multidrop lines. Most NCS systems work in conjunction with modems that perform testing and status queries via the use of a secondary channel. This permits data transmission to continue on the primary channel, with testing and responses to the testing flowing concurrently at a lower data rate on the secondary channel. In comparison, a few vendors offer NCS systems that work in conjunction with intelligent modems designed to recognize certain data patterns as a request to perform testing and/or issue status responses. When used with this type of modem, one's actual information flow of data is interrupted by the testing.

To allow users to implement network control functions in a data communications network without having to replace existing modems, many vendors now offer what is commonly called a "modem wraparound unit" or modem network adapter. Such devices are cabled to both ends of the modem and in schematic diagrams appear to wraparound the modem, hence the term "wraparound unit". Figure 2.31 illustrates the utilization of an NCS manufactured by one vendor with local and remotely located modems manufactured by a second vendor by the employment of two modem wraparound units. Through the use of the wraparound units, the NCS can conduct modem and line tests, even when the modems are manufactured by different vendors.

Modem indicators

Most modems contain a series of light-emitting diode indicators on the front panel of the device that display the status of the modem's operation. Typical

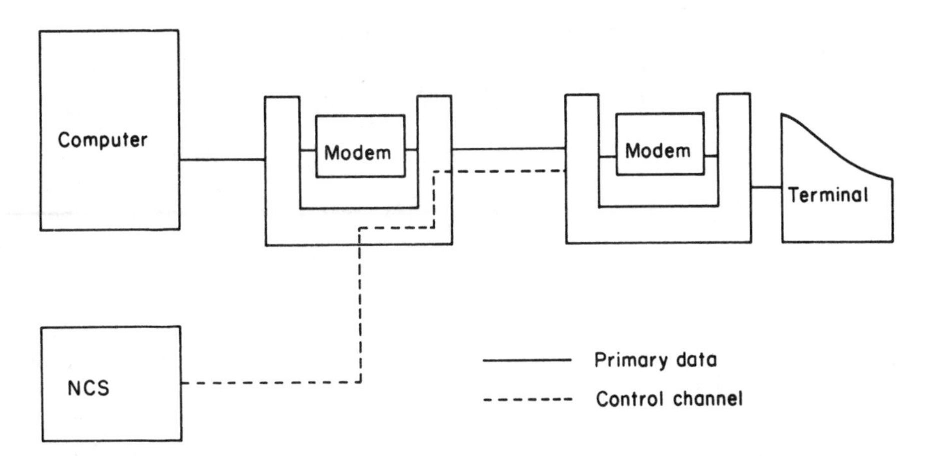

Figure 2.31 Modem wraparound unit utilization. The modem wraparound unit permits the use of a network control station produced by one vendor with modems manufactured by other vendors.

indicators include a power indicator which is illuminated when power to the modem is on, a terminal-ready indicator which is illuminated when an attached terminal is ready to send or receive data, a transmit-data indicator which illuminates when data is sent from the attached device to the modem, a receive-data indicator that illuminates when data is received from a distant computer or terminal, and a carrier-detect signal which illuminates when a carrier signal is received from a distant modem. Other indicators in some modems include an off-hook indicator that illuminates when the modem is using the telephone line and a high-speed indicator on dual speed modems that illuminates whenever the modem is operating at its high speed. Table 2.15 lists the most common modem indicator symbols, their meaning and the resulting status of the modem when the indicator is illuminated on the front panel of the modem.

2.3 INTELLIGENT MODEMS

Due to the popularity of the Hayes Microcomputer Products series of Smartmodems™, the command sets of those modems are the key to what the terms intelligent modems and "Hayes compatibility" mean. Since just about all modern personal computer communications software programs are written to operate with the Hayes command set, the degree of Hayes compatibility that a non-Hayes modem supports will affect the communications software that can be used with that modem. In some cases, non-Hayes modems will work as well or even better than a Hayes modem if the software supports the non-Hayes features of that device. In other cases, the omission of one or more Hayes Smartmodem features may require the personal computer user to reconfigure his or her communications software to work with a non-Hayes modem, usually resulting in the loss of a degree of functionality.

Table 2.15 Common modem indicator symbols.

Symbol	Meaning	Status
HS	High speed	ON when the modem is communicating with another modem at 2400 baud
AA	Auto answer/Answer	ON when the modem is in auto answer mode and when on-line in answer mode
CD	Carrier detect	ON when the modem receives a carrier signal from a remote modem. Indicates that data transmission is possible
OH	Off hook	ON when the modem takes control of the phone line to establish a data link
RD	Receive data	Flashes when a data bit is received by the modem from the phone line, or when the modem is sending result codes to the terminal device
SD	Send data	Flashes when a data bit is sent by the terminal device to the modem
TR	Terminal ready	ON when the modem receives a Data Terminal Ready signal
MR	Modem ready/Power	ON when the modem is powered on
AL	Analog loopback	ON when the modem is in analog loopback self-test mode

Hayes Command Set Modems

The Hayes command set actually consists of a basic set of commands and command extensions. The basic commands, such as placing the modem off-hook, dialing a number and performing similar operations are common to all Hayes modems. The command extensions, such as placing a modem into a specific operating speed are only applicable to modems built to transmit and receive data at that speed.

The commands in the Hayes command set are initiated by transmitting an attention code to the modem, followed by the appropriate command or set of commands one desires the modem to implement. The attention code is the character sequence AT, which must be specified as all uppercase or all lowercase letters. The requirement to prefix all command lines with the code AT has resulted in many modem manufacturers denoting their modems as Hayes AT compatible.

The command buffer in a Hayes Smartmodem holds 40 characters, permitting a sequence of commands to be transmitted to the modem on one command line. This 40-character limit does not include the attention code, nor does it include spaces included in a command line to make the line more readable. Table 2.16 lists the major commands included in the basic Hayes command set. Other modems, such as Telebit's Trailblazer employ a command set that can be considered to be a superset of the Hayes command set. While communications software that uses the Hayes command set will operate with

Table 2.16 Hayes command set. (Reprinted with permission from *Data Communications Management*, © 1988 Auerbach Publishers, New York.)

Major Commands	
Command	Description
A	Answer call
A/	Repeat last command
C	Turn modem's carrier on or off
D	Dial a telephone number
E	Enable or inhibit echo of characters to the screen
F	Switch between half- and full-duplex modem operation
H	Hang up telephone (on-hook) or pick up telephone (off-hook)
I	Request identification code or request check sum
M	Turn speaker off or on
O	Place modem on-line
P	Pulse dial
Q	Request modem to send or inhibit sending of result code
R	Change modem mode to "originate-only"
S	Set modem register values
T	Touch-tone dial
V	Send result codes as digits or words
X	Use basic or extended result code set
Z	Reset the modem

this modem, such software cannot utilize the full potential of the modem. This is because the Hayes command set in 1988 only supported a data rate selection up to 9600 bps, while the Trailblazer modem could be set to a data rate over 9600 bps by proprietary software commands.

The basic format required to transmit commands to a Hayes compatible intelligent modem is shown below.

AT Command[Parameter(s)]Command[Parameter(s)]. .Return

Each command line includes the prefix AT, followed by the appropriate command and the command's parameters. The command parameters are usually the digits 0 or 1, which serve to define a specific command state. As an example, H0 is the command that tells the modem to hang up or disconnect a call, while H1 is the command that results in the modem going off-hook, which is the term used to define the action that occurs when the telephone handset is lifted. Since many commands do not have parameters, those terms are enclosed in brackets to illustrate that they are optional. A number of commands can be included in one command line as long as the number of characters does not exceed 40, which is the size of the modem's command buffer. Finally, each command line must be terminated by a carriage return character.

To illustrate the utilization of the Hayes command set let us assume we desire to automatically dial New York City information. First, we must tell the modem to go off-hook, which is similar to one manually picking up the telephone handset. Then we must tell the modem the type of telephone system we are using, pulse or touch-tone, and the telephone number to dial. Thus, if we have a terminal or personal computer connected to a Hayes compatible modem, we would send the following commands to the modem:

AT H1
AT DT1,212-555-1212

In the first command, the 1 parameter used with the H command places the modem off-hook. In the second command, DT tells the modem to dial (D) a telephone number using touch-tone (T) dialing. The digit 1 was included in the telephone number since it was assumed we have to dial long-distance, while the comma between the long-distance access number (1) and the area code (212) causes the modem to pause for 2 s prior to dialing the area code This 2-s pause is usually of sufficient duration to permit the long-distance dial tone to be received prior to dialing the area code number.

Since a Smartmodem automatically goes off-hook when dialing a number, the first command line is not actually required and is normally used for receiving calls. In the second command line, the type of dialing does not have to be specified if a previous call was made, since the modem will then use the last type specified. Although users with only pulse dialing availability must specify P in the dialing command when using a Hayes Smartmodem, several vendors now offer modems that can automatically determine the type of dialing facility the modem is connected to and then use the appropriate dialing method without requiring the user to specify the type of dialing. For other non-Hayes modems, when the method of dialing is unspecified, such modems will automatically attempt to perform a touch-tone dial and, if unsuccessful, then redial using pulse dialing.

To obtain an appreciation of the versatility of operations that the Hayes command set provides, assume two personal computer users are communicating with one another. If the users wish to switch from modem to voice operations without hanging up or redialing, one user would send a message via the communications program he or she is using to the other user indicating that voice communications is desired. Then, both users would lift their telephone handsets and type +++(Return) ATH(Return) to switch from on-line operations to command mode (hang-up). This will cause the modems to hang-up, turning off the modem carrier signals and permitting the users to converse.

Result codes

The response of the Smartmodem to commands is known as result codes. The Q command with a parameter of 1 is used to enable result codes to be sent from the modem in response to the execution of command lines while a parameter of 0 inhibits the modem from responding to the execution of each command line.

If the result codes are enabled, the V command can be used to determine the format of the result codes. When the V command is used with a parameter of 0, the result codes will be transmitted as digits, while the use of a parameter of 1 will cause the modem to transmit the result codes as words. Table 2.17

Table 2.17 Smartmodern 1200 Basic Result Codes code set. (Reprinted with permission from *Data Communications Management*, © 1988 Auerbach Publishers, New York.)

Digit word	Word code	Meaning
0	OK	Command line executed without errors
1	CONNECT	Carrier detected
2	RING	Ring signal detected
3	NO CARRIER	Carrier signal lost or never heard
4	ERROR	Error detected in the command line

lists the Basic Results Codes set of the Hayes Smartmodem 1200. As an example of the use of these result codes, let us assume the following commands were sent to the modem:

AT Q0
AT V1

The first command, ATQ0, would cause the modem to respond to commands by transmitting result codes after each command line is executed. The second command, AT V1, would cause the modem to transmit each result code as a word code. Returning to Table 2.16, this would cause the modem to generate the word code "CONNECT" when a carrier signal is detected. If the command AT V0 was sent to the modem, a result code of 1 would be transmitted by the modem, since the 0 parameter would cause the modem to transmit result codes as digits.

By combining an examination of the result codes issued by a Smartmodem with the generation of appropriate commands, software can be developed to perform such operations as redialing a previously dialed telephone number to resume transmission in the event a communications session is interrupted and automatically answering incoming calls when a ring signal is detected.

Modem registers

A third key to the degree of compatibility between non-Hayes and Hayes Smartmodems is the number, use and programmability of registers contained in the modem. Hayes Smartmodems contain a series of programmable registers that govern the function of the modem and the operation of some of the commands in the modem's command set. Table 2.18 lists the functions of the first 12 registers built into the Hayes Smartmodem 1200, to include the default value of each register and the range of settings permitted. These registers are known as S registers, since they are set with the S command in the Hayes command set. In addition, the current value of each register can be read under program control, permitting software developers to market communications programs that permit the user to easily modify the default values of the modem's S registers.

Table 2.18 S register control parameters. (Reprinted with permission from *Data Communications Management*, © 1988 Auerbach Publishers, New York.)

Register	Function	Default value	Range
S0	Ring to answer on		0–255
S1	Counts number of rings	0	0–255
S2	Escape code character	ASC11 43	ASC11 0–127
S3	Carriage return character	ASC11 13	ASC11 0–127
S4	Line feed character	ASC11 10	ASC110–127
S5	Backspace character	ASC11 8	ASC11 0–127
S6	Dial tone wait time (s)	2	2–255
S7	Carrier wait time (s)	30	1–255
S8	Pause time caused by comma (s)	2	0–255
S9	Carrier detect response time (1/10 s)	6	1–255
S10	Time delay between loss of carrier and hang-up (1/10 s)	7	1–255
S11	Touch-tone duration and spacing time (ms)	70	50–255

To understand the utility of the ability to read and reset the values of the modem's S registers, consider the time period a Smartmodem waits for a dial tone prior to going off-hook and dialing a telephone number. Since the dial tone wait time is controlled by the S6 register, a program offering the user the ability to change this wait time might first read and display the setting of this register during the program's initialization. The reading of the S6 register would be accomplished by the program sending the following command to the modem:

AT S6?

The modem's response to this command would be a value between 2 and 255, indicating the time period in seconds that the modem will wait for a dial tone. Assuming the user desires to change the waiting period, the communications program would then transmit the following command to the modem, where n would be a value between 2 and 255.

AT S6 = n

Compatibility

For a non-Hayes modem to be fully compatible to a Hayes modem, command-set compatibility, result-codes compatibility and modem-register compatibility is required. Of the three, the modem-register compatibility is usually the least important and many users may prefer to consider only command-set and response-codes compatibility when acquiring intelligent modems.

The rationale for omitting register compatibility from consideration is the fact that many non-Hayes modem vendors manufacture "compatible" modems using the default values of the Hayes Smartmodem registers. This enables those manufacturers to avoid building the S registers into their modems, reducing the size, complexity and often also reducing the price of their modem. Thus, if the default values of the S registers are sufficient for the user and the modem under consideration is both command-set and result-code compatible the issue of register compatibility can normally be eliminated as an acquisition factor to consider.

Benefits of utilization

With appropriate software, intelligent modems can be employed in a variety of ways which may provide the user with the potential to reduce the cost of communications as well as to increase the efficiency of the user's data-processing operations. To obtain an understanding of the benefits that may be derived from the use of intelligent modems, let us assume your organization has a number of sales offices geographically dispersed throughout many states. Let us further assume that each sales office uses a personal computer to process orders, which are then mailed to company headquarters for fulfilment.

Due to postal delivery time or other factors, the delay between receiving an order at a sales office and its transmittal to company headquarters may be unacceptable. Since the order processing delay is making some customers unhappy, while other customers citing faster competitor delivery time have been cancelling or reducing their orders, management is looking for a way to expedite orders at a minimum cost to the organization.

Although a person in each sales office could be delegated to call a computer system at corporate headquarters at the end of each day and use a mainframe program to enter orders, this activity would operate at the speed of the person entering the data, communications costs would be high since the session would occur during the day and last-minute orders might not get processed until the next day unless the person performing the data entry activity agreed to stay late. Since each sales office is assumed to have a personal computer, another method you may wish to consider is the utilization of personal computers to expedite the transmittal of orders between the sales offices and the company headquarters.

Since the personal computer in each sales office is already used to process orders and prepare a report that is mailed to headquarters, one only has to arrange for the transmission of the order file each day, since the program that produces the report would only have to be sent to company headquarters once, unless the program was revised at a later date. Then, after a personal computer at company headquarters receives the order file, it would use that file as input to its copy of the order processing program, permitting the report to be produced at company headquarters.

Due to the desire to automate the ordering process at a minimal cost, it might be advisable to perform communications after 11 p.m. when rates for the use of the switched telephone network are at their lowest. Since it would defeat the purpose of communications economy to have an operator at each sales office late in the evening, a communications program that provides unattended operation capability would most likely be obtained. This type of communications program would require the use of an intelligent modem at the company headquarters location as well as at each sales office location. Then, the communications software program operating on a personal computer at company headquarters could be programmed through the use of macrocommands or menu settings to automatically dial each sales office computer at a predefined time after 11 p.m., request the transmission of the order file and then disconnect after that file had been received. The communications program would then dial the next sales office, repeating the file transfer procedure. At each sales office, a similar unattended communications program would be operating in the personal computer at that office. Upon receiving a call, the intelligent modem connected to the personal computer would inform the computer that a call had been received and the program would then answer the call, receive the request to transfer the order file, transmit that file to the distant computer and then hang up the telephone. To prevent anyone from dialing each sales office, most unattended communications programs permit password access to be implemented, enabling the user to assign appropriate passwords to the call program that will enable access to the files on the called personal computer.

Synchronous dialing language

In comparison to the Hayes command set that is a *de facto* standard for asynchronous modem operations there are several methods used to support synchronous modem operations. The earliest method of automating synchronous dialing was accomplished by the use of an automatic calling unit. In this method of establishing calls over the PSTN a computer or another DTE device is connected to an automatic calling unit via an RS-366 (parallel) or an RS-232 (serial) interface. The DTE controls the calling unit, which dials the telephone number and upon completion passes the call to the modem. Figure 2.32 illustrates this procedure.

Other methods that have been developed to obtain synchronous dialing capability include the incorporation of an RS-366 interface into some modems, the use of an integrated external keypad through which the operator stores dialing sequences which the modem dials when the attached DTE raises the Data Terminal Ready signal and the use of a synchronous dialing language

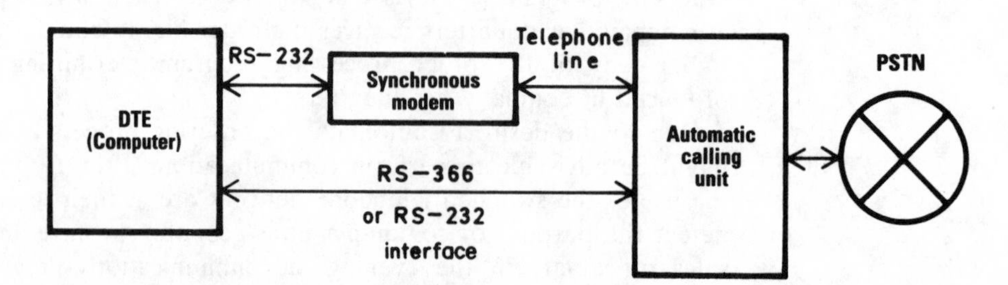

Figure 2.32 Using an automatic calling unit. Automatic calling units support both RS-232 (serial) and RS-366 (parallel) interfaces.

which combines dial control and data over a common RS-232 interface. Concerning the latter, there are currently several methods competing to become *de facto* industry standards.

Hayes Microcomputer Products proposed standard is the Hayes Synchronous Interface (HSI). This interface consists of a hardware independent connection between a user's application program and a synchronous communications driver which directly controls the hardware. Under the HSI method, dialing commands to a synchronous modem are issued asynchronously, with the modem directed to change into a synchronous mode of operation once a connection with a distant device is made. Obviously, this method is only applicable for use with modems that can work both synchronously and asynchronously; however, it does permit the ubiquitous Hayes AT command set to be used.

A second method being used to perform automatic dialing operations with synchronous modems is based upon the use of the Racal-Vadic Synchronous Auto-Dial Language (SADL). Under SADL special modem controlling messages in both bisynchronous and SDLC protocols are predefined. This enables software developers to modify application programs to support the use of SADL compatible modems for dialing on the PSTN.

The third method being used for synchronous dialing is similar to the previously described SADL technique. This method is called "SyncUp", which is sponsored by Universal Data Systems and is restricted to sending dialing commands under the bisynchronous protocol.

MNP Protocol

The Microcom Networking Protocol (MNP) was developed by the modem manufacturer Microcom, Inc., to provide a sophisticated level of error detection and correction as well as to enhance the data file transfer of intelligent modems. Microcom has licensed their MNP for use by other modem vendors, resulting in a large number of manufacturers incorporating this protocol into their products. Today, the MNP error-correcting protocol is considered a *de facto* industry standard, with over 500,000 modems having this feature when this book was published.

The MNP protocol was designed in a layered fashion like the OSI Reference Model developed by the International Standards Organization. MNP contains three layers instead of the seven layers in the OSI Reference Model. Figure 2.33 illustrates the correspondence between the OSI Reference Model and the MNP protocol.

The MNP link layer is responsible for establishing a connection between two devices. Included in the link layer is a set of negotiations that are conducted between devices to enable them to agree upon such factors as the transmission mode (full- or half-duplex), how many data messages can be transmitted prior to requiring a confirmation and how much data can be contained in a single message. After these values are established the link layer initiates the data transfer process as well as performing error detection and correction through the use of a frame checking scheme.

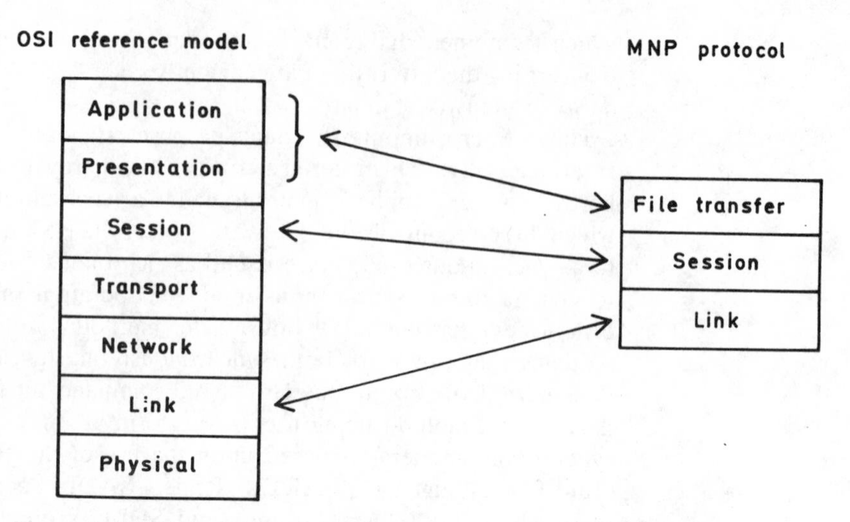

Figure 2.33 OSI reference model and MNP protocol: comparison.

Figure 2.34 illustrates the format of an MNP frame of information which has similarities to both bisynchronous and HDLC communications. Each frame contains three bytes which act as a "start flag". The SYN character tells the receiver that a message is about to arrive, the combination of data link escape (DLE) and start of text (STX) informs the receiver that everything following is part of the message. The first header describes the user data, such as the duplex setting, number of data messages before confirmation, etc. The session header defines additional information about the transmitted data which enables the automatic negotiation of the level of service that can be used between devices communicating with one another. Currently there are six versions or classes of the MNP protocol, with each higher level adding more sophistication and efficiency. When an MNP link is established the protocol assumes that the devices on both sides can only operate at the lowest level. Then, the devices

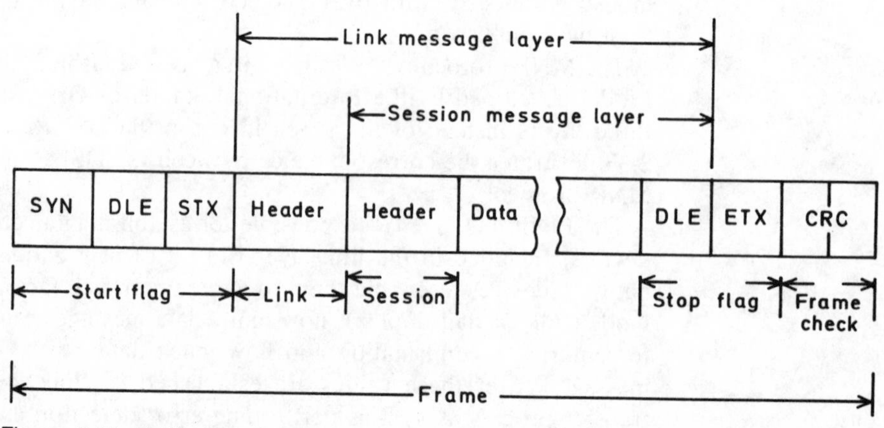

Figure 2.34 MNP block format.

Table 2.19 MNP protocol classes.

Protocol class	Description
Class 1	Data is exchanged in an asynchronous byte-oriented half-duplex block format
Class 2	Adds full-duplex capability to Class 1
Class 3	Data is exchanged using a synchronous bit-oriented full-duplex data packet format which eliminates the necessity of transmitting start and stop bits
Class 4	Adaptive Size Packet Assembly and Data Phase Packet Format Optimization is added to Class 3. Under Adaptive Size Packet Assembly the size of the data field is adjusted based upon the error rate encountered, with longer packets sent when the error rate is low while shorter packets are sent when the error rate is high. Under Data Phase Packet Format Optimization redundant administrative information during the data transfer phase of a connection is eliminated
Class 5	Data compression is added to Class 4
Class 6	Universal Link Negotiation and Statistical Duplexing is added to Class 5. Under Universal Link Negotiation an MNP modem can operate at a full range of speeds between 50 and 9600 bps. The Statistical Duplexing feature is similar to asymmetrical transmission, with the transmission direction shifted to simulate full-duplex operation.

negotiate with each other to determine the highest mutually supported class of MNP services they can support. If a non-MNP device is encountered the MNP device reverts to a "dumb" operating mode, providing an MNP modem with the ability to be used with non-MNP devices. Table 2.19 describes the six MNP protocol classes.

2.4 MULTIPORT MODEMS

The integration of modems and limited-function multiplexers into a device known as a multiport modem offers significant benefits to data communications users who require the multiplexing of only a few channels of data. Users who desired to multiplex a few high-speed data channels prior to the introduction of multiport modems were required to obtain both multiplexers and modems as individual units which were then connected to each other to provide the multiplexer and data transmission requirements of the user. Since multiplexers are normally designed to support both asynchronous and synchronous data channels, the cost of the extra circuitry and the additional equipment capacity was an excess burden for many user applications.

The recognition by users and vendors that a more cost-effective, less wasteful method of multiplexing and transmitting a small number of synchronous channels for particular applications led to the development of multiport modems. By the combination of the functions of a time division multiplexer (TDM) with the functions of a synchronous modem, substantial economies over the past data transmission methods can be achieved for certain applications.

Operation

A multiport modem is basically a high-speed synchronous modem with a built-in TDM that uses the modem's clock for data synchronization, rather than requiring one of its own, as would be necessary when separate modems and multiplexers are combined. In contrast with most traditional TDMs, a multiport modem multiplexes only synchronous data streams, instead of both synchronous and asynchronous data streams. An advantage of the built-in, limited-function multiplexer is that it is less complex and expensive, containing only the logic necessary to combine into one data stream information transmitted from as few as two synchronous data channels rather than the minimum capacity of four or eight channels associated with most separate multiplexers. The data channels in a multiport modem normally comprise a number of 2400-bps data streams, with the number of channels available being a function of the channel speed as well as the aggregate throughput of the multiport modem selected by the user.

Selection criteria

When investigating the potential use of multiport modems for a particular application, the user should determine the speed combinations and the number of selectable channels available, as well as the ability to control the carrier function (mode of operation) independently for each of the channels. One 9600-bps multiport modem now being marketed can have as many as six different modes of operation; however, only one mode can be functioning at any given time. As illustrated in Figure 2.35, operating speeds can range in combination from a single channel at 9600 bps through four 2400 bps channels.

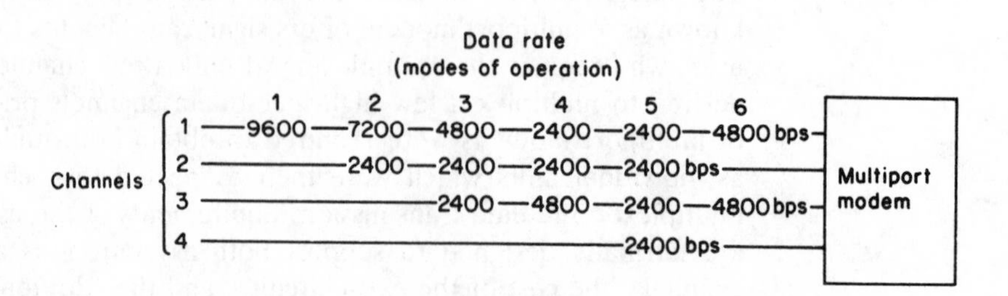

Figure 2.35 9600 multiport modem schematic utilization diagram. Multiport modem with six modes of operation is schematized here to show all possible data rate combinations for networking flexibility.

Application example

Using the fifth mode of operation shown in Figure 2.35 with four channels at 2400 bps, a typical application of a 9600-bps multiport modem is illustrated in Figure 2.36. This example shows a pair of 4-channel 9600 bps multiport modems servicing two interacting synchronous cathode ray tube (CRT) terminals, a synchronous printer operating at up to 300 characters per second or 2400 bps, and eight low-speed synchronous terminals connected by a traditional TDM. The output of the 8-channel TDM is a 2400-bps synchronous data stream, which is in turn multiplexed by the multiport modem. Here the multiport modem's multiplexer combines the eight asynchronous multiplexed 300 bps channels with the three synchronous unmultiplexed 2400-bps channels into a single multiplexed synchronous data stream. At the central site where the computer is located, the multiport modem at that end splits the 9600-bps stream into four 2400-bps data streams; one data stream is then channeled through another 8-channel, traditional TDM, whose eight output data streams in turn are connected to the computer. The 8-channel TDM takes the 2400-bps synchronous data stream from the multiport modem and demultiplexes it into eight 300-bps asynchronous data streams, which are passed to the appropriate computer ports. The remaining data streams produced by the demultiplexer in the multiport modem are connected to three additional computer ports. As this example demonstrates, the high-speed multiport modem's utilization in conjunction with other communications components permits a wide degree of flexibility in the design of a data communications network.

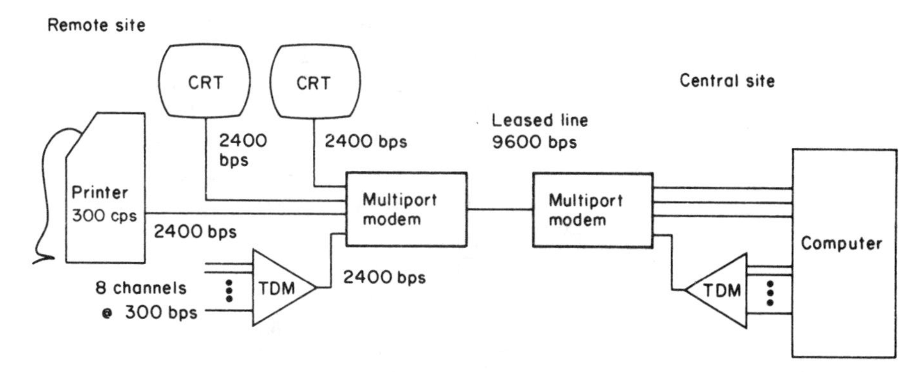

Figure 2.36 Multiport modem application example. A pair of 4-channel multiport modems services two CRT, a 300-cps printer, and eight teletypewriter terminals over a single transmission line.

Multiport modem and channel combinations available to the user are listed in Table 2.20, for modems whose data transfer rates range from 19,200 bps to 4800 bps. It should be noted that not all multiport modem channel combinations listed in Table 2.20 may be available for a particular vendor's modem that operates at the indicated aggregate throughput, since some vendors only offer a 4-port multiplexer with their modem, while other vendors may offer a 6-port or 8-port multiplexer with their device.

Table 2.20 Multiport modem channel combinations. The wide degree of flexibility that can be provided by multiport modems in a network configuration is a function of several factors: throughput, available modes and channels, and data rates.

Modem aggregate throughput	Operating mode	Multiport speed combinations							
		1	2	3	4	5	6	7	8
19,200	1	19,200							
	2	16,800	2,400						
	3	14,400	4,800						
	4	14,400	2,400	2,400					
	5	12,000	7,200						
	6	12,000	4,800	2,400					
	7	12,000	2,400	2,400	2,400				
	8	9,600	9,600						
	9	9,600	4,800	4,800					
	10	9,600	4,800	2,400	2,400				
	11	7,200	7,200	4,800					
	12	7,200	4,800	2,400	2,400	2,400			
	13	7,200	2,400	2,400	2,400	2,400	2,400	2,400	
	14	4,800	4,800	4,800	4,800				
	15	4,800	4,800	4,800	2,400	2,400			
	16	4,800	4,800	2,400	2,400	2,400	2,400		
	17	4,800	2,400	2,400	2,400	2,400	2,400	2,400	
	18	2,400	2,400	2,400	2,400	2,400	2,400	2,400	2,400
16,800	1	16,800							
	2	14,400	2,400						
	3	12,000	2,400	2,400					
	4	9,600	2,400	2,400	2,400				
	5	9,600	4,800	2,400					
	6	7,200	2,400	2,400	2,400	2,400			
	7	7,200	4,800	2,400	2,400				
	8	7,200	7,200	2,400					
	9	4,800	4,800	2,400	2,400	2,400			
	10	4,800	2,400	2,400	2,400	2,400	2,400		
14,400	1	14,400							
	2	12,000	2,400						
	3	9,600	4,800						
	4	9,600	2,400	2,400					
	5	7,200	7,200						
	6	7,200	4,800	2,400					
	7	7,200	2,400	2,400	2,400				
	8	4,800	4,800	4,800					
	9	4,800	4,800	2,400	2,400				
	10	2,400	2,400	2,400	2,400	2,400	2,400		
12,000	1	12,000							
	2	9,600	2,400						
	3	7,200	4,800						
	4	7,200	2,400	2,400					
	5	4,800	4,800	2,400					
	6	4,800	2,400	2,400	2,400				
	7	2,400	2,400	2,400	2,400	2,400			

Modem aggregate throughput	Operating mode	Multiport speed combinations 1	2	3	4	5	6	7	8
9600	1	9,600							
	2	7,200	2,400						
	3	4,800	2,400	2,400					
	4	2,400	2,400	4,800					
	5	2,400	2,400	2,400	2,400				
	6	4,800	4,800						
7200	1	7,200							
	2	4,800	2,400						
	3	2,400	2,400	2,400					
4800	1	4,800							
	2	2,400	2,400						

Although most manufacturers of multiport modems produce equipment that appears to be functionally equivalent, the system designer should exercise care in selecting equipment because of the differences that exist between modems but are hard to ascertain from vendor literature.

For modem aggregate throughput above 4800 bps, the modes of operation available for utilization by the system designer are quite similar regardless of manufacturer, with the major difference being the number of ports or data channels supported by the multiplexer.

However, at 4800 bps wide variances exist between equipment manufactured by different vendors. While most multiport modem manufacturers offer two-mode capability (one 4800-bps channel or two 2400-bps channels), some manufacturers have a built-in TDM which has the capability of servicing 1200-bps data streams. Another variance concerns the use of an independently controlled carrier signal in some multiport modems. By using multiport modems that permit an independently controlled carrier signal for each channel, data communications users can combine several polled circuits and further reduce leased line charges.

Standard and optional features

A wide range of standard and optional data communications features are available for users of multiport modems, including almost all of the features available in regular non-multiplexing modems; also available are unique multiport modem features such as multiport configuration selection, individual port testing, individual port display, and a data communications equipment (DCE) interface.

The multiport selector feature permits the user to alter the multiport configuration simply by throwing a switch into a new position. This feature

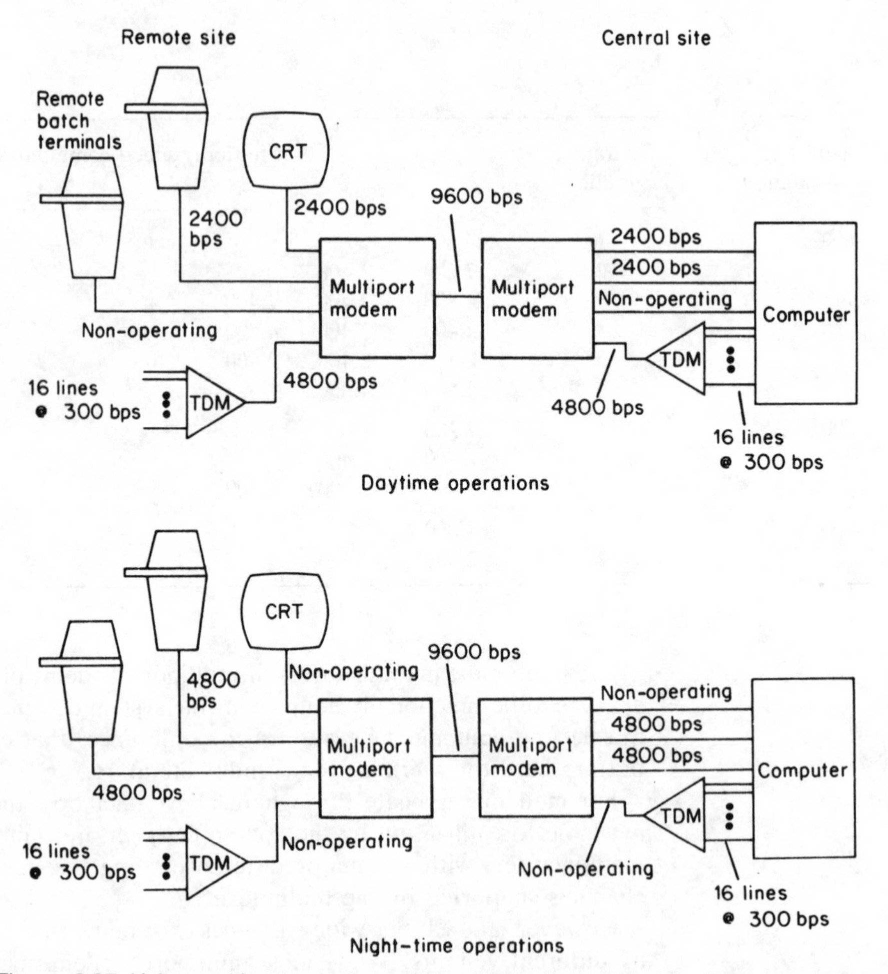

Figure 2.37 Using multiport selector switches. Day (top) and night (bottom) configurations for networks with multiport modems can be varied according to the requirements of different operations.

can be especially useful for an installation such as the one shown in Figure 2.37 (top), where daytime operations require the servicing of a large number of interactive time–sharing users; and operations at night (Figure 2.37 (bottom)) require the servicing of only two high-speed remote batch terminals. During daytime operations, 16 low-speed asynchronous, 300 bps, terminals with a composite speed of 4800 bps are serviced at this installation by one channel of the multiport modem. One remote batch terminal and a CRT are serviced by two additional channels, each of which operates at 2400 bps. Because of daytime load requirements, the second batch terminal cannot be operated since the modem's maximum aggregate speed of 9600 bps has been reached. On the assumption that the installation does not require the servicing of interactive time-sharing users at night, one possible reconfiguration is shown in Figure 2.37 (bottom). The multiport selector permits both remote batch terminals to be serviced until the start of the next business day by two 4800-bps channels while everything else in this network is shut down.

Numerous multiport modems contain a built-in test pattern generator and an error detector which permit users of such modems to determine if the device is faulty without the need of an external bit error rate tester. The use of this feature normally permits the individual ports of the modem to be tested.

Another option offered by some multiport modem manufacturers is a data communications equipment (DCE) interface. This option can be used to integrate remotely located terminals into a multiport modem network. Whereas the standard data terminal equipment (DTE) interface may require data sources to be collocated and within a 50 ft radius of the multiport modem, the DCE interface permits one or more data sources to be remotely located from the multiport modem. Installation of a multiport modem with a DCE option on one port permits that modem's port to be interfaced with another modem. This low-speed conventional modem can be used to provide a new link between the multiport modem's location and terminals located at different sites. As shown in Figure 2.38, the installation of a multiport modem with a DCE option on port 3 permits the port on that modem to be interfaced with another modem. This new modem can then be used to provide a new link between the multiport modem at location 1 and an additional remote batch terminal which is located at a second site.

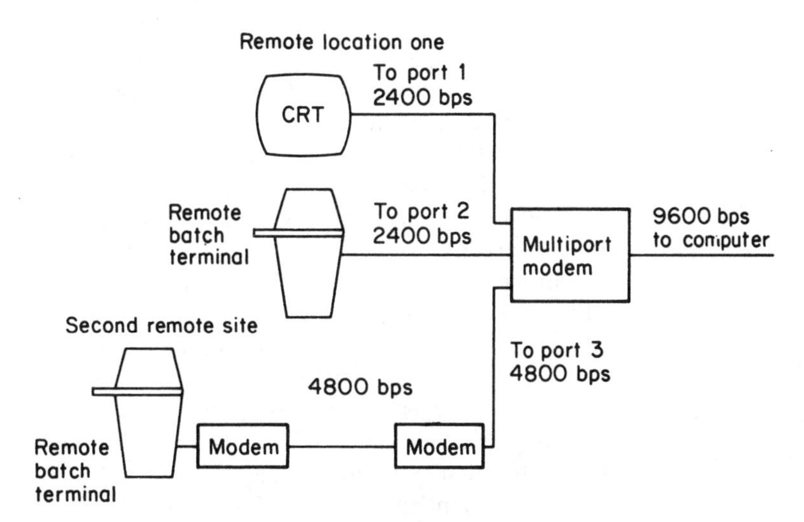

Figure 2.38 Multiport modem data communications equipment option. Using a data communications equipment (DCE) interface on port 3 of the multiport modem permits a second remote site to share the communications line from the first site to the computer.

2.5 MULTIPOINT MODEMS

To alleviate substantial confusion, it should be noted that a multiport modem contains a built-in multiplexer which enables two or more separate data streams to be combined for transmission over a single circuit. In contrast, a multipoint modem is basically a modem designed to achieve fast polling acquisition times on multipoint lines.

Multipoint lines, or multidrop lines as they are also referred to, are usually installed for applications that require interactive terminal access from a number of geographically dispersed locations into a central computer facility. This type of line may link a number of CRTs used for programming, debugging, and executing time-sharing jobs, or they may be installed to provide remote terminal access to a centralized database for one particular application.

An airline reservation system in which dispersed terminals randomly access the computer's database to determine flight information and seat availability is a representative application that uses multipoint lines. Another example would be an inventory control system where terminals are located at many warehouses and are used to report shippings and arrivals so that company inventories are continuously updated. For either application, the key item of interest is that each terminal only uses a small fraction of the total time available to all terminals connected to the line to complete a transaction, and the terminal is addressable and can recognize messages for which it is a recipient.

Although most terminals connected to multipoint circuits contain a buffer, this buffer primarily serves to enhance data throughput and is not a necessity. For example, consider the situation where an operator is entering data on the screen of a CRT through an attached keyboard. If the CRT does not have a buffer, the time it takes to transmit the data depends upon the operator's speed in typing it. During that time communications with all other terminals on the line are suspended. If the terminal has a buffer, the transmission speed from the terminal to the computer can be at a much higher rate than the operator's typing rate. Thus, once the operator has filled the CRT screen with data, the depression of a transmit key will permit the computer to select and receive the data from the terminal at a higher transfer speed and in a shorter time interval.

During the time the operator is entering data the computer is free to select other terminals. Therefore, the wait time per terminal on the common line is reduced. Since a slow operator on a multidrop line without buffered terminals could obtain an unjustified proportion of the total transmission time, some multidrop systems incorporating unbuffered terminals have a built-in time-out feature. This time-out feature permits another operator to gain control of the line if the first operator pauses for a time greater than the time-out feature permits. When either type of terminal is used in the previously described working environment, then many terminals can share the same communications circuit on an interleaved basis. The polling and selecting protocol used will make it appear to each terminal operator as if a private connection existed for his or her exclusive use for transmitting and receiving data from the computer. A typical multipoint circuit used to connect three terminals is illustrated in Figure 2.39.

Factors that affect multipoint circuits

When the applicability of a multipoint circuit is being investigated, several parameters warrant careful investigation; two such parameters are the response time and the transaction rate of the terminals.

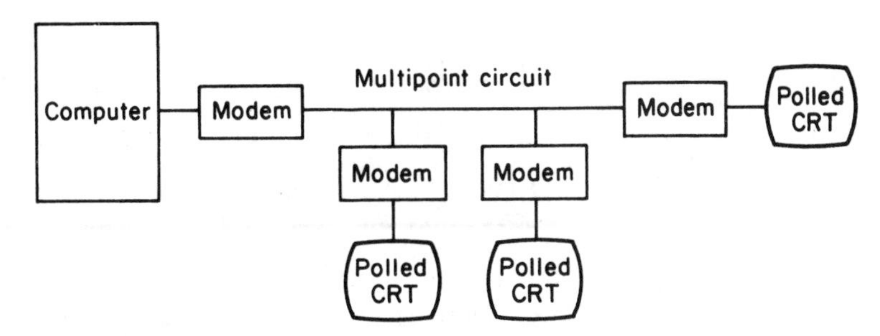

Figure 2.39 Typical multipoint circuit. On a multipoint circuit many polled terminals share the use of common communications facilities.

From a broad viewpoint, response time is the time interval from when an operator presses a transmit key at the terminal until the first character of the response appears back at that terminal. This response time consists of the many delay times associated with the components on the circuit, the time required for the message to travel down the circuit to the computer and back to the terminal, as well as the processing time required by the computer.

The transaction rate is a term used to denote the volume of inquiries and responses that must be carried by the circuit during a specific period of time. This rate is normally expressed in terms of a daily average and as a peak for a specific period of the day.

Additional factors that affect the data transfer rates on multipoint circuits include the line protocol used and its efficiency, the transmission rate of the modems, and the turnaround time of the line. While there are many factors that contribute to multipoint line efficiency utilization, the focus of this section will be on multipoint modem characteristics which should be investigated to obtain a more efficient transmission process on such circuits.

Delay factors

When a terminal connected to a computer via a multipoint circuit is polled, several factors, which by themselves may appear insignificant, accumulate to degrade response time.

In the transmission of a message or poll from the computer to a terminal on the multipoint line, the first delay encountered is caused by the modem's internal delay time (D_m). This is the delay that would be seen if the time between the first bit entering the modem and the first modulated tone put on the circuit were measured. For a poll transmitted from the computer to a terminal or a response transmitted from the terminal to the computer, total modem delay time is equal to two times the modem's internal delay time ($2D_m$), since the poll or message is transmitted through two modems. In this case, the internal delay time of the modem is equivalent to the delay one would measure if the modems were placed back to back and the time between the first bit entering the first modem and the first bit demodulated by the second modem were compared. This is shown in Figure 2.40. Depending upon

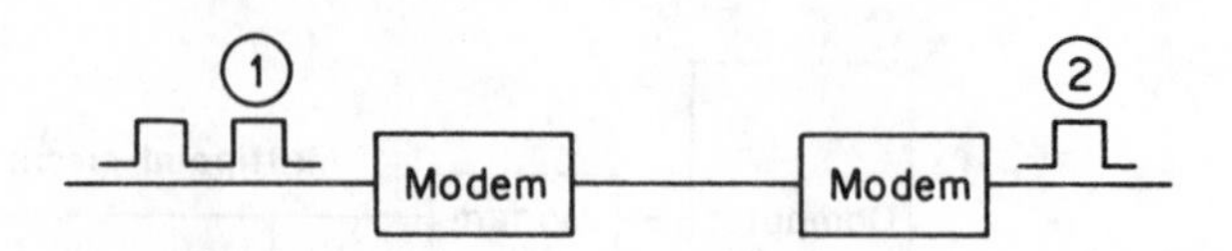

Figure 2.40 Internal modem delay time. Time difference from first bit transmitted (1) to first bit received (2) is denoted as total internal modem delay time ($2D_m$).

the type of modem used, this internal delay time can vary from a few milliseconds to 20 ms or more.

The second delay time on a multipoint circuit is a function of the distance between the computer and the terminal with which it is communicating. This delay time is called transmission delay and represents the time it takes for the signal to propagate down the line to the receiving location; hence, it is also referred to as propagation delay time (D_p). Although this delay time is insignificant for short transmission distances, coast-to-coast transmission can result in a transmission delay of approximately 30 ms on a terrestrial circuit or 250 ms if transmission occurs via a satellite.

When a response to a transmitted poll or message is returned from the remote terminal, several additional delay factors are encountered. First, the terminal itself causes a delay, since some time is required for the terminal to recognize the poll and initiate a response. This terminal delay (D_t) is usually a few milliseconds, but can be considerably longer, depending upon the design of the terminal and the software protocol used.

Since every modem on the multipoint line shares the use of this common circuit, only one modem at any point in time may have its carrier on. The carrier of each modem is turned on in response to the connected terminal or computer raising its Request to Send (RTS) signal, which indicates that a message is to be transmitted. The modem at the opposite end of the transmission path will then require some time interval to recognize this signal and adjust its internal timing. If the modems have automatic and adaptive equalization, additional time is spent adapting the modem to the incoming signal's characteristics.

Once these functions are completed, the modem located at the opposite end of the transmission path will raise its Clear to Send (CTS) signal which is required in response to the RTS signal if a return message is to be transmitted. This delay time is referred to as the Request to Send/Clear to Send delay time (RTS/CTS) and is usually denoted as $D_{R/C}$. Since the RTS/CTS delay time occurs twice in responding to a poll, the total RTS/CTS turnaround delay is $2D_{R/C}$. Disregarding terminal, propagation, and processing delays, the total turnaround delay attributable to the modems is M_D, where:

$$M_D = 2(D_{R/C} + D_m)$$

Although the RTS/CTS delay varies by the type of modem used, delays can range from about 5 to 100 ms or more and normally cause most of the line's turnaround delay.

When multipoint networks were initially implemented, modem operating rates were at speeds of 2400 bps or lower. These modems were manually equalized and were adjusted and set at installation time. For this category of modems, the RTS/CTS delay ranges from about 10 to 20 ms, depending upon manufacturer. Owing to the advancement in modem transmission rates as well as new applications which require higher traffic rates and shorter response times, multipoint networks with 4800-bps modems were implemented. At this data rate, two types of modem equalization are used, the manual type previously discussed and that which is automatically equalized. In comparison to the static nature of manual equalization, automatic equalization permits the modem to continuously monitor and compensate for changing line conditions. However, an initial period of time is required for the modem to "train" on the signal each time the transmission direction reverses. For manually equalized modems, the RTS/CTS delay time is normally between 10 and 20 ms, since they are static in nature. For automatically equalized modems, the training time necessary for the modem to adapt to the incoming signal adds significantly to the RTS/CTS delay time, with such delays increased to 50 ms or more.

Throughput problems

To illustrate the effect of the increased delay on data throughput, let us examine the transmission of a 50-character data block. If each character consists of 8 data bits, then the data block would contain 400 bits of information. If 2400-bps modems are used for transmission, the actual time required to transmit the block would be 400/2400 or 166 ms.

Let us assume that the RTS/CTS delay time of the 2400-bps modems was 10 ms and their internal delay time was 5 ms. Then, the total delay time associated with the transmission and acknowledgement of the data block would be 2 * (10 + 5) or 30 ms. Since the time to transmit the data block is 166 ms and the delay time associated with the block is 30 ms, the overhead attributable to the delay is 30/(166 + 30) or 15.3 percent. Given a modem operating rate of 2400 bps, the effective data transfer rate becomes:

$$2400 - (2400 * .153) = 2032 \text{ bps}$$

Now let us examine the effect of doubling the modems' operating rate to 4800 bps. At 4800 bps, modems employing automatic equalization originally had an RTS/CTS delay time of 50 ms. Assuming the internal delay time remained constant at 5 ms, the total delay time to transmit the data block and receive an acknowledgement would become:

$$2 * (50 + 5) = 110 \text{ ms}$$

When the modems operate at 4800 bps, the time to transmit a 400-bit block is 400/4800 or 83.3 ms. Then the overhead attributable to the delay times is 110/(83.3 + 110) or 56.9 percent. Given the modem operating rate of 4800 bps, the effective data transfer rate becomes:

$$4800 - (4800 * .569) = 2069 \text{ bps}$$

Based upon the preceding analysis, it is obvious that the doubling of a multipoint modem's operating rate only resulted in an insignificant increase in the effective data transfer rate. If this situation were allowed to persist, most data communications users would never benefit from upgrading the operating rate of modems used on multipoint circuits. Fortunately, several modem developments have resulted in an increase in the effective data transfer rate that has a high correspondence to the increase in a modem's operating rate.

Multipoint modem developments

Until the early 1970s, users were forced to trade off the benefits derived from automatic equalization with the longer RTS/CTS delay time that was obtainable through the use of a modem equipped with this feature. Fortunately, many modem manufacturers have incorporated techniques that reduce the RTS/CTS delay time while permitting automatic equalization.

One manufacturer uses a so-called "gearshift" technique where data transmission begins at a rate of 2400 bps, using a modulation technique that does not require extensive equalization. This reduces the RTS/CTS delay to a level of about 9 ms, which is the delay time normally associated with manually equalized modems transmitting at 2400 bps. Next, as transmission proceeds at 2400 bps, the receiving modem automatically equalizes on the incoming signal. After an initial transmission of 64 bits of data is received at the 2400-bps data rate, the training cycle is completed and both the sending and receiving modems "gearshift" up to the faster 4800 bps data rate to continue transmission. While this technique will reduce the RTS/CTS delay time, the actual number of data bits transmitted during an interval of time will depend upon the size of the message transmitted. This is because the first 64-bits of each message are transmitted at 2400 bps prior to the modem gearshifting to the 4800-bps data rate.

Another technique used to increase the number of bits of information transmitted has been obtained by incorporating a microprocessor into a modem. This microprocessor is used to perform equalization and provides a very fast polling feature which increases the data traffic transmitted during a period of time when compared to the standard Bell System equivalent 4800-bps modems or the gearshift-type modem. This comparison is illustrated in Table 2.21.

Table 2.21 Comparison of data traffic transmitted by 4800 bps multipoint modems.

	Data bits		
Time (ms)	Modem with microprocessor	Modem using "gearshift" technique	Bell System equivalent
9	0	0	0
20	24	24	0
26	58	41	0
36	106	65	48
50	173	133	115

Remote multipoint testing

Since it is much more difficult to ascertain the cause of problems on a multipoint line in comparison to a point-to-point circuit; modem manufacturers have developed several testing features that are either incorporated into multipoint modems or offered as an option for end-user selection. One of the more common testing features offered with multipoint modems is remote testing. This feature results in each remote multipoint modem containing a unique address, permitting a central site modem to send an address code and command signal to the remote modems on the multipoint circuit, which will then place each remote site modem into an analog or digital loop-back mode of operation. By incorporating an address, each remote modem can then be tested from the remote site.

2.6 SECURITY MODEMS

The goal in the development of the security modem is to identify authorized users based upon the telephone number from which they originate calls on the PSTN. To accomplish this goal the security modem is designed to receive calls originated on the PSTN, interrogate the user's identity based upon a code or password, disconnect the user and then initiate a call on the PSTN to a predefined telephone number stored in memory in the modem. Since the security modem initiates a callback to the user attempting access, another name commonly associated with this device is a callback modem.

Operation

A callback or security modem contains a battery powered buffer area into which the network administrator enters an access code and telephone number for users authorized to dial a PSTN telephone number that is connected to the modem. Depending upon the security modem used, between four and twelve characters are commonly available for the authorization code.

Once a network user receives the authorization code and dials the number assigned to the security modem, that device first prompts them to enter his or her code. Assuming the code is entered correctly, the modem then disconnects the user from the line and originates a call to the telephone number associated with the access code.

Since the security modem simply verifies the telephone number of the originator of the call it should not be confused with data encryption devices that scramble transmission based upon a predefined algorithm. Thus, the protection of the security modem is relegated to verifying that the data call originator has an authorized access code and is located at an authorized telephone number.

Memory capacity and device access

The memory capacity and network administrator access to a security modem differ considerably between devices. Most security modems have a buffer area that only permits the storage of three to five access codes and associated telephone numbers. A few security modems have the capacity to store up to approximately 20 access codes and associated telephone numbers, representing the maximum capacity of devices in this equipment category. Concerning network administrator access, most modems require the network administrator to first cable a terminal to the device to enter the access codes and telephone numbers assigned to each access code. Then, once this is accomplished, the security modem can be connected to the computer or multiplexer port it is designed to service. This type of modem is difficult to modify, since it requires the network administrator to place it out of service and recable a terminal to the device to alter previously entered data. Then, after modifications are made to the access codes and telephone numbers to add new users or delete previously authorized users, the network administrator can disconnect the terminal and recable the security modem to the device it services.

A second method of updating access codes and associated telephone numbers is available on a few security modems. This method requires the system administrator to first enter a master access code when they initially attach a terminal to the device. Thereafter, the network administrator can use any terminal attached to a compatible modem and dial the security modem via the PSTN. Once the master access code is entered, the network administrator can remotely enter or modify access codes and their associated telephone numbers.

Device limitations

Since it is a natural tendency of many persons to post access codes on terminals, the use of a security modem is only as good as the policy and enforcement of the policy conerning the use and distribution of access codes. In addition, since the security modem simply verifies the location of a terminal by a callback over the PSTN, data calls are still susceptible to illicit monitoring. Other limitations associated with security modems include the fixed telephone numbers stored in the modem's memory, the requirement and cost of a second call and the requirement of call originators to manually place some modems into an answer mode to receive the callback from the security modem.

Since a security modem requires the network administrator to associate fixed telephone numbers to access codes, this device is impractical for the traveling businessman to use. As an example of this, consider a salesperson who visits several customers during the day. Since a maximum of 20 access codes and associated telephone numbers can be entered into a security modem, even if the salesperson contacts the network administrator on a regularly scheduled basis it would be a demanding task to continuously adjust the security modem to correspond to one's sales calls.

The requirement of the security modem to initiate callbacks can considerably add to the cost of the organization's telephone bill. This is because the telephone companies in many metropolitan areas base the cost of a call upon message units. Since the initial call to the security modem results in a minimum of one message unit being used, even if the call only required 30 s for the user to enter his or her access code message units can rapidly build up. For long-distance calls the cost associated with the requirement for an initial call and a callback may be more pronounced. This is due to the tariff structure of long-distance calls, where the cost of the first minute is considerably higher than the cost per minute for succeeding minutes. As the initial call to the security modem is normally billed at the higher first-minute rate, this can considerably expand the communications cost of an organization over a period of time.

Most low-speed modems used to originate calls to a security modem have two modes of operation — originate and answer. When originating a call to the security modem that device is automatically in its answer mode of operation, while the originating modem must be placed into its originate mode to insure frequency compatibility between devices. After the user enters his or her access code and the security modem disconnects itself from the line the user must then place his or her modem into its answer mode of operation. This is necessary for frequency compatibility between devices, since the security modem automatically places itself into an originate mode of operation when it initiates the callback. If the user forgets to place his or her modem into an answer mode of operation it will not be able to answer the callback. Although the procedure to change modem operating modes is relatively simple, it is also easy to forget and can result in the user making an additional call to the security modem once the error is discovered. This in turn results in an additional callback, increasing the cost of using this security mechanism.

Another key limitation of security modems is the storage capacity of the device. If the modem, as an example, is limited to storing five access codes and associated telephone numbers a significant number of devices may be required to support a large number of users. As an example of this consider a geographical area that contains 50 terminals that will dial a multiplexer. If only six terminals are expected to be active at one time, six dial-in lines and conventional modems connected to the multiplexer would be sufficient. If security modems are required, 10 devices would be necessary, since each modem is assumed to store a maximum of five access codes and telephone numbers. This in turn would require 10 dial-in lines and 10 ports on the multiplexer, increasing the cost of communications facilities and equipment necessary when security modems are used. In addition, although a common rotary could be used with conventional modems, users dialing security modems would be restricted to accessing specific telephone numbers to which the modem with their access code is connected. This in turn further limits the flexibility associated with using security modems.

2.7 LINE DRIVERS

If one concentrates on conventional data transmission methods, eliminating such unconventional techniques as laser transmission through fiber optic bundles, four basic means of providing a data link between a terminal and a

Table 2.22 Terminal-to-computer circuit connections.

Direct connection of terminals through the use of wire conductors
Connection of terminals through the utilization of line drivers
Connection of terminals through the utilization of limited distance modems
Connection of terminals through the utilization of modems or digital service units

computer can be considered. These methods are listed in Table 2.22. It is interesting to observe that each of the methods listed in Table 2.22 provides for progressively greater distances of data transmission while incurring progressively greater costs to the users. In this section the limitations and cost advantages of the first two methods will be examined in detail.

Direct connection

The first and most economical method of providing a data circuit is to connect the terminal directly to a computer through the utilization of a wire conductor. Surprisingly, many installations limit such direct connections to 50 ft in accordance with terminal and computer manufacturers' specifications. These specifications are based upon EIA RS-232 and CCITT V.24 standards. If the maximum 50 ft standard is exceeded, manufacturers may not support the interface, yet terminals have been operated in a reliable manner at distances in excess of 1000 ft from a computer over standard data cables. This contradiction between operational demonstrations and usage and standard limitations is easily explained.

If one examines both the EIA RS-232-C/D and CCITT V.24 standards, such standards limit direct connections to 50 ft of cable for data rates up to and including 20 kbps. Since the data rate is inversely proportional to the width of the data pulses transmitted, taking capacitance and resistance into account, it stands to reason that slower terminals can be located further away than 50 ft from the computer without incurring any appreciable loss in signal quality. Simply stated, the longer the cable length the weaker the transmitted signal at its reception point and the slower the pulse rise time. As transmission speed is increased, the time between pulses is shortened until the original pulse may no longer be recognized at its destination. This becomes more obvious when one considers that a set amount of distortion will effect a smaller (less wide) pulse than a wider pulse.

In Figure 2.41, the relationship between transmission speed and cable length is illustrated for distances up to approximately 3400 ft and speeds up to 40,000 bps. This figure portrays the theoretical limits of data transfer speeds over an unloaded length of 22 American wire gauge (AWG) cable. Many factors can have an effect on the relationship between transmission speed and distance including noise, distortion introduced owing to the routing of the cable, and the temperature of the surrounding area where the cable is installed. The ballast of a fluorescent fixture, for instance, can cause considerable distortion of a signal transmitted over a relatively short distance.

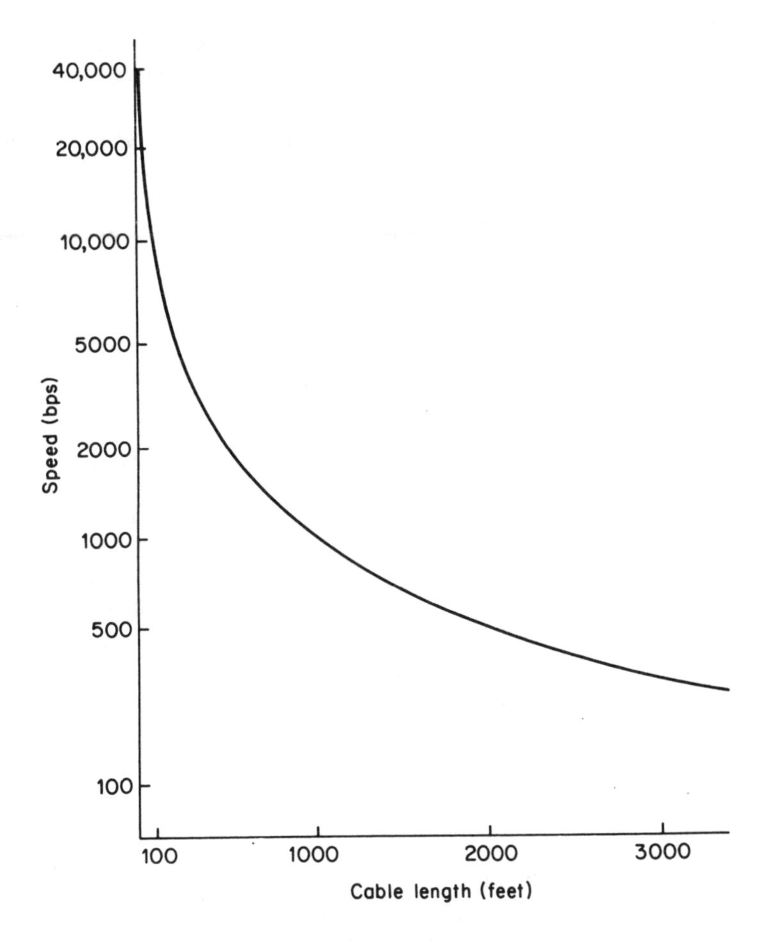

Figure 2.41 Speed and cable length relationship.

The diameter of the wire itself will affect total signal loss. If the cross-sectional area of a given length of wire is increased, the resistance of the wire to current flow is reduced. Table 2.23 shows the relationship between the dimensions and resistances of several types of commercially available copper wire denoted by gauge numbers. By increasing the gauge from 22 to 19, the resistance of the wire is reduced by approximately one-half.

Another method which can be utilized to extend the length of a direct wire connection is limiting the number of signals transmitted over the data link. After the connect sequence or handshaking is accomplished, only two signal leads are required for data transfer: the transmitted data and receive data leads. With some minor engineering at both ends of the data link and an available dc voltage source, the remaining signals can be held continuously high, permitting the use of a simple paired cable to complete the data link.

Cable length can be further extended by the use of commercially available low-capacitance shielded cable. The shield consists of a thin wrapping of lead foil around the insulated wires and is quite effective in reducing the overall capacitance of the data cable with a very modest increase in price over standard

Table 2.23 Relationship between wire diameter and resistance.

Gauge number	Diameter (in)	Ohms/1000 ft at 70°F
10	0.102	1.02
11	0.091	1.29
12	0.081	1.62
13	0.072	2.04
14	0.064	2.57
15	0.057	3.24
16	0.051	4.10
17	0.045	5.15
18	0.040	6.51
19	0.036	8.21
20	0.032	10.30
21	0.028	13.00
22	0.025	16.50
23	0.024	20.70
24	0.020	26.20
25	0.018	33.00
26	0.016	41.80
27	0.014	52.40
28	0.013	66.60
29	0.011	82.80

unshielded cables. The use of low-capacitance shielded cables is strongly recommended when several cables must be routed through the same limited-diameter conduit. Once the practical limitation of cable length has been reached, signal attenuation and line distortion can become significant and either reduce the quality of data transmission or prevent its occurrence. One method of further extending the direct interface distance between terminals and a computer is by incorporating a line driver into the cable connection.

Using line drivers

As the name implies, a line driver is a device which performs the function of extending the distance a signal can be transmitted down a line. A single line driver, depending upon manufacturer and transmission speed, can adequately drive signals over distances ranging from hundreds of feet up to a mile. One manufacturer has introduced a line driver capable of transferring signals at a speed of 100 kbps at a distance of 5000 ft and a 1-Mbps signal over a distance of 500 ft using a typical multipair RS-232/V.24 cable.

A multitude of names have been given to the various brands of line drivers to include local data distribution units and modem eliminator drivers. For the purpose of this discussion a line driver is a stand-alone device inserted into a digital transmission line in order to extend the signal distance.

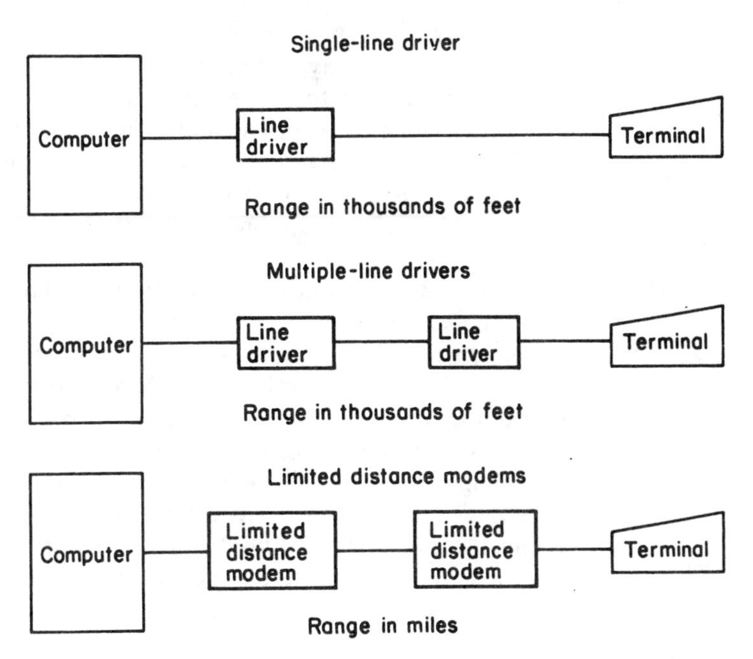

Figure 2.42 Line drivers and limited distance modems. When line drivers are used, the signal remains in its digital form for the entire transmission. Distances can be extended by the addition of one or more line drivers which serve as digital repeaters. For longer distances, limited distance modems can be utilized where the transmitted data is converted into an analog signal and then reconverted back into its original signal by the modem.

In Figure 2.42 the distinction between single and multiple line drivers is illustrated and contrasted to limited distance modems which will be explored in Section 2.8. The primary distinction between line drivers and limited distance modems is that two identical units must be used as limited distance modems to pass data in analog form over a conductor, whereas a line driver serves as a repeater to amplify and reshape digital signals.

Although some models of line drivers can theoretically have an infinite number of repeaters installed along a digital path, the cost of the additional units as well as the extra cabling and power requirements must be considered. Generally, the use of more than two line drivers in a single digital circuit makes the use of limited distance modems a more attractive alternative.

Applications

The characteristics of line drivers become important when one considers their incorporation into a data link. If the EIA RS-232 signals are accepted, amplified, regenerated, and passed over the same leads, they can be used as repeaters. If, however, the line driver also serves as a modem eliminator by providing a synchronous clock, inserting RTS/CTS delays and reversing transmit and receive signals, care should be taken when attempting to use them as repeaters. If this type of line driver is used and strapping options are provided for the RTS/CTS delay and such desirable features as internal/external clock, it is a simple matter to convert it to a repeater by setting the delay to zero, setting the clock to external, and using a short pigtail cable to reverse the signals. One manufacturer offers a single stand-alone device that performs the function of a pair of two synchronous modems along with a less expensive remote cable extender option which serves as a matching line driver similar to that previously described.

In Figure 2.43, a typical application where line drivers would be installed is illustrated. In this office building the computer system is located in the basement. The three terminals to be connected to the computer are located on different floors of the building. A remote terminal located on the second floor of the building is only 100 ft from the computer and is directly connected by the use of a low-capacitance shielded cable. The second terminal is located on the eighth floor, approximately 500 ft from the computer and is connected by the use of line drivers to extend the signal transmission range. A third terminal, located on the 30th floor, uses a pair of limited-distance modems for transmission since a large number of line drivers would be cost-prohibitive.

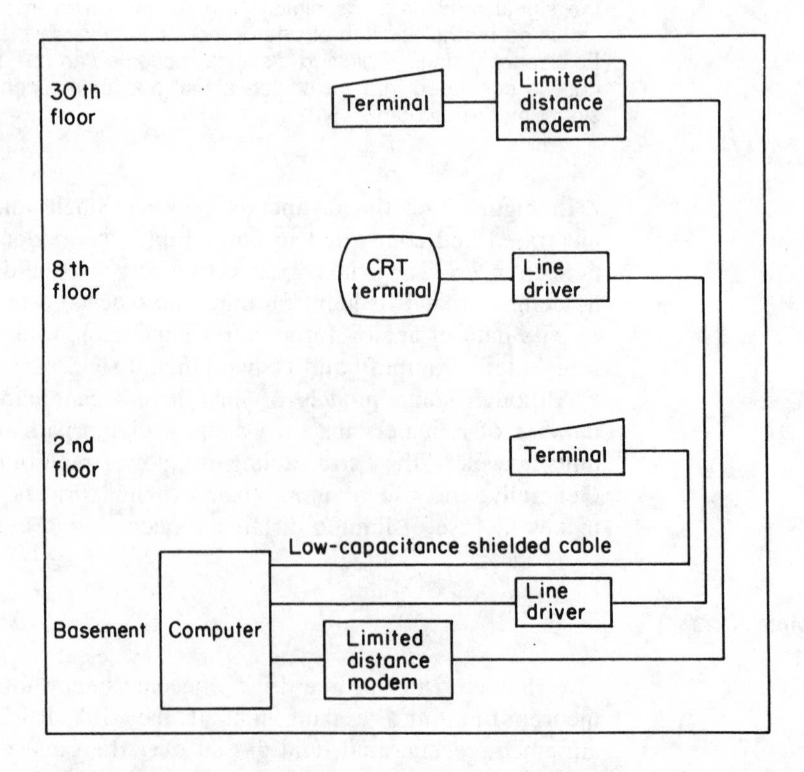

Figure 2.43 A variety of methods can be used to connect a terminal to a computer located in the same building.

2.8 LIMITED-DISTANCE MODEMS

Limited-distance modems are being employed more frequently in data communications networks. This increase in usage is the result of a number of factors, the cost of the device being a major consideration. As the name implies, limited-distance modems are designed for data transmission over relatively short distances when compared to traditional modems. The utilization of such devices can result in dramatic savings on the cost of using conventional modems for the transmission of data over short distances. Currently, more than a dozen manufacturers produce these devices, with purchase prices ranging from several hundred to several thousand dollars. These devices have operating rates ranging from 110 bps to over 1 million bps for distances ranging from 1 to 20 miles or more. The names given to these devices include not only limited-distance modems but such descriptive terms as modem bypass units, short-range data sets, short-haul modems, and wire line modems.

Contrasting devices

In contrast to line drivers where one or more such devices are used to regenerate digital data and extend transmission ranges, limited distance modems require two matching components, one at each end of the circuit. In most cases, limited distance modems convert digital data into analog signals for transmission. However, some devices on the market convert the terminal's serial binary bits into bipolar return to zero signals to maintain transmission entirely in digital form. The utilization of this type of device can provide one with a direct limited-distance digital extension to a DATAPHONE® digital service channel service unit (CSU).

Transmission mediums

In Figure 2.44, some of the distances that can be achieved with limited-distance modems are illustrated when transmission is over unloaded, twisted-pair metallic wire cables. The representative transmission distances illustrated represent a refined composite derived from manufacturers' data sheets and should be used as a guide to the variations in methods of signaling and sensitivity in receiving level between devices. As discussed in Section 2.7, the smaller the wire gauge the greater the diameter of the wire and the lower the resistance of the wire to passage of current. The 4-wire gauges selected for illustration in Figure 2.44 are used because they correspond to the common wire sizes used in the US telephone systems, and many users prefer to utilize the existing common carrier cables in lieu of routing their own private cables. Of course, there is no gauge restriction in routing a private cable, and in light of the differences shown in Figure 2.44 between 19- and 26-gauge cable, many users will install a lower gauge cable even when transmitting at a normal rate to alleviate the necessity of recabling if the data transmission rate should increase at a later date.

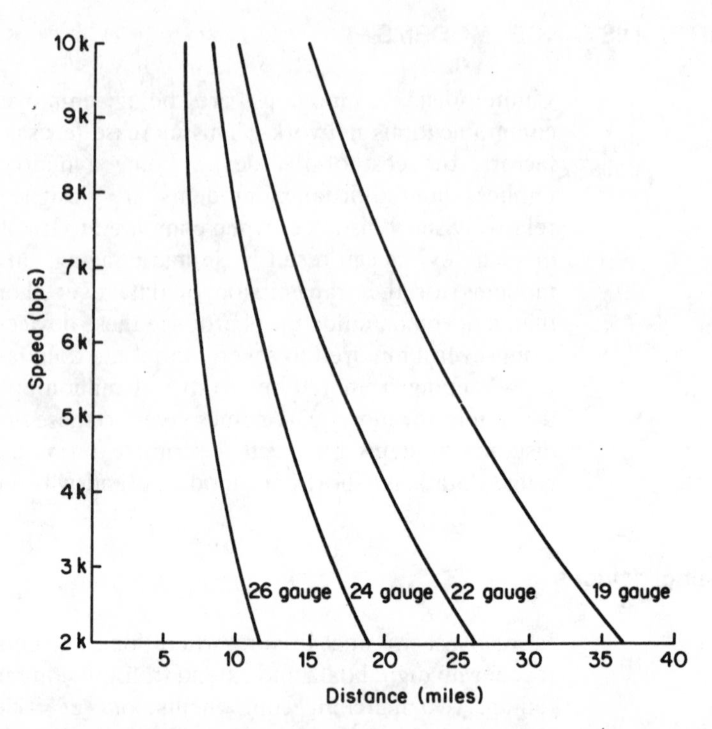

Figure 2.44 Representative transmission distance — miles per twin pair wire gauges (unloaded).

In Table 2.24, common telephone wire sizes and associated measured resistances are listed. In this table, a loop mile is a term which is used to describe two wires connecting two points which are physically located 1 mile apart, an important measurement when one considers using carrier facilities for supporting their limited-distance transmission requirements.

Table 2.24 Telephone wire sizes and resistances

Circuit type	Gauge (AWG)	Diameter (mm)	Resistance	
			1000 ft	Loop mile
Station wire	22	1.00	16	
Station wire	24	0.79	25	
Station wire	26	0.625	40	
Toll wire	19	1.42	8	
Interoffice wire	19	1.42	8	
Open wire lines				
Copper	10	4.00		6.7
Copper	12	3.00		10.2
Copper-clad steel	12	3.00		25
Copper-clad steel	14	2.00		44

If the use of existing carrier facilities is being contemplated, a close liaison with the local telephone company should be established. Common carriers and their operating companies in some areas may not be completely familiar with this particular type of hardware or the tariff structure for the service to support limited-distance transmission. It is also important that the proposed limited-distance modem conforms to the specifications set forth in the Bell System publication 43401 entitled "Transmission Specifications for Private Line Metallic Circuits." This publication describes the signal level criteria objectives for private line metallic circuits (cable pairs without signal battery or amplification devices). In addition, the publication notes that the telephone companies have no obligation to provide private line channels on a metallic basis. Most manufacturers of limited-distance modems clearly specify that their equipment operate in accordance with the previously mentioned publication. If it is not explicitly stated, one may encounter delays and additional cost to insure that the transmitter of the device is modified to comply with the specifications.

Operational features

Most limited-distance modems utilize a differential diphase modulation scheme and permit internal transmit timing or externally derived timing from the associated data terminal. These devices act similarly to a line driver, and most will accommodate 4-wire half/full duplex and 2-wire half-duplex/simplex data transmission. Data rate switches on some asynchronous units provide selectable data rates ranging from 110 through 1800 bps, while the selectable rates of most synchronous units include 2400, 4800, 7200, 9600 and 19,200 bps. Users often select a transmission speed only to realize that by the time the equipment is installed changing requirements may indicate a different speed; therefore, the ease of adjustment of the unit should be investigated. This is especially true when an installation goes operational for the first time and the user starts operation with leased terminals which may be replaced at a later date.

Some manufacturers state that their limited-distance modems can be inserted into a transmission line to serve as a repeater, as illustrated in Figure 2.45, to further extend the transmission distance. An obvious limitation to inserting such devices to serve as repeaters or data regenerators is the fact that they must be sheltered, have an available power source, and be readily accessible for diagnostics and maintenance. Since these combinations are difficult to achieve at locations between buildings, normal utilization of limited-distance modems has their locations fixed at each end of a transmission medium.

A very desirable feature that is offered on some units is a multiport or split stream feature. This feature permits several collocated terminals to utilize a single limited-distance transmission link at a considerable cost saving over the less expensive but more limited capability single-channel device. The key advantage to the employment of a multiport limited-distance modem is that only one cable instead of many cables can be used to service the transmission requirements of multiple terminals, as illustrated in Figure 2.46. For additional information on the advantages of multiport operations, the reader is referred to Section 2.4.

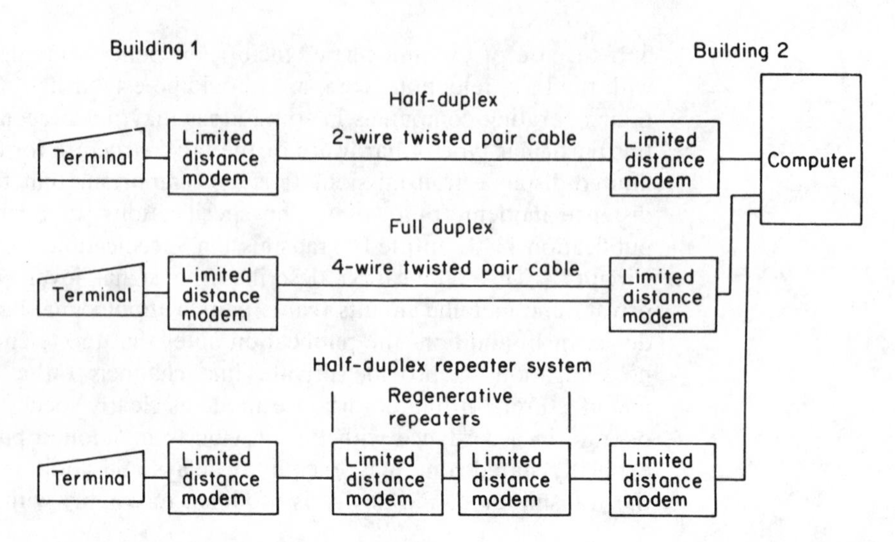

Figure 2.45 Typical point-to-point applications. If a limited distance modem is planned to be used as a repeater, care should be taken to insure that shelter, a power source, and access for diagnostic testing and maintenance is available.

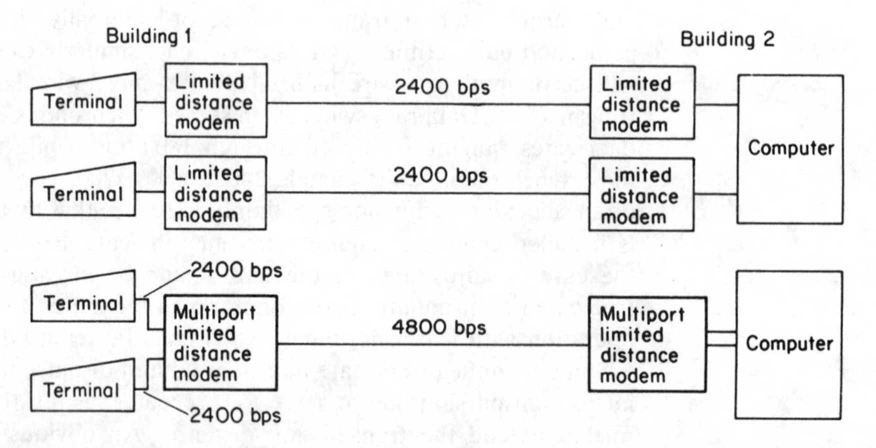

Figure 2.46 Multiport operation reduces cabling requirements. Using a limited distance modem with a multiport feature, one cable may be utilized to provide access to a computer from several terminals.

Diagnostics

Diagnostic capabilities vary both by the model produced by a manufacturer and between manufacturers of these units. Some limited-distance modems have self-testing circuitry which permits the user to easily determine if the unit is operating correctly; or, if it is defective, it notifies the user of the operational status of the unit through the display of one or more lights which indicate equipment status and alarm conditions. Most self-testing features available with limited-distance modems involve some form of loopback testing. In such a test, the transmitter output of the modem is looped back or returned to the receiver of the same unit so that the transmitter signal can be checked for errors. Other tests which are available on some models include dc busback, in which the received data and clocking is transmitted back to the limited distance modem at the opposite end of the line to provide an end-to-end test and a remote loop-back test which can be used to trigger a dc busback at a remote location.

2.9 DIGITAL SERVICE UNITS

At the beginning of the 1970s, communications carriers began offering communication systems designed exclusively for the transmission of digital data. Specialized carriers, including the now-defunct DATRAN, performed a considerable service to the information-processing community through their pioneering efforts in developing digital networks. Without their advancements, major communications carriers may have delayed the introduction of an all-digital service.

In December of 1974, the FCC approved the Bell System's DATAPHONE® digital service (DDS), which was shortly thereafter established between five major cities. Since then the service has been rapidly expanded to the point where, by the late 1980s, more than 100 cities had been added to the DDS network. Western Union International set another milestone in February of 1975 by applying to the FCC for authority to offer their International Digital Data Service (IDDS) from New York to Austria, France, Italy, and Spain. Digital data transmission by major carriers had become a reality.

In Europe, several countries have implemented digital networks, with British Telecom's KiloStream service currently being the most comprehensive network. Equipment was placed into production in the United Kingdom in September of 1982, and the first customer was commercially connected to KiloStream in January of 1983. By 1988, over 600 exchanges in the United Kingdom were equipped to offer KiloStream services.

Comparison of facilities

When data is transmitted on analog or voice-grade facilities by using FSK, the data stream is commonly modulated into two distinct frequencies representing marks and spaces or binary ones and zeros, respectively, as illustrated in the center of Figure 2.47. In so-called "voice-grade" telephone circuits, the usable bandwidth extends from 300 Hz to 3300 Hz, and the power transmitted at the higher frequency is significantly lower than the lower frequency. This bandwidth limitation not only causes a loss of distinction between the vocal "S" and "F" sounds, but limits the amount of information which can be transmitted via modulated digital forms.

In the case of the switched telephone network, the characteristics of a data link cannot be exactly determined because each new call may take a different path. Over long distances, multiple voice-grade lines are often combined into 3600 channels of 4000 Hz each and sent by microwave transmission. In the combining or multiplexing process, an original 2225-Hz signal may be shifted

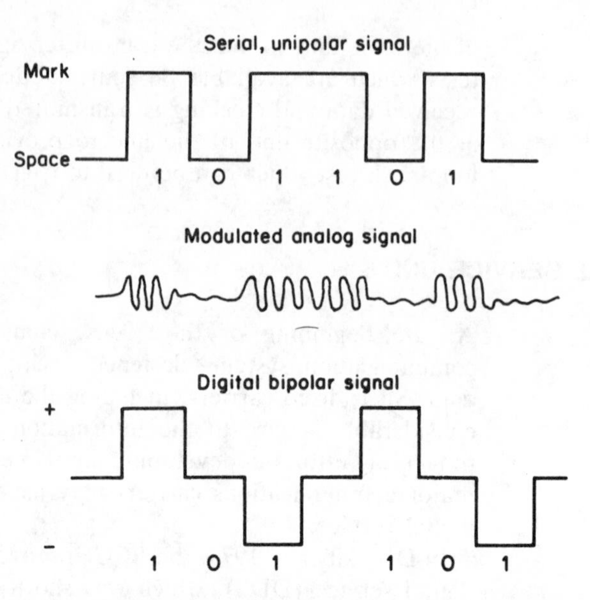

Figure 2.47 Analog versus digital signaling. When data is transmitted over an analog medium, the digital data is first modulated at the source and then demodulated at the destination back into its original form. On a digital network the digital data is transmitted as a digital bipolar signal.

to 19225 Hz for transmission and end up as a 2220-Hz or 2230-Hz signal at the receiver. This transmission over the switched network normally occurs at data rates up to 9600 bps. By obtaining a leased line, employing automatic equalization, and conditioning the line, data rates of 19,200 bps can be achieved.

With the voice-grade type of analog transmission, the data travels in a continuous manner; although it is easily amplified, any noise or distortion along the link is also amplified. In addition, the data signals become highly attenuated or weakened by the telephone characteristics originally geared to voice transmission. For the analog transmission of data, expensive and complex modems must be employed at both ends of the link to shape (modulate) and reconstruct (demodulate) the digital signals.

When digital transmission facilities are used, the data travels from end to end in digital form with the digital pulses regenerated at regular intervals as simple values of one and zero. Inexpensive digital service units are employed at both ends of the link to condition the digital signals for digital transmission.

AT&T's DDS is strictly a synchronous facility providing full-duplex, point-to-point, and multipoint service limited to speeds of 2.4, 4.8, 9.6, and 56 kbps. In addition to these leased-line services, AT&T introduced a switched 56 kbps digital service in 1985 and at the time this book was prepared several communications carriers were considering providing 19.2 kbps point-to-point and multipoint digital transmission facilities. Access to AT&T's Switched 56 service is obtained by dialing a "700" number, which was available in over 60 cities in early 1989.

Rates for leased-line digital services are based primarily upon distance and transmission speeds. This type of digital service is normally cost effective for high-volume users that could justify the expense associated with a dedicated communications facility. In comparison, pricing for Switched 56 kbps service is usage sensitive, based both on connection duration and distance between calling and called parties. Due to the cost of this service essentially corresponding to usage, it is attractive for such applications as the backup of critical DDS and T1 lines, peak-time overload usage to eliminate the necessity of installing additional leased digital circuits as well as for infrequent activities that may require a high data rate, such as still-frame videoconferencing and facsimile transmission.

Terminal access to the DDS network is accomplished by means of a digital service unit which alters serial unipolar signals into a form of modified bipolar signals for transmission and returns them to serial unipolar signals at the receiving end. The various types of service units will be discussed in detail later in this section.

Digital signaling

It is important to understand what modified bipolar signaling is and why it is necessary, since this form of signaling is the cornerstone of digital transmission. Figure 2.47 shows a comparison of how a serial unipolar signal from a teletype is transmitted as a modulated analog "voice" signal and as a digital bipolar signal. In normal return to zero bipolar signaling, a binary zero is transmitted as zero volts and binary ones are alternately transmitted as positive and negative pulses. Since DDS incorporates its own network codes (to include zero suppression, idle, and out-of-service) the original digital format is "violated" in that two successive binary ones could have the same polarity. To avoid any highly undesirable dc build-up in the line, each "violation" is in turn given an alternating polarity which will again return the voltage sum to zero.

Figure 2.48 illustrates six common types of digital signals that have been developed to meet various communications and electronic device operational requirements.

Unipolar non-return to zero is the simplest type of signaling and was used for early key telegraphy and is currently used with private line teletypewriter systems as well as representing the signal pattern used by RS-232 and V.24 interfaces. In this signal scheme a dc current or voltage represents a mark, while the absence of current or voltage represents a space. When used with transmission systems, line sampling determines the presence or absence of current, which is then translated into an equivalent mark or space.

A variation of unipolar non-return to zero signaling is unipolar return to zero. Here the signal always returns to zero after every "1" bit. While this signal is easier to sample, it requires more circuitry to implement and is not commonly used.

In place of having current for a mark and no current for a space, it is often desirable to provide a definite current for both pieces of information. The polar non-return to zero signal uses positive current to represent a mark and

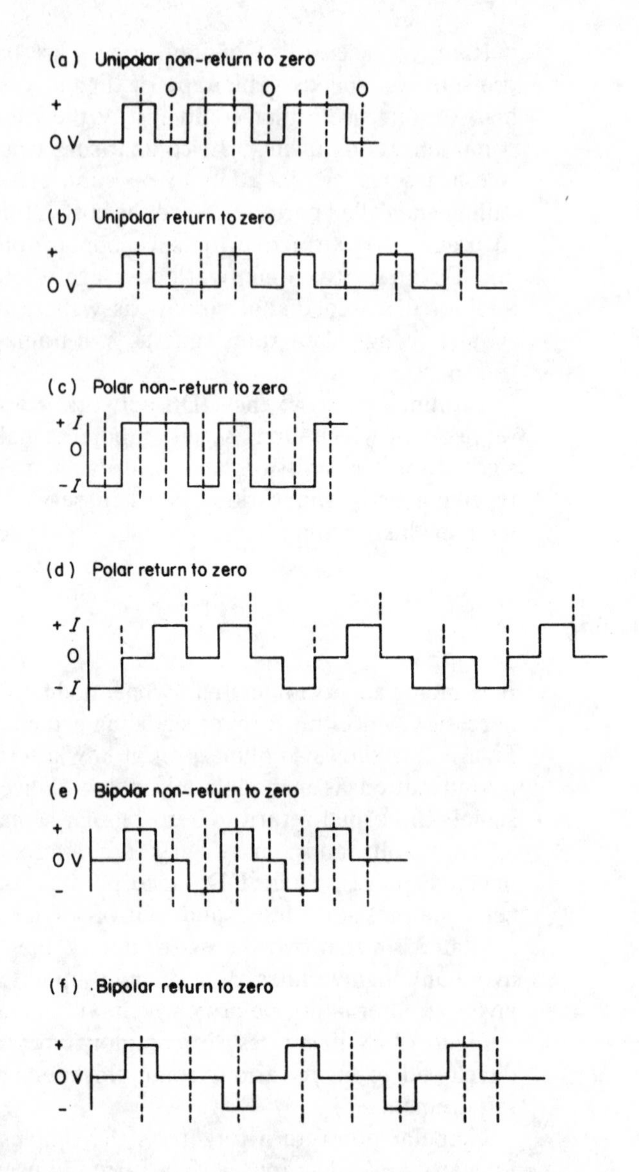

Figure 2.48 Digital signal characteristics.
(a) Unipolar non-return to zero — signal does not return to zero after each '1' bit.
(b) Unipolar return to zero — signal always returns to zero after every '1' bit.
(c) Polar non-return to zero — instead of current or no current, opposite directions or polarities of current can be used to denote the'1' and '0' conditions. In polar non-return to zero the current alternates between a positive and negative value.
(d) Polar return to zero — using opposite polarities of current but returning to zero after each '1' bit.
(e) Bipolar non-return to zero — a signal in which alternating polarity pulses are used for 1 symbols and no pulse for 0 symbols and the signal does not return to zero after each '1' bit.
(f) Bipolar return to zero — bipolar signaling with the signal returning to zero after each '1' bit.

negative current to represent a space. As no transition occurs between two consecutive bits of the same value the signal must be sampled to determine the value of each received bit.

The polar return to zero signal also uses opposite polarities of current, however, the signal returns to zero after each bit is transmitted. Since there is a pulse that has a discrete value for each bit, sampling of the signal is not required. Thus, the circuitry required to determine whether a mark or a space has occurred is reduced.

Two signals similar to the return to zero signal are the bipolar non-return to zero and the bipolar return to zero signals. In bipolar non-return signaling, alternating polarity pulses are used to represent marks while a zero pulse is used to represent a space. In bipolar return to zero signaling, the bipolar signal returns to zero after each mark. As this signal ensures that there is no dc voltage buildup on the line, repeaters can be placed relatively far apart in comparison to other signaling techniques and this signaling technique is employed in modified form on digital networks. Since two or more spaces cause no change from a normal zero voltage, sampling is required to determine the value of bits.

An interesting aspect of the network code insertion, which modify and violate the bipolar signal, is the use of a zero-suppression code. Since a long succession of binary zeros would not provide the necessary transitions to maintain proper timing recovery, strings of more than six zeros are replaced with zero suppression codes to maintain synchronization. Figure 2.49 shows how a bipolar signal undergoes violation insertion. In this example, a zero-suppression sequence is inserted into the binary channel signal. The resultant signal, as shown, returns the voltage sum to zero.

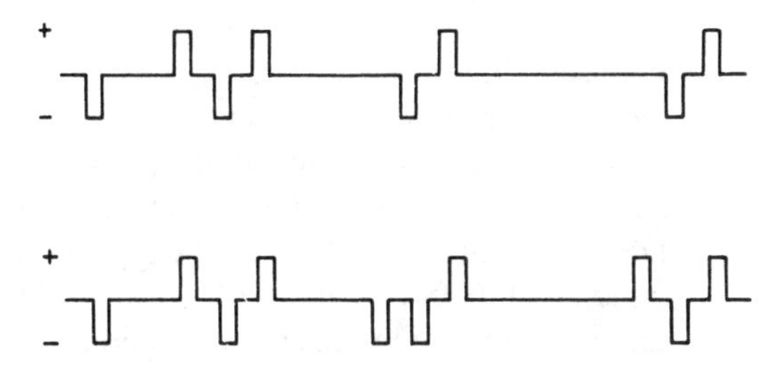

Figure 2.49 Modified bipolar signaling. Top: Original bipolar signal. Bottom: Modified bipolar signal 'violated' by zero-suppression codes. Since a long succession of binary zeros does not provide the necessary translations to maintain proper timing recovery, strings of zeros are replaced with zero suppression codes to maintain synchronization.

Clocking

For digital transmission, precise synchronization is the key to success of an all-digital network. It is essential that the data bits be generated at precise intervals, interleaved in time, and read out at the receiving end at the same interval to prevent loss or garbling of data sequences. To accomplish the necessary clock synchronization on the AT&T digital network, a master

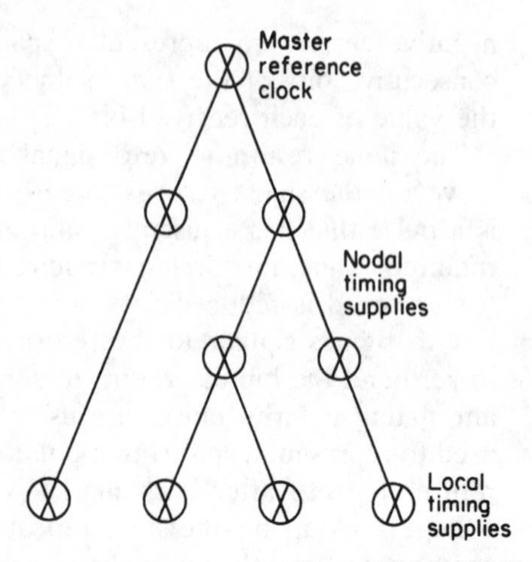

Figure 2.50 DDS timing subsystem. In AT&Ts DATAPHONE® digital service network a hierarchy of timing is provided to effect network synchronization.

reference clock is used to supply a hierarchy of timing in the network. Should a link to the master clock fail, the nodal timing supplies can operate independently for up to 2 weeks without excessive slippage during outages. In Figure 2.50, the hierarchy of timing supplies as linked to AT&T's master reference clock is illustrated. As shown, the subsystem is a treelike network containing no closed loops.

Service units and network integration

In discussing the characteristics of service units which interface terminals to digital networks, it is important to understand the functional differences between channel service units (CSU) and data service units (DSU). Figure 2.51 illustrates a simplified schematic diagram of the Bell System 500A-type DSU compared to their 550A-type CSU. Originally DSUs and CSUs were bundled into the DDS tariff, however, today each device can be obtained as separate items that users can acquire from both communications carriers or third party vendors. The DSU performs the actual conversion of unipolar data into a bipolar format for transmission over a digital facility. In comparison, the CSU terminates the digital circuit at the end-user's premises and performs such functions as line conditioning, remote loopback testing, signal regeneration and monitoring of the incoming digital signal to detect violations of rules that govern the transmission of data on a digital facility. On the left of Figure 2.52, the reader will note the relationship between a DSU and a CSU. It should also be noted that combined DSU/CSU units are commercially available and can be used to replace two separate units.

The CSU is designed to accept nominal 50 percent duty cycle bipolar pulses from the customer on the transmit and receive data leads. The pulses, synchronized with the DDS, are amplified, filtered, and passed on to the 4-wire metallic telephone company cable. The signals on the receive pair are

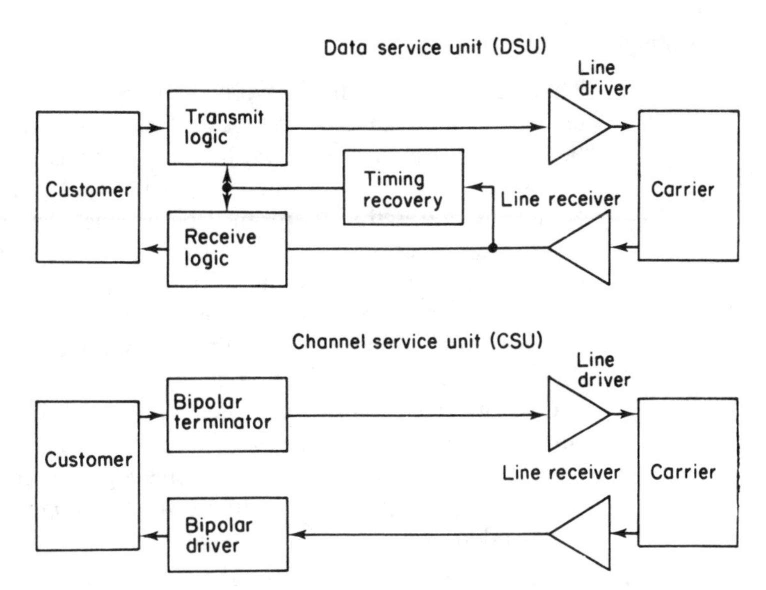

Figure 2.51 Service units for digital transmission. The DSU contains all of the circuitry necessary to make the device plug compatible with existing modems and terminals. When using a CSU, the customer must provide timing recovery and detect or generate DDS network control codes.

amplified, equalized, and sliced by the line receiver. The resultant bipolar pulses are then passed to the customer, whose DSU must recover the synchronous clock used for timing the transmitted data and sampling the received data. The customer's DSU must further detect DDS network codes, enter appropriate control states, and remove bipolar "violations" from the data stream.

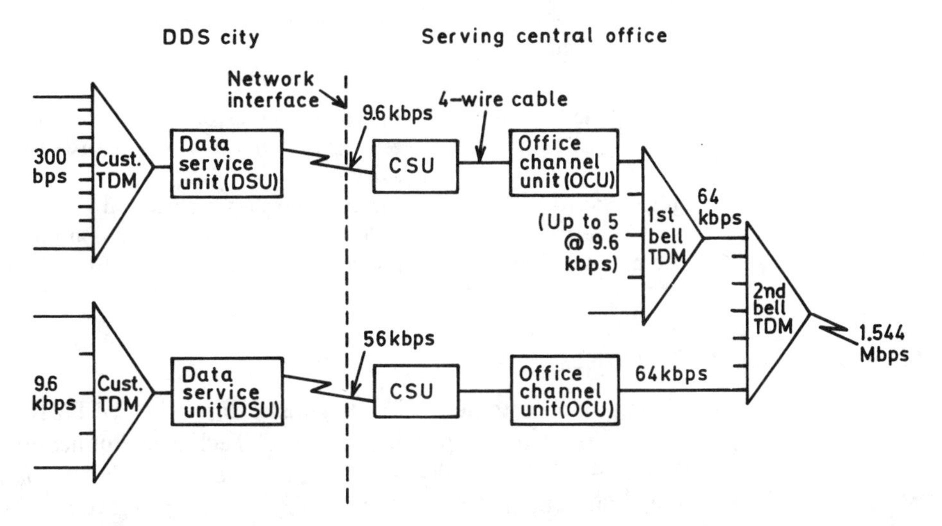

Figure 2.52 DDS multiplexing arrangement. Signals from a DSU are terminated into a complementary office channel unit in the serving central office. From there they enter into a multiplexing hierarchy which may carry voice as well as data signals.

Violation detection

There are two rules that govern the transmission of data on DDS that a CSU checks for compliance. First, AT&T specifies that no more than 15 zeros can be sent in a row on its digital facilities. This is because successive zeros result in a steady, zero voltage that cannot be used to generate timing information. A second rule is that there must be at least three ones in every 24 bits. This insures that necessary timing information can be obtained, since it is the ones that provide this information on the DDS network. As data is transmitted the CSU checks for the occurrence of 15 zeros or less than three ones in every 24 bits and produces a warning if either condition occurs. Some CSUs also correct such conditions, incorporating circuitry to perform timing recovery.

CSU interfacing is accomplished by use of a 15-pin female D-type connector which utilizes the first 6 pins: where pin 1 is signal ground, pin 2 is status indicator, pins 3 and 4 are the receive signal pair, while pins 5 and 6 are the transmit signal pair. In addition to the communications carriers, several independent vendors offer compatible CSU for customer interconnection to digital networks.

In comparison to CSUs, DSUs incorporate all the circuitry necessary to make the device plug compatible with existing terminals. The unit incorporates circuit similar to that described in the CSU, plus a digital circuit which performs bipolar to unipolar and unipolar to bipolar conversion, bipolar signal violation detection (two pulses of the same polarity transmitted in succession) and control code processing. By the transmission of an intentional bipolar violation the DDS network indicates to the DSU that a control code follows. Table 2.25 lists the four control codes that can be inserted into a received data stream and their meaning to the DSU.

Table 2.25 DSU control codes.

Control code	Meaning
Zero code	Indicates a series of six or seven zeros that is used to establish a fixed level of density of ones to maintain clocking
Idle code	Indicates no data transferred
Out of service code	Indicates an abnormal condition on the DDS circuit
Remote loop code	Indicates a request to the DSU to perform a remote test

Secondary Channel Diagnostics

To provide customers with diagnostic capability, AT&T defined DDS with a secondary channel specification in its Technical Publication 62120 in 1984. In 1987, AT&T formally tariffed a DDS Secondary Channel Option, with secondary channel data rates of 133 bps at a primary rate of 2400 bps, 266 bps at a primary rate of 4800 bps, 533 bps at a primary rate of 9600 bps and 2666 bps at a primary rate of 56 kbps. The secondary channel operates asynchronously and independent of the primary channel. Due to this, users of

DDS can continue transmitting data on the primary channel while performing secondary channel operations.

DSU interfacing is accomplished by use of a standard 25-pin EIA RS-232 female connector on the 2.4- through 9.6-kbps units, using 10 pins for signaling. The wideband, 56-kbps device utilizes a 34-pin CCITT, V.35 (Winchester-type) connector using 14 pins for signaling. Several independent suppliers also manufacture DSU-type units which offer even more flexibility in the form of multiport options.

Current AT&T 550A CSU and 550A-type DSU are listed in Table 2.26. Both the DSU and CSU incorporate properly balanced and equalized terminations for the 4-wire loop as well as circuitry to permit rapid, remote testing of the channel. The signals on the 4-wire loop are the same for both devices and are terminated in the servicing central office of the communications carrier into a complementary unit called an office channel unit (OCU). From here, the time division multiplexing hierarchy begins as illustrated in Figure 2.52.

Table 2.26 Dataphone digital service interface units.

	Speed (kbps)	List code
500A-type DSU	2.4	500A-L1/2
	4.8	500A-L1/3
	9.6	500A-L1/4
	56	500A-L1/5
500A-type CSU	2.4	550A-L1/2
	4.8	550A-L1/3
	9.6	550A-L1/4
	56	550A-L1/5

Signals from the OCU are fed into the first stage of multiplexing, which combines up to twenty 2.4-, ten 4.8-, or five 9.6-kbps signals into a single 64-kbps channel which is the digital capacity of a voice channel in the T1 digital transmission system. A second stage of multiplexing takes the 64-kbps bit streams and efficiently packs them into a T1-bit stream operating at 1.544 Mbps, which may carry voice as well as data signals over existing long-line facilities. Using this scheme, future expansion of DDS may be accomplished at a very rapid pace which could at a later date relegate analog transmission of data to history. In the early 1970s, AT&T was expanding their T1 system by as much as 10,000 channel miles per day. However, this expansion rate substantially decreased after their initial network was established.

Analog extensions to DDS

AT&T provides an 831A data auxiliary set which allows analog access to DDS for customers located outside the DDS servicing areas. The 831A connects the EIA RS-232 interfaces between a data service unit (500A-type) and a voice-band data set. The 831A contains an 8-bit elastic store, control, timing, and test circuits which allow loop-back tests toward the digital network. The elastic store is a data buffer that is required by the DSU to receive data from the modem in time with the modem's receive clock. The data is then held in the elastic store until the DSU's transmit clock requests it. Thus, the buffer serves as a mechanism to overcome the timing differences between the clocks of the two devices. In the reverse direction, no buffer is required when the DSU's receive clock is used as the modem's external transmit clock. When the modem cannot be externally clocked or when one DSU is connected to a second DSU or a DTE that cannot accept an external clock, a second elastic store will be required. Figure 2.53 illustrates a typical analog extension to a DDS servicing area.

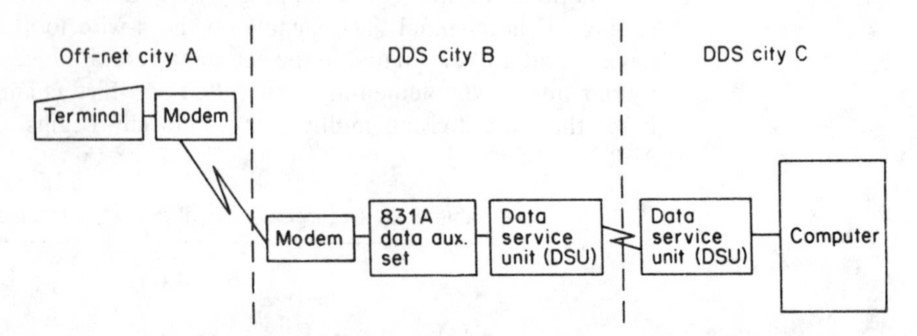

Figure 2.53 Analog extension to DDS. In order to obtain an analog extension to a digital network, a device known as a data auxiliary set, which provides an interface between a modem and a service unit, must be installed.

Applications

As discussed in previous sections, the requirement for expensive modems in an off-net analog extension could negate any real savings, gained in utilizing DDS. A network arrangement in the form of Figure 2.54, on the other hand, could easily achieve the high-performance characteristics inherent in DDS while reducing the overall costs of creating two independent data links.

In this example, we have incorporated a device marketed by AT&T which is called a split stream unit (SSU). The SSU is similar to the muliplexer incorporated into Bell System 209 data sets, in that the user can select various combinations of data transfer rates on up to four individual channels, up to the maximum capability of the DDS line. The SSU plugs directly into the DSU, with operational settings at one-half or one-quarter the specified DDS data rate. The unit provides local loop and remote loop testing of each individual channel and very effectively lifts the four speed restrictions of DDS service. Table 2.27 lists the operational modes of SSUs designed for use on 2400-, 4800-, and 9600-bps DDS circuits.

Of course, one can also obtain a variety of multiplexers manufactured by independent communications vendors that can be used in place of SSUs. Since DDS is a synchronous network that operates at a series of fixed data rates, one can also employ multiplexers as a means to concentrate data for communications over a DDS line. One can also use multiplexers to convert asynchronous data into a synchronous data stream that is compatible with the data format for which DSUs and CSUs are designed.

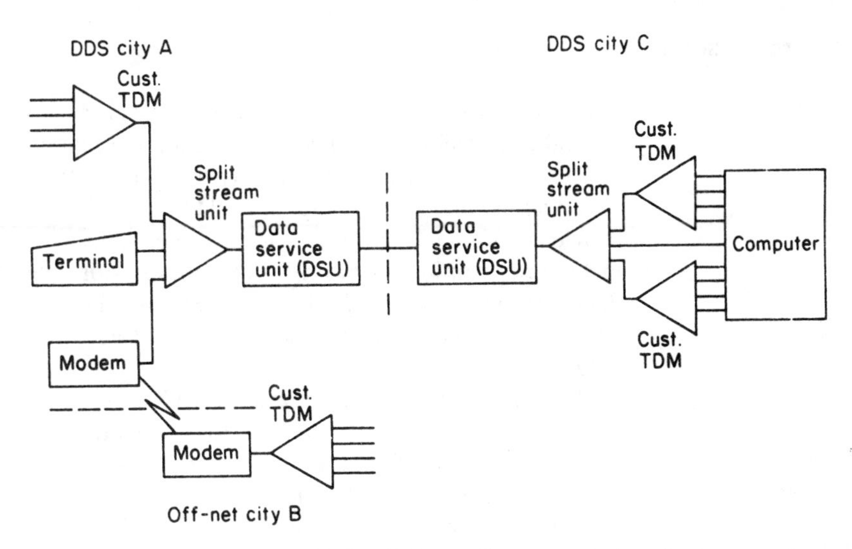

Figure 2.54 Multiplexing over DDS utilizing split stream units. An inexpensive split stream unit, or limited-function synchronous multiplexer, can offer considerable flexibility when interfacing into the DDS network.

Table 2.27 Split stream unit modes of operation.

DDS rate	Mode	SSU	Channel	Data	Rate
		1	2	3	4
9600	1	9600			
	2	2400	7200		
	3	7200	2400		
	4	4800	4800		
	5	2400	4800	2400	
	6	4800	2400	2400	
	7	2400	2400	2400	2400
4800	1	4800			
	2	1200	3600		
	3	3600	1200		
	4	2400	2400		
	5	1200	2400	1200	
	6	2400	1200	1200	
	7	1200	1200	1200	1200
2400	1	2400			
	2	600	1800		
	3	1800	600		
	4	1200	1200		
	5	600	1200	600	
	6	1200	600	600	
	7	600	600	600	600

KiloStream service

Although British Telecom's KiloStream service is similar to DDS, there are several significant differences that warrant discussion.

The British Telecom customer is provided with an interface device which is called a network terminating unit (NTU), which is similar to a DSU. The NTU provides a CCITT interface for customer data at 2.4, 4.8, 9.6, or 48 kbps to include performing data control and supervision, which is known as structured data. At 64 kbps, the NTU provides a CCITT interface for customer data without performing data control and supervision, which is known as unstructured data.

The NTU controls the interface via CCITT recommendation X.21, which is the standard interface for synchronous operation on public data networks. An optional V.24 interface is available at 2.4, 4.8, and 9.6 bps while an optional V.35 interface can be obtained at 48 kbps. The X.21 interface is illustrated in Figure 2.55. Here the control circuit (C) indicates the status of the transmitted information – data or signaling, while the indication circuit (I) signals the status of information received from the line. The control and indication circuits control or check the status bit of an 8-bit envelope used to frame six information bits.

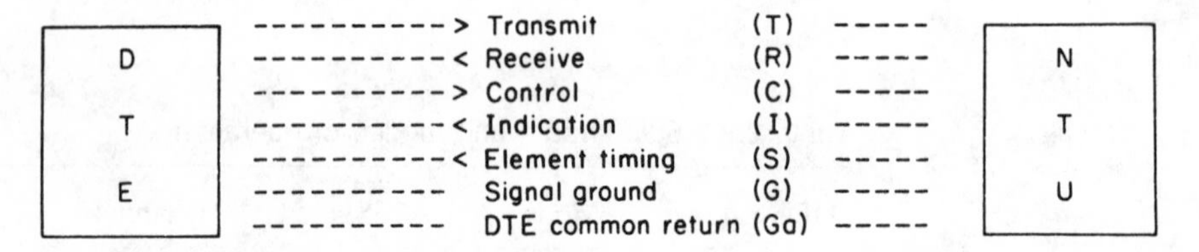

Figure 2.55 CCITT X.21 interface circuits.

Customer data is placed into a 6 + 2 format to provide the signaling and control information required. This is known as envelope encoding and is illustrated in Figure 2.56.

The NTU performs signal conversion, changing unipolar non-return to zero signals from the V.21 interface into a di-phase WAL 2 encoding format. This ensures that there is no dc content in the signal transmitted to the line, provides isolation of the electronic circuitry from the line, and provides transitions in the line signal to enable timing to be recovered at the distant end. Table 2.28 lists the NTU operational characteristics of KiloStream.

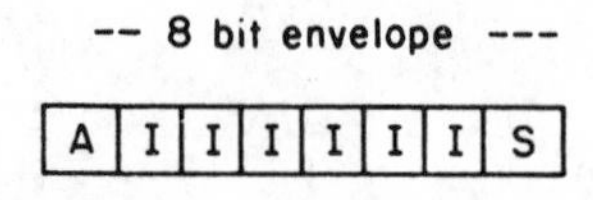

Figure 2.56 KiloStream envelope encoding.
A= Alignment bit which alternates between '1' and '0' in successive envelopes to indicate the start and stop of each 8-bit envelope.
S = Status bit which is set or reset by the control circuit and checked by the indicator circuit.
I = Information bits.

Table 2.28 KiloStream NTU operational characteristics.

Customer data rate (kbps)	DTE/NTU interface	Line data rate (kbps)	NTU operation
2.4	X.21	12.8	6+2 envelope encoding
4.8	X.21	12.8	6+2 envelope encoding
9.6	X.21	12.8	6+2 envelope encoding
48	X.21	64	6+2 envelope encoding
64	X.21	64	No envelope encoding
48	X.21 bis/V.35	64	6+2 envelope encoding
2.4	X.21/V.24	12.8	6+2 envelope encoding
4.8	X.21/V.24	12.8	6+2 envelope encoding
9.6	X.21/V.24	12.8	6+2 envelope encoding

The KiloStream network

In the KiloStream network, the NTUs on a customer's premises are routed via a digital local line to a multiplexer operating at 2.048 Mbps. This data rate is the European equivalent of the T1 line in the United States that operates at 1.544 Mbps. The multiplexer can support up to 31 data sources and may be located at the local telephone exchange or on the customer's premises if traffic justifies. It is connected via a digital line or a radio system into the British Telecom KiloStream network as illustrated in Figure 2.57.

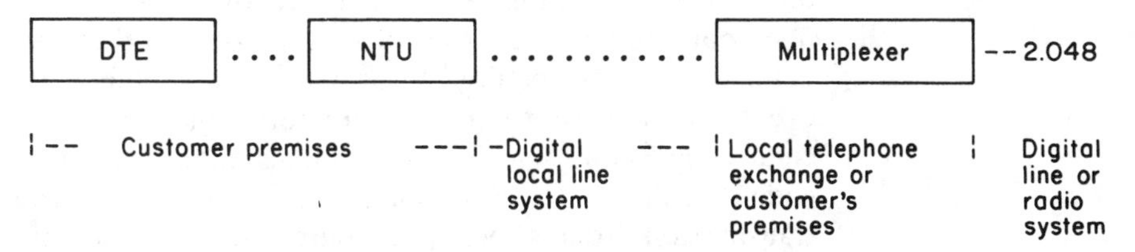

Figure 2.57 The KiloStream connection.

2.10 PARALLEL INTERFACE EXTENDERS

When an application arises that will require batch processing at a site remotely located from the computer, several approaches can be considered to satisfy this requirement.

A traditional approach is the establishment of a remote batch processing operation. The establishment of this type of operation normally requires the procurement of several communications components in addition to the remote batch terminal. First, a communications controller must be installed and interfaced to the computer if such a device does not already exist at the computational facility. This controller, in conjunction with the computer, performs such tasks as character assembly and disassembly, transmission error checking by generating check characters from the received data blocks, and comparing the check character to the check character generated by the remote batch terminal, as well as performing numerous traffic management functions. Next, a teleprocessing software module will be added and integrated to the computer's operating system to perform and control the transmission discipline. This software may not only be costly but may affect computer performance since it typically requires between 10,000 to 40,000 or more memory locations, depending upon complexity. Last, a transmission medium and either a high-speed modem or DSU to translate the signals into an acceptable form for transmission must be installed. This traditional remote batch processing operation is illustrated in Figure 2.58.

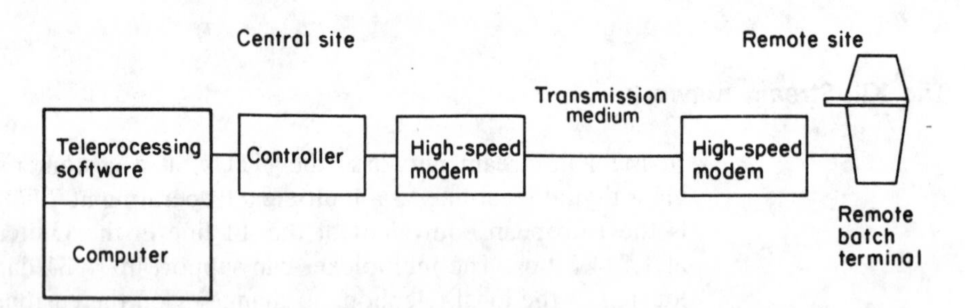

Figure 2.58 Traditional remote batch processing. A large portion of the computer's memory may be reserved for teleprocessing software.

Another problem which may arise is the compatibility of the remote batch terminal to the computer system already installed. If the terminal obtained does not support the protocol of the computer's teleprocessing software an emulator may be required to provide an acceptable transmission link. In any event, transmission from the computer to the terminal will most likely require code conversion, since most terminals cannot accept the computer's native code. Realizing these problems, a device was introduced which permits transmission from selected computers to a variety of computer peripherals without the necessity of the addition of special teleprocessing software or a communications controller. This device is called a parallel interface extender.

Extender operation

A parallel interface extender is a device which translates the parallel protocol of an input/output (I/O) channel from such devices as a computer or selected computer peripheral units into a serial protocol which is suitable for transmission over a normal serial communications link. At the other end of the link, another parallel interface extender or similar operating device translates the serial protocol formed by the first parallel interface extender back into the original parallel protocol transmitted by the I/O channel for reception by devices similar to the standard peripherals used for local data processing at a computer center.

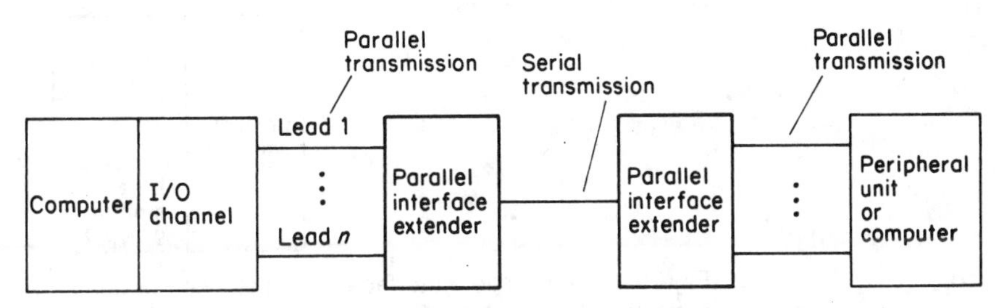

Figure 2.59 Parallel interface extender operation. Operation is similar to a multiplexer, combining the data from a number of leads from a computer or peripheral I/O channel into a single bit stream for transmission.

From a broad viewpoint, a parallel interface extender can be compared as being similar in performance to a multiplexer, combining the data from a number of leads of the parallel I/O channel into a single bit stream for transmission over a serial communications link, as shown in Figure 2.59.

By permitting a computer to utilize its regular I/O channel when communicating with a remotely located peripheral device, the necessity of obtaining specialized communications software is alleviated, and communications with remote peripheral units can then take place with the same software which is regularly used for communicating with the local peripherals at the computer center. Depending upon the computer configuration in use, a parallel interface extended may reduce operating system software requirements from 10 to 30 percent or more, when compared with communicating to remote devices by using a line controller and the required teleprocessing software modules. Another advantage obtained through the use of parallel interface extenders is the ability to program remote applications in Fortran, Cobol, other higher level languages, as well as in assembly language, using the READ, WRITE statements of Fortran and Cobol or the GET and PUT macros of assembly language to perform input and output remote peripheral functions. In addition to the translation of parallel to serial and serial to parallel data, to perform remote peripheral functions the device encodes commands and status information into a serial bit stream at the transmission end of the link while the device at the receiving end of the link performs a decoding function to reconstruct the original commands, status, and data before passing such information to the remote peripheral unit.

Extender components

A parallel interface extender consists of a control unit and one or more line module groups, as shown in Figure 2.60. The control unit of the extender connects to the multiplexer channel of a computer and emulates the functions of several peripheral control units, such as card readers, card punches, magnetic tapes, and line printers, thus supporting the computer's standard software, as previously mentioned. Each line module group is connected to the control unit of the extender on one side and provides an interface for the connection of dedicated, switched, or leased lines on the opposite side. The line module

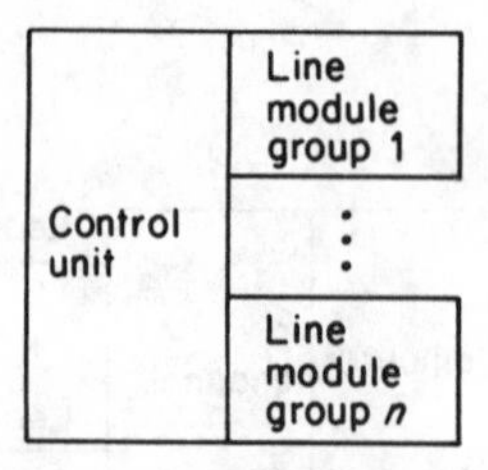

Figure 2.60 Parallel interface extender components. A parallel interface extender consists of a control unit and one or more line module groups. In addition, some manufacturers provide modems built in to the device.

contains all necessary line control and error control components, as well as a built-in modem which alleviates the necessity of having special communications software such as IBM's basic telecommunications access method (BTAM) or synchronous data link control (SDLC), a communications controller, line adapter, and a separate modem. For alternate high-speed data transfer in either direction, half-duplex line modules can be used while full-duplex line modules provide data transfer in both directions at the same time. For data transfer over the switched network, half-duplex line modules are normally used while either half-duplex or full-duplex line modules can be used with dedicated or leased 4-wire lines.

Applications examples

The parallel interface extender manufactured by Paradyne Corporation can accommodate sixteen half-duplex or eight full-duplex line modules. This unit connects to either an IBM system 360 or an IBM system 370 central processing unit via the system's byte multiplexer channel and can be used for such diverse applications as permitting two computers to communicate with each other in their native code, as in Figure 2.61; or it can be used to connect a computer to a variety of local or remotely located peripheral units, as shown in Figure 2.62.

With data transmission rates of 4800, 7200, 9600, 19,200 and 56,000 bps, the parallel interface extender permits a level of data transfer that can be matched to the operating speed of most peripheral devices. Parallel interface extenders permit peripherals to be transferred to remote locations for other applications than originally required; hence, the use of such equipment warrants further examination.

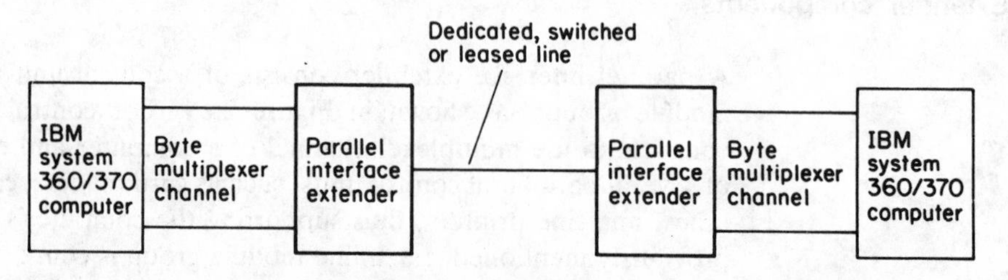

Figure 2.61 Intercomputer communications using a parallel interface extender. Using a parallel interface extender, the parallel transmission of the byte multiplexer channel of a computer is converted into a serial data stream for transmission.

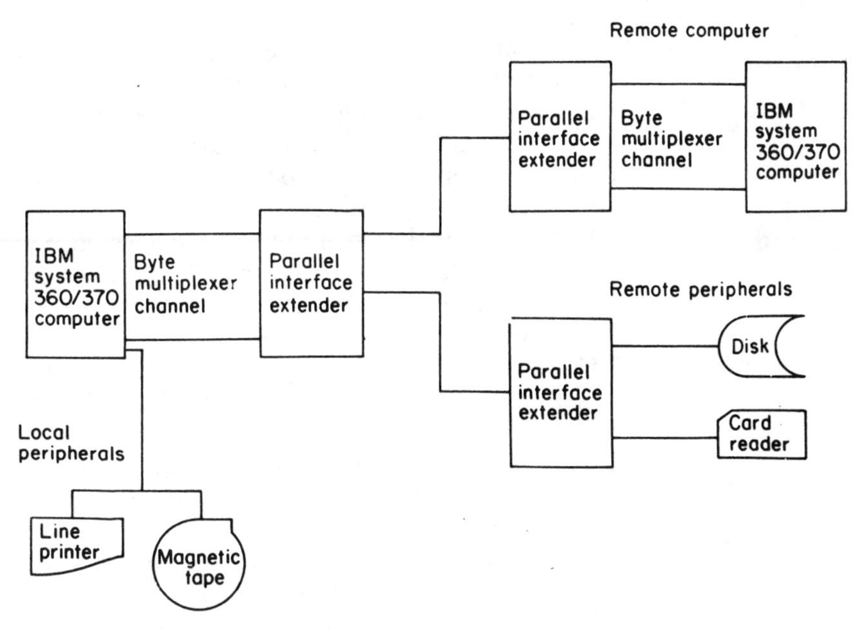

Figure 2.62 Local and remote peripherals as well as computers can be serviced. Through the utilization of a parallel interface extender, remote peripherals as well as remote computers are referenced by the central computer as if they were local devices.

REVIEW QUESTIONS

2.1 What is the difference between an acoustic coupler and a modem with respect to their line connection?

2.2 What does the term "Bell System" compatibility mean when discussing the operational characteristics of a modem?

2.3 If an originate mode acoustic coupler transmits a mark at f_1 and a space at f_2 and receives a mark at f_3 and a space at f_4, what would be the corresponding frequencies of an answer mode coupler to ensure communications compatibility?

2.4 Discuss the conventional utilization of originate and answer mode couplers and modems.

2.5 Why is it more likely than not that an American using his or her portable personal computer in Europe would not be able to communicate with a computer located in Europe?

2.6 What is a carrier signal? By itself, does it convey any information? Why?

2.7 How can the characteristics of a carrier signal be altered?

2.8 What is the difference between a bit per second and a baud? When can they be equivalent? When are they not equivalent?

2.9 What is the Nyquist relationship and why does it require modem designers to develop multilevel phase-shift keying modulation schemes for modems to operate at high data rates?

2.10 What does the signal constellation pattern of a modem represent? What is the normal relationship between the density of the signal constellation pattern and the susceptibility of a modem to transmission impairments?

2.11 Discuss the difference between Trellis coded modulation and conventional quadrature amplitude modulation with respect to the density of the signal constellation and the susceptibility of a modem employing each modulation technique to transmission impairments?

2.12 How could you use a reverse channel? What is the difference between a reverse channel and a secondary channel?

2.13 What is a multiport modem and under what circumstances should you consider using this device?

2.14 Why are Bell System 212-type modems operating at 1200 bps sometimes compatible with V.22 modems while at other times they are incompatible?

2.15 Why are Bell System 202 type modems incompatible with CCITT V.23 modems?

2.16 Discuss the compatibility of a CCITT V.26 modem employing a pattern A phase change with a similar modem using the pattern B phase change.

2.17 Explain why the V.29 signal constellation pattern forms a mirror image.

2.18 Under what conditions in the United States can a modem be connected to the switched telephone network without requiring the use of a Data Access Arrangement?

2.19 Assume two modems connected to a leased line have both local and digital loop-back self-testing features. If communications were disabled, discuss the tests you would perform to determine if the line or one or both modems caused the communications failure.

2.20 What is a wraparound unit? When would it be used?

2.21 What are the advantages and disadvantages in using multiport modems instead of separate modems and multiplexers?

2.22 What function does a DCE option perform? How would you obtain a DCE option through the use of a cable?

2.23 What is the key difference between multipoint modems and conventional modems? Discuss the differences in throughput obtained on a multipoint circuit as the average block size transmitted increases.

2.24 Why is the effect of multipoint modems on satellite circuits minimal?

2.25 Discuss the relationship between the data rate, wire gauge, and transmission distance.

2.26 What is the difference between a line driver and a limited-distance modem?

2.27 Assume your organization has eight terminals that require a connection to the computer system located in the same building 600 feet distant. If the terminals operate at 2400 bps and can only transmit pulses 400 feet before the pulses become distorted, determine the most economic method to connect the terminals given the following cost for equipment and cable:

cable cost per foot	$ 0.50
line driver	$100.00
multiport limited-distance modem – (4 port)	$250.00

2.28 Why is the probability of a transmission error occurring on a digital transmission facility less than an analog transmission facility?

2.29 Why do digital transmission facilities employ bipolar signaling?

2.30 What is the advantage of using a parallel interface extender with respect to teleprocessing software? How could you use a central computer facility to distribute reports to several remotely located printers connected to the central computer via parallel interface extenders?

CHAPTER **3**

DATA CONCENTRATION EQUIPMENT

Although the communications components included in this chapter perform a variety of functions which govern their utilization for selected application areas, their inclusion here is based upon their common function of concentrating data. In this chapter, the operation and utilization of components designed primarily to accomplish data concentration will be covered. Specific devices to be investigated in this chapter include a variety of multiplexing equipment, concentrators, and front-end processors, as well as components which permit modems, lines, and the ports of computers and other devices to operate on a shared use basis. In addition, a device which splits a data stream into two streams for transmission to take advantage of the difference in the tariff between wideband and voice-band leased lines will also be covered.

EQUIPMENT SIZING

One of the most frequent problems in data communications involves determining the appropriate capacity of a communications device. This usually involves selecting the number of ports or channels to be installed in the communications device to satisfy one's communications requirements in an efficient manner. If too few ports are installed, users will encounter busy signals or may be placed in queues, resulting in a loss of productivity, if the users are within one's organization, or the possible loss of customers, if users are from outside the organization. If too many ports are installed, many ports will be underutilized, probably resulting in the expenditure of funds for unnecessary hardware.

Owing to the importance of equipment sizing, a review of the mathematics associated with such problems is presented in Appendix A. The material presented in that appendix can be used for sizing such devices as multiplexers, concentrators, and front-end processors.

3.1 MULTIPLEXERS

With the establishment of distributed computing, the cost of providing the required communications facilities became a major focus of concern to users. Numerous network structures were examined to determine the possibilities of using specialized equipment to reduce these costs. For many networks where geographically distributed users accessed a common computational facility, a central location could be found which would serve as a hub to link those users to the computer. Even when terminal traffic was low and the cost of leased lines could not be justified on an individual basis, quite often the cumulative cost of providing communications to a group of users could be reduced if a mechanism was available to enable many terminals to share common communications facilities. This mechanism was provided by the utilization of multiplexers whose primary function is to provide the user with a reduction of communications costs. This device enables one high-speed line to be used to carry the formerly separate transmissions of a group of lower speed lines. The use of multiplexers should be considered when a number of data terminals communicate from within a similar geographical area or when a number of leased lines run in parallel for any distance.

Evolution

From the historical perspective, multiplexing technology can trace its origination to the early development of telephone networks. Then, as today, multiplexing was the employment of appropriate technology to permit a communications circuit to carry more than one signal at a time.

In 1902, twenty-six years after the world's first successful telephone conversation, an attempt to overcome the existing ratio of one channel to one circuit occurred. Using specifically developed electrical network terminations, three channels were derived from two circuits by telephone companies.

The third channel was denoted as the phantom channel, hence the name "phantom" was applied to this early version of multiplexing. Although this technology permitted two pairs of wires to effectively carry the load of three, the requirement to keep the electric network finely balanced to prevent crosstalk limited its practicality.

Comparison with other devices

In the past, differences between multiplexers and concentrators were pronounced, with multiplexers being prewired, fixed logic devices; they produced a composite output transmission by sharing frequency bands (frequency division multiplexing) or time slots (time division multiplexing) on a predetermined basis, with the result that the total transmitted output was equal to the sum of the individual data inputs. Multiplexers were also originally transparent to the communicator, so that data sent from a terminal through a multiplexer to a computer was received in the same format and code by the computer as its original form.

In comparison, concentrators were developed from minicomputers by the addition of specialized programming and originally performed numerous tasks that could not be accomplished through the use of a multiplexer. First, the intelligence provided by the software in concentrators permits a dynamic sharing technique to be employed instead of the static sharing technique used in traditional multiplexers. If a terminal connected to a concentrator is not active, then the composite high-speed output of the concentrator will not automatically reserve a space for that terminal as will a traditional multiplexer.

This scheme, commonly known as dynamic bandwidth allocation, permits a larger number of terminals to share the use of a high-speed line through the use of a concentrator than when such terminals are connected to a multiplexer, since the traditional multiplexer allocates a time slot or frequency band for each terminal, regardless of whether the terminal is active. For this reason, statistics and queuing theory plays an important role in the planning and utilization of concentrators. Next, owing to the stored program capacity of concentrators, these devices can be programmed to perform a number of additional functions. Such functions as the preprocessing of sign-on information and code conversion can be used to reduce the burden of effort required by the host computer system.

The advent of statistical and intelligent multiplexers, which are discussed later in this section, has closed the gap between concentrators and multiplexers. Through the use of built-in microprocessors, these multiplexers can now be programmed to perform numerous functions previously available only through the use of concentrators. The reader should refer to Section 3.5, "Concentrators and front-end processors", as well as the portion of this section which covers statistical and intelligent multiplexers, for additional information on these devices.

Device support

In general, any device that transmits or receives a serial data stream can be considered a candidate for multiplexing. Data streams produced by the devices listed in Table 3.1 are among those that can be multiplexed. The intermix of devices as well as the number of any one device whose data stream is considered

Table 3.1 Candidates for data stream multiplexing.

Analog network private line modems
Analog switched network modems
Digital network data service units
Digital network channel service units
Data terminals
Data terminal controllers
Minicomputers
Concentrators
Computer ports
Computer–computer links
Other multiplexers

for multiplexing is a function of the multiplexer's capacity and capabilities, the economics of the application, and cost of other devices which could be employed in that role, as well as the types and costs of high-speed lines being considered.

Multiplexing techniques

Today, two basic techniques are commonly used for multiplexing: frequency division multiplexing (FDM) and time division multiplexing (TDM). Within the time division technique, two versions are available – fixed time slots which are employed by traditional TDMs and variable use of time slots which are used by statistical and intelligent TDMs.

FDM

In the FDM technique, the available bandwidth of the line is split into smaller segments called data bands or derived channels. Each data band in turn is separated from another data band by a guard band which is used to prevent signal interference between channels, as shown in Figure 3.1. Typically, frequency drift is the main cause of signal interference and the size of the guard bands are structured to prevent data in one channel drifting into another channel.

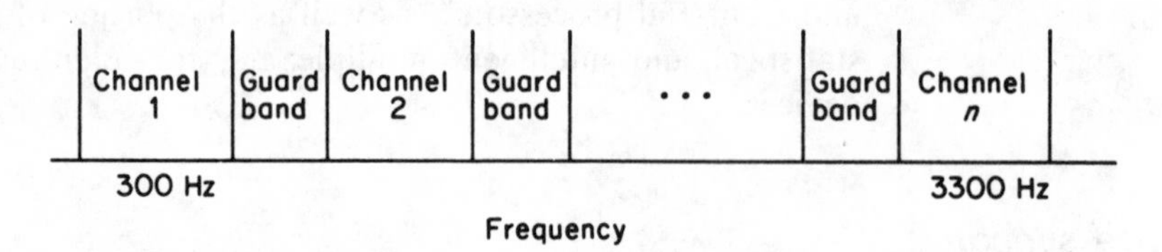

Figure 3.1 FDM channel separations. In frequency division multiplexing the 3-kHz bandwidth of a voice-grade line is split into channels or data bands separated from each other by guard bands.

Physically, an FDM contains a channel set for each data channel as well as common logic, as shown in Figure 3.2. Each channel set contains a transmitter and receiver tuned to a specific frequency, with bits being indicated by the presence or absence of signals at each of the channel's assigned frequencies. In FDM, the width of each frequency band determines the transmission rate capacity of the channel, and the total bandwidth of the line is a limiting factor in determining the total number or mix of channels that can be serviced. Although a multipoint operation is illustrated in Figure 3.2, FDM equipment can also be utilized for the multiplexing of data between two locations on a point-to-point circuit. Currently, data rates up to 1200 bps can be multiplexed by FDM. Typical FDM channel spacings required at different data rates are listed in Table 3.2.

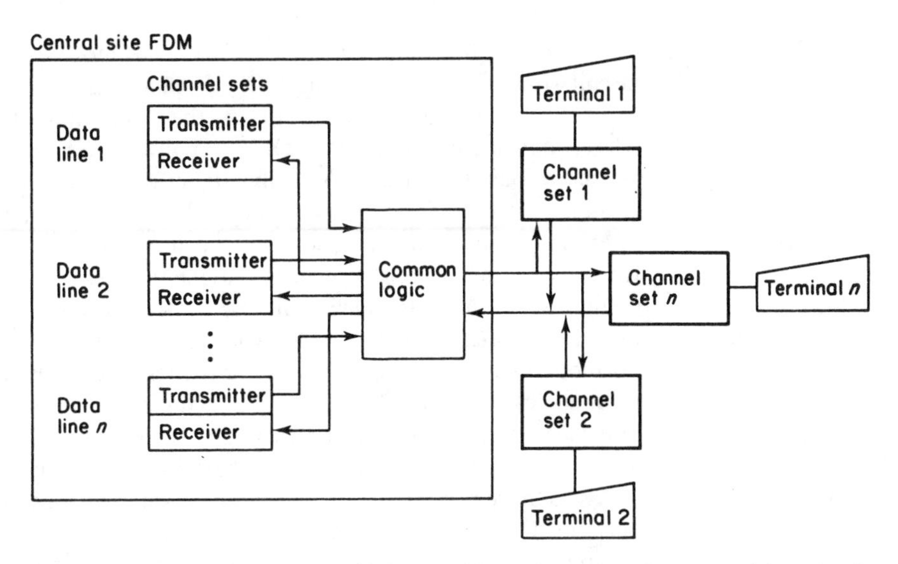

Figure 3.2 Frequency division multiplexing. Since the channel sets modulate the line at specified frequencies, no modems are required at remote locations.

Table 3.2 FDM channel spacings.

Speed (bps)	Spacing (Hz)
75	120
110	170
150	240
300	480
450	720
600	960
1200	1800

The overall FDM's aggregate data handling capacity depends upon the mixture of subchannels as well as the type of line conditioning added to the circuit. A chart of FDM subchannel allocations is illustrated in Figure 3.3. This chart can be used to compute the mixture of data subchannels that can be transmitted via a single voice-grade channel when frequency division multiplexing is employed. The referenced chart is based upon data subchannel spacing standards formulated by the CCITT. As illustrated, 17 75 bps subchannels can be multiplexed on an unconditioned (C0) circuit, 19 subchannels on a C1 conditional circuit, 22 subchannels on a C2 circuit and 24 channels on a C4 conditioned circuit. For higher data rates the CCITT standards allocate a fractional proportion of the bandwidth allocated to the previously discussed data rate of 75 bps.

With the development of terminals operating at speeds that were not multiples of CCITT frequencies, such as 134.5-bps teleprinters, a number of vendors developed FDM equipment tailored to make more efficient use of voice-grade circuits than permitted by the CCITT standards.

Since the physical bandwidth of the line limits the number of devices which may be multiplexed, FDM is mainly used for multiplexing low-speed asynchronous terminals. An advantage obtained through the use of such equipment is its code transparency. Once a data band is set, any terminal operating at that speed or less can be used on that channel without concern for the code of the terminal. Thus, a channel set to carry 300-bps transmission could also be used to service an IBM 2741 terminal transmitting at 134.5 bps or a Teletype 110-bps terminal. Another advantage of FDM equipment is that no modems are required since the channel sets modulate the line at specified frequencies, as shown in Figure 3.2. At the computer site, the FDM multiplexer interfaces the computer ports through channel sets. The common logic acts as a summer, connecting the multiplexer channel sets to the leased line.

At each remote location, a channel set provides the necessary interface between the terminal at that location and the leased line. When using FDM equipment, individual data channels can be picked up or dropped off at any point on a telephone circuit. This characteristic permits the utilization of multipoint lines and can result in considerable line charge reductions based upon developing a single circuit which can interface multiple terminals. Each remote terminal to be serviced only needs to be connected to an FDM channel set which contains bandpass filters that separate the line signal into the individual frequencies designated for that terminal. Guard bands of unused frequencies are used between each channel frequency to permit the filters a degree of tolerance in separating out the individual signals.

Although FDM normally operates in a full-duplex transmission mode on a 4-wire circuit by having all transmit tones sent on one pair of wires and all receive tones return on a second pair, FDM can also operate in the full-duplex mode on a 2-wire line. This can be accomplished by having the transmitter and receiver of each channel set tuned to different frequencies. For example, with 16 channels available, one channel set could be tuned for channel 1 to transmit and channel 9 to receive while another channel set would be tuned to channel 2 to transmit and channel 10 to receive. With this technique, the number of data channels is halved. However, the cost differential between a 4-wire and a 2-wire circuit may justify its use if one has only a small number of terminals to service.

FDM utilization

As mentioned previously, one key advantage in utilizing FDM equipment is the ability afforded the user in installing multipoint circuits for use in a communications network. This can minimize line costs since a common line, optimized in routing, can now be used to service multiple terminal locations. An example of FDM equipment used on a multipoint circuit is shown in Figure 3.4, where a 4-channel FDM is used to multiplex traffic from terminals located in four different cities. Although the entire frequency spectrum is transmitted on the circuit, the channel set at each terminal location filters out the preassigned bandwidth for that location, in effect producing a unique individual

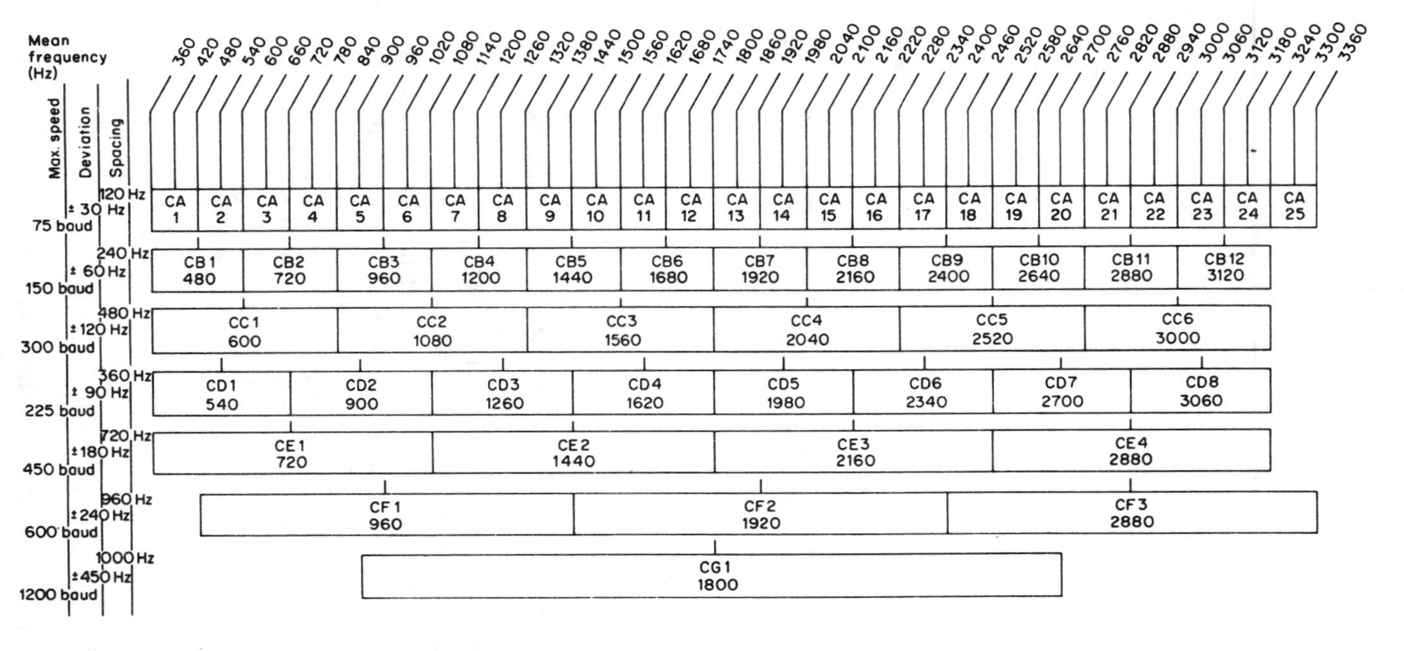

Figure 3.3 CCITT FDM subchannel allocations.

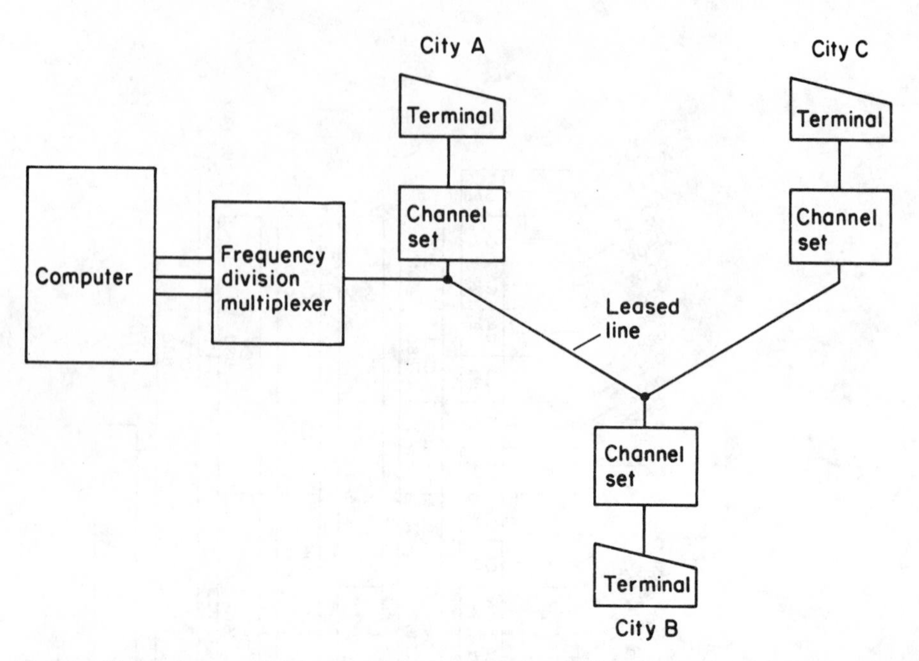

Figure 3.4 Frequency division multiplexing permits multipoint circuit operations. Each terminal on an FDM multipoint circuit is interfaced through the multiplexer to an individual computer port.

channel that is dedicated for utilization by the terminal at each location. This operation is analogous to a group of radio stations transmitting at different frequencies and setting a radio to one frequency so as always to be able to receive the transmission from a particular station.

In contrast to poll and select multipoint line operations where one computer port is used to transmit and receive data from many buffered terminals connected to a common line, FDM used for multipoint operations as shown requires one computer port for each terminal. However, such terminals do not require a buffer area to recognize their addresses nor is poll and select software required to operate in the computer. When buffered terminals and poll and select software are available, polling by channel can take place, as illustrated in Figure 3.5. In this example, channels 1 and 2 are each connected to a number of relatively low-traffic terminals which are polled through the multiplexer system. Terminals 3 and 6 are presumed to be higher traffic stations and are thus connected to individual channels of the FDM or to individual channel sets.

Time division multiplexing

In the FDM technique, the bandwidth of the communications line serves as the frame of reference. The total bandwidth is divided into subchannels consisting of smaller segments of the available bandwidth, each of which is used to form an independent data channel. In the TDM technique, the

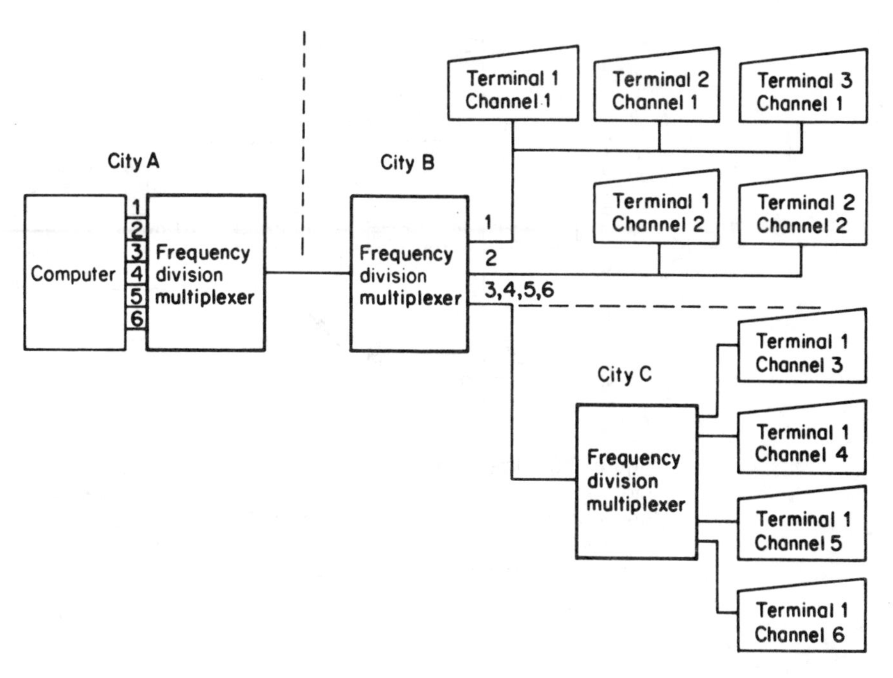

Figure 3.5 FDM can intermix polled and dedicated terminals in a network. Of the 6 channels used in this network, channels 1 and 2 service a number of polled terminals, while channels 3 through 6 are dedicated to service individual terminals.

aggregate capacity of the line is the frame of reference, since the multiplexer provides a method of transmitting data from many terminals over a common circuit by interleaving them in time. The TDM divides the aggregate transmission on the line for use by the slower-speed devices connected to the multiplexer. Each device is given a time slot for its exclusive use so that at any one point in time the signal from one terminal is on the line. In the FDM technique, in which each signal occupies a different frequency band, all signals are being transmitted simultaneously.

The fundamental operating characteristics of a TDM are shown in Figure 3.6. Here, each low- to medium-speed terminal is connected to the multiplexer through an input/output (I/O) channel adapter. The I/O adapter provides the buffering and control functions necessary to interface the transmission and reception of data from the terminals to the multiplexer. Within each adapter, a buffer or memory area exists which is used to compensate for the speed differential between the terminals and the multiplexer's internal operating speed. Data is shifted from the terminal to the I/O adapter at different rates (typically 110 to 9600 bps), depending upon the speed of the terminal; but when data is shifted from the I/O adapter to the central logic of the multiplexer, or from central logic to the composite adapter, it is at the much higher fixed rate of the TDM. On output from the multiplexer to each data terminal the reverse is true, since data is first transferred at a fixed rate from central logic to each adapter and then from the adapter to the terminal at the data rate

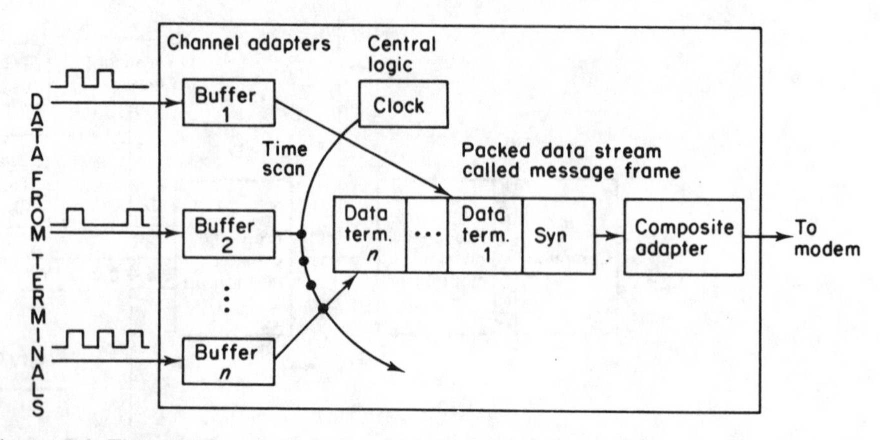

Figure 3.6 Time division multiplexing. In time division multiplexing, data is first entered into each channel adapter buffer area at a transfer rate equal to the device to which the adapter is connected. Next, data from the various buffers are transferred to the multiplexer's central logic at the higher rate of the device for packing into a message frame for transmission.

acceptable to the terminal Depending upon the type of TDM system, the buffer area in each adapter will accommodate either bits or characters.

The central logic of the TDM contains controlling, monitoring, and timing circuitry which facilitates the passage of individual terminal data to and from the high-speed transmission medium. The central logic will generate a synchronizing pattern which is used by a scanner circuit to interrogate each of the channel adapter buffer areas in a predetermined sequence, blocking the bits of characters from each buffer into a continuous, synchronous data stream which is then passed to a composite adapter. The composite adapter contains a buffer and functions similar to the I/O channel adapters. However, it now compensates for the difference in speed between the high-speed transmission medium and the internal speed of the multiplexer.

The multiplexing interval

When operating, the multiplexer transmits and receives a continuous data stream known as a message train, regardless of the activity of the terminals connected to the device. The message train is formed from a continuous series of message frames which represents the packing of a series of input data streams. Each message frame contains one or more synchronization characters followed by a number of basic multiplexing intervals whose number is dependent upon the model and manufacturer of the device. The basic multiplexing interval can be viewed as the first level of time subdivision which is established by determining the number of equal sections per second required by a particular application. Then, the multiplexing interval is the time duration of one section of the message frame.

When TDMs were first introduced the section rate was established at 30 sections per second, which then produced a basic multiplexing interval of 0.033

second or 33 ms. Setting the interval to 33 ms made the multiplexer directly compatible to a 300-baud asynchronous channel which transmits data at up to 30 characters per second (cps). With this interval, the multiplexer was also compatible with 150-baud (15-cps) and 110-baud (10-cps) data channels, since the basic multiplexing interval was a multiple of those asynchronous data rates. Later TDMs had a section rate of 120 sections per second, which then made the multiplexer capable of servicing a range of asynchronous data streams up to 1200 bps.

TDM techniques

The two TDM techniques available are bit interleaving and character interleaving. Bit interleaving is generally used in systems which service synchronous terminals, whereas character interleaving is generally used to service asynchronous terminals. When interleaving is accomplished on a bit-by-bit basis, the multiplexer takes 1 bit from each channel adapter and then combines them as a word or frame for transmission. As shown in Figure 3.7 (top), this technique produces a frame containing one data element from each channel adapter. When interleaving is accomplished on a character-by-character basis, the multiplexer assembles a full character into one frame and then transmits the entire character, as shown in Figure 3.7 (bottom). Although a frame containing only one character of information is illustrated in Figure 3.7, to increase transmission efficiency most multiplexers transmit long frames containing a large number of data characters to reduce the synchronization overhead associated with each frame. Thus, while a frame containing one character of information has a synchronization overhead of 50 percent, a frame containing four data characters has its overhead reduced to 20 percent, and a frame containing nine data characters has a synchronization overhead of only 10 percent, assuming constant slot sizes for all characters.

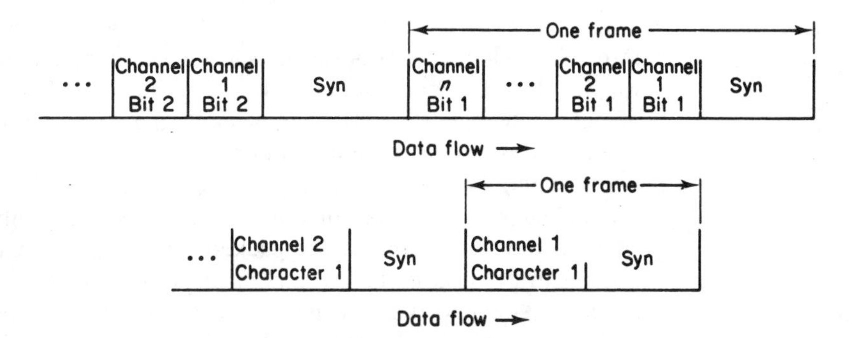

Figure 3.7 Time division interleaving bit-by-bit and character-by-character. When interleaving is accomplished bit-by-bit (top), the first bit from each channel is packed into a frame for transmission. Bottom: Time division multiplexing character-by-character. When interleaving is conducted on a character-by-character basis, one or more complete characters are grouped with a synchronization character into a frame for transmission.

For the character-by-character method, the buffer area required is considerably larger; and although memory costs have declined, TDM character interleaving is still slightly more expensive than TDM bit interleaving systems. Since the character-by-character interleaved method preserves all bits of a character in sequence, the TDM equipment can be used to strip the character of any recognition information that may be sent as part of that character. Examples of this would be the servicing of such terminals as a Teletype Model 33 or an IBM Personal Computer transmitting asynchronous data, where a transmitted character contains 10 or 11 bits which include a start bit, 7 data bits, a parity bit, and 1 or 2 stop bits. When the bit-interleaved method is used, all 10 or 11 bits would be transmitted to preserve character integrity, whereas in a character-interleaved system, the start and stop bits can be stripped from the character, with only the 7 data bits and on some systems the parity bit warranting transmission.

To service terminals with character codes containing different numbers of bits per character, two techniques are commonly employed in character interleaving. In the first technique, the time slot for each character is of constant size, designed to accommodate the maximum bit width or highest level code. Making all slots large enough to carry American Standard Code for Information Interchange (ASCII) characters makes the multiplexer an inefficient carrier of a lower level code such as 5-level Baudot. However, the electronics required in the device and its costs are reduced. The second technique used is to proportion the slot size to the width of each character according to its bit size. This technique maximizes the efficiency of the multiplexer, although the complexity of the logic and the cost of the multiplexer increases. Owing to the reduction in the cost of semiconductors, most character-interleaved multiplexers marketed are designed to operate on the proportional assignment method.

While bit interleaving equipment is less expensive, it is also less efficient when used to service asynchronous terminals. On the positive side, bit-interleaved multiplexers offer the advantage of faster resynchronization and shorter transmission delay, since character-interleaved multiplexers must wait to assemble the bits into characters; whereas, a bit-interleaved multiplexer can transmit each bit as soon as it is received from the terminal. Multiplexers, which interleave by character use a number of different techniques to build the character, with the techniques varying between manufacturers and by models produced by manufacturers.

A commonly utilized technique is the placement of a buffer area for each channel adapter which permits the character to be assembled within the channel adapter and then scanned and packed into a data stream. Another technique which can be used is the placement of programmed read only memory within the multiplexer so that it can be used to assemble characters for all the input channels. The second technique makes a multiplexer resemble a concentrator since the inclusion of programmed read only memory permits many additional functions to be performed in addition to the assembly and disassembly of characters. Such multiplexers with programmed memory are referred to as intelligent multiplexers and are discussed later in this chapter.

TDM applications

The most commonly used TDM configuration is the point-to-point system, which is shown in Figure 3.8. This type of system, which is also called a two-point multiplex system, links a mixture of terminals to a centrally located multiplexer. As shown, the terminals can be connected to the multiplexer in a variety of ways. Terminals can be connected by a leased line running from the terminal's location to the multiplexer, by a direct connection if the user's terminal is within the same building as the multiplexer and a cable can be laid to connect the two, or terminals can use the switched network to call the multiplexer over the dial network. For the latter method, since the connection is not permanent, several terminals can share access to one or more multiplexer channels on a contention basis.

As shown in Figure 3.8, the terminals in cities B and C use the dial network to contend for one multiplexer channel which is interfaced to an automatic answer unit on the dial network. Whichever terminal accesses that channel maintains use of it and thus excludes other terminals from access to that particular connection to the system. As an example, one might have a network which contains 50 terminals within a geographical area wherein between 10 to 12 are active at any time; and one method to deal with this environment would be through the installation of a 12-number rotary interfaced to a 12-channel multiplexer. If all of the terminals were located within one city, the only telephone charges that the user would incur in addition to those of the leased line between multiplexers would be local call charges each time a terminal user dialed the local multiplexer number.

Series multipoint multiplexing

A number of multiplexing systems can be developed by linking the output of one multiplexer into a second multiplexer. Commonly called series multipoint multiplexing, this technique is most effective when terminals are distributed at

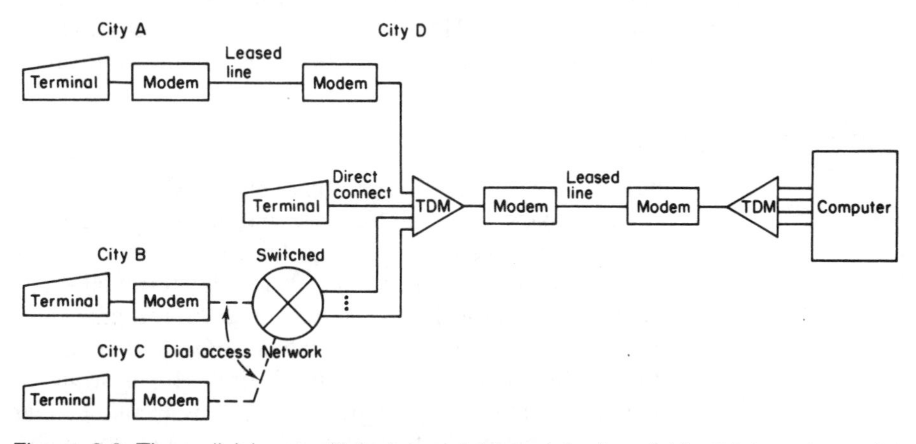

Figure 3.8 Time division multiplexing point-to-point. A point-to-point or two-point multiplexing system links a variety of data users at one or more remoted locations to a central computer facility.

two or more locations and the user desires to alleviate the necessity of obtaining two long-distance leased lines from the closer location to the computer. As shown in Figure 3.9, four low-speed terminals are multiplexed at city A onto one high-speed channel which is transmitted to city B where this line is in turn multiplexed along with the data from a number of other terminals at city B. Although the user requires a leased line between city A and city B, only one line is now required to be installed for the remainder of the distance from city B to the computer at city C. If city A is located 50 miles from city B, and city B is 2000 miles from city C, 2000 miles of duplicate leased lines are avoided by using this multiplexing technique.

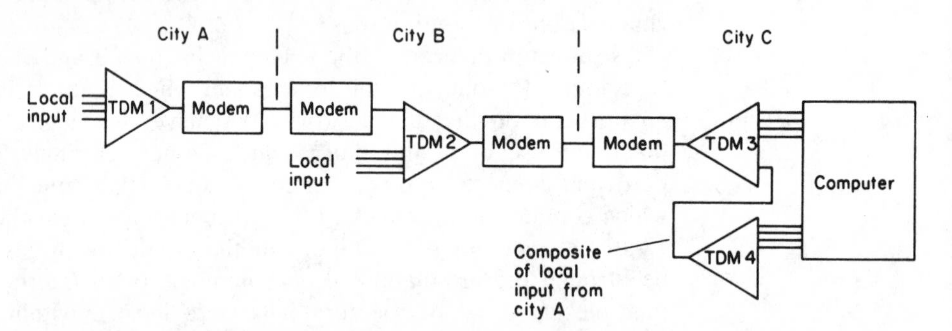

Figure 3.9 Series multipoint multiplexing. Series multipoint multiplexing is accomplished by connecting the output of one multiplexer as input to a second device.

Multipoint multiplexing requires an additional pair of channel cards to be installed at multiplexers 2 and 3 and higher-speed modems to be interfaced to those multiplexers to handle the higher aggregate throughput when the traffic of multiplexer 1 is routed through multiplexer 2; but, in most cases the cost savings associated with reducing duplicated leased lines will more than offset the cost of the extra equipment. Since this is a series arrangement a failure of either TDM2 or TDM3 or a failure of the line between these two multiplexers will terminate service to all terminals connected to the system.

Hub-bypass multiplexing

A variation of series multipoint multiplexing is hub-bypass multiplexing. To be effectively used, hub-bypass multiplexing can occur when a number of remote locations have the requirement to transmit to two or more locations. To satisfy this requirement, the remote terminal traffic is multiplexed to a central location which is the hub, and the terminals which must communicate with the second location are cabled into another multiplexer which transmits this traffic, bypassing the hub. Figure 3.10 illustrates one application where hub bypassing might be utilized. In this example, eight terminals at city 3 require a communications link with one of two computers; six terminals always communicate with the computer at city 2, while two terminals use the facilities of the computer at city 1. The data from all eight terminals are multiplexed over a common line to city 2 where the two channels that correspond to the

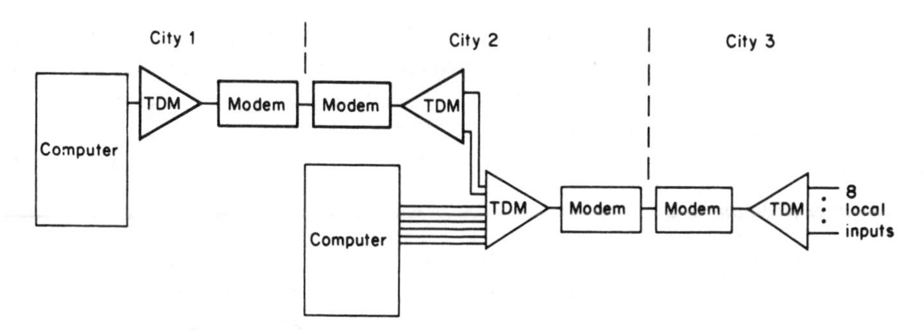

Figure 3.10 Hub-bypass multiplexing. When a number of terminals have the requirement to communicate with more than one location, hub-bypass multiplexing should be considered.

terminals which must access the computer at city 1 are cabled to a new multiplexer, which then remultiplexes the data from those terminals to city 1. When many terminal locations have dual location destinations, hub-bypassing can become very economical. However, since the data flows in series, an equipment failure will terminate access to one or more computational facilities, depending upon the location of the break in service.

Although hub-bypass multiplexing can be effectively used to connect collocated terminals to different destinations, if more than two destinations exist a more efficient switching arrangement can be obtained by the employment of a port selector or a multiplexer that has port selection capability. The reader is referred to Section 3.9 for information concerning port selection.

Front-end substitution

Although not commonly utilized, a TDM may be installed as an inexpensive front end for a computer, as shown in Figure 3.11. When used as a front end, only one computer port is then required to service the terminals which are connected to the computer through the TDM. The TDM can be connected at the computer center, or it can be located at a remote site and connected over a leased line and a pair of modems. Since demultiplexing is conducted by the computer's software, only one multiplexer is necessary.

However, owing to the wide variations in multiplexing techniques of each manufacturer, no standard software has been written for demultiplexing; and, unless multiple locations can use this technique, the software development

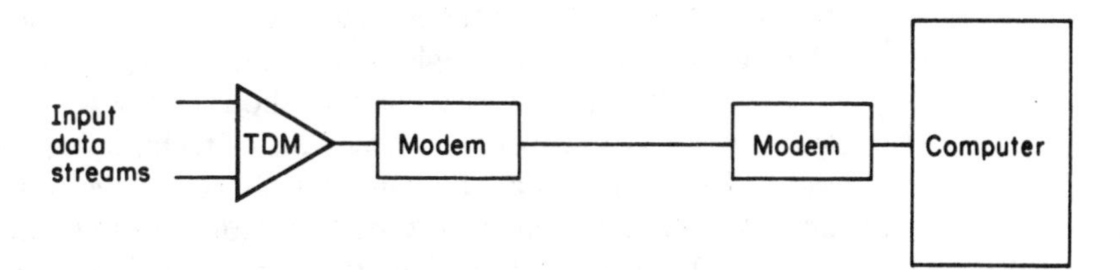

Figure 3.11 TDM system used as a front end. When a TDM is used as a front-end processor, the computer must be programmed to perform demultiplexing.

costs may exceed the hardware savings associated with this technique. In addition, the software overhead associated with the computer performing the demultiplexing may degrade its performance to an appreciable degree and must be considered.

Inverse multiplexing

A multiplexing system which is coming into widespread usage is the inverse multiplexing system. As shown in Figure 3.12, inverse multiplexing permits a high-speed data stream to be split into two or more slower data streams for transmission over lower-cost lines and modems.

Because of the tariff structure associated with wideband facilities, the utilization of inverse multiplexers can result in significant savings in certain situations. As an example, their use could permit 38,400 bps transmission over two voice-grade lines at a fraction of the cost which would be incurred when using wideband facilities. The reader should refer to Section 3.3 for additional information on these devices.

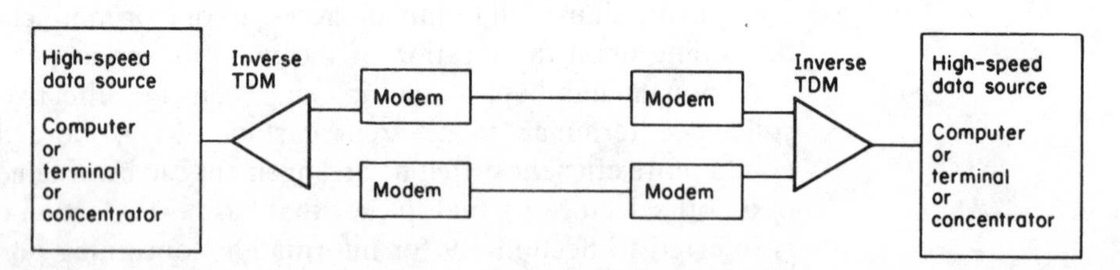

Figure 3.12 Inverse multiplexing. An inverse multiplexer splits a serial data stream into two or more individual data streams for transmission at lower data rates.

Multiplexing economies

The primary motive for the use of multiplexers in a network is to reduce the cost of communications. In analyzing the potential of multiplexers, one should first survey terminal users to determine the projected monthly connect time of each terminal. Then, the most economical method of data transmission from each individual terminal to the computer facility can be computed. To do this, direct dial costs should be compared with the cost of a leased line from each terminal to the computer site.

Once the most economical method of transmission for each individual terminal to the computer is determined, this cost should be considered the "cost to reduce". The telephone mileage costs from each terminal city location to each other terminal city location should be determined in order to compute and compare the cost of utilizing various techniques, such as line dropping and the multiplexing of data by combining several low- to medium-speed terminals' data streams into one high-speed line for transmission to the central site.

In evaluating multiplexing costs, the cost of telephone lines from each terminal location to the "multiplexer center" must be computed and added to the cost of the multiplexer equipment. Then, the cost of the high-speed line from the multiplexer center to the computer site must be added to produce the total multiplexing cost. If this cost exceeds the cumulative most economical method of transmission for individual terminals to the central site, then multiplexing is not cost-justified. This process should be reiterated by considering each city as a possible multiplexer center to optimize all possible network configurations. In repeating this process, terminals located in certain cities will not justify any calculations to prove or disprove their economic feasibility as multiplexer centers, because of their isolation from other cities in a network.

An example of the economics involved in multiplexing is illustrated in Figure 3.13. In this example, assume the volume of terminal traffic from the devices located in cities A and B would result in a dial-up charge of $3000 per month if access to the computer in city G was over the switched network. The installation of leased lines from those cities to the computer at city G would cost $2000 and $2200 per month, respectively. Furthermore, let us assume that the terminals at cities C, D, and E only periodically communicate with the computer, and their dial-up costs of $400, $600, and $500 per month,

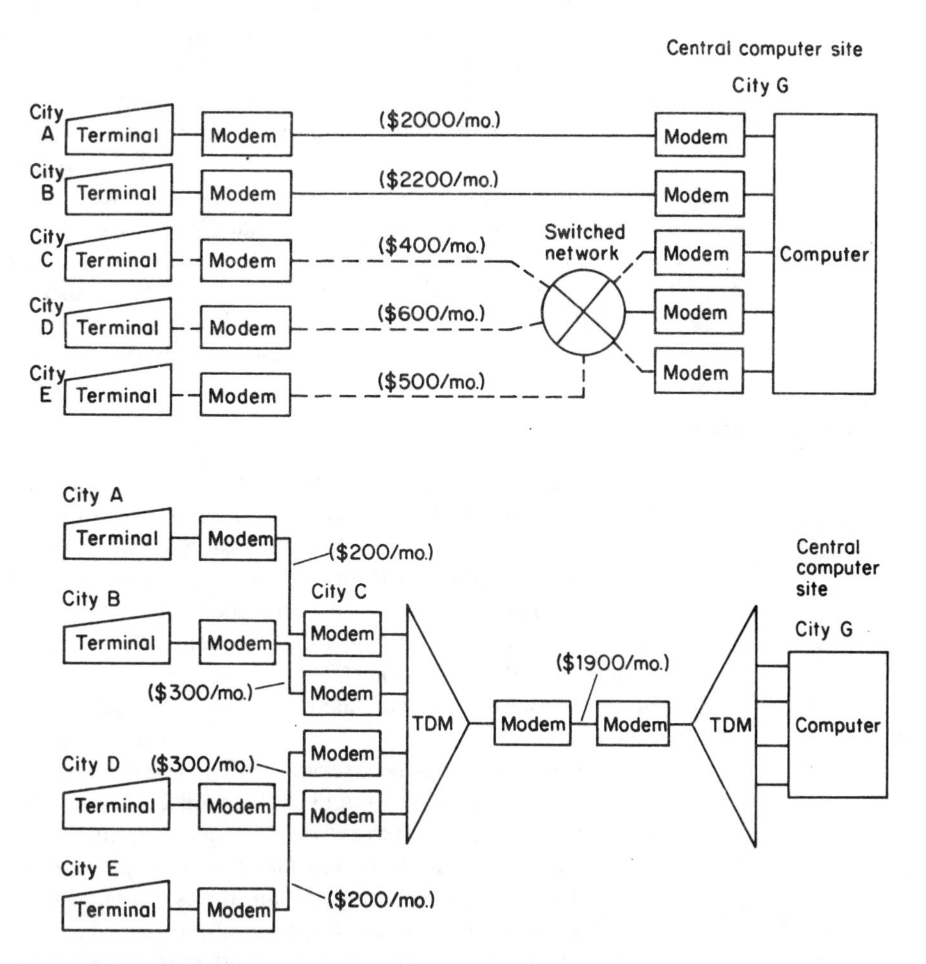

Figure 3.13 Multiplexing economics. On an individual basis, the cost of five terminals accessing a computer system (top) can be much more expensive than when a time division multiplexer is installed (bottom).

respectively, are much less than the cost of leased lines between those cities and the computer. Then, without multiplexing, the network's most economical communications cost would be:

Location	Cost per month
city A	$2000
city B	$2200
city C	400
city D	600
city E	500
Total cost	$5700

Let us further assume that city C is centrally located with respect to the other cities so we could use it as a homing point or multiplexer center. In this manner, a multiplexer could be installed in city C, and the terminal traffic from the other cities could be routed to that city, as shown in the bottom portion of Figure 3.13. Employing multiplexers would reduce the network communications cost to $2900 per month which produces a potential savings of $2800 per month, which should now be reduced by the multiplexer costs to determine net savings. If each multiplexer costs $500 per month, then the network using multiplexers will save the user $1800 each month. Exactly how much saving can be realized, if any, through the use of multiplexers depends not only on the types, quantities, and distributions of terminals to be serviced but also on the leased line tariff structure and the type of multiplexer employed.

Mixing multiplexers

While FDM equipment is limited by the telephone lines' 3-kHz bandwidth, the main limitation on a TDM system is the transmission capability of the high-speed modem attached to the multiplexer. FDM service, as an example, is usually limited to 16 100-bps or 8 150-bps channels, while TDM systems can service a mixture of low- and high-speed terminals whose composite speed is less than or equal to the attached modem's rated speed. Thus, a single TDM system could service 64 150-bps terminals when interfaced to a 9600-bps modem, whereas eight 8-channel FDM systems might be necessary to provide equivalent service.

Although FDM systems service only low-speed terminals, TDM systems can service a mixture of low- and high-speed lines providing the user with more flexibility in both network design and terminal selection. As mentioned previously, while TDM systems favor point-to-point applications, FDM systems are well suited for multidrop configurations where a number of widely separated terminals can be serviced most economically over a single multipoint line. Although terminal quantities, locations, and transmission rates will frequently dictate which type of system to use, a mixture of systems should often be considered to provide an optimum solution to network problems. While FDM equipment is often regarded as obsolete technology because of its limited data speed capacity, its inherent capability of providing a multipoint line connection without requiring poll and select software or addressable terminals should keep this equipment in limited use through the 1980s.

So, in many networks, TDM systems will be used for transmission at high data rates between two widely separated locations, while FDM will often be used in the same system to provide multidrop servicing to terminals, which because of transmission rates and locations are best serviced by an FDM system. Such a combined system is shown in Figure 3.14. In developing this system, the terminal requirements were first examined and denoted as follows:

City	Terminal quantity	Terminal speed	Aggregate terminal speed
A	8	300	2400
A	2	1200	2400
B	2	1200	2400
C	1	300	300
D	1	300	300
E	1	300	300
F	1	300	300

In examining the terminal requirements of city A, the total aggregate throughput of the eight 300-bps terminals and the two 1200-bps terminals becomes 4800 bps. Since this exceeds the typical capacity of FDM systems, city A becomes a candidate for TDM. Although the two terminals at city B may lie within the upper range of servicing by an FDM system, further examination of the terminals located in cities C, D, E, and F makes them ideal candidates for FDM. If an FDM system is installed to service those cities, based upon geographical distances, we may then desire to install a TDM at city B which we can use to service the two terminals at that city, as well as servicing the output of the TDM in city A and the FDM system.

Thus, instead of three long-distance lines, city B can act as a homing point for all the multiplexers in the system. Since the aggregate throughput of TDM3 will be 8400 bps, an extra channel card(s), depending upon manufacturer, may be required to make the system run at a standard 9600-bps data rate. Although this channel will not be utilized by any terminal, the extra 1200-bps capacity is available for servicing additional terminals at a later date. This extra channel card can be one 1200-bps card (as shown in Figure 3.14) or a number of channel cards whose total capacity adds up to 1200 bps.

At city G only two TDMs are necessary. Since the output of the FDM system is four 300-bps lines which are input to TDM3, TDM4 now demultiplexes this data into four 300-bps channels which are then interfaced to the computer. TDM4, in addition, separates a 4800-bps channel from the 9600-bps composite speed, and this channel is further demultiplexed by TDM2 into its original 10 channels which were provided by TDM1. In addition, TDM4 separates the

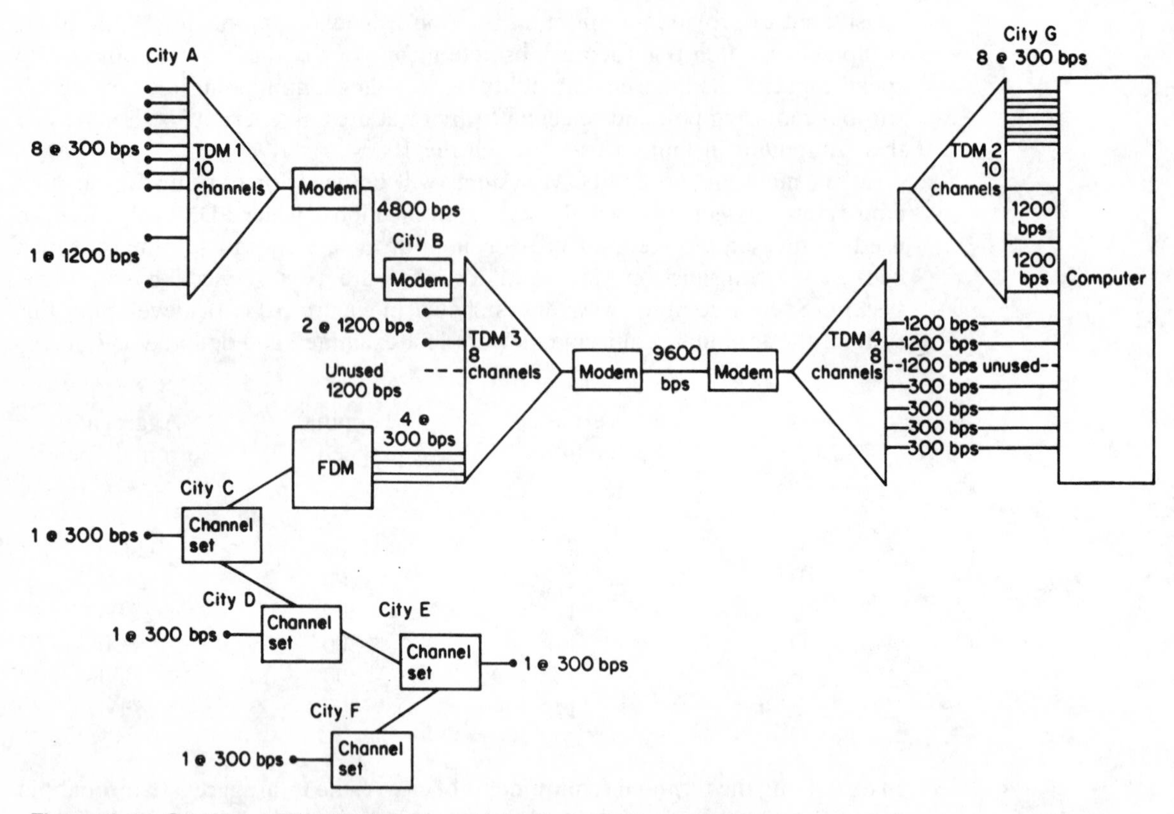

Figure 3.14 Combined FDM-TDM system. Using both FDMs and TDMs in a network permits the capabilities of both devices to be used more advantageously.

two 1200-bps channels which were multiplexed by TDM3. Thus, the 10 data channels of TDM1 and the four channels of the FDM system are serviced by the 8-channel TDM3 (which includes one 1200-bps unused channel for timing). At city G, TDM4 contains eight channels (again, one 1200-bps channel is unused and provides timing) of which the 4800-bps channel is further demultiplexed by the 10-channel TDM2 system. Owing to the similarity in channels, TDM4 can be considered the mirror image of TDM3, and TDM2 would be the mirror image of TDM1. With the advent of new families of multiplexers produced by many vendors, only one multiplexer may be required at city G to demultiplex data from all remote sites.

Statistical and intelligent multiplexers

In a traditional TDM, data streams are combined from a number of devices into a single path so that each device has a time slot assigned for its use. While such TDMs are inexpensive and reliable, and can be effectively employed to reduce communications costs, they make inefficient use of the high-speed transmission medium. This inefficiency is due to the fact that a time slot is reserved for each connected device, whether or not the device is active. When the device is inactive, the TDM pads the slot with nulls and cannot use the slot for other purposes.

These pad characters are inserted into the message frame since demultiplexing occurs by the position of characters in the frame. Thus, if these pads are eliminated, a scheme must then be employed to indicate the origination port or channel of each character. Otherwise, there would be no way to correctly reconstruct the data and route it to its correct computer port during the demultiplexing process.

A statistical multiplexer is in many respects very similar to a concentrator since both devices combine signals from a number of connected devices in such a manner that there is a certain probability that a device will have access to the use of a time slot for transmission. Whereas a concentrator may require user programming and always requires special software in the host computer to demultiplex its high-speed data stream, statistical multiplexers are built around a microprocessor that is programmed by the vendor, and no host software is required for demultiplexing since another statistical multiplexer at the computer site performs that function.

By dynamically allocating time slots as required, statistical multiplexers permit more efficient utilization of the high-speed transmission medium. This permits the multiplexer to service more terminals without an increase in the high-speed link as would a traditional multiplexer. The technique of allocating time slots on a demand basis is known as statistical multiplexing and means that data is transmitted by the multiplexer only from the terminals that are actually active.

Depending upon the type of TDM, either synchronization characters or control frames are inserted into the stream of message frames. Synchronization characters are employed by conventional TDMs, while control frames are used by TDMs which employ a high-level data link control (HDLC) protocol between multiplexers to control the transmission of message frames.

The construction technique used to build the message frame also defines the type of TDM. Conventional TDMs employ a fixed frame approach as illustrated in Figure 3.15. Here, each frame consists of one character or bit for each input channel scanned at a particular period of time. As illustrated, even when a particular terminal is inactive, the slot assigned to that device is included in the message frame transmitted since the presence of a pad or null character in the time slot is required to correctly demultiplex the data. In the lower portion of Figure 3.15 the demultiplexing process which is accomplished by time slot position is illustrated. Since a typical terminal may be idle 90 percent of the time, this technique contains obvious inefficiencies.

Statistical frame construction

A statistical multiplexer employs a variable frame building technique which takes advantage of terminal idle times to enable more terminals to share access to a common circuit. The use of variable frame technology permits previously wasted time slots to be eliminated, since control information is transmitted with each frame to indicate which terminals are active and have data contained in the message frame.

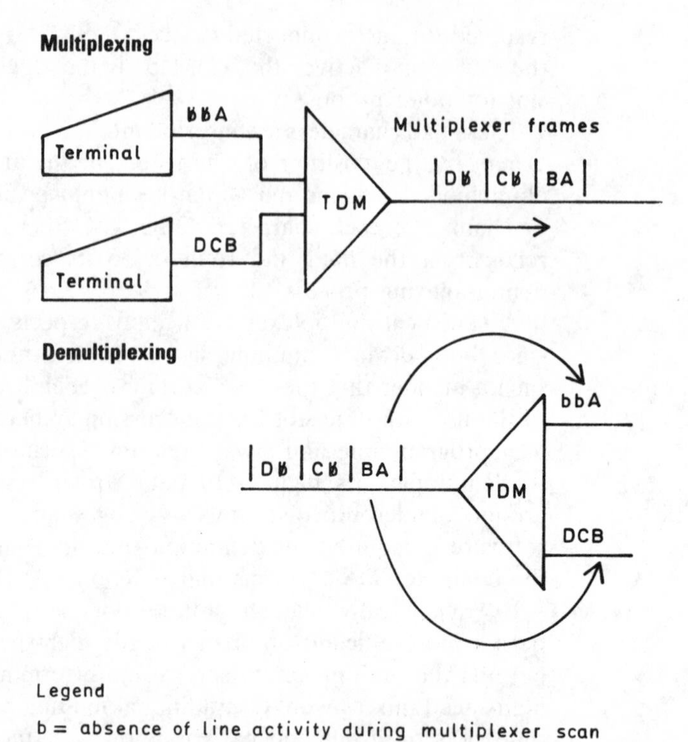

Figure 3.15 Multiplexing and demultiplexing by TDMs.

One of many techniques that can be used to denote the presence or absence of data traffic is the activity map which is illustrated in Figure 3.16. When an activity map is employed, the map itself is transmitted before the actual data. Each bit position in the map is used to indicate the presence or absence of data from a particular multiplexer time slot scan. The two activity maps and data characters illustrated in Figure 3.16 represent a total of 8 characters which would be transmitted in place of 16 characters that would be transmitted by a conventional multiplexer.

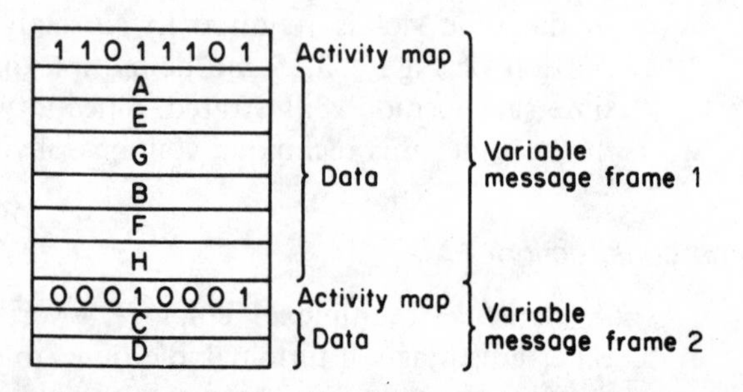

Figure 3.16 Activity mapping to produce variable frames. Using an activity map where each bit position indicates the presence or absence of data for a particular data source permits variable message frames to be generated.

Another statistical multiplexing technique involves buffering data from each data source and then transmitting the data with an address and byte count. The address is used by the demultiplexer to route the data to the correct port, while the byte count indicates the quantity of data to be routed to that port. Figure 3.17 illustrates the message frame of a 4-channel statistical multiplexer employing the address and byte count frame composition method during a certain time interval. Note that since channels 3 and 4 had no data traffic during the two particular time intervals, there was no address and byte count nor data from those channels transmitted on the common circuit. Also note that the data from each channel is of variable length. Typically, statistical multiplexers employing an address and byte count frame composition method wait until either 32 characters or a carriage return is encountered prior to forming the address and byte count and forwarding the buffered data. The reason 32 characters was selected as the decision criterion is that it represents the average line length of an interactive transmission session.

A few potential technical drawbacks of statistical multiplexers exist which users should note. These problems include the delays associated with data blocking and queuing when a large number of connected terminals become active or when a few terminals transmit large bursts of data. For either situation, the aggregate throughput of the multiplexer's input active data exceeds the capacity of the common high-speed line, causing data to be placed into buffer storage.

Another reason for delays is when a circuit error causes one or more retransmissions of message frame data to occur. Since the connected terminals may continue to send data during the multiplexer-to-multiplexer retransmission cycle, this can also fill up the multiplexer's buffer areas and cause time delays.

If the buffer area should overflow, data would be lost which would create an unacceptable situation. To prevent buffer overflow, all statistical multiplexers employ some type of technique to transmit a traffic control signal to attached terminals and/or computers when their buffers are filled to a certain level. Such control signals inhibit additional transmission through the multiplexer until the buffer has been emptied to another predefined level. Once this level has been reached, a second control signal is issued which permits transmission to the multiplexers to resume.

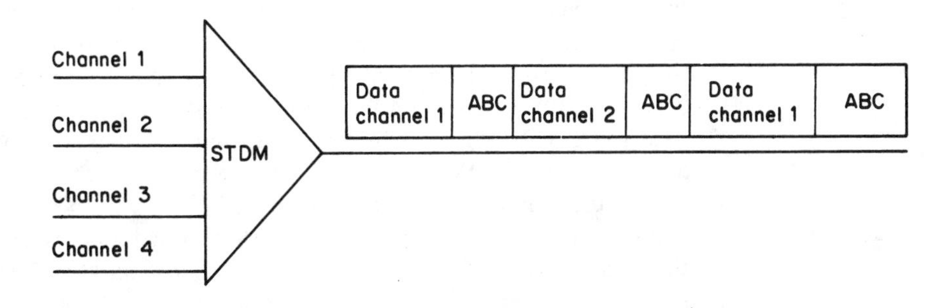

Figure 3.17 Address and byte count frame composition.

Buffer control

The three major buffer control techniques employed by statistical multiplexers include inband signaling, outband signaling, and clock reduction. Inband signaling involves transmitting XOFF and XON characters to inhibit and enable the transmission of data from terminals and computer ports that recognize these flow control characters. Since many terminals and computer ports do not recognize these control characters, a second common flow control method involves raising and lowering the Clear to Send (CTS) control signal on the RS-232 or CCITT V.24 interface. Since this method of buffer control is outside the data path where data is transmitted on pin 2, it is known as outband signaling.

Both inband and outband signaling are used to control the data flow of asynchronous devices. Since synchronous devices transmit data formed into blocks or frames, one would most likely break a block or frame by using either inband or outband signaling. This would cause a portion of a block or frame to be received, which would result in a negative acknowledgement when the receiver performs its cyclic redundancy computation. Similarly, when the remainder of the block or frame is allowed to resume its flow to the receiver, a second negative acknowledgement would result.

To alleviate these potential causes of decrease of throughput, multiplexer vendors typically reduce the clocking speed furnished to synchronous devices. Thus, a synchronous terminal operating at 4800 bps might first be reduced to 2400 bps by the multiplexer halving the clock. Then, if the buffer in the multiplexer continues to fill, the clock might be further reduced to 1200 bps.

Service ratio

The measurement used to denote the capability of a statistical multiplexer is called its service ratio, which compares its overall level of performance in comparison to a conventional TDM. Since synchronous transmission by definition denotes blocks of data with characters placed in sequence in each block, there are no gaps in this mode of transmission. In comparison, a terminal operator transmitting data asynchronously may pause between characters to think prior to pressing each key on the terminal. Thus, the service ratio of STDMs for asynchronous data is higher than the service ratio for synchronous data. Typically, STDM asynchronous service ratios range between 2 : 1 and 3.5 : 1, while synchronous service ratios range between 1.25 : 1 and 2 : 1, with the service ratio dependent upon the efficiency of the STDM as well as its built-in features to include the stripping of start and stop bits from asynchronous data sources. In Figure 3.18, the operational efficiency of both a statistical and a conventional TDM are compared. Here we have assumed that the STDM has an efficiency of twice that of the TDM.

Assuming four 1200-bps and ten 300-bps data sources are to be multiplexed, the conventional TDM illustrated in the top part of Figure 3.18 would be required to operate at a data rate of at least 7800 bps, thus requiring a 9600-bps modem if transmission was over an analog facility. For the STDM shown in the lower portion of this illustration, assuming a two-fold increase in

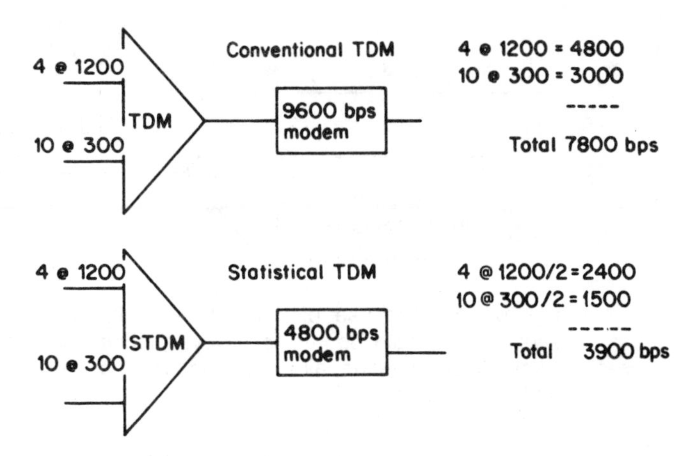

Figure 3.18 Comparing statistical and conventional TDMs. An STDM typically has an efficiency of two to four times a conventional TDM. Using an efficiency level twice the conventional TDM results in a composite operating data rate requirement of 3900-bps which is serviced by the use of a 4800-bps modem.

efficiency over the conventional TDM, the composite data rate required will be 3900 bps. This permits the employment of a lower operating rate modem which is also a lower-cost modem. In addition, the STDM can accept a combined average input data rate increase of 1800 bps from additional sources prior to requiring the 4800-bps modem to be upgraded. If an input data rate increase of over 1800 bps were directed to the conventional TDM, a significantly more expensive wideband line would probably be required to replace the voice-grade facility whose data rate in many locations is limited to 9600 bps.

Data source support

Some statistical multiplexers only support asynchronous data while other multiplexers support both asynchronous and synchronous data sources. When a statistical multiplexer supports synchronous data sources it is extremely important to determine the method used by the STDM vendor to implement this support.

Some statistical multiplexer vendors employ a band pass channel to support synchronous data sources. When this occurs, not only is the synchronous data not multiplexed statistically, but the data rate of the synchronous input limits the overall capability of the device to support asynchronous transmission. Figure 3.19 illustrates the effect of multiplexing synchronous data via the use of a band pass channel. When a band pass channel is employed, a fixed portion of each message frame is reserved for the exclusive multiplexing of synchronous data, with the portion of the frame reserved proportional to the data rate of the synchronous input to the STDM. This means that only the remainder of the message frame is then available for the multiplexing of all other data sources.

As an example of the limitations of band pass multiplexing, consider an STDM that is connected to a 9600-bps modem and supports a synchronous

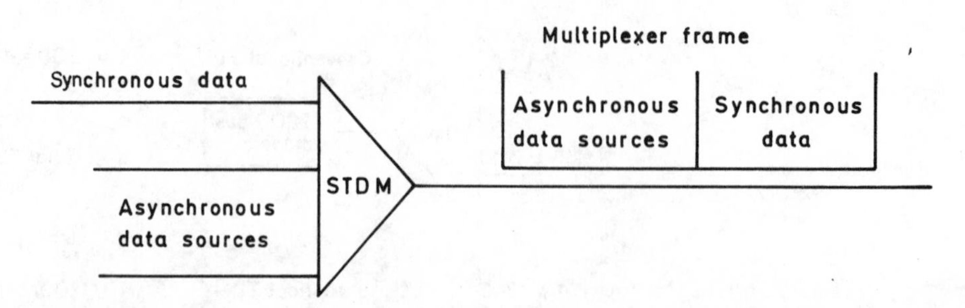

Figure 3.19 The use of a band pass channel to multiple synchronous data. The synchronous data source is always placed into a fixed location on the multiplexer frame. In comparison, all asynchronous data sources contend for the remainder of the multiplexer frame.

terminal operating at 7200 bps. If band pass multiplexing is employed, only 2400 bps is then available in the multiplexer for the multiplexing of other data sources. In comparison, assume another STDM statistically multiplexes synchronous data. If this STDM has a service ratio of 1.5 : 1, then a 7200-bps synchronous input to the STDM would on the average take up 4800 bps of the 9600-bps operating line. Since the synchronous data is statistically multiplexed, when that data source is not active other data sources serviced by the STDM will flow through the system more efficiently. In comparison, the band pass channel always requires a predefined portion of the high-speed line to be reserved for synchronous data, regardless of the activity of the data source.

Switching and port contention

Two features normally available with more sophisticated statistical multiplexers are switching and port contention. Switching capability is also referred to as alternate routing and requires the multiplexer to support multiple high-speed lines whose connection to the multiplexer is known as a node. Thus, switching capability normally refers to the ability of the multiplexer to support multiple nodes. Figure 3.20 illustrates how alternate routing can be used to compensate for a circuit outage. In the example shown, if the line connecting locations 1 and 3 should become inoperative an alternate route through location 2 could be established if the STDMs support data switching.

Port contention is normally incorporated into large capacity multinodal statistical multiplexers that are designed for installation at a central computer facility. This type of STDM may demultiplex data from hundreds of data channels, however, since many data channels are usually inactive at a given point in time, it is a waste of resources to provide a port at the central site for each data channel on the remote multiplexers. Thus, port contention results in the STDM at the central site containing a lesser number of ports than the number of channels of the distant multiplexers connected to that device. Then, the STDM at the central site contends the data sources entered through remote multiplexer channels to the available ports on a demand basis. If no ports are available, the STDM may issue a "NO PORTS AVAILABLE" message and

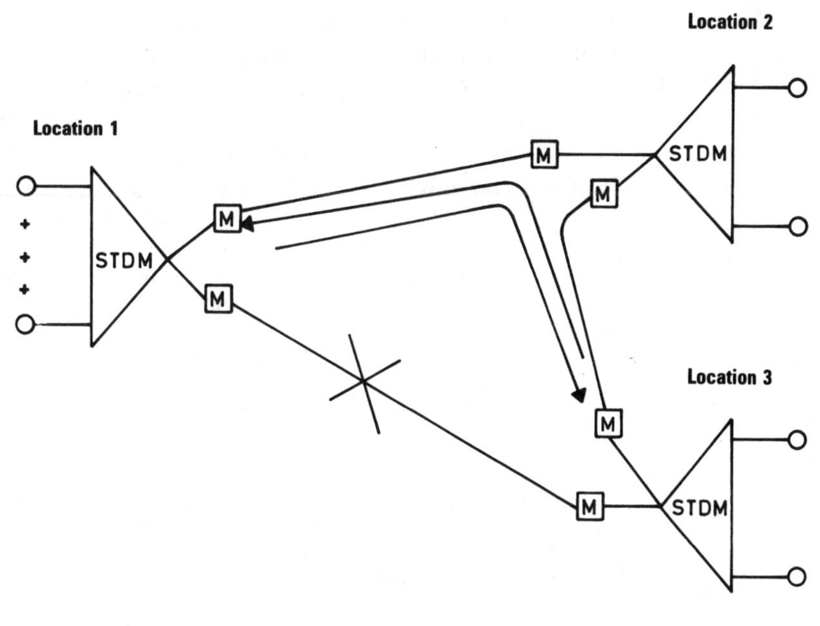

Figure 3.20 Switching permits load balancing and alternate routing if a high-speed line should become inoperative.

disconnect the user or places the user into a queue until a port becomes available.

ITDMs

One advancement in statistical multiplexer technology resulted in the introduction of data compression into a few STDMs. Such devices intelligently examine data for certain characteristics and are known as intelligent time division multiplexers (ITDM). These devices take advantage of the fact that different characters occur with different frequencies and use this quality to reduce the average number of bits per character by assigning short codes to frequently occurring characters and long codes to seldom-encountered characters.

The primary advantage of the intelligent multiplexer lies in its ability to make the most efficient use of a high-speed data circuit in comparison to the other classes of TDMs. Through compression, synchronous data traffic which normally contains minimal idle times during active transmission periods can be boosted in efficiency. Intelligent multiplexers typically permit an efficiency four times that of conventional TDMs for asynchronous data traffic and twice that of conventional TDMs for synchronous terminals.

STDM/ITDM statistics

Although the use of statistical and intelligent multiplexers can be considered on a purely economic basis to determine if the cost of such devices is offset by the reduction in line and modem costs, the statistics that are computed and made available to the user of such devices should also be considered. Although many times intangible, these statistics may warrant consideration even though an economic benefit may at first be hard to visualize. Some of the statistics normally available on statistical and intelligent multiplexers are listed in Table 3.3 Through a careful monitoring of these statistics, network expansion can be preplanned to cause a minimum amount of potential busy conditions to users. In addition, frequent error conditions can be noted prior to user complaints and remedial action taken earlier than normal when conventional devices are used.

Table 3.3 Intelligent multiplexer statistics.

Multiplexer loading: % of time device not idle
Buffer utilization: % of buffer storage in use
Number of frames transmitted
Number of bits of idle code transmitted
Number of negative acknowledgements received

$$\text{Traffic density} = \frac{\text{non-idle bits}}{\text{total bits}}$$

$$\text{Error density} = \frac{\text{NAKs received}}{\text{frames transmitted}}$$

$$\text{Compression efficiency} = \frac{\text{total bits received}}{\text{total bits compressed}}$$

$$\text{Statistical loading} = \frac{\text{number of actual characters received}}{\text{maximum number which could be received}}$$

$$\text{Character error rate} = \frac{\text{characters with bad parity}}{\text{total characters received}}$$

Features to consider

In Table 3.4, the reader will find a list of the primary selection features one should consider when evaluating statistical multiplexers. Although many of these features were previously discussed, a few features were purposely omitted from consideration until now. These features include auto baud detect, flyback delay and echoplex, and primarily govern the type of terminal devices that can be efficiently supported by the statistical multiplexer.

Auto baud detect is the ability of a multiplexer to measure the pulse width of a data source. Since the data rate is proportional to the pulse width, this feature enables the multiplexer to recognize and adjust to different speed terminals accessing the device over the switched telephone network.

On electromechanical printers, a delay time is required between sending a carriage return to the terminal and then sending the first character of the next line to be printed. This delay time enables the print head of the terminal to

Table 3.4 Statistical multiplexer selection features.

Feature	Parameters to consider
Auto baud detect	Data rates detected
Flyback delay	Settings available
Echoplex	Selectable by channel or device
Protocols supported	2780/3780, 3270, HDLC/SDLC, other
Data type supported	Asynchronous, synchronous
Service ratios	Asynchronous, synchronous
Flow control	XON-XOFF, CTS, clocking
Multinodal capability	Number of nodes
Switching	Automatic or manual
Port contention	Disconnect or queued when all ports in use
Data compression	Stripping bits or employs compression algorithm

be repositioned prior to the first character of the next line being printed. Many statistical multiplexers can be set to generate a series of fill characters after detecting a carriage return, enabling the print head of an electromechanical terminal to return to its proper position prior to receiving a character to be printed. This feature is called flyback delay and can be enabled or disabled by channel on many multiplexers.

Since some networks contain full-duplex computer systems that echo each character back to the originating terminal the delay from twice traversing through statistical multiplexers may result in the terminal operator obtaining the feeling that his or her terminal is non-responsive. When echoplexing is supported by an STDM, the multiplexer connected to the terminal immediately echoes each character to the terminal, while the multiplexer connected to the computer discards characters echoed by the computer. This enables data flow through the multiplexer system to be more responsive to the terminal operator. Since error detection and correction is built into all statistical multiplexers, a character echo from the computer is not necessary to provide visual transmission validation and is safely eliminated by echoplexing.

The other options listed in Table 3.4 should be self-explanatory, and the user should check vendor literature for specific options available for use on different devices.

Utilization considerations

Although one's precise network requirements govern the type of multiplexer that will result in the best price-performance, several general comparisons can be made between devices. In comparison to FDM, the principal advantages of TDM include the ability to service high input data sources, the capacity for a greater number of individual inputs, the performance of data compression (intelligent multiplexers), the detection of errors, and the request for retransmission of data (statistical and intelligent multiplexers). The key differences between multiplexers are tabulated in Table 3.5.

Table 3.5 Multiplexer comparisons.

	FDM	TDM	STDM	ITDM
Efficiency	poor	good	better	best
Channel capacity	poor	good	better	best
High-speed data	very poor	poor	better	best
Configuration change				
Data rate	good	fair	good	good
Number of channels	poor	good	better	better
Installation ease	poor	poor	good	good
Problem isolation	poor	poor	good	good
Error detection/ retransmission	n/a	n/a-good	automatic	automatic
Multidrop capability	good	n/a	possible	possible

3.2 GROUP-BAND AND T1 MULTIPLEXERS

A group-band multiplexer is a TDM which has been specifically designed to permit channel usage of group-band (wideband) data circuits. Although the capabilities and capacities of group-band multiplexers vary by manufacturer, basically they provide the user with the ability to intermix different speed data streams in order to obtain a composite synchronous data stream that is suitable for transmission over group-band (wideband) facilities. Since one of the most commonly used circuits employed to transport the composite output of a type of group-band multiplexer is the T1 carrier; thus, another name for a device similar to a group-band multiplexer is the T1 multiplexer.

Multiplexer differences

The major differences between a group-band and a T1 multiplexer are in the areas of line interface, input speeds supported, voice/data capability, the use of synchronization channels, and the method of subdivision of the composite output channel.

Group-band multiplexers were developed for use on group-band (wideband) analog facilities; therefore, they were designed to operate with wideband or group-band modems. In comparison, T1 multiplexers are connected to T1 digital facilities via a channel service unit, although they can also be interfaced to analog facilities. As group-band multiplexers predate the more modern T1 multiplexers, they are usually fixed logic devices while T1 multiplexers are microprocessor-based devices. Thus, the group-band multiplexer is usually limited to accepting only synchronous inputs while the T1 multiplexer normally can accept both asynchronous and synchronous data input sources.

Owing to the tariff structure of T1 facilities, it is most advantageous from an economic perspective to transmit multiplexed voice and data. Thus, most T1 multiplexer vendors offer plug-in voice modules or channel cards that digitize voice so it can be multiplexed. In comparison, group-band multiplexers normally do not offer this capability.

Another difference between the two devices is the requirement of the group-band multiplexer to use one of its channels as a synchronization channel, while a T1 multiplexer does not require this type of channel. Finally, since a T1 multiplexer normally accepts both asynchronous and synchronous data inputs, the method of subdivision of the composite channel capacity is usually slightly different from a group-band multiplexer.

Since a T1 multiplexer can normally perform all the functions of a group-band multiplexer, we will first examine the latter device. Then, using the operational characteristics of the group-band multiplexer for comparison purposes, we will discuss the operation of T1 multiplexers.

Group-band multiplex operation

Typical group-band multiplexer servicing speeds can range from 2400 bps through 56 kbps, with such intermediate speeds as 4800, 7200, 9600, and 19,200 bps being acceptable to most devices. The composite transmission speed developed by the group-band multiplexer for transmission over wideband facilities can range from 19.2 kbps through 1.544 Mbps, with intermediate data rates of 38.4, 40.8, 50.0, 56.0, 64.0, and 230.4 kbps commonly used.

Similar in operation to a traditional TDM developed to service synchronous data streams, a group-band multiplexer uses a bit interleaving process to combine two or more channels of synchronous digital data onto a single synchronous wideband facility. Although techniques vary by manufacturer, one or more channels are usually utilized as a synchronization channel within the group-band multiplexer. Owing to the utilization of channels for synchronization, a portion of the high-speed data rate in effect becomes reserved for system overhead. While this overhead is not significant when compared to the total data transmission rate of the system, it must be taken into consideration when configuring the channel utilization of a group-band multiplexer.

Thus, while a traditional TDM would develop a 9600-bps output data rate when it multiplexes two 4800-bps input data streams, a group-band multiplexer used to service 31 2400-bps data streams would produce a composite speed in excess of 74,200 bps because of the synchronization overhead. This is shown in Figure 3.21, where a group-band multiplexer is used to combine 31 2400-bps data streams. In this example, one additional channel is used only for synchronization, which makes the composite speed rise to 76.8 kbps although the usable data transmission capacity of the circuit is 74.2 kbps.

Since synchronization techniques vary, manufacturers' literature should be consulted to determine the number of synchronization channels which must be used at different data rates. In addition, the user should check vendor literature to determine the number and speed of the synchronous data channels that are supported. Some devices have a program pin on each channel module so that the user can select the needed speed, as long as it is some multiple of 2400 bps, up to and including 19,200 bps. On other devices a change in channel rates may require a return of the equipment to the factory or service by field engineers.

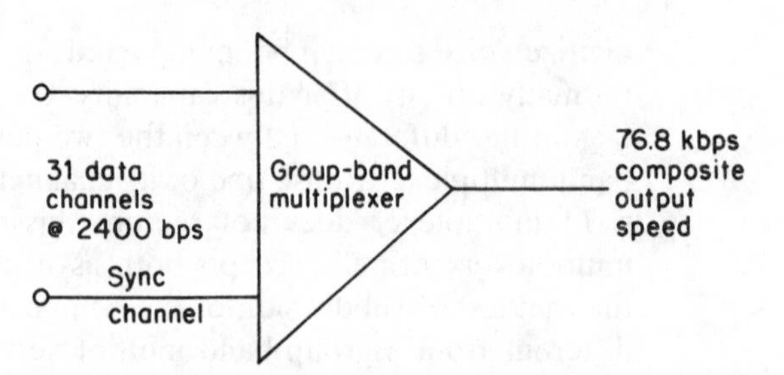

Figure 3.21 Group-band multiplexer overhead. An allowance for one or more synchronization channels must be considered when group-band multiplexers are employed.

Applications

In addition to providing a link between two computer centers where a requirement may exist for multiple computer-to-computer communications, group-band multiplexers are adaptable for a number of communication network applications. An example of their application would be in a communications network where one or more regional areas are distant from the computer center so that the charges for leased lines connecting individual terminals and multiplexers to the computer center cumulatively exceed the cost of providing wide-band service between a central point in that region and the computer center. An example of this type of situation is shown in Figure 3.22.

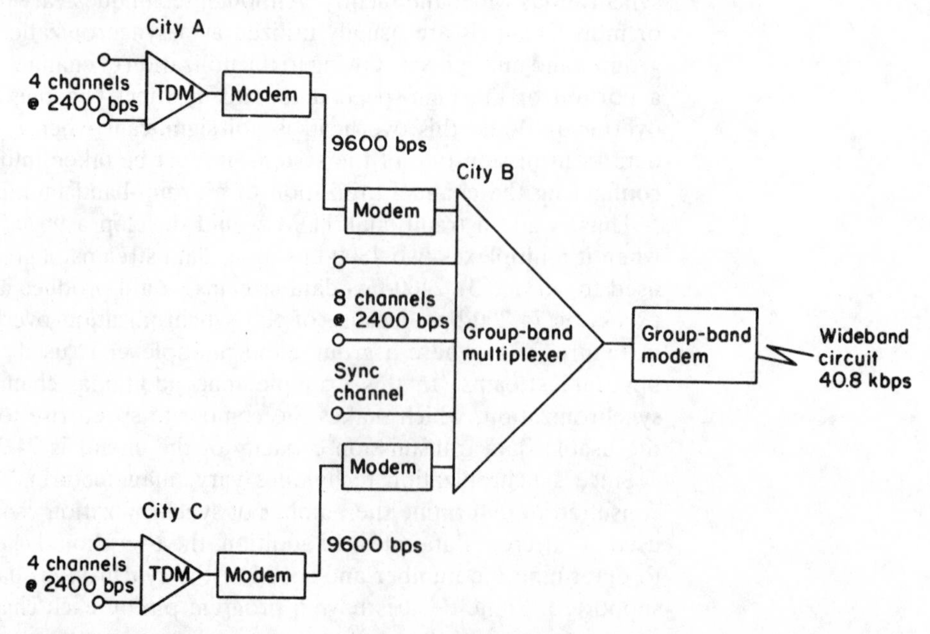

Figure 3.22 Group-band multiplexer servicing regional area. The capacity of a group-band multiplexer can be utilized by employing the device to serve as a hub or data transmission focal point at regional locations.

Here, TDMs are installed in cities A and C to multiplex four 2400-bps synchronous data streams in each city and transmit the multiplexed data over voice-grade lines at 9600 bps to the group-band multiplexer which is installed in city B. At city B, the group-band multiplexer services eight additional 2400-bps data streams which originate at that location, in addition to the two 9600-bps data streams from cities A and C. If the group-band multiplexer being utilized requires one 2400-bps synchronization channel as shown, although data is transmitted on the wideband facility at 40,800 bps, only 38,400 bps is actually used for data transmission with the difference being allotted to synchronization overhead.

Although not commonly used, group-band multiplexers can be pyramided at one location to make it possible for the user to combine the transmissions from 1000 or more low-speed terminals for transmission over a circuit at a speed of 1.544 Mbps. Owing to the limited number of organizations which have 1000 terminals in a regional area, let alone in a city, a more common pyramiding technique is to use a mixture of TDM and group-band multiplexers to service a few hundred terminals within a regional area as well as providing a computer-to-computer link over the same circuit used to service the terminals.

An example of this type of application is shown in Figure 3.23. In this example, a TDM services eight 1200-bps terminals in city A and produces a 9600-bps synchronous data stream which is input to the group-band multiplexer

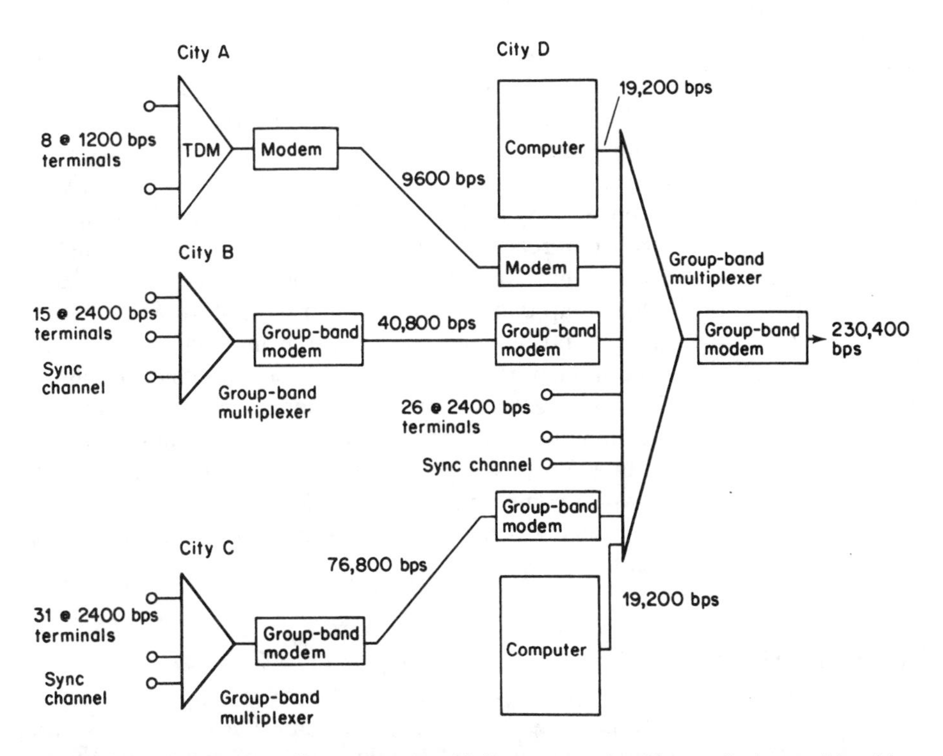

Figure 3.23 Pyramiding. Group-band multiplexers can simplify monitoring and trouble-shooting by permitting functionally identical units to be installed at many locations.

at city D. This group-band multiplexer acts as a pyramid, servicing the 40.8-kbps data stream of the group-band multiplexer in city B and the 76.8-kbps data stream of the group-band multiplexer in city C. At city D two computers transmit at 19,200 bps and 26 2400-bps terminals are serviced by the group-band multiplexer. With one channel used for synchronization, this group-band multiplexer produces a composite output of 230.4 kbps which now contains the traffic of 8 1200-bps terminals, 73 2400-bps terminals, two computers at 19.2 kbps, and three synchronization data streams at 2400 bps. Although it may be hard to visualize, one advantage of pyramiding group-band multiplexers is that monitoring, cabling, and troubleshooting can be simplified through the installation of functionally identical units at cities B, C, and D.

Options

Like most communications equipment, group-band multiplexers can be obtained with either a data terminal equipment (DTE) interface, a data communications equipment (DCE) interface, or a combination of the two interfaces. Thus, if the terminals to be serviced by the group-band multiplexer are within the cable length specifications of the RS-232 or CCITT V.24 interface, they can be directly connected to the group-band multiplexer and the user does not have to obtain modems. By using the data communications equipment interface, the group-band multiplexer can service distant data streams via modems, as shown in the previous example.

Group-band modems

While group-band multiplexers multiplex data streams for transmission over group-band facilities, group-band modems provide the mechanism for the transmission of that data over the facilities at speeds ranging from 19.2 kbps through 1.544 Mbps. Usually, the group-band multiplexer derives its wideband send and receive clock from the attached modem and by integer division divides the clock into the individual lower data rates. Some manufacturers offer group-band modems which have strap selectable rates, whereas on some group-band modems the data transmission rate is fixed. In selecting a group-band modem the user should investigate what options are available, as well as their cost. Some common options include strap selectable rates as previously mentioned, simultaneous voice channel capability where a plug-in card allows simultaneous usage of the 104- to 108-kHz segment of the available bandwidth for voice communications, Bell System 303 data set interface adapters, as well as buffers which will permit the group-band modem to operate on a digital network.

T1 multiplexing

The primary utilization of T1 multiplexers is to concentrate both data and digitized voice on a "T series" digital facility. To obtain an appreciation of the utilization of T1 multiplexers it is necessary to examine the process by which voice conversations can be digitized as well as the relationship between voice digitization and the data formats of the T1 carrier.

Although T1 transmission popularity is a relatively recent phenomenon, the technology for this transmission method dates back to the early 1960s. At that time, the T-carrier was used exclusively by telephone companies and was based upon the time division multiplexing of digitized voice.

PCM

One of the earliest methods employed to digitize voice was pulse code modulation (PCM). Under this modulation method an analog signal, such as the human voice, is sampled. Although PCM conversion from an analog to a digital signal can be done in one step by a single integrated circuit chip, due to economics it is often done by two chips in a two-step process. First, the analog signal which represents the intensity of a voice signal is sampled at predefined time increments as illustrated in Figure 3.24, which is known as Pulse Amplitude Modulation Sampling. Next, the height of each sample is converted into a series of binary digits that represents the analog signal height at the time sample occurred.

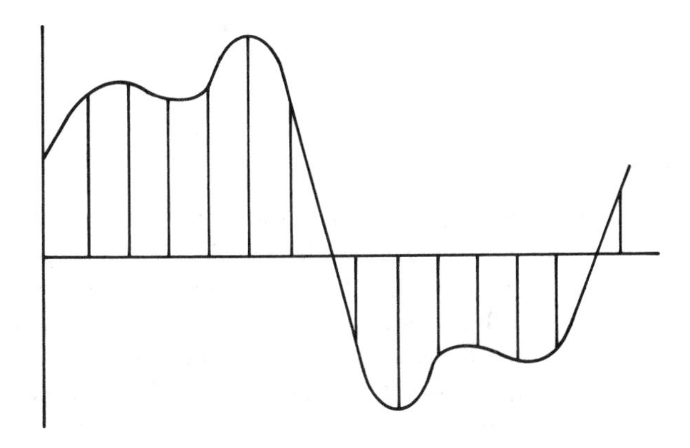

Figure 3.24 Pulse amplitude modulation.

DS1 framing

Under the PCM technique, a voice conversation is sampled 8000 times per second. In the United States, AT&T developed a 24-channel PCM system known as DS1, which resulted in the multiplexing of 24 voice conversations onto one 1.544 Mbps line. In the DS1 system, the analog signal of each voice channel is quantized through the use of a 7-level code, permitting 2^7 or 128 quantizing steps. To every 7 bits that represent a coded quantum step an additional bit was added for signaling, resulting in a total of 8 bits used to represent one sample of the analog signal. In addition, for each frame representing 24 analog signal samples, an additional bit was added. This bit is known as the framing bit and results in the following composition of the DS1 frame, which is illustrated in Figure 3.25. Since each voice signal is sampled 8000 times per second under PCM, 8000 frames per second are transmitted on a DS1 system. This results in a data rate of 193×8000 or 1.544 Mbps.

$$(7 + 1) \times 24 + 1 = 193 \text{ bits}$$

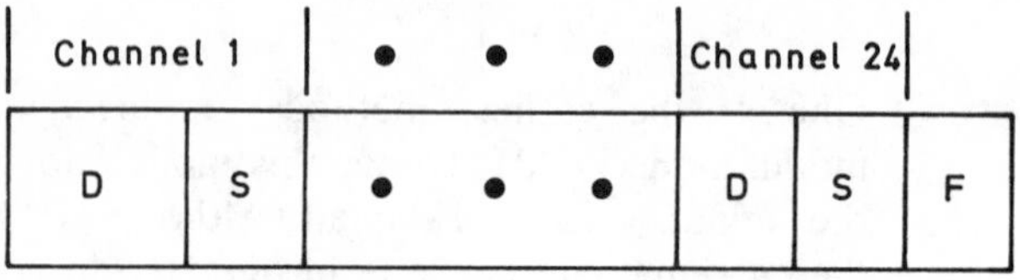

Channel 1		• • •	Channel 24		
D	S	• • •	D	S	F

Legend:

D = 7 data bits representing nearest digital value of the analog signal at the time of sampling

S = signaling bit

F = framing bit

Figure 3.25 The DS1 frame.

Digital signal levels

In the United States, the telephone network's digital hierarchy contains several digital signal (DS) levels. With the exception of the first level known as DSO signaling, each succeeding level is made up of a number of lower level signals.

Table 3.6 lists the four most widely used DS levels, their data rate and use in a telephone network. In such networks, 24 DS0 signals form one DS1 signal, while four DS1s make up a DS2 and seven DS2s form one DS3 signal.

If the reader multiplies the data rate of the lower level signal by the number of lower level signals used to make up a higher level signal, the sum will be slightly less than the data rate of the higher level signal. This difference is used for framing the lower level signals onto the higher level signal and is not employed for actual data transmission. Thus, multiplying the data rate of a lower level signal by the number of signals that make up a higher level signal can be used to obtain the actual data carrying rate of a DS level while the data rate in Table 3.6 is the signaling rate of data and framing information.

Framing changes

The name of the service represented by the original DS1 system has changed several times since the early 1960s. In addition, the composition of the framing structure was changed, in that in 5 out of every 6 frames a full 8 bits are now used for quantizing, while in the sixth frame, 7 bits are used for quantizing

Table 3.6 Digital transmission hierarchy.

Digital signal level	Data rate	Telephone company network use
DS0	64 kpbs	Basic voice bandwidth data channel encoded via PCM; digital data service (DDS), analog/digital channel-bank inputs
DS1	1.544 Mbps	This is the well-known T1 carrier which consists of 24 DS0 signals. Used for point-to-point communications between telephone company offices
DS2	6.312 Mbps	Used between telephone company central offices as well as inter- and intra-building communications
DS3	44.736 Mbps	Used in high-capacity digital radio, coaxial cable and fiber optic transmission systems for communications between telephone company offices

and 1 bit is used for signaling. The resulting T1 frame structure is illustrated in Figure 3.26. This frame structure change permits 256 quantizing steps to be used to represent the voice signal in 5 out of every 6 frames, with the sixth frame using 128 steps to represent the signal. Although the composition of the frame was changed, it has an equivalent bit structure of 8×24+1 or 193 bits

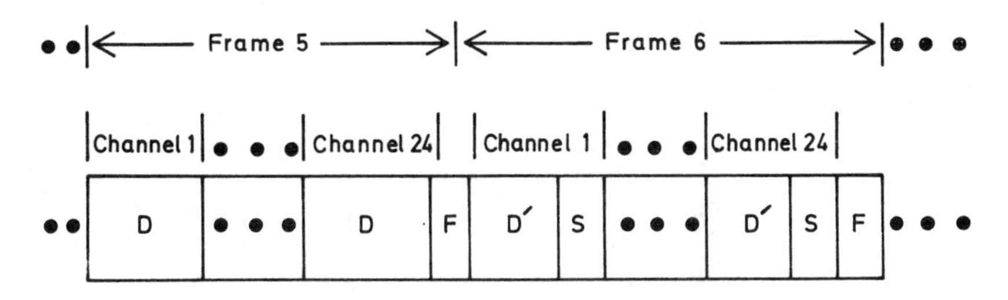

Legend:

D = 8 data bits representing nearest digital value of the analog signal at the time of sampling, permitting 256 quantizing steps

D′ = 7 data bits representing nearest digital value of the analog signal at the time of sampling, permitting 128 quantizing steps

S = signalling bit

F = framing bit

Figure 3.26 T1 frame structure.

per frame, producing a data rate of 193 bits × 8000 samples per second, or 1.544 Mbps. Today, this data rate is known as a T1 carrier in the United States and represents a communications facility capable of operating at 1.544 Mbps. Common T1 services include AT&T's Accunet T1.5, Accunet Reserved and DACS/CCR, which is an acronym for Digital Access Cross-Connected System/ Customer Controlled Reconfiguration.

T1 signal characteristics

The 1.544 Mbps T1 bit stream is a bipolar signal which is also called alternate mark inversion (AMI). Under this signaling format, the one pulses have an alternating polarity. Thus, if the *n*th one bit is represented by a positive pulse, the *n*th+1 one will be represented by a negative pulse. In comparison to alternating polarities for one bits, all zero bits are represented by a 0 voltage. In addition to bipolar signaling a T1 data stream must conform to the same signaling requirements as a DDS facility. That is, there can be no more than 15 consecutive zero bits present in the stream and at least 3 one bits in every 24 information bits.

Until 1985, suppliers of T1 multiplexing equipment only had to develop products compatible with the previously discussed signal characteristics for their devices to work on T1 facilities. Since then, two new framing patterns have been added to the T1 frame structure illustrated in Figure 3.26. The first frame structure is known as D4 framing, while the second structure is called the Extended SuperFrame (ESF). Currently, all T1 multiplexers must be compatible with the D4 pattern to operate on T1 facilities and eventually such equipment will have to support the ESF pattern.

D4 framing

In D4 framing a sequence of 12 frame bits is used to develop a precise pattern employed by T1 multiplexers to keep the bit stream in synchronization. Figure 3.27 illustrates the D4 framing pattern, showing the value of the 193rd bit in each of the first 12 frames transmitted on a T1 circuit. This 12-bit frame pattern continuously repeats itself, providing the synchronization signal used by equipment attached to a T1 circuit.

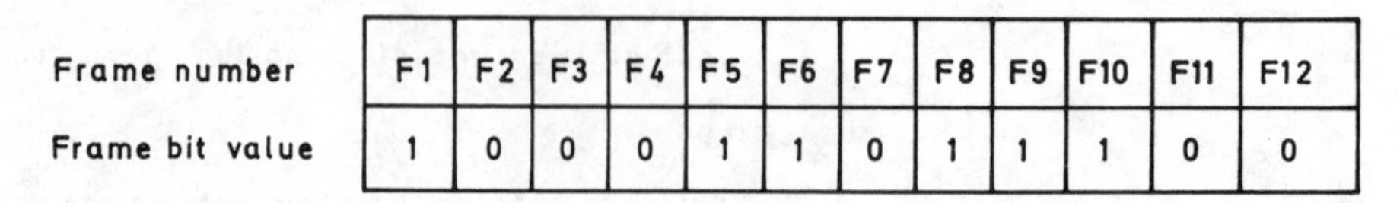

Frame number	F1	F2	F3	F4	F5	F6	F7	F8	F9	F10	F11	F12
Frame bit value	1	0	0	0	1	1	0	1	1	1	0	0

Figure 3.27 D4 framing pattern.

ESF framing

A more recently introduced T1 framing pattern called the Extended SuperFrame (ESF) format can be expected to result in an increase in the performance of T1 networks. This framing pattern contains 24 frame bits. Unlike the D4 pattern which repeats itself, the ESF consists of three types of frame bits, two of which can vary. These three types of frame bits include line control, cyclic redundancy checking, and a frame pattern for synchronization.

In the ESF frame the bits in the odd frames from 1 through 23 are used by the telephone company to perform network monitoring, set alarm conditions, and perform other control operations. The frame bits in frames 2, 6, 10, 14, 18, and 22 are used for cyclic redundancy checking. Finally, the frame bits in frames 4, 8, 12, 16, 20, and 24 are used to form the repeating pattern 001011, which is used for synchronization.

Accunet T1.5 is AT&T's regular, dedicated private line T1 service which is a full-duplex, digital transmission facility operating at 1.544 Mbps. Accunet Reserved is a switched T1 service which permits the transmission of digital signals at a data rate of 1.544 or 3.0 Mbps, with the higher data rate achieved by the use of two 1.544 Mbps channels in parallel.

A relatively recent offering, DACS/CCR, permits users to cross-connect digital subchannels of 64 kbps. Through the use of this service, organizations can control the allocation of the bandwidth of their T1 facilities by issuing commands to AT&T central offices which, in effect, results in the central office equipment functioning like a T1 matrix switch.

European T1 facilities

In Europe, a 32-channel system was developed to encode and multiplex voice signals in comparison to the 24-channel system used in the United States and other North American countries. Under the 32-channel system, 30 channels digitize voice signals resulting from incoming telephone lines, while the remaining 2 channels are used to provide signaling and synchronization information. Since each channel operates at 64 kbps which represents 8 bits used for each sample at a sample rate of 8000 samples per second, 64 kbps times 32 channels results in a composite data rate of 2.048 Mbps on a European T1 facility, which is technically referred to as a G703/732 channel.

Table 3.7 lists the composition of the 32 channels that are used to establish the G703/732 frame structure. In comparing the North American T1 frame structure to the European G703/732 structure, there are several key differences between the two in addition to the different data rates. First, North American T1 systems derive the 1.544 Mbps data rate from the use of 24 channels

Table 3.7 European G703/732 frame structure.

Time slot	Type of information
0	Synchronizing
1–15	PCM encoded speech
16	Signaling
17–31	PCM encoded speech

while the European G703/732 system uses 30 voice channels plus separate synchronization and signaling channels, with each channel operating at 64 kbps to produce a 2.048 Mbps data rate. Secondly, the North American T1 system uses the 193rd bit in each frame for synchronization, whereas the G703/732 system provides a separate 64 kbps channel for this function. Finally, the T1 system uses the eighth bit in every sixth frame for signaling, whereas the G703/732 system also uses a separate 64 kbps channel for this function. Table 3.8 compares the signaling characteristics of the T1 and G703/732 systems. Due to these major differences between systems, many T1 multiplexers designed for operation in North America will not perform properly if used in Europe, while other T1 multiplexers can operate correctly on both sides of the Atlantic.

Table 3.8 Signaling characteristics comparison.

	T1	G703/732
Composite data rate (Mbps)	1.544	2.048
Number of channels	24	32
Channel data rate (kbps)	64	64
Synchronization	Frame bit	Channel 0
Signaling	Eighth bit in sixth frame	Channel 16

T1 MULTIPLEXERS

When T1 became available for subscriber use in 1980, it was initially relegated to use in voice networks. This limited utilization was based upon the lack of equipment to efficiently combine voice and data on one T1 facility as well as the availability of telephone company equipment designed exclusively for the combination of 24 voice channels onto one T1 circuit. Due to the cost of less than 20 analog circuits used for transmitting data equaling the cost of one T1 facility while providing approximately 6 percent of the capacity of a T1 facility, many communications equipment manufacturers developed T1 multiplexers designed to economically multiplex both voice and data onto a composite T1 channel.

In addition to performing data multiplexing, most T1 multiplexer manufacturers offer users a variety of optional voice digitization modules that can be used to digitize voice signals at data rates ranging from the standard 64 kbps PCM rate to data rates as low as 9.6 kbps. Such modules can be employed to increase the number of voice channels that can be transmitted on a T1 line by a factor of two to four or more over the normal 24 or 30 channels obtainable when PCM digitization is employed. The use of these multiplexers enables organizations to effectively integrate voice, data and video information onto a common T1 facility. Prior to examining how T1 multiplexers can be employed in a network, a discussion of the various voice digitization modules commonly available is warranted to obtain an understanding of the benefits and limitations of these modules.

Voice digitization modules

In a conventional PCM digitization process the height of the analog signal is converted into an 8-bit word which represents the analog signal at the time sampling occurred. Since sampling occurs 8000 times per second, an analog voice signal is converted into a 64 kbps digital data stream, resulting in a maximum of 24 voice channels that can be carried on a North American T1 facility when PCM is employed. To increase the number of voice signals that can be carried on this facility, a variety of voice digitization techniques have been developed to include Adaptive Differential Pulse Code Modulation (ADPCM) and Continuous Variable Slope Delta Modulation (CVSD), as well as several less widely employed schemes known as Time Assigned Speed Interpolation (TASI) and Differential PCM (DPCM).

ADPCM

When Adaptive Differential Pulse Code Modulation is employed, a transcoder is utilized to reduce the 8-bit samples normally associated with PCM into 4-bit words, retaining the 8000 samples-per-second PCM sampling rate. This technique results in a voice digitization rate of 32 kbps, which is one-half the PCM voice digitization data rate.

Under the ADPCM technique, the use of 4-bit words permits only 15 quantizing levels, however, instead of representing the height of the analog signal each word contains information required to reconstruct the signal. This information is obtained by circuitry in the transcoder which adaptively predicts the value of the next signal based upon the signal level of the previous sample. This technique is known as adaptive prediction and its accuracy is based upon the fact that the human voice does not significantly change from one sampling interval to the next. Unfortunately, until 1988 there were no standard ADPCM algorithms and most T1 multiplexer vendors offering such modules employ proprietary transcoder algorithms. As a result of this lack of ADPCM algorithm standardization, T1 multiplexers from one vendor will probably be incapable of directly passing through digitized voice into a T1 multiplexer manufactured by a different vendor, requiring ADPCM data channels to be first reconverted to voice and then redigitized prior to remultiplexing. Fortunately, a pending American National Standards Institute (ANSI) standard known as ANSI T1Y1 should eliminate this incompatibility problem if it is adopted by vendors.

CVSDM

In the Continuously Variable Slope Delta Modulation (CVSD) digitization technique, the analog input voltage is compared to a reference voltage. If the input is greater than the reference a binary "1" is encoded, while a binary "0" is encoded if the input voltage is less than the reference level. This permits a 1-bit data word to represent each sample.

At the receiver, the incoming bit stream represents changes to the reference voltage and is used to reconstruct the original analog signal. Each "1" bit causes the receiver to add height to the reconstructed analog signal, while each

"0" bit causes the receiver to decrease the analog signal by a set amount. If the reconstructed signal was plotted, the incremental increases and decreases in the height of the signal will result in a series of changing slopes, resulting in the naming of this technique: Continuously Variable Slope Delta Modulation.

Since only changes in the slope or steepness of the analog signal are transmitted, a sampling rate higher than the PCM sampling rate is required to recognize rapidly changing signals. Typically, CVSD samples the analog input at 16,000 or 32,000 times per second. With a 1-bit word transmitted for each sample, the CVSD data rate normally is 16 kbps or 32 kbps. Other CVSD data rates are obtainable by varying the sampling rate. Some T1 multiplexer manufacturers offer a CVSD option which permits sampling rates from 9600 to 64,000 samples per second, resulting in a CVSD data rate ranging from 9.6 kbps to 64 kbps, with the lower sampling rates reducing the quality of the reconstructed voice signal. Normally, voice signals are well recognizable at 16 kbps and above, while a data rate of 9.6 kbps will result in a marginally recognizable reconstructed voice signal.

T1 multiplexer employment

Modern T1 multiplexers are microprocessor-based time division multiplexers designed to combine the inputs from a variety of data, voice and video sources onto a single communications circuit that operates at 1.544 Mbps in North America and 2.048 Mbps in Europe. Table 3.9 lists the typical input channel rates accepted by most T1 multiplexers. It should be noted that, although digitized voice is treated as synchronous input, its digitized data rate can vary considerably based upon the type of optional voice digitization modules offered by the T1 multiplexer manufacturer.

Table 3.9 Typical T1 multiplexer channel rates.

Type	Data rates (bps)
Asynchronous	110; 300; 600; 1200; 1800; 2400; 3600; 4800; 7200; 9600; 19,200
Synchronous	2400; 4800; 7200; 9600; 14,400; 16,000; 19,200; 32,000; 38,400; 40,800; 48,000; 50,000; 56,000; 64,000; 112,000; 115,200; 128,000; 230,400; 256,000; 460,800; 700,000; 756,000
Voice	9600; 16,000; 32,000; 48,000; 64,000

In most applications, input to the T1 multiplexer's voice channels results from an interface to an organization's Private Branch Exchange (PBX), resulting in one or more tie lines being obtained through the use of two T1 multiplexers and a T1 carrier facility. Figure 3.28 illustrates a typical T1 multiplexer application where voice, video, and data are combined onto one T1 carrier facility. In this example, it was assumed that PCM digitization channel modules were selected for use in the T1 multiplexer, resulting in the

10 voice channels on the PBX interface using 640 kbps of the available 1.544 Mbps T1 operating rate.

Since digitized video normally requires a data rate of 700 kbps to be effective, it was assumed that the organization using the T1 multiplexer has a conference room to connect to a distant location for video conferencing. Thus, 700 kbps input to the T1 multiplexer in Figure 3.28 represents a digitized video conferencing signal. Similarly, it is assumed that the organization has two data centers, one at each location where there is also a PBX. This permits computer-to-computer transmission to occur at 128 kbps. Finally, it was assumed that 12 data terminals, each operating at 4.8 kbps at one site, required access to the computer located at the other end of the T1 link. Since the cost of 10 tie lines and a few leased lines to support the data terminal traffic would normally equal the cost of a T1 carrier facility, in effect, the employment of T1 multiplexers provides the organization with no cost for the bandwidth required for video conferencing and wideband computer-to-computer transmission.

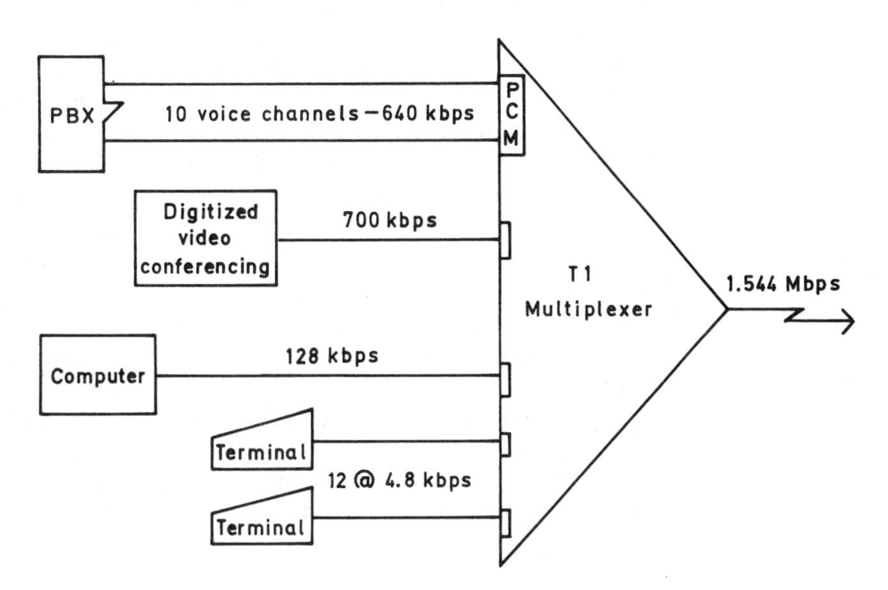

Figure 3.28 Typical T1 multiplexer application.

Multiplexer features to consider

Table 3.10 lists some of the more important features that users contemplating the acquisition of a T1 multiplexer should consider. As previously indicated in Table 3.9, the data rates supported for asynchronous, synchronous and voice transmission can vary considerably from multiplexer to multiplexer and must be examined to insure that the proposed equipment can support the user's requirements. Concerning Channel Interfaces, although most vendors support the Electronic Industry Association (EIA)RS-232-C interface as well as the North American T1 interface, other interfaces may not be supported by some vendors. Thus, the more modern RS-422, RS-423 interface which is equivalent

Table 3.10 Multiplexer features to consider.

Multiplexer features
Multiplexer channel rates: asynchronous; synchronous; voice
Channel interfaces: EIA RS-232-C (CCITT V.24); RS-422, RS-423 (CCITT V.10, V.11); V.35; G703/732; MIL-STD-188; T1
Number of channels supported and channel mixture
Multinode capability

to the CCITT V.10 and V.11 interface and which is not commonly included in communications equipment manufactured in the United States may not be supported by some T1 multiplexer vendors. Similarly, the V.35 wideband interface, the MIL-STD-188 current loop interface and the European T1 interface known as G703/732 may not be supported.

Even if all the user's interface requirements are supported, one must determine the total number of channels supported by the T1 multiplexer as well as the manufacturer's constraints regarding the mixture of such channels. Some T1 multiplexers cannot support more than 48 independent channels whereas other T1 multiplexers can support up to 512 or 1024 channels. Similarly, some T1 multiplexers are designed so that asynchronous data channels actually use a 64 kbps synchronous channel time slot on the aggregate T1 link regardless of the asynchronous data rate. In comparison, other T1 multiplexers may more efficiently service asynchronous data traffic, permitting a larger number of channels to be supported by the multiplexer.

Concerning multinode capability, some T1 multiplexers offering this feature permit users to network the multiplexers to service several locations. Figure 3.29 illustrates an example where three T1 multinodal multiplexers are employed to interconnect three distinct locations via three T1 lines. In comparison, uninodal T1 multiplexers can only be employed on point-to-point lines and would not provide the user with the routing flexibility illustrated in Figure 3.29.

Similar functional devices

Prior to the selection of group-band or T1 multiplexers for a particular application, a number of similar functional devices should be checked to determine their applicability and cost of usage.

First, inverse multiplexers present an economic alternative to the use of wide-band transmission at speeds in a range between 19,200 and 115,200 bps. Next, if a large number of asynchronous terminals are to be multiplexed, another potential device that should be considered is an intelligent multiplexer. This device will strip away the start and stop bits associated with asynchronous transmission, perform selected data compression for transmission, and reverse the compression and reinstitute the start and stop bits at the destination. Through these techniques, between four and eight terminals can effectively share one time slot that would normally require four to eight such slots on a

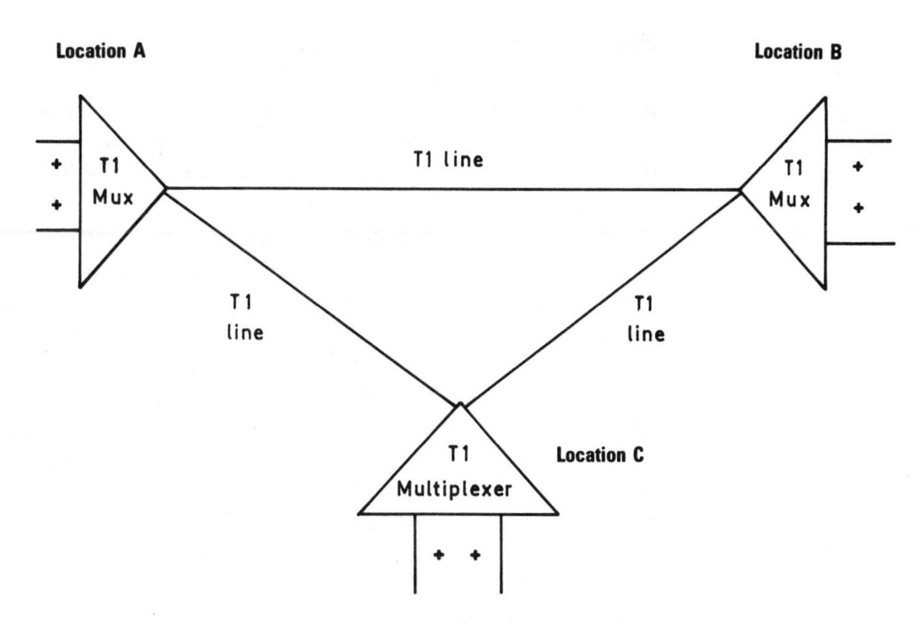

Figure 3.29 Multinodal T1 multiplexers can be networked.

traditional multiplexer. The reader should refer to Sections 3.1 and 3.3 for additional information on these devices.

3.3 INVERSE MULTIPLEXERS

Through the introduction of a new class of data communications equipment, network users can now obtain transmission at wideband rates through the utilization of two to six voice-grade lines. These devices also provide network configuration flexibility and provide reliable back-up facilities during leased line outage situations. In addition, since wideband transmission is not available at many locations, these devices have the extra advantage of extending wideband service to every location where the more available leased line service can be obtained.

Initially in this section we will focus our discussion of inverse multiplexers to devices that split a data stream into two substreams and recombine those substreams at the other end of the communications link. Then, using the previously discussed material as a foundation, we will examine the use of inverse multiplexers that have four and six subchannel capacities.

Operation

An inverse multiplexer splits a data stream at the transmitting station, and two to six substreams travel down different paths to a receiving station. Such a data communications technique has several distinct advantages over single-channel wideband communications lines. Using leased lines increases network routing flexibility and permits the use of the direct distance dialing (DDD) network as a back-up in the event of the failure of one or more leased lines.

Similar in design and operation, devices produced by several manufacturers permit data transmission at speeds up to 115,200 bps by combining the transmission capacity of two to six voice-grade circuits. Their operation can be viewed as reverse time division multiplexing. Input data streams are split into two to six paths by the unit's transmitting section. Although this chapter will concentrate in greater detail on the general operation of these devices, two specific units are described in more detail in Figures 3.30 and 3.31.

The most basic inverse multiplexer produces two serial bit streams by dividing all incoming traffic into two paths. In its simplified form of operation, all odd bits are transmitted down one path and all even bits down the other. At the other end, the receiver section continuously and adaptively adjusts for differential delays caused by 2-path transmission and recombines the dual bit streams into one output stream, as illustrated in Figure 3.32.

Each inverse multiplexer contains a circulating memory that permits an automatic training sequence, triggered by modem equalization, to align the memory to the differential delay between the two channels. This differential delay compensation allows, for example, the establishment of a 19,200-bps circuit consisting of a 9600-bps satellite link, and a 9600-bps ground or undersea cable, as shown in Figure 3.33. Since the propagation delay on the satellite circuit would be 250 ms or more while the delay on the cable link would be less than 30 ms, without adjusting for the difference in transmission delays the bit stream could not be reassembled correctly. In this type of application, the failure of one circuit can be compensated by transmitting the entire data load over the remaining channel at one-half the normal rate.

Typical applications

As stated previously, one key advantage obtained by using inverse multiplexers is the cost savings associated with two voice-grade lines in place of wideband facilities. Another advantage afforded by these devices gives the user the ability to configure and reconfigure their network based upon the range of distinct speeds available to meet changing requirements. If synchronous 16,800-bps speed-selectable modems are installed with the inverse multiplexers, up to ten possible throughput bit rates may be transmitted, as shown in Table 3.11. For transmission over two voice-grade circuits at 9600 bps or above for a composite rate of at least 19,200 bps, utilization of inverse multiplexers at this speed may be considered to provide a computer-to-computer transmission path. This is illustrated in Figure 3.34. Another application for such a device could be to connect a high-speed remote batch terminal to a computer at a lower but still wideband transmission rate, as shown in Figure 3.35.

Another possible use of such devices is to permit time-sharing operations and computer-to-computer or remote batch terminal-to-computer transmissions to occur over common leased lines. This sharing of the transmission medium

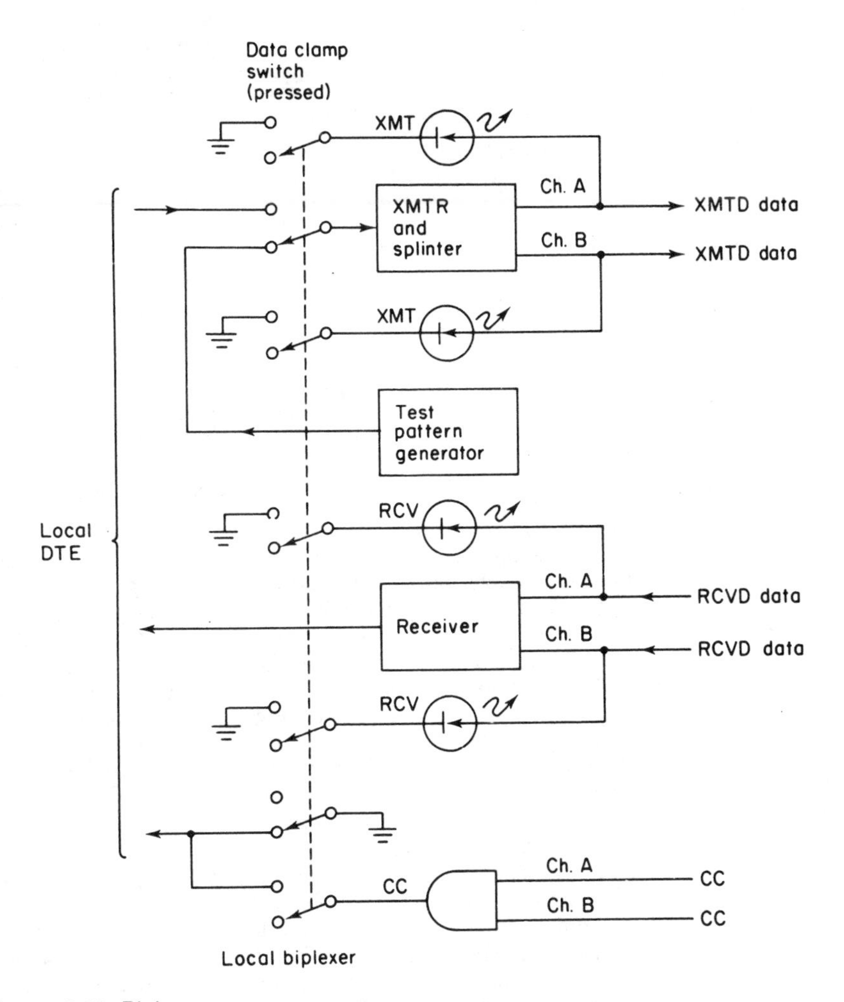

Figure 3.30 Biplexer.

The Codex Biplexer provides full-duplex transmission at speeds as high as 38,400-bps (composite) over two conditioned voice-grade lines by combining the capacity of two independent channels. These channels may be diversely routed with differential delays of up to 800 ms. The Biplexer contains monitor circuits that detect modem degradations and/or failures so that the Biplexer can always maintain at least one-half the normal data rate, which is based upon the type of modem connected to each subchannel.

Features include: automatic compensation for differential transmission paths; adaptation to changing line delays; automatic restoral over the direct distance dialing (DDD) network in the event of dedicated line failure; remote and local loop-backs; complete fault isolation; and system-monitoring capabilities.

Interface. For high-speed terminals, four interface configurations are available — Bell 303B, EIA RS-232-C/CCITT V.24, CCITT V.35 for standard European wideband modems, and MILSTD 188 for military specifications.

Physical specifications. Power supply: ac inputs of either 115 or 230 V ± 10 percent; 47–63 Hz, switch-selectable. The internal oscillator provides stability within 1 part in 10^6 per year. The Biplexer unit is 7 in high, 19 in wide, and 21 in deep.

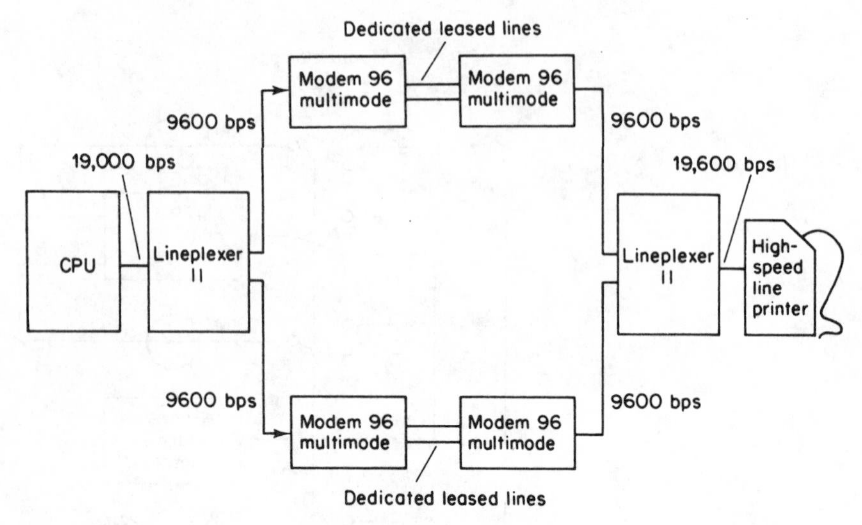

Figure 3.31 Lineplexer II.

The Racal-Milgo, Inc. Lineplexer II operates in a full-duplex mode to transmit and receive odd bits down one type 3002 voice-grade channel at speeds up to 14,400-bps while it simultaneously uses a second voice-grade channel to transmit and receive even bits at the same rate. Recombination of 'plexed' data results in a composite speed that is double the rate on the individual channels.

The Lineplexer II features a total backup system over the direct distance dialing (DDD) network, reconfiguration for day–night operations that allows system rearrangement of port configuration and operating speed via front-panel switches and displays, unattended operation through a self-monitoring system that causes the associated modem to fall back to a lower speed if the line degrades, automatic alternate-route selection, and diagnostic capabilities.

Specifications. Technical specifications include: data-transfer rate of 28,800-bps on high-speed channels interfacing with RS-232-C, MIL STD 188C, Bell 300 series, CCITT V.24, and CCITT V.35. Voice-band channels at speeds from 4800-bps to 14,400-bps for composite speeds of 9600-bps to 28,800-bps over leased lines are available with interfaces to RS-232-C, MIL STD 188C, and CCITT V.24. Modes of operation include channel A only, channel B only, channel A and B (normal operation), and channel A or B (both channels handle identical data, but only one is used, according to signal quality). Primary power is from an ac input at 105–125 or 210–250 V ac, 47–63 Hz.

Signal restoration. Speed is reduced after receipt of a low signal quality indication from either modem by external command, by using the manual Lineplexer switch, or through the operation of the Lineplexer syn monitor, special circuitry that stops transmission completely until both channels are resynchronized. Reestablishment of the high rate is automatic.

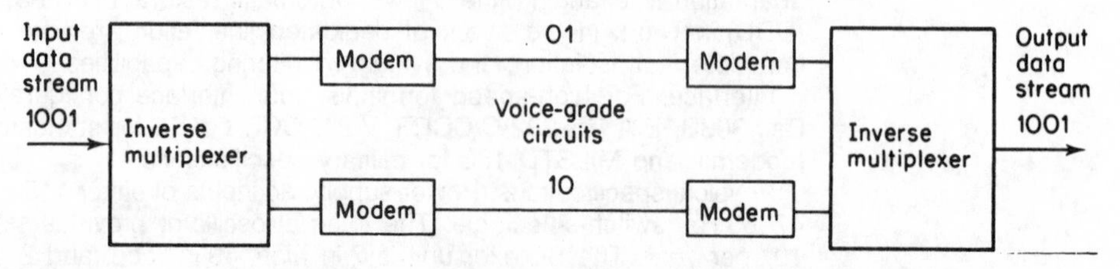

Figure 3.32 Inverse multiplexer operation. An inverse multiplexer splits a transmitted data stream into odd and even bits which are transmitted over two voice-grade circuits and recombined at the distant end.

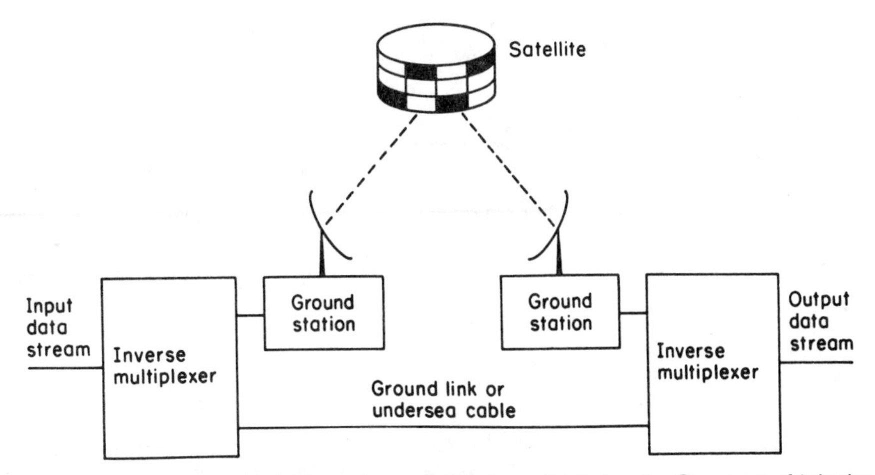

Figure 3.33 Inverse multiplexing using satellite-terrestrial circuits. One type of 'plexing' configuration could involve the use of a satellite link for one data stream with a ground or undersea cable link for the other.

Table 3.11 Wideband transmission combinations using 16,800 bps modems.

Subchannel A	Subchannel B	Transmission speed
16,800	16,800	33,600
14,400	16,800	31,200
14,400	14,400	28,800
14,400	12,000	26,400
12,000	12,000	24,000
12,000	9,600	21,600
9,600	9,600	19,200
9,600	7,200	16,800
7,200	7,200	14,400
7,200	4,800	12,000

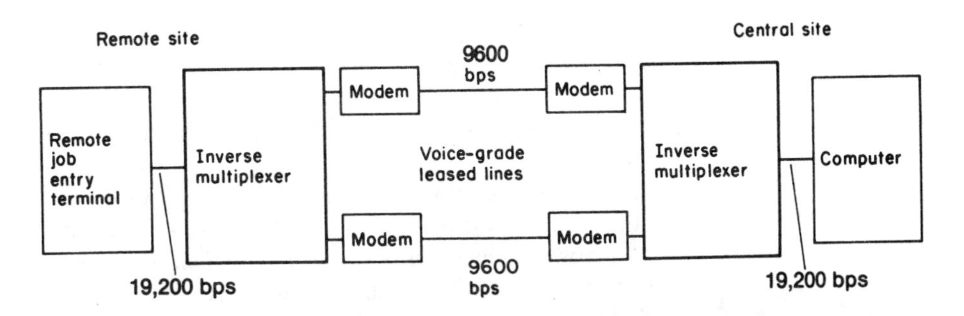

Figure 3.34 Computer-to-computer transmission. Inverse multiplexer at remote computer site splits data stream into odd and even bit streams that travel at data rates up to 19,200-bps over conditioned voice-grade lines and recombine at the CPU located at the central site. In the example illustrated, four 9600-bps modems provide 19,200-bps pseudo-wideband transmission.

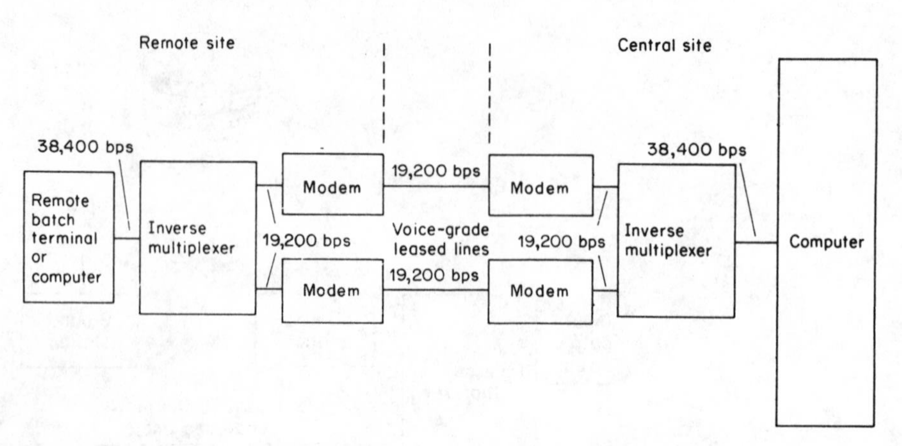

Figure 3.35 Reduced but still wideband transmissions. Lower transmission rates at wideband speeds can be maintained by the use of inverse multiplexers.

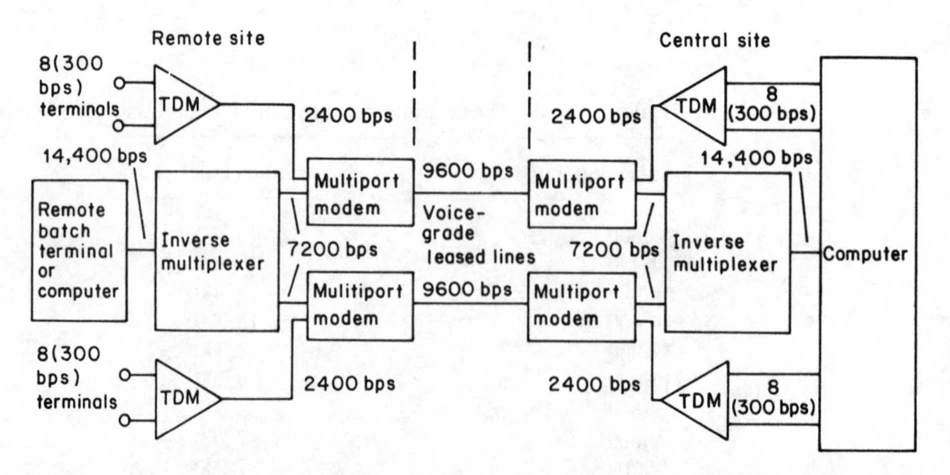

Figure 3.36 Shared use of wideband facilities. A number of configurations can be developed by using inverse multiplexers and multiport modems to permit shared use of wideband facilities.

can be accomplished through the use of multiport modems, as illustrated in Figure 3.36.

In the example shown in Figure 3.36, 16 300-bps terminals have their data multiplexed by two 8-channel TDMs at the remote site. Each TDM produces a 2400-bps synchronous data stream which is connected to one port on a multiport modem. The remote batch terminal or computer transmits data at 14,400 bps which is split into two 7200-bps data streams by the inverse multiplexer. Each 7200- and 2400-bps data stream is then multiplexed by the multiport modem for transmission at 9600 bps to the central site. At the central site the process is reversed, with the inverse multiplexer recombining the 7200-

bps output from one port of each modem into a single 14,400-bps data stream that is transferred to the computer. Each 2400-bps output is demultiplexed by a TDM into eight 300-bps channels and connected to the computer to provide a total of 16 low-speed terminal paths that are transmitted over the same facility as the computer or remote job entry (RJE) data.

Contingency operations

If the user is faced with the problem of channel degradation rather than a total outage, an automatic fallback capability allows the pair of channels to be switched to a slower speed until the return of good signal quality permits the resumption of higher speed transmission. Another important capability not available on wideband facilities is that service can be restored to one or both voice channels by dialing over the voice network during a dedicated line outage. This restoral capability provides for the configuraton of the network in such a way as to develop a wide range of contingency plans, in addition to the range of inherent equipment speeds and the combination of line speed available. If one or both of the leased lines should fail or transmission degrades on a circuit to a point where the continued use of that circuit is unacceptable, circuit compensation can be achieved in a variety of ways. Figure 3.37 details two possible methods of circuit compensation, when using 9600-bps modems on each subchannel. In the first example, data throughput may be maintained at 19,200 bps by the utilization of the dial-up or switched voice network. If the quality of the dial-up connection is such that 9600 bps transmission cannot

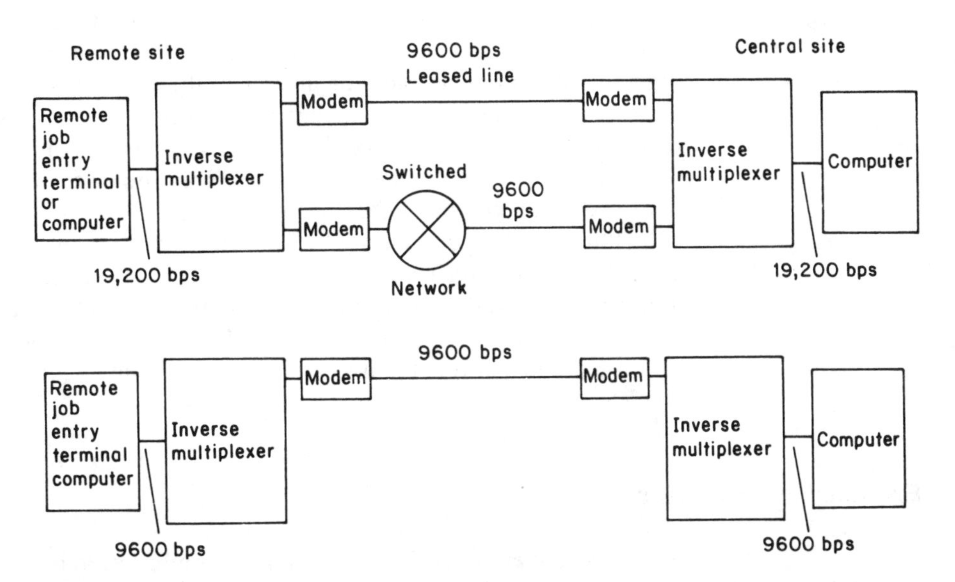

Figure 3.37 Methods of circuit compensation. Top: Dial restore to compensate for failure of leased line. Bottom: Continued operations on one line at a data rate of 9600-bps or less. Leased-line failure can be remedied directly over the DDD network. If this is not possible, transmission can continue at half-speed on the remaining channel.

be maintained, a drop in the transmission to 7200 bps may become necessary. Even at this speed, total aggregate throughput will be 16,800 bps. In the second example, it is postulated that because of high cost the DDD network would be unavailable, as would be the case in the failure of a satellite channel. The remaining circuit permits the maintenance of a data transmission rate as high as one-half the normal 19,200 bps.

Economics of use

One advantage in the use of an inverse multiplexer is the cost savings associated with using voice-grade lines in comparison to wideband facilities. Although both voice-grade and wideband circuit tariffs follow a sliding scale of monthly per mile fees based upon distance, the cost of wideband facility is often up to 16 times more expensive than a similar-distance voice-grade line. This means that the cost of four modems, two inverse multiplexers, and two voice-grade lines will normally be less expensive than one wideband circuit and two wideband modems.

Since tariffs follow a sliding scale, with the monthly cost per mile of a short circuit much more expensive than a long circuit, there are certain situations where the preceding does not hold true. This economic exception is usually encountered when connecting locations 80 to 100 miles or less distant from one another. Normally, if the two locations to be connected are over 80 to 100 miles apart, inverse multiplexing is a more economic method of performing communications, while wideband is more economically advantageous at distances under that mileage range.

The economics of inverse multiplexing are illustrated in Figure 3.38. Note that the 80- to 100-mile economic decision point range results from the fact that the monthly lease cost of inverse multiplexers varies between vendors.

A second advantage of using inverse multiplexers is the availability of service. Since wideband transmission is generally available only in major cities, two voice-grade circuits transmitting at a composite rate of up to 38,400 bps, in effect, extend wideband capacity nationwide.

A third advantage associated with using inverse multiplexing is the reduction in ordering time such devices may allow. Unfortunately, in many locations a notification period ranging up to 12 months in duration may be required for the installation of a wideband circuit. In comparison, the telephone company can usually install one or more voice-grade circuits within 60 days of receiving one's order.

Extended subchannel support

As initially mentioned in this section, a number of vendors have extended the capabilities of their inverse multiplexers to the support of four and six subchannels. Using four subchannel inverse multiplexers with eight 19,200-bps modems would permit a composite data transfer rate of 76,800-bps, while six subchannel inverse multiplexers with twelve 19,200-bps modems would permit a pseudo-wideband data rate of 115,200 bps to be obtained.

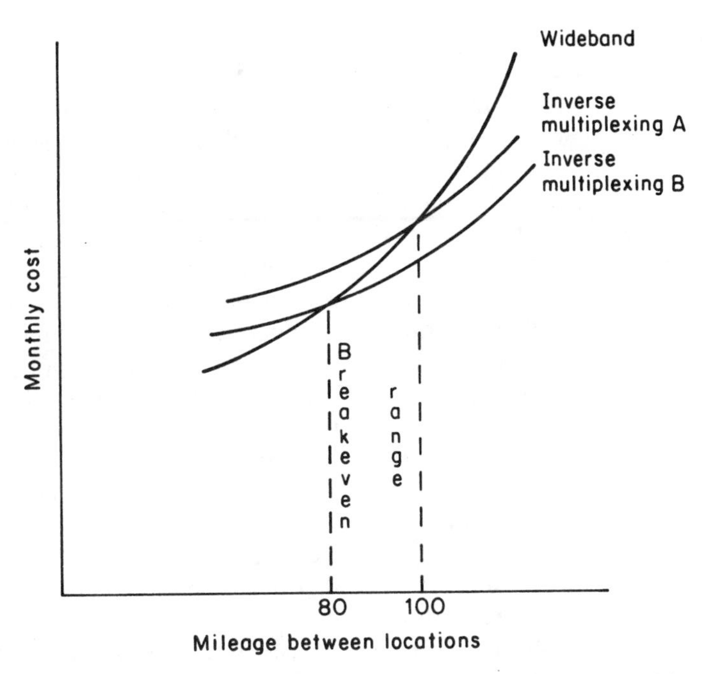

Figure 3.38 Economics of inverse multiplexing. In comparing the monthly cost of one wideband circuit and two wideband modems to the cost of two voice-grade lines, two inverse multiplexers, and four voice-grade modems, the breakeven range will be between 80 and 100 miles.

Although the utilization of two subchannel inverse multiplexers is fairly common, the employment of four and six subchannel devices is far less frequently encountered, primarily owing to economics. This is because the monthly cost of the additional pairs of modems and extra voice-grade lines will rapidly approach and then exceed the cost of one wideband circuit and two wideband modems.

Thus, the use of four and six subchannel inverse multiplexers is usually governed by the unavailability of wideband facilities and not by the economics associated with their use. In addition, because of the ease of channel speed selection and line restoral capability, the use of inverse multiplexers can provide more efficient line utilization and better backup capabilities than are currently available on wideband transmission facilities. Therefore, the utilization of inverse multiplexers can be expected to continue, probably without regard to future tariff changes.

3.4 PACKET ASSEMBLER/DISASSEMBLER

The Packet Assembler/Disassembler (PAD) is a specialized type of multiplexer originally developed to permit multiple terminals at one location to obtain access to a packet switching network via a common circuit. Originally PADs were designed to convert asynchronous start–stop protocols to an X.25 protocol. Later, many vendors expanded the functionality of PADs by adding additional protocol support to these devices.

Applications

The primary use of PADs is to enable non-X.25 networking devices to be connected to a packet switching network. There are many kinds of PADs, with their features and operation governing where they can be used to attach devices to an X.25 network. Some PADs support only asynchronous terminals, while other PADs support bisynchronous devices, SNA/SDLC devices or a mixture of terminals using different protocols.

PADs are used in both public and private X.25 networks. A public X.25 network is owned and operated by a communications carrier, which may sell or lease PADs to companies using the network. In addition, companies may purchase PADs for use on a public network once the device is certified by the communications carrier for use on their network. A private X.25 network can be constructed by an organization purchasing or leasing PADs and other network equipment and circuits or it can be formed by a communications carrier subdividing a portion of their network for the exclusive use of an organization. Regardless of the method used, the company that has a private X.25 network can either purchase or lease PADs from communications carriers or third-party vendors.

Figure 3.39 illustrates two of the main uses of PADs for connecting non-X.25 devices to an X.25 network. In the top left corner of Figure 3.39 several non-X.25 terminals are assumed to require a connection to a packet network. Although the terminals could either dial a port on the packet network or be routed to the network via individual leased lines, it is normally more economical to install a PAD at the terminal location. Then, the terminals can be connected to the PAD and the PAD in turn can be connected to the packet network by the use of a single circuit over which data is transmitted according to the X.25 protocol.

Once the X.25 data enters the packet network it will flow through the network based upon its destination code and the activity on different trunks in the network. As indicated in Figure 3.39, most X.25 networks include a number of packet switching processors interconnected to one another by trunk lines which form a mesh structure. If any trunk should become overloaded or fail, the network control center can issue commands through the network which will result in the alternate routing of data. Once data reaches a destination node on the network, it can be transmitted to a host computer in one of two ways. If the host computer directly supports the X.25 protocol a leased line can be used to connect the computer to the packet network. If the host computer does not support the X.25 protocol, a PAD can be used as a conversion device between the multiplexed X.25 protocol and individual ports supporting specific protocols on the computer. This conversion process is illustrated at the lower right corner of Figure 3.39, where a PAD is used to interface a non-X.25 host computer to a packet network.

Types of PADs

The most common type of PAD supports asynchronous terminals and its operation is defined by a set of CCITT recommendations. Recommendation

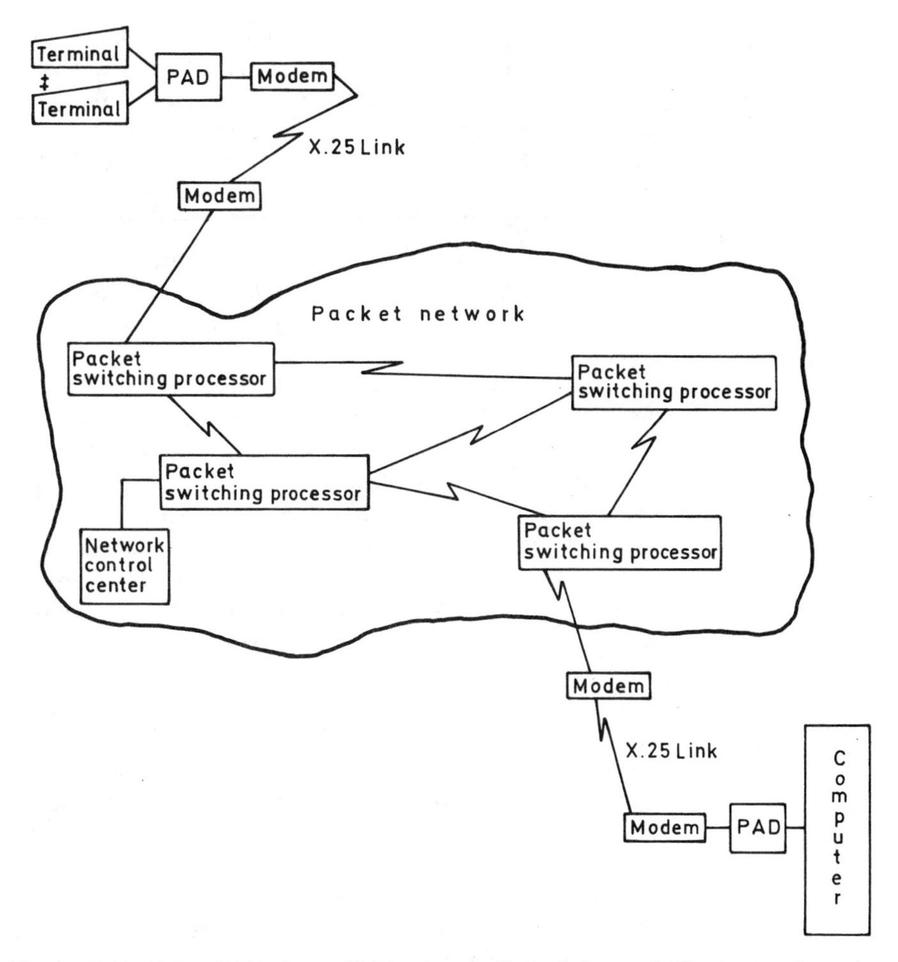

Figure 3.39 Using PADs in an X.25 network. By installing a PAD at a location where two or more terminals requiring access to an X.25 network are located the cost of communications to that network can be reduced. At a computer site, a PAD can be used to convert the X.25 packet network protocol to the protocol supported by a host computer.

X.3 defines the functions and operating parameters of the PAD and asynchronous terminals attached to that device. Included in the X.3 recommendation is a set of 22 parameters. Table 3.12 lists the meaning of each of the PAD parameters defined by the CCITT X.3 recommendation.

Each of the PAD parameters has two or more possible values. Ten of the more commonly altered parameters and their possible values are listed below.

2:m Echo. This parameter controls the echoing of characters on the user's screen as well as their forwarding to the remote DTE. Possible values are: 0 — no echo; 1 — echo.

3:m Selection of data forwarding signal. This parameter defines a set of characters that act as data forwarding signals when they are entered by the user. Coding of this parameter can be a single function or the sum of any combination of the functions listed below. As an example, a 126 code represents the functions 2 through 64, which results in any character to include control characters being forwarded. Possible values of parameter 3 are: 0 — no data forwarding character; 1 — alphanumeric characters; 2 — character CR; 4 — characters ESC, BEL, ENQ, ACK; 8 — characters DEL, CAN, DC2; 16 — characters ETX, EOT; 32 — characters HT, LF, VT, FF; 64 — all other characters: X'00' to X'1F'.

4:m Selection of idle timer delay. This parameter is used to specify the value of an idle timer used for data forwarding. Possible values of this parameter include: 0 — no data forwarding on time-out; 1 — units of 1/20 second, maximum 255.

7:m Procedure on receipt of break signal. This parameter specifies the operation to be performed upon entry of a break character. Possible values of this parameter include: 0 — nothing; 1 — send an interrupt; 2 — reset; 4 — send an indication of break PAD; 8 — escape from data transfer state; 16 — discard output. Similar to parameter 3, parameter 7 can be coded as a single function or as the sum of a combination of functions.

12:m Flow control of the terminal PAD. This parameter permits flow control of received data using X-ON and X-OFF characters. Possible values are: 0 — no flow control; 1 — flow control.

13:m Line feed insertion after carriage return. This parameter instructs the PAD to insert a line feed (LF) into the data stream following each carriage return (CR). Possible values are: 0 — no LF insertion; 1 — insert an LF after each CR in the received data stream; 2 — insert an LF after each CR in the transmitted data stream; 4 — insert an LF after each CR in the echo to the screen. The coding of this parameter can be as a single function or a combination of functions by summing the values of the desired options.

15:m Editing. This parameter permits the user to edit data locally or at the host. If local editing is enabled, the user can correct any data buffered locally, otherwise it must flow to the host for later correction. Possible values of this parameter include: 0 — no editing in the data transfer state; 1 — editing in the data transfer state.

16:m Character delete. This parameter permits the user to specify which character in the ASCII (International Alphabet Number 5) character set will be used to indicate that the previously typed character should be deleted from the buffer. Possible values for this parameter include: 0 — no character delete; 1–127 — character delete character.

17:m Line delete. This character is used to enable the user to specify which character in the character set denotes that the previously entered line should be deleted. Possible values are: 0 — no line deleted; 1–127 — line delete character.

18:m Line display. This parameter enables the user to define the character which will cause a previously typed line to be redisplayed. Possible values are: 0 — no line display; 1–127 — line display character.

20:m Echo mask. This parameter is only applicable when parameter 2 is set to 1. When this occurs parameter 20 permits the user to specify which characters will be echoed. Possible values include: 0 — no echo mask (all characters echoed); 1 — no echo of character CR; 2 — no echo of character

LF; 4 — no echo of characters VT, HT, FF; 8 — no echo of characters BEL, BS; 16 — no echo of characters ESC, ENQ; 32 — no echo of characters ACK, NAK, STX, SOA, EOT, ETB, and ETX; 64 — no echo of characters defined by parameters 16, 17, and 18; 128 — no echo of all other characters in columns 0 and 1 of International Alphabet Number 5 and the character DEL.

Table 3.12 CCITT X.3 PAD parameters

X.3 parameter number	X.3 meaning
1	PAD recall
2	Echo
3	Selection of data forwarding signal
4	Selection of idle timer delay
5	Ancillary device control
6	Control of PAD service signals
7	Procedure on receipt of break signal
8	Discard output
9	Padding after carriage return
10	Line folding
11	Binary speed
12	Flow control of the terminal pad
13	Line feed insertion after carriage return
14	Padding after line feed
15	Editing
16	Character delete
17	Line delete
18	Line display
19	Editing PAD service signals
20	Echo mask
21	Parity treatment
22	Page wait

In addition to supporting asynchronous terminals a variety of PADs are marketed to support vendor-specific protocols. Most PADs in this category support IBM bisynchronous or SNA/SDLC terminals, although a few vendors manufacture equipment to support other vendor-specific protocols.

The physical location of the PAD can be in one of three places. First, the PAD can reside on an X.25 network node, with terminals required to access the node via the PSTN or by leased lines. This is the most common PAD location, since the majority of persons accessing packet networks use asynchronous terminals and dial a node via the PSTN. Secondly, a stand-alone PAD can be installed at the end-user's terminal or host computer location as previously illustrated in Figure 3.39. The use of this type of PAD permits the multiplexed X.25 protocol to be routed on one circuit between the packet node and the end-user location, in effect serving to minimize the communications cost of the end-user. The third PAD location is inside a computer or terminal

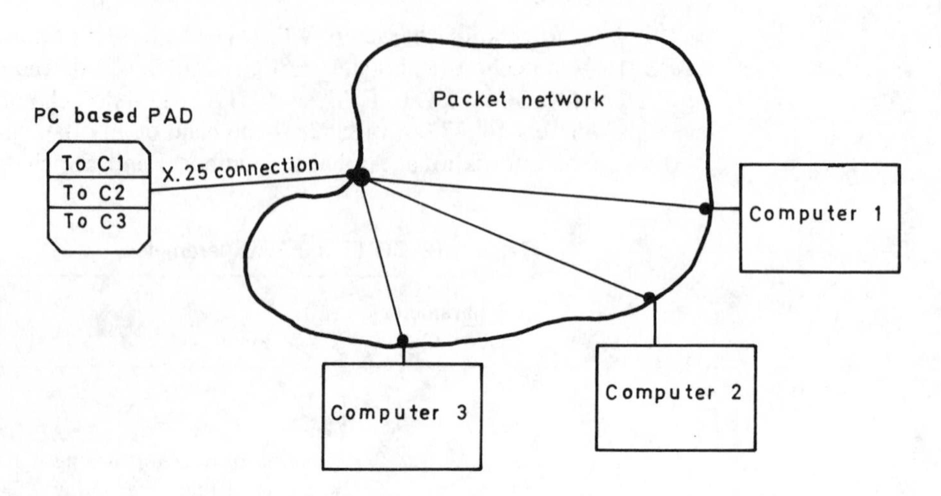

Figure 3.40 PC based PAD utilization.

device. This type of PAD is normally a special adapter card that can be installed within a personal computer or on the channel adapter of a front-end processor. Most PADs built on an adapter card permit the personal computer user to establish simultaneous communications with two or more remote computers on one X.25 connection as illustrated in Figure 3.40. This capability permits the user to switch from one host session to another without requiring the log-off from one computer system prior to signing onto a second system.

3.5 CONCENTRATORS AND FRONT-END PROCESSORS

Integral to almost every data communications network are two devices which, although consisting of many similar hardware components, must be recognized and utilized as distinct entities whose performance is designed for particular applications. These devices are communications concentrators and front-end processors. Substantial confusion concerning the utilization of these devices can occur because of the multitude of functions they perform. A front-end processor in effect performs concentration functions by concentrating a number of lines into a few data transfer paths between that processor and a host computer.

Likewise, a remote network processor can be viewed as performing the functions of a front-end processor when its high-speed data link is used to transmit data directly into another processor. To alleviate some of the existing confusion about the utilization of these processors, this section will examine the basic components of these devices, the functions they perform, the characteristics that should be investigated for evaluation purposes, and their placement within a data communications network.

Since the front-end processors offered by IBM under the name communication controller can be remotely located from a host computer, in effect serving as a remote concentrator, we will examine one member from the family of IBM devices at the end of this section.

Concentrators

As a general statement, a concentrator is a device which concentrates *m* incoming lines to *n* outgoing lines, where the number of incoming lines is greater than or equal to the number of outgoing lines. The incoming lines are usually referred to as concentrator-to-terminal links, whereas the outgoing lines are normally called concentrator-to-concentrator or concentrator-to-host links, as illustrated in Figure 3.41. Although the concentrator-to-host link implies such lines from the concentrator are connected to the host processor, in actuality they can terminate at a front-end processor which is in turn connected to a host processor or main computer. Depending upon the hardware components and operating software one selects, the concentrator can be used to perform a number of functions. These functions can include concentration, pure contention, store and forward concentration, concentration and message switching, and remote network processing.

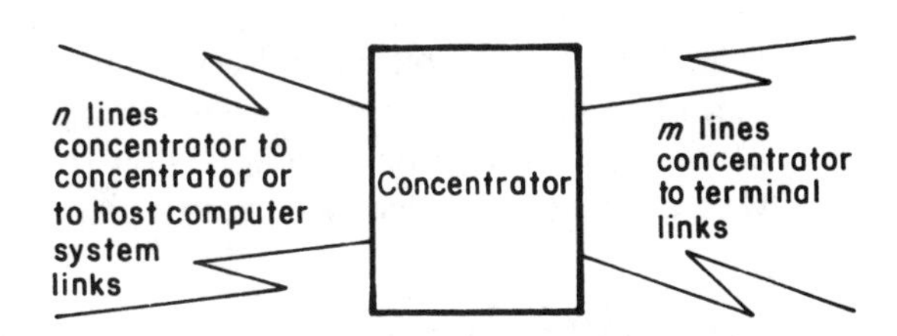

Figure 3.41 Concentrator links. In a concentration role *m* incoming lines (concentrator-to-terminal links) are concentrated onto one or more high-speed lines (concentrator-to-host computer system links).

When functioning as a concentrator, data from a large number of low- to medium-speed lines are combined for retransmission over one or more high-speed lines. The high-speed output can be transmitted directly to a front-end processor or to another concentrator. In a pure contention role, the concentrator serves as a programmed switch and is also referred to as a port selector or contention unit. Through the addition of on-line storage capacity and appropriate software, the concentrator can be configured to store and forward messages or to perform message switching alone. Another function rapidly gaining acceptance is that of remote network processing, which is a term used to denote concurrent concentration and remote batch processing on one processor.

In a concentration role, concentrators combine *m* low- to medium-speed lines onto one or more high-speed lines, similar to the function performed by a conventional multiplexer. However, concentrators can be programmed to transmit data only from terminals that are active as opposed to a conventional TDM, in which a fixed fraction of the multiplexed channel is reserved for each terminal regardless of whether or not the terminal is active. By using the memory areas in concentrators for program execution, changes in terminal speeds, data formats, communications procedures and the number of terminals to be serviced can be accommodated; whereas a traditional multiplexer is primarily an inflexible hardware device configured to accept predetermined data inputs and produce a predetermined data output. Concentrators can also be programmed to make more efficient utilization of the high-speed data link through data compression, making use of the different frequencies of characters to cause a reduction in the average number of bits transmitted per character through the utilization of codes to denote different characters or groups of characters.

If the host computer's native code is different from the terminal's code, code conversion can be performed by the concentrator, relieving the host processor of this function. While concentrators provide a significant advantage over the use of traditional multiplexers, users should also explore the capabilities of statistical and intelligent multiplexers. These devices are built around microprocessors and perform many communications functions previously available only through the use of concentrators.

Table 3.13 lists the major differences between concentrators and statistical and intelligent multiplexers. Although the differences in the type of processors used by each device may appear insignificant, since minicomputers operate faster and are capable of addressing more memory than microprocessors, they permit more sophisticated programming to be performed on concentrators. Thus, sophisticated data compression routines are more frequently employed with concentrators than with statistical and intelligent multiplexers.

Table 3.13 Comparing concentrators to statistical/intelligent multiplexers.

Hardware/Function	Concentrator	Statistical/intelligent multiplexer
Processor	Minicomputer	Microprocessor
Attached peripheral devices	Can include tape, disk, card reader, line printer	Usually limited to RS-232 port for monitoring
Programming	Loaded into RAM	ROM-based
Communications capability	Supports standard and vendor specific protocols	Usually supports vendor-specific protocol

In the area of attached peripheral devices one can normally add magnetic tape units, disk drives, and card readers, as an example, to concentrators. In comparison, the statistical or intelligent multiplexer is usually limited to an RS-232 port for the attachment of a terminal to monitor statistics computed by that device. This means that a concentrator is more readily tailored for such activities as journalization and flow control by the addition of software modules and such peripheral devices as magnetic tape units and disk drives.

The programming loaded into the multiplexer resides in the form of read-only memory (ROM) chips. In comparison, programs are loaded into random access memory (RAM) in a concentrator. This means that the concentrator is a much more flexible device for users who desire to modify and tailor software parameters to their specific requirements.

The last major difference between concentrators and statistical and intelligent multiplexers concerns their communications compatibility. Most concentrators support such standard protocols as bisynchronous and higher-level data link control, in addition to vendor-unique protocols. In comparison, most multiplexer vendors use a proprietary link control protocol, although several vendors have implemented an HDLC link control structure. This means that concentrators can usually be interfaced to other vendor products in a network while multiplexers frequently are limited to operating in pairs manufactured by the same vendor. The reader is referred to Section 3.1 for additional information on statistical and intelligent multiplexers.

Concentrator components

The typical hardware components included in a concentrator in a concentration role are shown in Figure 3.42. The single-line controllers provide the necessary control and sensing signals to interface the concentrator to individual data communications circuits. While single-line controllers can be asynchronous or synchronous, the majority are used to transmit synchronous data since each controller is used to provide only one high-speed transmission link from the concentrator to another concentrator or host computer (front-end processor).

Since the support of numerous lines would be expensive and would consume a large amount of physical space if implemented with single-line controllers, most communications support for the concentrator-to-terminal links are implemented through the use of multiline controllers (MLC). These can be categorized by capacity (number of lines and speed of lines supported), type of operation (hardware- or software-controlled), and the type of data transfer employed (character at a time or block transfer). Hardware-controlled MLCs place no additional burden on the concentrators' central processing unit (CPU), whereas programmed controllers place a large burden on the processor although they achieve the lowest cost per line by reducing the hardware in the interfaces

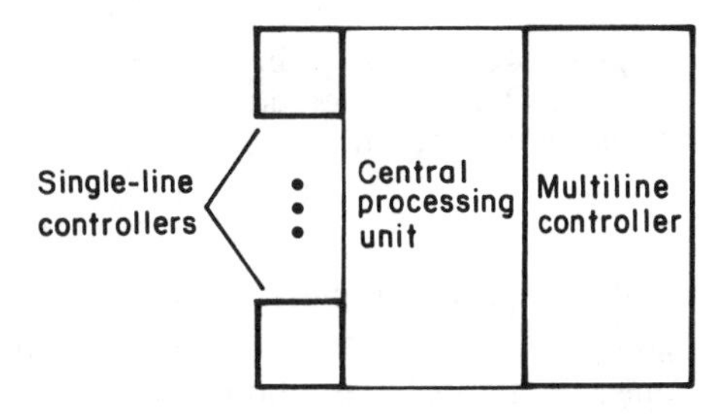

Figure 3.42 Concentrator hardware components. The single- and multiline controllers provide the necessary control and sensing signals to interface the concentrator to the communications circuits.

and the controller to a minimum. For a programmed controller, all sampling control, bit detection, and buffering is performed by the processor with the amount of processing time required by the operational program being the delimiting factor determining the amount of lines that can be connected to the concentrator.

To reduce the complexity of circuits in hardware MLCs as well as to reduce the software overhead of programmed controllers, incoming lines must normally be grouped on some controllers. These groupings are by data rate, code level, and the number of stop bits for asynchronous terminal support. Figure 3.43 illustrates a typical grouping by channel for an MLC that requires a minimum of four groups per class with any mixture of classes until the number of groups multiplied by 4 equals the total number of channels supported by the controller.

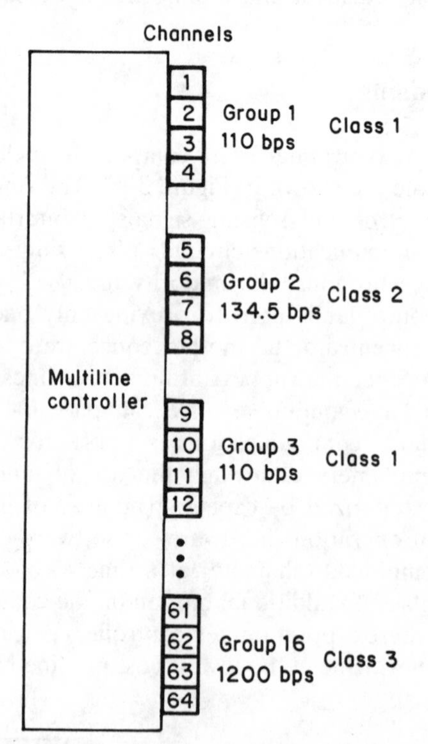

Figure 3.43 Groupings by channel. The number of groups and classes permitted by the multiline controller are important constraints that must be determined prior to equipment selection.

Although a complete examination of controllers might warrant the investigation of up to 50 parameters, Table 3.14 denotes the key types of information one should ascertain about the different controllers that may be supported by a concentrator. Another area that can have a major bearing on communications costs is the type of interfaces that the controller supports, as illustrated in Figure 3.44. If the controller supports directly connected terminals and the concentrator can be located within the vicinity of a number of those terminals,

Table 3.14 Controller selection factors.

Control type: hardware or software
Number of lines supported (MLC only)
Number of lines per group (MLC only)
Number of classes (MLC only)
Maximum throughput (MLC only)
Data codes and speeds supported
Number of bits per character
Number of stop bits
Syn characters support
Number per concentrator
Full-duplex/half-duplex support
Automatic dial/automatic answer support
Modem interface
Line-type serviced
Leased
Switched
Direct connect
Asynchronous/synchronous support
Protocol supported
Parity checking
Expansion capability (MLC only)

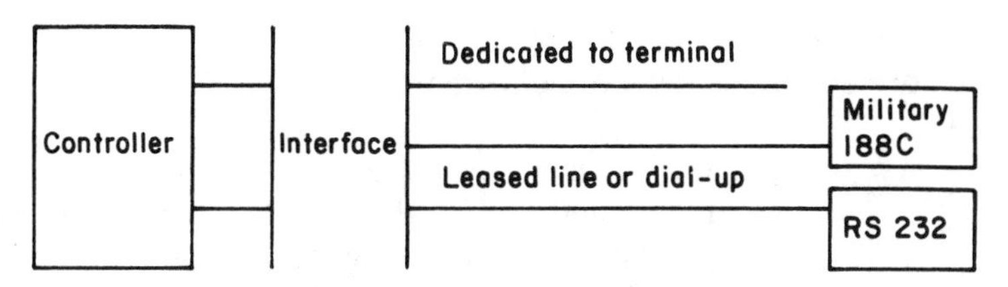

Figure 3.44 Controller interfaces. The types of interfaces that the controller supports may have a major bearing on communications costs.

directly connecting those terminals to the concentrator will alleviate the cost of installing line drivers or pairs of modems between the concentrator and each terminal.

Pure contention concentrator applications

In essence, a pure contention concentrator is a programmed switch which is also referred to as a port selector. In performing this function, any of m input lines are connected to any of n output lines as one of the n output lines becomes available. The m input lines are commonly called the line side of the concentrator, whereas the n output lines are referred to as the port side of the concentrator, since a concentrator used for this type of function usually interfaces the ports of a front-end processor on the output side of the device.

The basic hardware components of a contention concentrator are illustrated in Figure 3.45. Incoming data on each line of the line side of the device is routed through the processor which searches for a non-busy line on the port side to which the data can be transmitted. Priorities can be programmed so that groups of incoming lines can be made to contend for one or a group of lines on the port side of the device. When all ports are in use messages can be generated to new terminals attempting to access the system; and through the addition of peripheral devices jobs can be batched to await the disconnection of a user from the system and then to use the newly available port side line to gain entry to the computational facility. Additional information about port selectors will be found in Section 3.9.

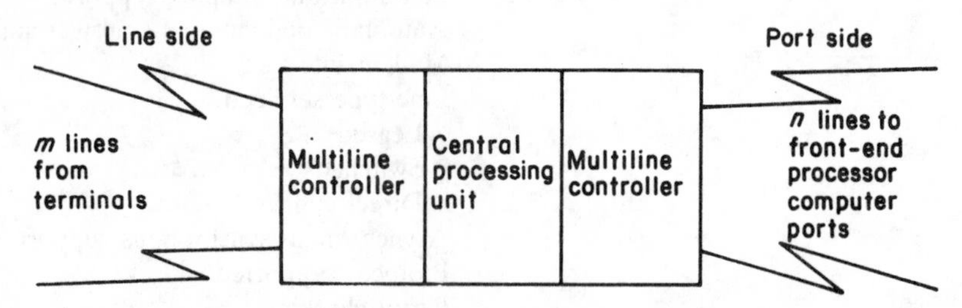

Figure 3.45 Contention concentrator components. A pure contention concentrator functions as a port selector, with m input lines contending for access to n output lines.

Store and forward message concentration

Although some vendors offer a complete concentration package to include controllers and CPU, other vendors permit the customers to select the CPU that is to be used for the concentrator. The evaluation and selection of the CPU of the concentrator should be accomplished similarly to the evaluation and selection of any stand-alone computer. Both the hardware and software should be evaluated according to user requirements. Thus, if one has the requirements for a store and forward message concentration system, emphasis should be placed upon peripheral equipment, data transfer rates, and software appropriate for this particular application.

In addition, if the application is critically time dependent, examining hardware reliability may not suffice by itself, and the user will most likely want to consider a redundancy configuration. Figure 3.46 illustrates a redundant store and forward message concentration system, in which both systems are directly connected to each other by an intercomputer communications unit and share access to incoming and outgoing lines and peripherals via electronic switches. During operation, one system is considered the operational processing or master system while the other system is the slave or standby system. Upon a hardware failure or power interruption, the master system signals the slave system to take over processing via the intercomputer communications unit, generates an alarm message, and conducts an orderly shutdown. Since the slave system has been in parallel processing, it resets all switches and becomes the master system, holding the potential of losing data to a minimum. This

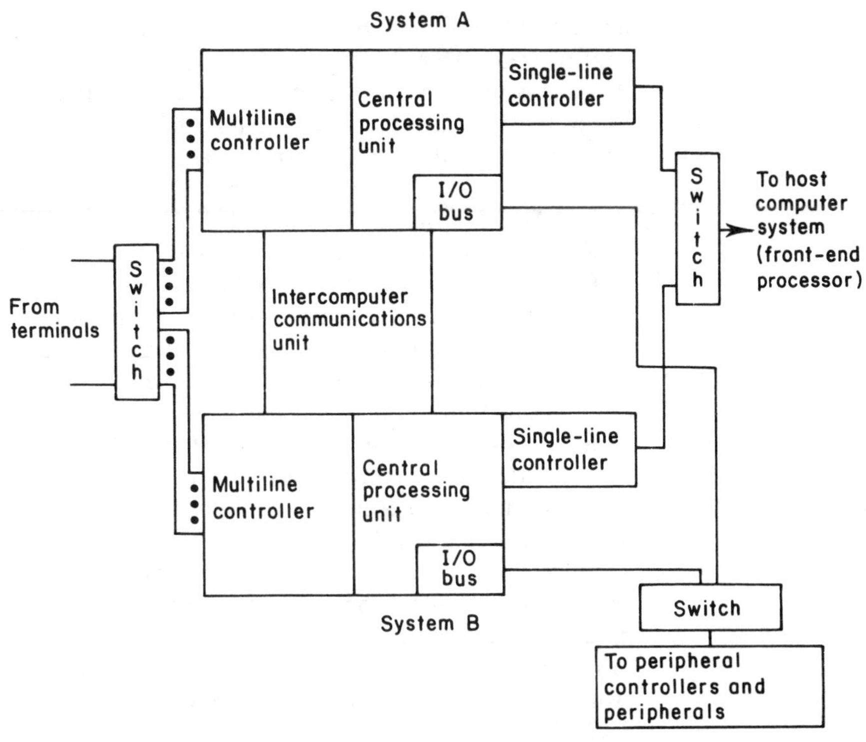

Figure 3.46 Redundant store and forward message concentration hardware. Upon a hardware failure or power interrupt the master system signals the slave system to take over processing via the intercomputer communications unit.

procedure can usually be completed within 500 ms for processors with a cycle time of 750 ns (nanosecond) or less, since 666 cycles (500 ms/750 ns) or more are then available within that time period to execute the required instructions to transfer control and effect the orderly shutdown. Since instructions normally require two to three processor-cycles, this time permits about 220 instructions to be executed to cause the switchover and shutdown.

Message switching

To effect message switching, incoming data is routed to a central point where messages are concentrated for processing, and based upon some processing criteria, messages are then routed out over one or more lines connected to the system. In a message switching system, all terminals connected to the system can communicate with every other terminal connected to the system once the message has been processed and the destination data is acted upon.

The hardware required for a message switching system is quite similar to that required by a store and forward message concentration system. The primary difference between them is the application software, wherein incoming messages are processed and then routed out over one or more of the incoming lines; and access to peripherals might be accomplished via a data-multiplexed control (DMC) or direct memory access (DMA) option in the concentrator instead of through the use of the lower data transfer rates obtained with the use of the I/O bus. The interface used to transfer data to and from controllers and peripheral devices is normally a function of the required data transfer rate.

The DMC provides peripheral devices with high-speed access to the computer's memory. It is a passive device that responds to the requirements of the devices connected to it. When a particular device has data to input or is ready to accept data, it uses the DMC control lines to request service. The DMC then sends a break request to the CPU, and a DMC cycle is executed when the current instruction is completed. During the DMC cycle the appropriate transfer between the device and the computer's memory occurs with the DMC using the computer's standard I/O bus for data transfer to and from the device. A DMA interface allows data transfers to bypass the I/O bus and thus allows external devices such as line controllers to access memory without program control.

All received characters enter a prescribed table in memory where they are then operated on by the program in the processor. In the DMC and DMA modes, data transfers are effected independent of program control so that other processing may be accomplished during data transfers. A message switching system configured for redundancy of operation is shown in Figure 3.47.

In addition to examining the controller parameters listed in Table 3.14, the line and station characteristics of the message switching system which are a

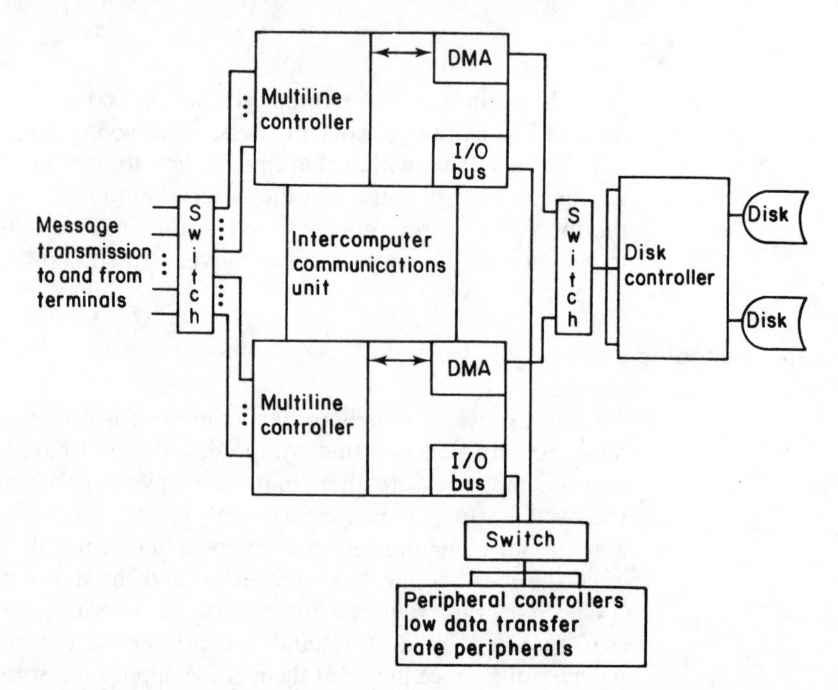

Figure 3.47 Redundant message switching system.

function of both hardware and software should be determined. Table 3.15 lists some of the message switching system characteristics that should be investigated prior to selecting such equipment. The number of stations supported many times is far more important than the number of lines supported since, as an example, a number of foreign exchange (FX) lines in diverse cities could each be connected to a channel on the MLC of the message switching system. This situation would enable a large number of terminals to contend for the FX line in each city, with each terminal having a unique station code. Station type refers to whether full-duplex, half-duplex, or simplex transmission is supported, while message addressing refers to the number of addresses per message as well as the availability of broadcasting a message to a predetermined group of stations or to all stations by using a group addressing scheme.

Table 3.15 Message switching system characteristics.

Number of lines supported
Number of stations supported
Line speeds and codes supported
Station types
Message addressing permitted
Levels of message priority
Message length
Input error checking
Alternate routing
Line and station holds and skips
Journalization capability
Retrieval capability
Intercept and recovery provisions
Alarms and reports

Most message switching software permits message priority of several levels. An example of priorities would be *expedite*, *normal*, and *deferred*, where a *deferred* priority message is transmitted to its destination after normal working hours, whereas a message of *expedite* priority would terminate any message being transmitted to the addressee of the expedite message until the expedite message transmission is completed. Depending upon vendors, some software packages permit a station to assign any message a priority while other systems can lock out terminal stations from assigning one or more priorities to a message. Normally, the maximum length of any message is a function of disk space and system throughput, but for all practical purposes users can transmit messages without worrying about their length.

Software for a good message switching system will preprocess the message header as it is entered and inform the user of invalid station codes, invalid group addresses, garbled transmission for both header and text, and other errors prior to routing the message. On some systems, a message denoting these errors will also be routed to the message switching systems' operator's console so that person will be able to ascertain user problems as they occur and furnish assistance, if required.

Alternate routing permits supervisory personnel to route all traffic destined to one or a group of stations to some other station. This capability is important if for some reason a communications component or terminal becomes inoperative and there is a station nearby that can handle messages destined for the inoperative terminal. Line and station holds and skips supplement alternate routing as they permit polling to any line to be skipped and permit traffic to any station or line to be held from delivery.

Journalization permits every message transmitted from the system to be recorded on journal storage. This permits message retrieval from the journal. However, the type of retrieval permitted varies from system to system. Some systems permit any station to retrieve messages delivered to itself, while other systems permit any station to retrieve any message by sequence number. On some systems, only the operator can access the journal by either station number, sequence number, or time of day.

Intercept permits the operator to have traffic that is routed to specific lines or stations rerouted to intercept storage and to be delivered to the addressed destinations by using a recovery function. These characteristics are especially important if a large number of lines become inoperative at one time and messages cannot be alternately routed to other nearby terminals. Since line drops are of critical importance to a message switching system, events of this type can normally be expected to cause an alarm message to be generated to an operator console so that he or she can take appropriate action.

Remote network processing

A remote network processor (RNP) is in effect a concentrator. However, it also performs the additional function of remote batch processing, thus providing two distinct functions in one package. RNPs vary in capabilities ranging from basic, single-job stream remote batch processing plus remote message concentration to multiple job stream remote batch processing combined with remote message concentration. Owing to the addition of remote batch processing to the concentration function, line utilization to the host computer is extremely high since the RNP may be servicing a variety of devices to include card readers, magnetic tape units, and line printers, in addition to concentrating the data from remote terminals for transmission to the host computer. A typical RNP configuration is illustrated in Figure 3.48.

As shown in Figure 3.48, several concurrent remote and local batch processing jobs can be accomplished in addition to the standard concentration function. Remote batch jobs can be entered for transmission to the host computer while completed jobs are printed on the line printer and other jobs are being punched out on the card punch.

Some RNPs have a read-only memory module which permits down-line loading of operational software from the host computer to the RNP. This is a valuable feature since it permits programming changes for new batch and remote terminal equipment to be performed at the central site and alleviates the necessity of employing programmers at every RNP installation to effect equipment configuration changes.

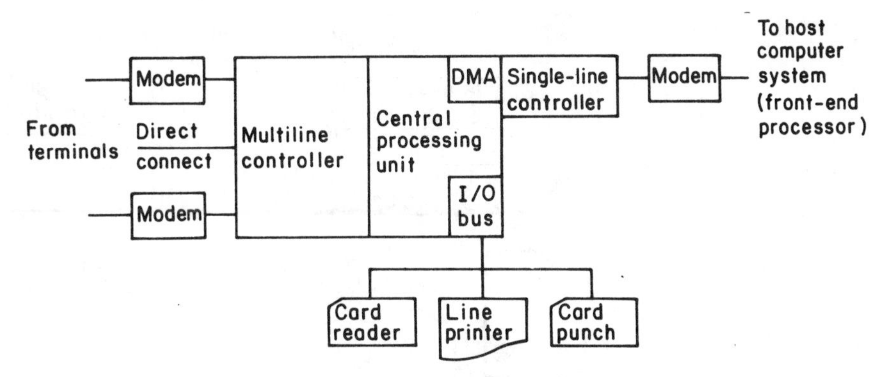

Figure 3.48 Remote network processor. Both remote batch processing and the concentration of terminal data for transmission are performed concurrently by a remote network processor.

By offloading work from the host computer by blocking the characters transmitted from each terminal into messages, the RNP permits users to better load balance their computational equipment; and in many instances this can obviate a costly host processor upgrade or the threat of encountering degraded service.

Down-line loading

Most concentrators being marketed today can be obtained with a read-only memory (ROM) module. This feature can be used to permit dynamic concentrator configuration of the types of terminals serviced or to change processing functions by down-line loading of the concentrator. An example of this is shown in Figure 3.49.

Although the multiline programmed controller is interfaced to 22 lines, for daytime operations it only services sixteen 300-bps time-sharing terminal lines and four 1200-bps cathode ray tube (CRT) lines. The data from the lines serviced are transmitted at 9600 bps to the central computer facility. For night-time operations, eight of the time-sharing lines and two of the CRT lines are removed from service in order to permit two 2400-bps printers to be connected via the concentrator to the central computer. This dynamic configuration can be controlled by the central computer by having the central computer down-line load the concentrator. During the down-line loading procedure, instructions are sent to the concentrator which activates the concentrator's read-only memory module.

The read-only memory module contains a fixed program which then takes the incoming data from the central computer, stores it in a predetermined location as a new program, and then initiates the program into operation. The incoming data can be an entirely new program or new tables and commands which alter the previous software configuration and in effect can become a new or revised program. This new program then determines which lines and what types of terminals will now be serviced by the concentrator. The key

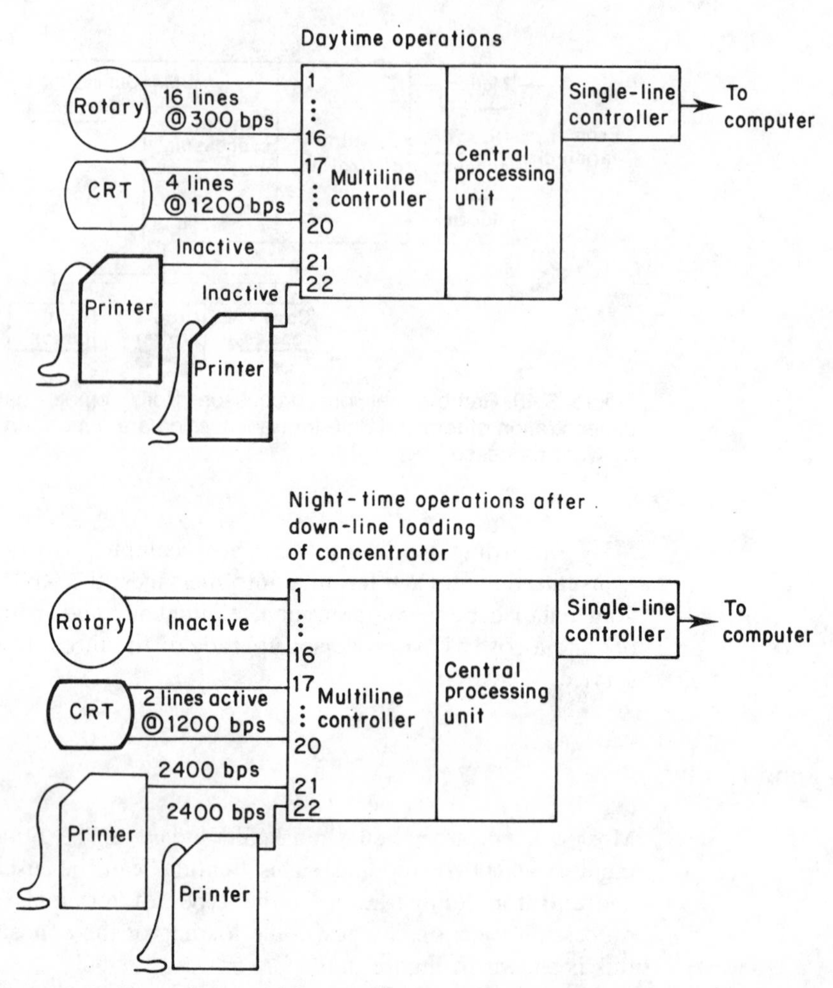

Figure 3.49 Dynamic reconfiguration by down-line loading. Dynamic reconfiguration of terminals serviced by a concentrator can be controlled by the host computer system through the process of down-line loading the concentrator.

advantage of this type of operation is that no operator intervention is necessary and the concentrator can also be down-line loaded in the morning to resume daytime requirements before anyone reports to work.

Alternate routings

Since a concentrator can have several high-speed output lines (single-line controllers), it is possible to configure a network so that one concentrator can be utilized to provide an alternate link routing capability to a second concentrator, as shown in Figure 3.50. In this example, the remote terminals which are serviced by the concentrator at location A are linked to the computer center on the primary remote concentrator A to computer center line. If this

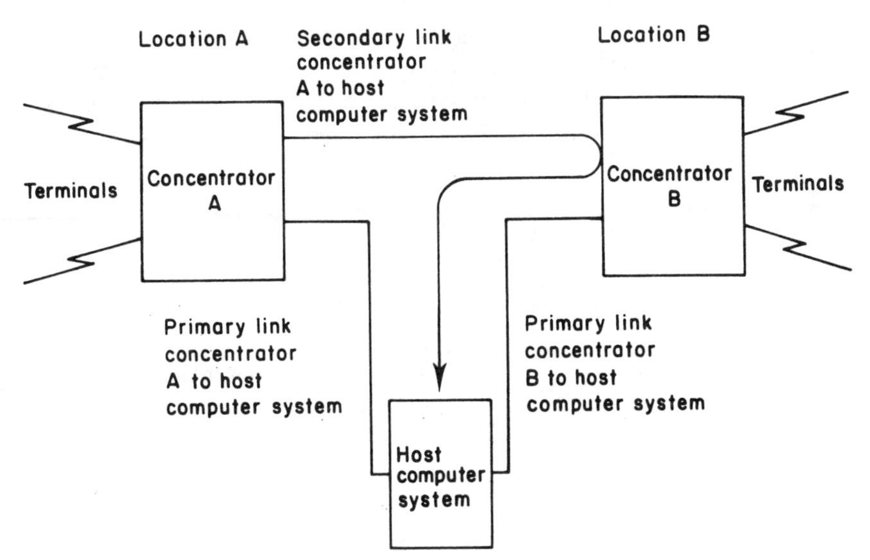

Figure 3.50 Alternate link routing using concentrators. One concentrator can be used to provide an alternate routing capability by permitting the sharing of its primary link to the host computer system.

primary link should become inoperative, the central computer can communicate with the remote terminals attached to concentrator A through the secondary link; this link consists of a high-speed line from concentrator A to concentrator B and the sharing of concentrator B's primary link from concentrator B to the computer center.

Satisfying average data transfer requirements

Since a concentrator can be used to store input data from the terminals to which it is connected until memory is full or by transferring the information to secondary storage such as a disk or tape unit, the high-speed lines from the concentrator to another concentrator or to the computer center only need the capacity to handle the average data rate of the terminals. This permits the condition to occur where the concentrator service's terminal input exceeds the maximum line transmission rate from the concentrator to the computer center. Thus, information is stored in memory or on secondary storage until such time as the concentrator can retrieve the information and transmit it to its destination.

For a traditional multiplexer, the absence of terminal activity still requires the multiplexer to take up signal space on the high-speed line, although the activity slot for that terminal is padded with nulls. Since numerous terminal applications have data transmission conducted in burst mode with the terminal activity reduced to a minimum between each burst, a portion of the high-speed line is usually idle for a majority of the time, with the multiplexer transmitting mostly nulls and thus wasting a large portion of the line's capacity. This type of situation can be alleviated to a large degree when concentrators are used.

As shown in Figure 3.51, 16 1200-bps terminals with an average data rate of 600 bps can be serviced by a single concentrator and, depending upon the line overhead (protocol) and buffering techniques employed, transmission from the concentrator to the computer center can occur over one 9600-bps line.

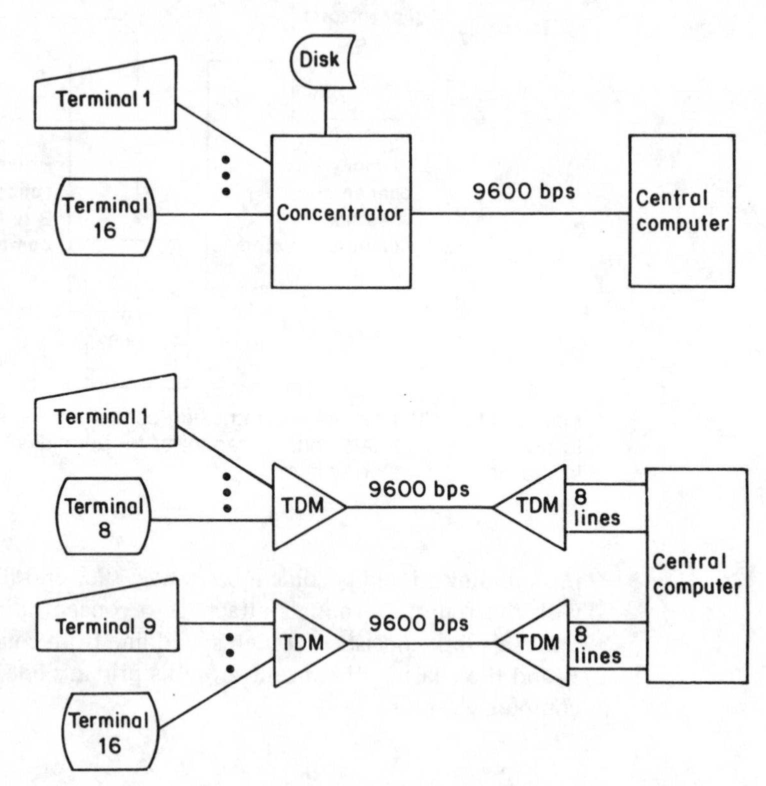

Figure 3.51 Using concentrators to satisfy average data rates. Sixteen 1200-bps terminals with 600-bps average data rate. In addition to servicing more terminals than time division multiplexers, concentrators only require one computer port which may substantially reduce hardware costs at the central computer site.

In reality, because of buffering and protocol overhead, only 70 to 90 percent of the high-speed link capacity is normally available for data transfer, reducing the effective data transfer on a 9600-bps line to between 6700 and 8600 bps. Through the addition of a disk, traffic peaks and valleys can be smoothed out, since the disk can be used to provide intermediate storage for the condition where seven or more terminals become active at the peak data rate of 1200 bps and thus exceed the 9600-bps transmission capacity of the line from the concentrator to the central computer. To service the same number of terminals with a conventional multiplexer would require the installation of two such devices at the remote location as well as two demultiplexers at the central computer site, with each multiplexer transmitting over a 9600-bps line. Since multiplexers must be configured to service a terminal's peak data rate, the two multiplexers require double the line capacity which the concentrator needs in transmitting to the central computer.

In addition, since a concentrator can be programmed to communicate directly with a high-speed channel of the central computer, no additional hardware may be necessary at the central site, whereas two multiplexers may be necessary to split the high-speed multiplexer traffic back into its original form so that the computer can talk to the remote terminals. This requirement for demultiplexers is due to the fact that multiplexer protocols are usually vendor-unique and computers cannot communicate with the high-speed multiplexer traffic without extensive software modifications.

In determining which type of device to use, the savings in line charges and modem costs achieved by the concentrator's capability must be balanced by the higher cost of the concentrator and its disk subsystem as well as the queuing delay when the average data rate exceeds the transmission line capacity.

Front-end processors

Front-end processors provide a large volume of network communications power in support of a particular computer system. Although they are similar in design and have components in common with concentrators, normally they have larger word sizes, faster cycle time, and larger memory, and permit the interfacing of more communications devices to the processor. In addition, front-end processors are usually more "closely coupled" to a particular host computer and may be specifically programmed to operate with that computer and its operating system. A typical front-end processor configuration is illustrated in Figure 3.52.

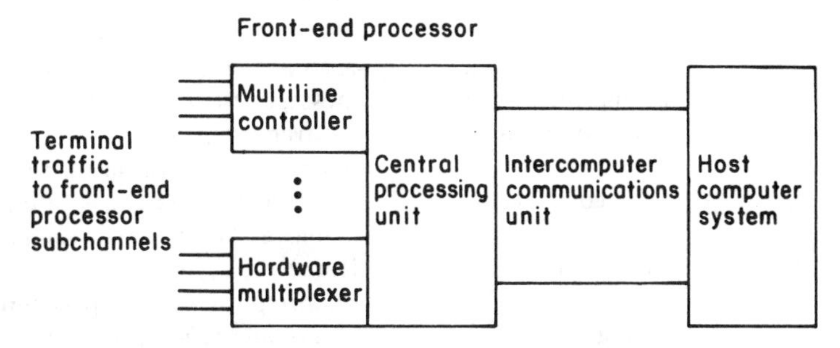

Figure 3.52 Front-end processors. A front-end processor can be considered the heart of a computer network, relieving the host computer system of many software burdens, performing such functions as code conversion, character blocking, and character deblocking.

Not only are more MLCs available for connection to a front-end processor than for a concentrator, the MLCs are close to being universal in their ability to service a mixture of synchronous and asynchronous data at speeds ranging from 50 to over 50,000 bps. Another device encountered on some front-end processors but normally not used on concentrators is a local communications multiplexer which provides for time division multiplexing by character, to and from the front-end processor, for a variety of low-speed terminals at transmission rates up to 300 bps. These local multiplexers can handle terminals with differing communications speeds and code settings, with the character demultiplexing performed by software in the front-end processor. In addition to performing network and communications processing activities that one normally associates with front-end processors, owing to the larger memory and word size, quite often they can be used to perform message switching functions by the addition of modular software.

The operating system which supervises the overall control and operation of all system functions is the key element of a front-end processor. Although numerous software elements must be evaluated, major consideration should be given to determining supported line protocols as well as supported processor communications. In the first area, most vendors divide their supported line protocols into several categories or classes of support. Normally, category one refers to vendor-developed and tested software to support certain line protocols. Category two usually refers to vendor-developed but non-qualified tested software; while category three refers to customer-developed interfaces designed to support certain terminal line protocols.

The functionality of front-end processors is primarily obtained through software. Table 3.16 lists the major front-end processor software modules offered by most vendors of this device, followed by a brief functional description of the module.

Table 3.16 Front-end processor software modules.

Module	Functional description
Character/message assembly/disassembly	Assemble the serial bit stream into characters and messages for transfer to the host computer and vice versa
Data formatting	Restructure incoming data to a more compact format to permit an increase in host computer efficiency
Code conversion	Convert the data codes of different terminals into the code employed by the host computer
Message switching	Route messages from one terminal to another without requiring the data to pass through the host computer
Polling	Query input/output devices to see if they have information to transfer or are available to accept information
Error checking	Check incoming data to insure transmission errors did not occur; reject data blocks and request retransmission if errors are computed to have occurred
Protocol support	Ability to communicate with different terminal devices according to a standard method of interface
Automatic operation	Handle automatic answering and/or outward calling on the public switched telephone network
Statistics	Compute such important parameters as traffic density and circuit availability as well as other statistics which may be essential for the effective management of a data communications network

While the features of front-end processors are similar to concentrators, the reliability and redundancy as well as diagnostics should be more extensive since the front-end processor is the heart of a communications network. Figure 3.53 illustrates a typical data communications network consisting of several different types of concentrators and a front-end processor.

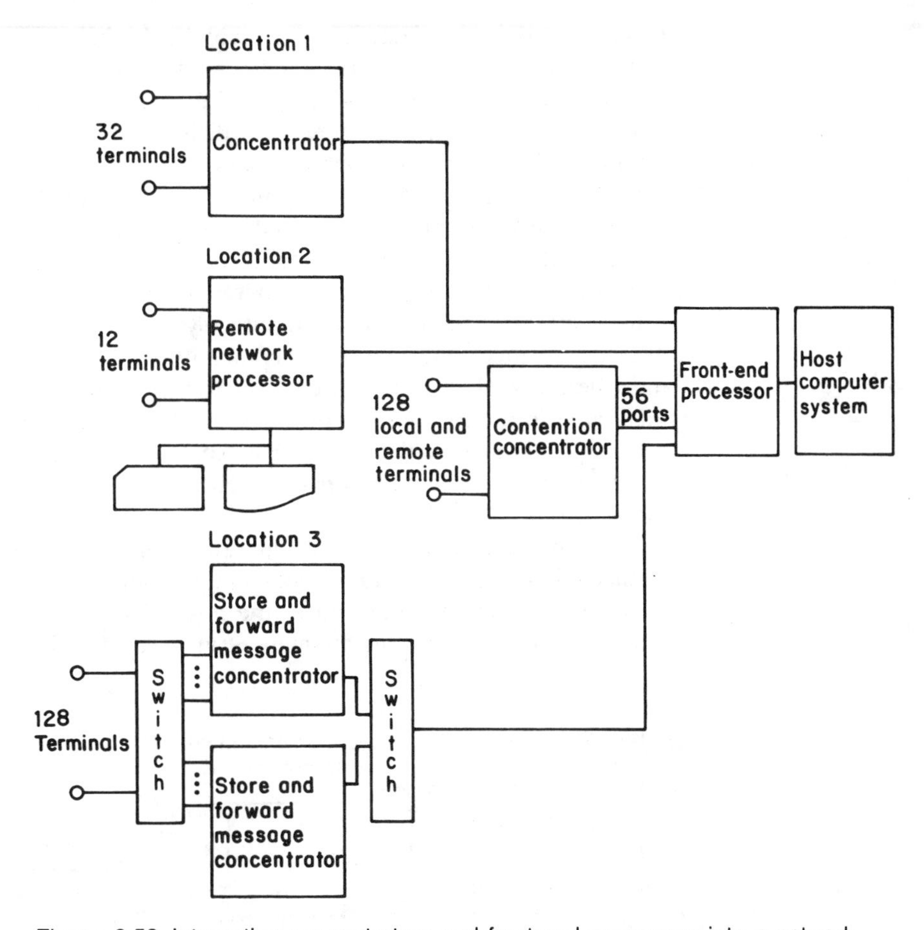

Figure 3.53 Integrating concentrators and front-end processors into a network.

At location 1, a standard concentrator is used to focus the traffic from 32 terminals onto a high-speed line for transmission to the front-end processor. Since location 2 has a requirement for remote batch processing as well as connecting 12 terminals to the host computer, a remote network processor has been installed to perform these two functions. Since location 3 has a significant number of terminals doing an important application, a redundant store and forward message concentrator was installed at that location. The remainder of the terminals in the network totals 128. However, it was felt that at most only 56 would ever become active at any given time. Therefore, to economize on front-end processor ports, a contention concentrator was "front-ended" to the front-end processor, making 128 lines connected to 128 terminals contend for the use of 56 front-end processor ports.

Communication controller

In place of the term front-end processor IBM labeled its equivalent hardware as a communication controller. Members of the IBM communication controller family include the IBM 3704, 3705, 3720, 3725, and 3745 which are designed to operate with the IBM System/370, 30XX, and 4300 series computer systems. Although there are numerous hardware differences between members of the IBM communication controller family, the examination of the IBM 3725 presented in this section can be used as a reference for reviewing the hardware functionality and operational capability of the earlier models in this family.

The IBM 3725 is a modular, programmable communication controller that can be attached to an IBM computer system in one of two ways: channel-attached or link-attached. In a channel-attached mode of operation the 3725 must be physically located in close proximity to the host computer system as illustrated in the top portion of Figure 3.54. In this method of attachment the communication controller is physically connected to a data channel on the host computer system and data is transferred in parallel between the computer and the controller.

In a link-attached mode of operation the communication controller is connected via a telecommunications line to another communication controller, which can be channel-attached to the host computer or may be another link-attached controller. In this attachment method, data is transferred serially. In

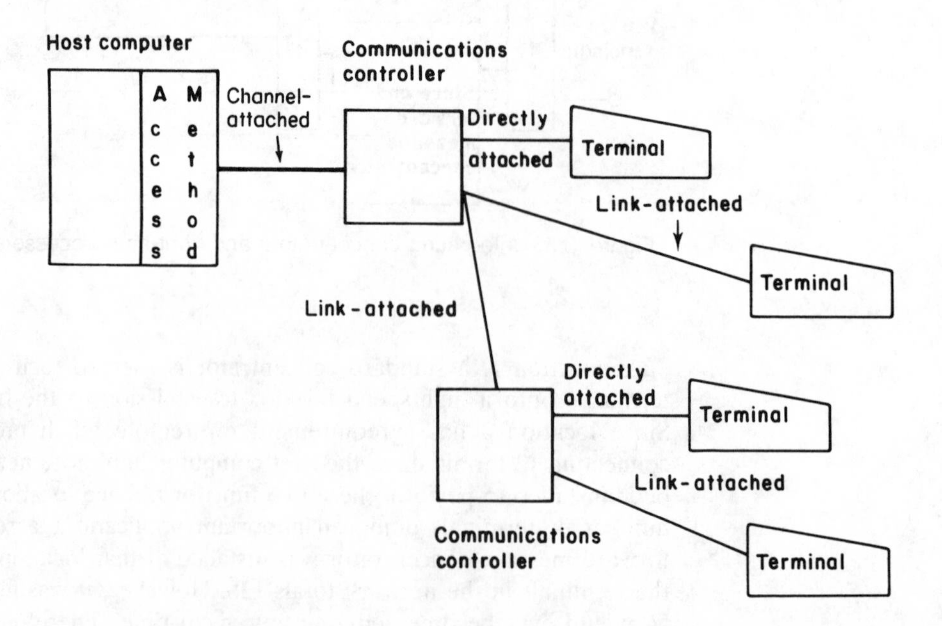

Figure 3.54 IBM communication controller attachments.

the lower portion of Figure 3.54, a link-attached controller connected to the channel-attached controller previously discussed is illustrated.

Terminals can be connected both directly and remotely via a communications facility to either a link-attached or channel-attached communication controller. Note that for the sake of simplicity, such communications equipment as modems or digital service units are omitted from Figure 3.54, even though they would be required for link-attached controllers and terminals to operate.

The 3725 communication controller contains three logical subsystems as illustrated in Figure 3.55.

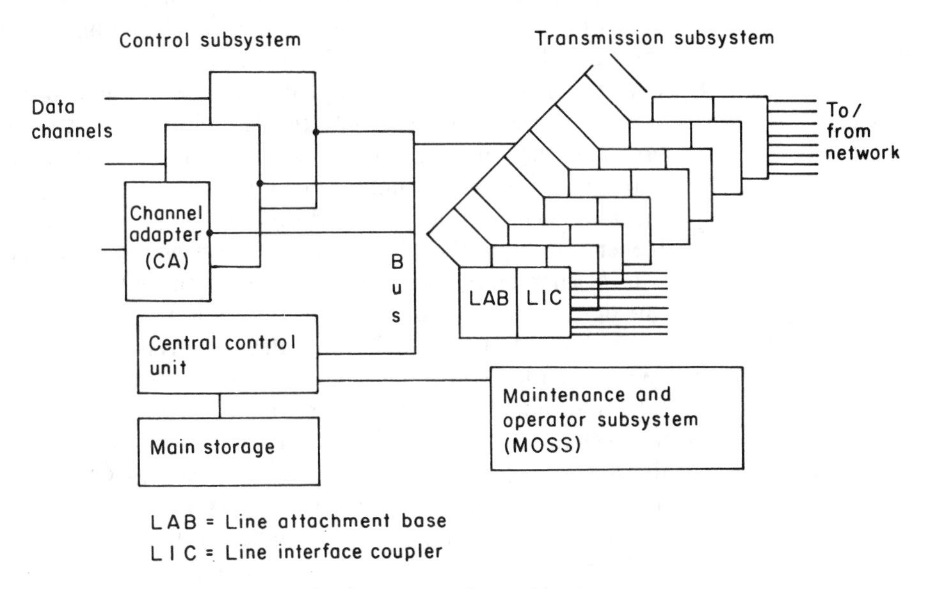

Figure 3.55 3275 communications controller subsystems.

The control subsystem consists of a central control unit, main storage, and channel adapters. The transmission subsystem consists of up to eight line attachment bases (LABs), each containing one or more communication scanners and line attachment hardware called line interface couplers (LICs). The maintenance and operator subsystem (MOSS) consists of a processor and storage, a diskette drive, control panel, and up to two externally connected operator consoles.

The channel adapters provide the physical interface between the communication controller and its direct attachment to a host computer system, permitting the controller to be attached to a byte-multiplexer, block multiplexer, or selector channel on the host computer. Up to two channel adapters can be installed in a 3725, while four additional channel adapters can be installed in the 3725's expansion unit, which is formally called the 3726 Communication Controller Expansion.

The line attachment bases house communication scanners and line interface couplers. Two types of LABs can be selected by the end-user based upon the data transmission rate of the connected line. The type A line attachment base

(LABA) contains one communication scanner and supports up to 32 low- and medium-speed lines, operating at or under 9600 bps. These lines can be connected to up to eight LICs, with each LIC supporting the physical connection of up to four lines.

The second type of LAB is the type B (LABB) device which contains two communication scanners and can support up to eight high-speed lines or 32 medium-speed lines, attached through a maximum of eight LICs. Up to three LABs can be installed in the basic 3725 controller, permitting it to service up to 96 lines. Five additional LABs can be installed in the 3726 Communication Controller Expansion, providing support for a total of 256 lines. The first two LABs must be type A devices, while the remaining six line attachment bases can be eight type A or type B, as required.

The scanner in each line attachment base is a hardware device that contains a microprocessor, storage for the scanner microcode, and a buffer area for servicing the lines. Each scanner is near universal in terms of support, as scanners can service a mixture of line speeds and communication protocols to include asynchronous, bisynchronous, and synchronous data link-control. Depending upon the type of line connection, the scanner will insert or delete control characters, perform error detection and correction, and perform such functions as bisynchronous code translation.

The line interface coupler provides the physical connection between the IBM communication controller and a directly connected terminal or communications facility terminated by a modem or digital service unit. Several types of LICs are available for selection, with their actual use based upon such factors as the line speed, type of line interface (RS-232/CCITT V.24 or CCITT V.35), and line protocol. Some LICs can be installed in either LABA or LABB bases, while other LICs can only be installed in type B line attachment bases.

The third major subsystem of the 3725 is the maintenance and operator subsystem (MOSS). This subsystem has its own microcode and is used to detect and isolate failures within the controller. Since the microcode in MOSS operates independently of the rest of the controller, it can function even when the controller is inoperative. Up to two external operator consoles can be connected to the 3725 maintenance and operator subsystem.

3.6 MODEM- AND LINE-SHARING UNITS

Cost-conscious company executives are always happy to hear of ways to save money on the job. One of the things a data communications manager can best do to make his or her presence felt is to produce a realistic plan for reducing expenses. It may be evident that a single communication link is less costly than two or more. What is sometimes less obvious is the most economical and effective way to make use of even a single link.

Multiplexing is usually the first technique that comes to mind. But there are many situations where far less expensive, albeit somewhat slower, equipment is quite adequate. Here, terminals are polled one by one through a "sharing device" that acts under the instructions of the host computer. Typically, the applications where this method would be most useful and practical would be those where messages are short and where most traffic between host computer and terminal moves in one direction during any one period of time. The technique which can be called "line-sharing" (as distinct from multiplexing) may work in some interactive situations, but only if the overall response time can be kept within tolerable limits. The technique is not as a rule useful for remote batching or RJE, unless messages can be carefully scheduled so as not to get in each other's way because of the long run-time for any one job. Although line-sharing is inexpensive, it has some limits to its usefulness, particularly in situations where a multiplexer, most likely a TDM, can be used to produce additional economic leverage.

A similar device

Another device which functions similarly to modem- and line-sharing units, although its operation and usage is different, is a control unit. This device is built to work with a specific type of terminal device, usually manufactured by the vendor that produced the control unit. In addition, while modem- and line-sharing units operate basically transparent to the data flow, control units scan and interpret the data, looking for and operating upon orders received in the information flow. For additional information on control units the reader is referred to Section 3.8.

Operation

A TDM operates continuously to sample in turn each channel feeding it, either bit by bit or character by character; and this produces an aggregate transmission at a speed equal to the sum of the speeds of all its terminals.

A conventional TDM operation is illustrated in Figure 3.56 (top). For example, a multiplexer operating character by character assembles its first frame by taking the letter A from the first terminal, the letter E from the second, and the letter I from the third terminal. During the next cycle, the multiplexer takes the second character of each message (B, F, and J, respectively) to make up its second frame. And the sampling continues in this way until traffic on the line is reversed to allow transmission from the computer to the terminals. The demultiplexing side of the TDM (operating on the receiving side of the network) assembles incoming messages and distributes them to their proper terminals or computer ports.

An FDM divides up the transmission link's total bandwidth into a number of distinct frequency bands, each of which is able to carry a low-speed channel. The FDM accepts and moves transmissions from all of its terminals and ports simultaneously and continuously.

A line-sharing network is connected to the host computer by a local link, through which the host polls the terminals one by one. The central site transmits the address of the terminal to be polled throughout the network by way of the sharing unit. This is illustrated in Figure 3.56 (bottom). The terminal assigned this address (01 in the diagram) responds by transmitting a Request to Send (RTS) signal to the computer, which returns a Clear to Send (CTS),

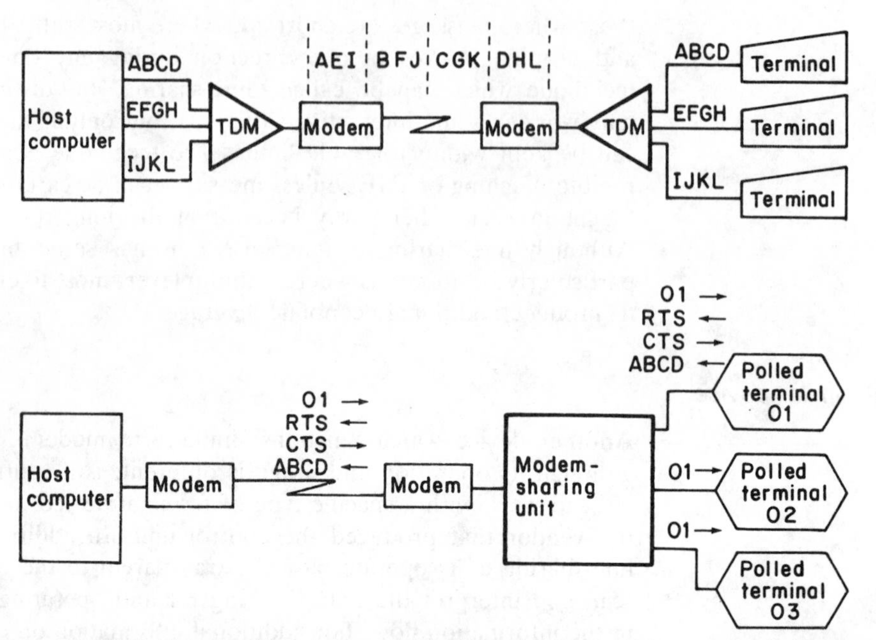

Figure 3.56 Multiplexing versus line-sharing. Top: Time division multiplex network. Bottom: Modem-sharing network. Multiplexer needs: A time or frequency division multiplex system (top) requires one computer port for each terminal and a multiplexer at each end. A sharing system (bottom) needs only one computer port. Because it requires terminals to be polled, a sharing system can be cost-effective for interactive operation, but may not be so for long messages such as are likely to move in remote job entry or remote batch types of applications.

to prompt the terminal to begin transmitting its message (ABCD in diagram). When the message is completed, the terminal drops its RTS signal, and the computer polls the next terminal.

Throughout this sequence, the sharing device continuously routes the signals to and from the polled terminal and handles supporting tasks, such as making sure the carrier signal is on the line when the terminal is polled and inhibiting transmission from all terminals not connected to the computer.

Device differences

There are two types of devices that can be used to share a polled line: modem-sharing units and line-sharing units. They function in much the same way to perform much the same task — the only significant difference being that the line-sharing unit has an internal timing source, while a modem-sharing unit gets its timing signals from the modem it is servicing.

A line-sharing unit is mainly used at the central site to connect a cluster of terminals to a single computer port, as shown in Figure 3.57. It does, however, play a part in remote operation, when a data stream from a remote terminal cluster forms one of the inputs to a line-sharing unit at the central site to make it possible to run with a less expensive single-port computer.

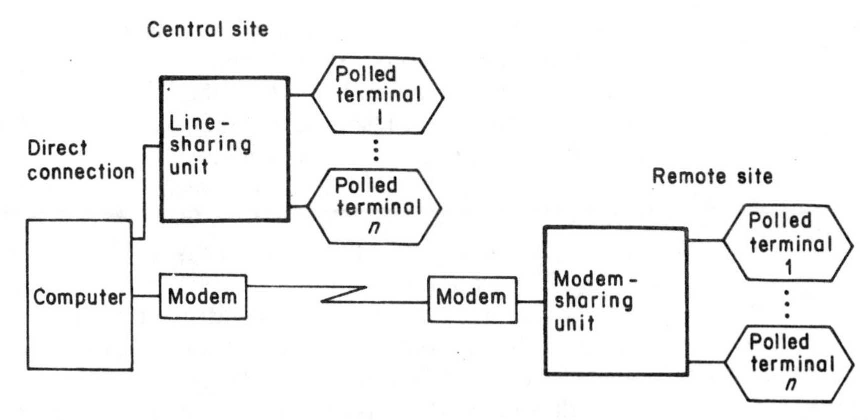

Figure 3.57 Line-sharing and modem-sharing use compared. Line-sharing units tie central site terminals to the computer, but modem-sharing units handle all the remote terminals. A line-sharing unit requires internal timing, whereas a modem-sharing unit gets its timing from the modem to which it is connected. In either case, access to the host is made through a single communications link — either a 2-wire or 4-wire — and a single port at the central site computer.

In a modem-sharing unit, one set of inputs is connected to multiple terminals or processors, as shown in Figure 3.57. These lines are routed through the modem-sharing unit to a single modem. Besides needing only one remote modem, a modem-sharing network needs only a single 2-wire (for half-duplex) or 4-wire (for full-duplex) communications link. A single link between terminals and host computer allows all of them to connect with a single port on the host, a situation that results in still greater savings.

If multiplexing were used in this type of application the outlay would likely be greater, because of the cost of the hardware and the need for a dedicated host computer port for each remote device. A single modem-sharing unit, at the remote site, is all that is needed for a sharing system, but multiplexers come in pairs, one for each end of the link.

The polling process makes sharing units less efficient than multiplexers. Throughput is cut back because of the time needed to poll each terminal and the line turnaround time on half-duplex links. Another problem is that terminals must wait their turn. If one terminal sends a long message others may have to wait an excessive amount of time, which may tie up operators if unbuffered terminals are used; but terminals with buffers to hold messages waiting for transmission will ease this situation.

Sharing unit constraints

Sharing units are generally transparent within a communications network. There are, however, four factors that should be taken into account when making use of these devices: the distance separating the data terminals and the sharing unit (generally set at no more than 50 ft under RS-232-C/D interface specifications); the number of terminals that can be connected to the unit; the various types of modems with which the unit can be interfaced; and whether the terminals can accept external timing from a modem through a sharing unit. Then, too, the normal constraints of the polling process, such as delays arising from turnaround and response and the size of the transmitted blocks, must be considered in designing the network.

The 50 ft limit on the distance between terminal and sharing unit (RS-232/ CCITT V.24 standard) can cause problems if terminals cannot be clustered closely. A way to avoid this constraint is to obtain a sharing unit with a DCE option. This option permits a remote terminal to be connected to the sharing unit through a pair of modems, as illustrated in Figure 3.58. This in turn allows the users the economic advantage of a through connection out to the farthest point. Since the advantage of modem-sharing units over a multipoint line is the reduction in the total number of modems when terminals are clustered, only one or at most a few DCE options should be used with a modem-sharing unit, as it could defeat the economics of clustering the terminals to utilize a common modem.

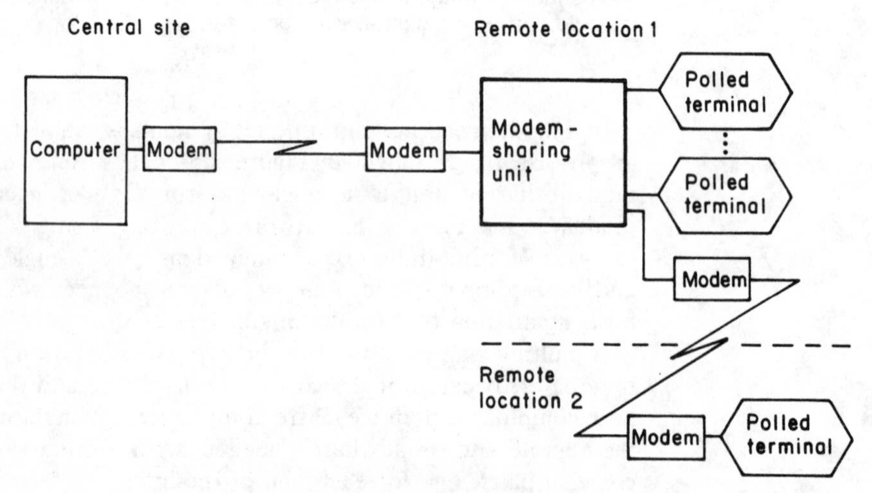

Figure 3.58 Extending the connection. Line- or modem-sharing units form a single link between a host computer and terminals. This system contains a modem-sharing unit with inputs from the terminals at its own site as well as from remote terminals. A line-sharing unit at the central site can handle either remote site devices or local devices more than 50 ft away from the host computer, which is the maximum cable length advisable under the RS-232-C/D/CCITT V.24 standards.

It is advisable to check carefully into what types of modem can be supported by modem-sharing units, since some modems permit a great deal more flexibility of network design than others. For instance, if the sharing unit can work with a multiport modem, the extra modem ports can service remote batch terminals or dedicated terminals that frequently handle long messages. An example of this flexibility of design is shown in Figure 3.59. Some terminals that cannot accept external timing can be fitted with special circuitry through which the timing originates at the terminal itself instead of at the modem.

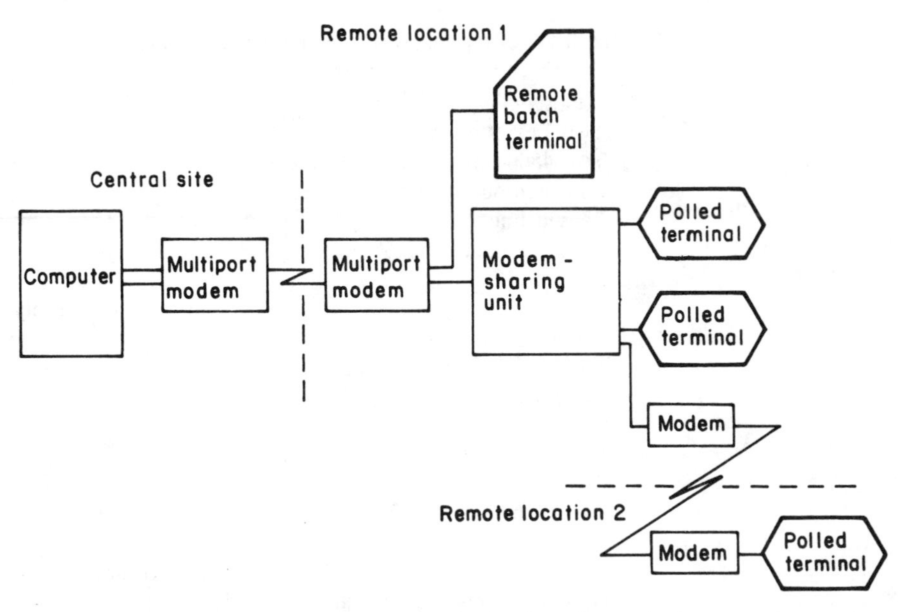

Figure 3.59 Multiple applications can share the line. Through the use of a modem-sharing unit with a data communications equipment interface, a terminal distant from the cluster (location 2) can share the same line segment (computer to location 1) that is used to transmit data to those terminals at location 1. With a second application that requires a remote batch terminal at location 1, additional line economies can be derived by installing multiport modems so both the polled terminals and the RBT can continue to share the use of one leased line from the computer to location 1.

Economics of utilization

The prices of sharing units range from $500 to $3000, depending mainly upon the number of terminals that can be connected through the unit. At present this number varies, the most versatile units being able to handle up to 32 terminals.

As shown in Table 3.17, a typical multiplexing system containing a line leased at $1000 a month and designed to service four 1200-bps terminals (Part A of the table) might cost the user $2160 a month; a system with a modem-sharing unit designed to service four polled 4800-bps terminals (Part B of the table) would cost $1900 a month, or 12 percent less. Because the leased line contributes the largest part of the system's cost, the percentage saved with a less expensive line can be even greater. For instance (Part C of the table), a line leased for $500 a month would increase the overall saving to 16 percent.

Other sharing devices

Sections 3.7 through 3.9 cover other sharing devices. Section 3.7 explores the use of port-sharing units, which are used in polled networks for the programmed

Table 3.17 Comparison of monthly rental costs.

A. Multiplexing		Monthly cost
Two TDMs at $90 each		$ 180
Two 4800-bps modems at $120 each		$ 240
Four computer ports at $35 each		$ 140
Leased line		$1000
Four terminals at $150 each		$ 600
	Total monthly cost	$2160
B. Using a modem-sharing unit		Monthly cost
Two 4800-bps modems at $120 each		$ 240
Computer port		$ 35
Modem-sharing unit		$ 25
Modem-sharing unit		$1000
Leased line		$ 600
Four terminals at $150 each		
	Total monthly cost	$1900
C. How percentage savings increase as leased line cost decreases		
	$1000 a month leased line	$500 a month leased line
Multiplexing	$2160	$1660
Modem-sharing unit	$1900	$1400
Percentage savings	12%	16%

selection of the computer port. Section 3.8 examines control units which although functioning similarly to sharing units are different in operation and usage. Section 3.9 discusses port selection (or port contention) units, where the unit provides random access to the computer ports under its control. The most significant differences in the various types of sharing units lies in their placement and function in the options available to them. Unlike line-, modem-, and port-sharing units, the port selection devices operate by time-sharing or contention. Access to any one port is provided on a first-come, first-served basis whenever a port is available. With port selection, therefore, a large number of lines contend for a small number of ports. Users let go of a port by signing off in a way similar to that found in time-sharing and RJE applications. The reader should refer to these sections for additional information.

3.7 PORT-SHARING UNITS

An alternative to the utilization of modem- and line-sharing units in a communications network can be obtained through the employment of devices known as port-sharing units. In addition, the proper employment of such devices can be used to complement or supplement modem- and line-sharing units and in certain situations may result in large economies being realized.

When to consider their use

Port-sharing units are devices that are installed between a host computer and modem and that control access to and from the host for up to six terminals with the number of terminals limited by the capabilities of current hardware. In this way, port-sharing units are able to cut down on the number of computer ports needed for these terminals. The port-sharing unit is versatile, inexpensive (about $500), and available from many modem and terminal manufacturers. Its utilization can save the cost of one or more relatively expensive computer channels that does essentially the same job but may have more capabilities than are needed. Port-sharing units can be used to service both local and remote terminals and so expand the job that can be done by a single port of the host computer.

To put the concept of port-sharing into perspective, the user should be aware of related devices designed to cut networking costs. Modem-sharing units and line-sharing units are available to minimize modem and line costs at remote locations, but they do not deal with the problem of overloading the host computer's ports. Modem- and line-sharing units are partial solutions to the high cost of data communications networking, but they are limited to the types of modems they can handle; and they can, by themselves, complicate the life of the network designer.

A problem that surfaces when either modem- or line-sharing units are used by themselves is the distribution of polled terminals within the network. For either kind of sharing unit to be effective, the terminals should be placed so that several are grouped close together. A typical modem-sharing unit employed to connect a number of terminals at a remote location for access to one computer port via a single pair of modems is illustrated in Figure 3.60. Some modem-sharing units can be obtained with a DCE interface option which is an RS-232-C/D interface by which remote terminals at two or more locations can be connected to the modem-sharing unit through the installation of a pair of modems between the terminal and the modem-sharing unit. Such a configuration is shown in Figure 3.61. Although the use of the interface shown in Figure 3.61 permits the network to have a more flexible configuration, with a number of terminals remote from the sharing unit, the number of such terminals that can be served by any one unit is usually limited to one or two.

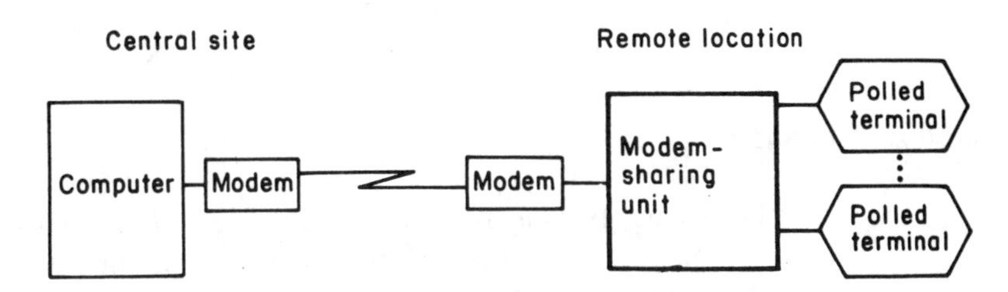

Figure 3.60 Modem-sharing unit usage. A modem-sharing unit permits a number of polled terminals to share the usage of a single line, one pair of modems, and a single computer port.

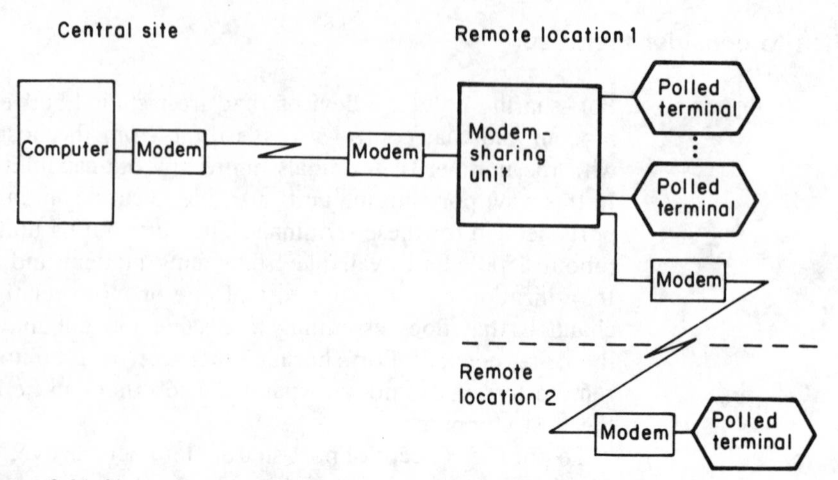

Figure 3.61 Network expansion using modem-sharing units with data communications equipment (DCE) interface. Remote terminals at two or more locations can share a common polled circuit when a DCE interface is present on the modem-sharing unit.

Another disadvantage of this arrangement is that it is rather like putting all your eggs in one basket. If either modem on the high-speed link between the computer and the modem-sharing unit should fail, or if the circuit itself goes down, all the remote terminals become inoperative. Multiplexed terminals can use the dial-up network to restore data communications if the dedicated line fails. Polled terminals, however, do not have this advantage, since the host computer software is set up to seek and recognize the addresses of specific terminals in a certain order on the line. Therefore polled terminals must stay in their respective places, relative to each other, along the communications route. Any change in route necessitates changes in hardware at the terminals as well as software at the host computer.

As new applications develop and the number of remote terminals connected to the host computer increases, a situation can arise where the network designer runs out of ports to service the network. If no additional ports are available, a costly computer upgrade or the installation of a second computer system would represent a major economic burden. A method to obviate or postpone these types of equipment upgrades is through the utilization of port-sharing units.

Operation and usage

Port-sharing, then, is presented either as an alternative or as a supplement to modem- and line-sharing, in networks without multiplexers. A port-sharing unit is connected to a computer port and can transmit and receive data to and from two to six either synchronous or asynchronous modems, as shown in Figure 3.62.

Data from the computer port is broadcast by the port-sharing unit, which passes the broadcast data from the port to the first modem that raises a receiver

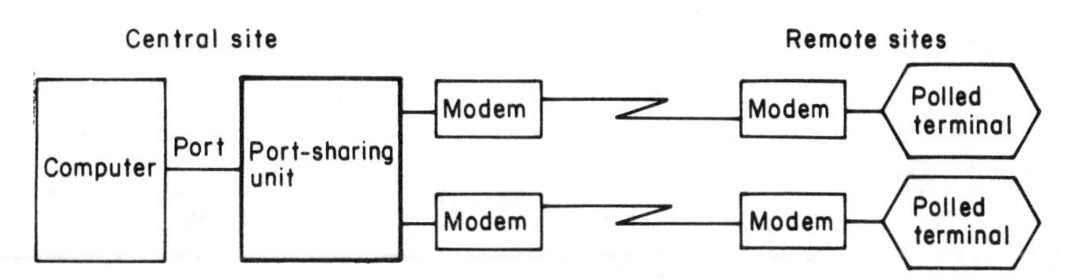

Figure 3.62 Using a port-sharing unit. The port-sharing unit lies between the host computer and the modems at the central site. One advantage is that if a failure occurs on the communications line or at the modem, only the terminal on that particular line goes down. On a polled, multidrop line, a line failure renders all terminals beyond it inaccessible.

— carrier detect (RCD) signal. Data for any other destination will be blocked by the unit until the first modem stops receiving. The port-sharing unit thus provides transmission by broadcast and reception by contention for the port connected to it. Like a modem-sharing unit, a port-sharing unit is transparent with respect to data transmission. Data rates are limited only by the capabilities of the terminal, modem, and computer port.

But in order to gain the same results without a port-sharing unit would require a multidrop configuration. Both the port-sharing unit and a multidrop network allow a large number of terminals to be served by one computer port; but in a multidrop network the failure of any part of the circuit will put all terminals beyond the failure out of action. In the configuration in Figure 3.62, however, failure of modem or outage on the line will only cut out a terminal on that segment. Failure of a computer port or of the port-sharing unit would of course bring down the entire network, but these devices are stable and such failures are fairly unusual.

Port-sharing as a supplement

Port-sharing units may also be used alongside modem-sharing units. If modem-sharing units alone are used, a situation can arise where there are not enough ports to serve the network, as in the top of Figure 3.63. If each modem-sharing unit serves its full complement of terminals and all the computer ports are in use, expansion of the network, even by just one port, may require a second mainframe computer.

This problem can also be dealt with by the use of a port-sharing unit at the central site which by cutting down the number of ports currently needed allows a network to expand without additional computer ports. This is illustrated at the bottom of Figure 3.63 which shows how one port-sharing unit with a two-modem interface can free a computer port from the configuration shown at the top of that illustration.

One versatile feature of port-sharing units is an option that allows the unit to accept a local interface instead of the normal RS-232-C/D interface, so that up to two local terminals may be operated without modems at the central site, as shown in Figure 3.64.

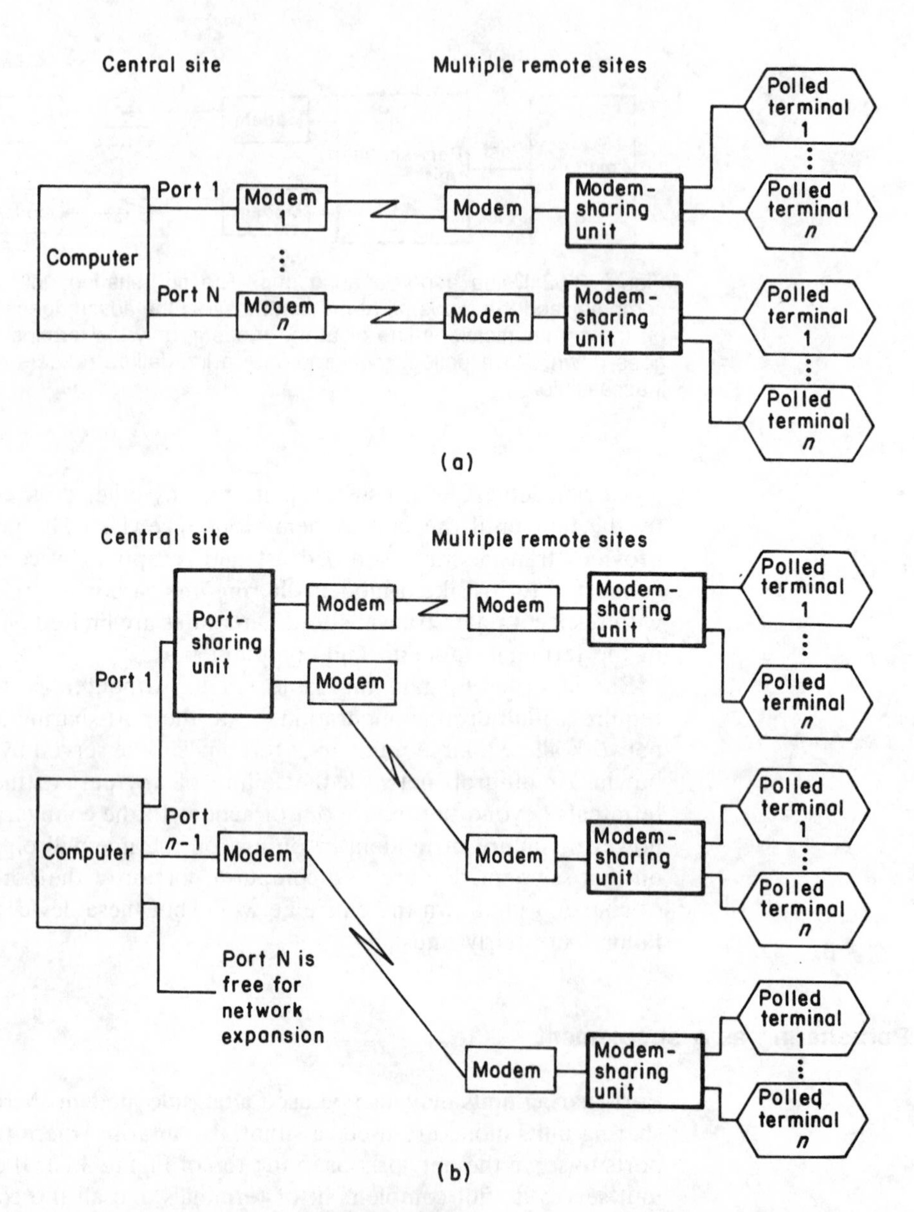

Figure 3.63 Two sharing techniques combined. When only modem-sharing units are used (a) a time may come when every port is in use and no further expansion on the network is possible. Rather than add another computer to serve a single new terminal, the user may prefer to invest in a port-sharing unit (b) that will solve the problem for about $500.

While both modem-sharing units and port-sharing units are similar in the way they are used, there is an important difference in the normal placing of their interfaces. In Table 3.18, a comparison of the characteristics of a port-sharing unit with those of modem- and line-sharing units will be found. For additional information on the latter two devices the reader is referred to Section 3.6.

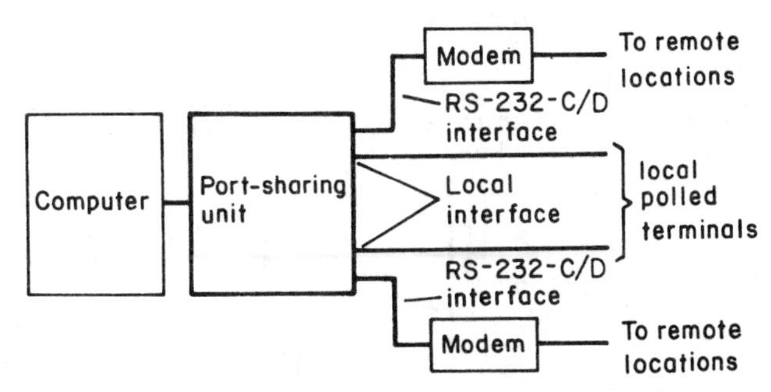

Figure 3.64 Connecting local peripherals. A local interface option to port-sharing lets local and remote sites to be served by the same port. Peripherals are polled as if they were at remote sites.

Table 3.18 Features of sharing units.

Feature	Modem-sharing/line-sharing unit	Port-sharing unit
Transmit mode	Broadcast	Broadcast
Receive mode	Contention	Contention
Number of modems interfaced	2–32	2–6
Terminals supported	Polled	Polled
Options	RS-232-C interface (MSU to modem)	Local interface (PSU to terminal)
Normal interface placement	Between modem and terminal	Between computer port and modem

Comparing cost with alternate components

In Figure 3.65, the reader is presented with two alternate means of connecting four 2400-bps remotely located terminals to a central computer facility. In the top portion of that illustration, a port-sharing unit and four individual lines and four pairs of modems are installed to enable the terminals to communicate on a poll and select basis with the computer. For this configuration, the failure of any modem or line will only render the terminal connected to the failing modem or line inoperative. In the lower portion of Figure 3.65, a pair of 4-channel synchronous multiplexers has been installed to service the four terminals which are now restricted to being collocated unless other equipment is installed in addition to the devices shown in that illustration. Table 3.19 compares the monthly cost of a four-terminal network in multiplexed and port-sharing configurations. The breakeven point comes when the cost of each leased line (x) reaches \$103 a month. This figure can be arrived at by taking the total known costs (for multiplexers, computer ports, modems, and four terminals, at an average of \$150 a month apiece), and adding an unknown cost — for the leased line — which remains constant whether the user employs port-

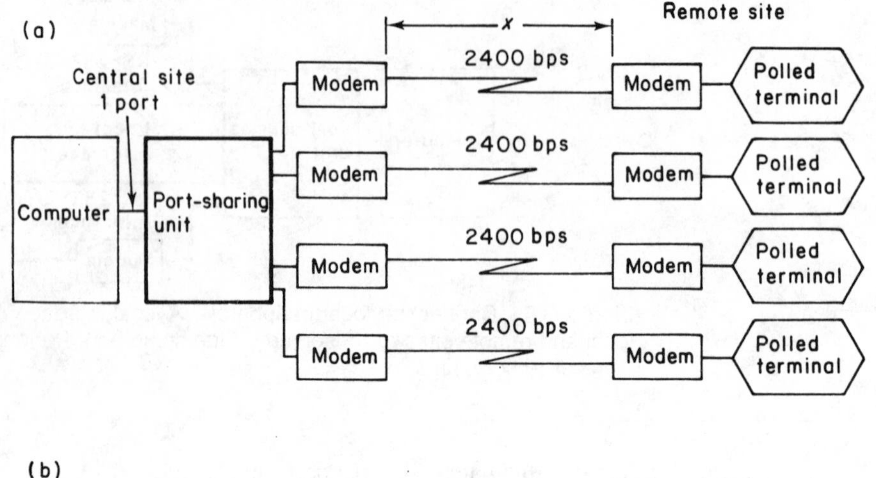

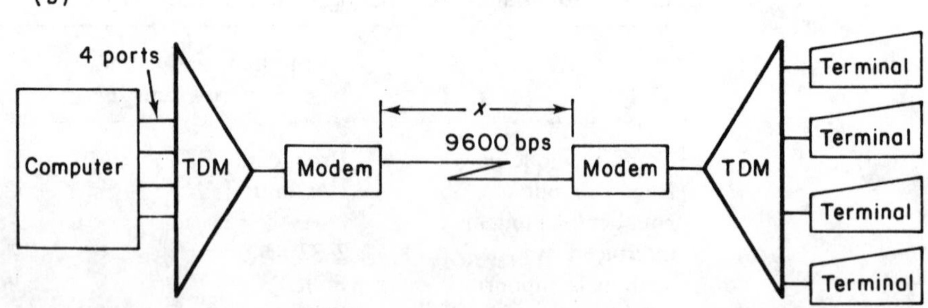

Figure 3.65 Comparing port-sharing units with multiplexers. (a) Port-sharing unit servicing four polled terminals. (b) Using multiplexers to service the terminals. When a leased line is equidistant from all terminals to the computer, a cost comparison becomes simple as listed in Table 3.19.

sharing units or chooses a multiplexed network configuration. Balanced against this is the somewhat smaller total rental amount for a single port-sharing unit, one computer port, four lower-speed modems, and four terminals, all at the same rate of $150 a month.

More leased lines

The lower part of Table 3.19, however, shows an increase in the number of leased lines from one to four. The upper part of the table, therefore, gives a fixed cost of $1420 with a variable amount on a one-time basis for a single leased line, while the lower part shows a significantly lower equipment cost but gives a variable cost for leased lines four times that for a single-line multiplexed configuration.

In this example, $1420 less $1110 equals $310, or 3*x* (with *x* still representing the variable cost of the leased line). Dividing $310 by 3 gives a breakeven point of $103. Therefore, until monthly leased line charges total $103 for each line, the use of a port-sharing unit is more economical. Although the preceding

Table 3.19 Comparison of monthly costs.

Multiplexed configuration costs	
Two 4-channel TDM at $120	$ 240
Four computer ports at $35	140
Two 9600-bps modems at $220	440
Four terminals at $150	600
One leased line	x
	$1420 + x
Port-sharing unit configuration costs	
One port-sharing unit at $25	$ 25
One computer port at $35	35
Eight 2400-bps modems at $55	440
Four terminals at $150	600
Four leased lines	$4x$
	$1110 + $4x$
equating: $1420 + x = 1110 + 4x$, $x = 103$	

x = monthly cost of a leased line

cost comparison assumed that the cost of the terminals used on the port-sharing and multiplexer networks illustrated in Figure 3.57 were equivalent, this may not necessarily be true. Since terminals used on a port-sharing network generally require a buffer area, these terminals may be more expensive than non-buffered ones that could be used on the multiplexer network. If, owing to the application requirement, buffered terminals become necessary, then the terminal cost can be eliminated from consideration since they would be necessary regardless of the network employed.

In addition, without increasing network costs users can add up to two more local or remotely located terminals to a configuration based on a port-sharing unit, since the table is based on the costs related to four terminals and the average port-sharing unit can support up to six. The only additional cost would be for the rental of the terminal units; none would be incurred for additional modems or leased lines when local terminals are connected.

In order to support two additional terminals in a multiplexed configuration, however, the user would have to pay substantially more. Not only is an additional $35 per computer port required, but since the multiplexer is operating at the transmission limit of a leased voice-grade circuit, substantially more expensive wideband facilities would have to be installed to service the upgraded multiplexers. In addition, as explained earlier, the cost of adding two computer ports can be further aggravated if all ports are already in use on the computer, so that any extra load requires another entire processing unit.

The port-sharing unit, therefore, is most evidently a cost-saving tool when the user is already straining his computer system to its limits. Saving money is a constant preoccupation for all cost-conscious data communications managers and port-sharing should be considered in any polled-terminal situation where instantaneous response is not the most important network condition; yet there are times, such as when the computer runs out of capacity, when the cost of any further network expansion takes a leap from a few hundred dollars to perhaps tens of thousands for another computer.

A similar device

Another device which has a function similar to port-sharing units, although its operation and usage differ, is a port selector (or port contention) unit. This unit provides access to a computer port on a first-come, first-served basis. Instead of up to six lines contending for a port, a larger number of lines can reach the same port when it opens up. The availability of a port occurs when a current user of the port disconnects from the system. For additional information, the reader is referred to Section 3.9.

3.8 CONTROL UNITS

Although similar in functionality to modem- and line-sharing units, a control unit is a much more sophisticated device. Basically the control unit functions like a modem- or line-sharing unit, since it enables many devices to share the use of a common line facility. However, apart from a degree of similarity with respect to their functionality, a control unit both operates and can be utilized differently from modem- and line-sharing units.

Control unit concept

The concept of the control unit originated in the 1960s with the introduction of the IBM 3270 Information Display System which consisted of three basic components — a control unit, display station, and printer. The control unit enabled many display stations and printers to share access to a common port on IBM's version of a front-end processor, marketed as a communication controller, or via a common channel on a host computer, when the control unit was directly attached to a host computer channel. Here the term display station represents a CRT display with a connected keyboard.

Since the introduction of the 3270 Information Display Unit, most computer vendors have manufactured similar products, designed to connect their terminals and printers to their computer systems. Control units vary considerably in configuration, ranging from a single control unit built into a display station to a configuration in which the control unit directs the operation of up to 32 attached display stations and printers. These attached display stations and printers are usually referred to by the term cluster; hence, another popular name for a control unit is cluster controller.

The major differences between IBM and other vendor control units is in the communications protocol employed for data transmission to and from the control unit, the types of terminals and printers that can be attached to the control unit, and the physical number of devices the control unit supports. In this section we will focus our attention upon IBM control units; however, when applicable, differences in the operation and utilization of IBM and other vendor devices will be discussed.

Attachment methods

IBM control units can be connected to a variety of IBM computer systems, ranging in scope from System/360 and System/370 processors to 43XX computers as well as certain System/3 models. The control unit can be connected either directly to the host computer via a channel attachment to a selector, multiplexer, or byte multiplexer channel on the host; or it can be link-attached to IBM's version of the front-end processor known as a communication controller as illustrated in Figure 3.66. Once connected, the control unit directs the operation of attached display stations and printers, with such devices connected via a coaxial cable to the control unit. Note that a generic type of control unit is indicated in Figure 3.66 by the use of the term 327X. Several control unit models are marketed by IBM, with numerous types of each model offered. The actual control unit installed depends upon the communications protocol (bisync or synchronous data-link control) used for communications between the control unit and the host computer, the type of host system to which the control unit is to be connected, and the method of attachment (channel- or link-attached), and the number of devices the control unit is to support.

Two of the most popularly used IBM control units are the 3274 and 3276 devices. The 3274 enables up to 32 devices to be connected while the 3276 is a table-top control unit with an integrated display station that permits up to seven additional devices to be connected to it. In 1987 IBM introduced its 3174 control unit which can be obtained in several different configurations and which like the 3274 also supports a maximum of 32 devices. The major difference between the 3174 and 3274 is the availability of a token-ring option and a protocol conversion option for the newer 3174 control unit. The token-ring option permits the control unit to be connected to IBM's token-ring network, while the protocol conversion option permits asynchronous ASCII

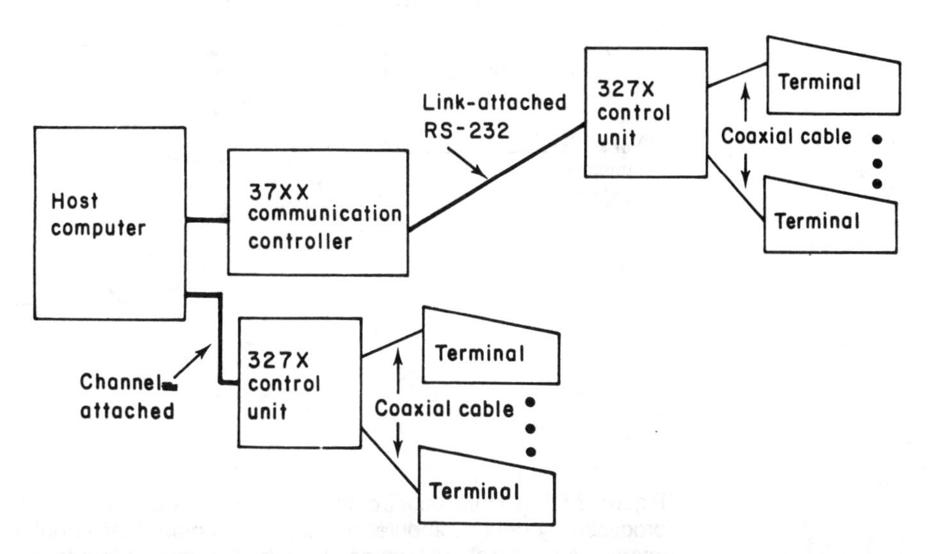

Figure 3.66 IBM 3270 Information Display System. Control units used in the IBM 3270 Information Display System can be either link-attached to a communication controller or channel-attached to a host computer system.

terminals to be connected to the control unit in place of normally more expensive EBCDIC devices.

Since the attachment of terminal devices to an IBM control unit requires constant polling from the control unit to the terminal, either dedicated or leased lines are normally used to connect the two devices together. Due to this, the 3270 Information Display System is considered to be a "closed system." In addition, many IBM control units require terminals to be connected by the use of coaxial cable. Thus, this normally precludes the use of asynchronous terminals using the PSTN to access a 3270 Information Display System.

The devices marketed by other computer vendors vary considerably in comparison to the method of attachment employed by IBM. Most non-IBM control units permit terminal devices to be connected via an RS-232/CCITT V.24 interface while this capability is only available on certain 3174 and 3274 models. Thus, terminal devices remotely located from the main cluster of terminals can be connected to such control units via tail circuits as illustrated in Figure 3.67. Another difference between IBM and other computer vendor control units concerns the method of attachment of the control unit to the host computer system. IBM control units can be channel-attached or link-attached, whereas most non-IBM control units are link-attached to the vendor's front-end processor, even when the control unit is located in close proximity to the computer room.

Unit operation

One of the major differences between a control unit and modem- and line-sharing units is the fact that the former is not a passive device. The 3270 data

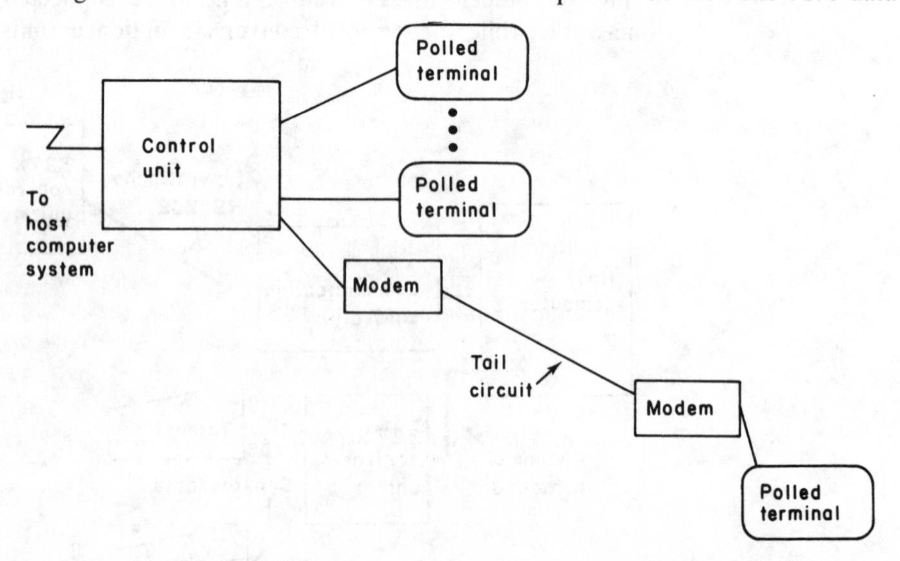

Figure 3.67 Non-IBM control units can connect terminals via tail circuits. Control units produced by many computer manufacturers other than IBM have RS-232/CCITT V.24 interfaces for attaching terminal devices. This type of interface permits distant terminals to be connected to the controller via tail circuits.

stream that flows between the control unit and the host computer system contains orders and control information in addition to the actual data. Both buffer control and printer format orders are sent to the control unit, where they are interpreted and acted upon. Buffer control orders are used by the control unit to perform such functions as positioning, defining, modifying, and assigning attributes and formatting data that is then written into a display character buffer that controls the display of information on the attached terminal. Thus, the terminal display information flows from the host computer to the control unit in an encoded, compressed format, where it is interpreted by the control unit and acted upon. Similarly, printer format orders are stored in the printer character buffer in the control unit as data and are interpreted and executed by the printer's logic when encountered in the print operation. Since a network can consist of many control units, with each control unit having one or more attached terminal devices, both a control unit address and terminal address is required to address each terminal device.

In comparison to a sharing unit where the host polls each terminal, the control unit performs the polling operation. This enables a higher degree of data throughput, since the computer system can then transmit and receive data during what would normally be a polling interval if sharing units were employed.

Protocol support

Both binary synchronous communications and synchronous data link control (SDLC) communications are supported by most members of the IBM family of 327x control units. Some control units are soft-switch selectable, incorporating a built-in diskette drive that permits a communications module incorporating either communication protocol to be loaded into the control unit; other models may require a field modification to change the communications discipline.

By employing a communications discipline, end-to-end error detection and correction is accomplished between the control unit and the host computer system. In comparison, line- and modem-sharing units are essentially transparent to the data flow and do not perform error checking, which either is performed by the individual terminals connected to the sharing device or is not performed.

Another difference between IBM control units and devices manufactured by other computer vendors is the communications protocol employed to link the control unit to the host computer. As previously mentioned, IBM control units communicate employing BISYNC or SDLC, the latter being IBM's version of the International Standards Organization's Higher-level Data Link Control (HDLC). Most control units manufactured by other computer vendors communicate with that vendor's host computer system using a proprietory protocol that is unique to that vendor or a standard protocol. One example of the former is Honeywell's V.I.P. 7705 protocol which, although similar to BISYNC, is also different and thus incompatible with IBM's BISYNC. An example of the latter would be the HDLC protocol offered by several computer vendors which, while closely matching the standard HDLC protocol, is sufficiently different from IBM's SDLC version as to be incompatible. Thus, although there are numerous vendors that manufacture IBM plug compatible control units, such control units normally will not operate with non-IBM computer systems. Similarly, most control units manufactured by other computer vendors for use with their computer systems will not work in an IBM environment.

Breaking the closed system

One of the key constraints of the IBM family of control units is the fact that it is a closed system, requiring terminal devices to be directly connected to the control unit. Without the introduction of a variety of third party products, this cabling restriction would preclude the utilization of personal computers and asynchronous terminals in a 3270 network. Fortunately, a variety of third party products have reached the communications market that enable users operating 3270 networks to overcome the cable restriction of control unit connections. Foremost among such products are protocol converters and terminal interface units.

Protocol converters

When used in a 3270 network, a protocol converter will function as either a terminal emulator or a control unit/terminal emulator, depending upon the method employed to attach a non-3270 terminal into a 3270 network. When functioning as a terminal emulator, the protocol converter is, in essence, an asynchronous to synchronous converter that converts the line by line transmission display of an asynchronous terminal or personal computer into the screen-oriented display upon which the control unit operates. Similarly, the protocol converter changes the full screen-oriented display image transmitted by the control unit into the line-by-line transmission which the terminal or personal computer was built to recognize.

During this translation process cursor positioning, character attributes, and other control codes are mapped from the 3270 format into the format which the terminal or personal computer was built to recognize, and vice versa. A protocol converter functioning as a terminal emulator is illustrated at the top of Figure 3.68. It should be noted that some protocol converters are stand-alone devices while other protocol converters are manufactured as adapter boards that can be inserted into a system expansion slot within a personal computer's system unit.

In the lower portion of Figure 3.68, a protocol converter functioning as a control unit/terminal emulator is illustrated. Since all terminals access a 3270 network through control units, stand-alone terminals or personal computers must also emulate the control unit function if they are to be directly attached to the host computer system.

Terminal interface unit

Another interesting device that warrants attention prior to concluding our discussion of control units is the terminal interface unit. This device is basically a coaxial cable to RS-232 converter, which enables coaxial cabled terminals to

Terminal emulation protocol converter

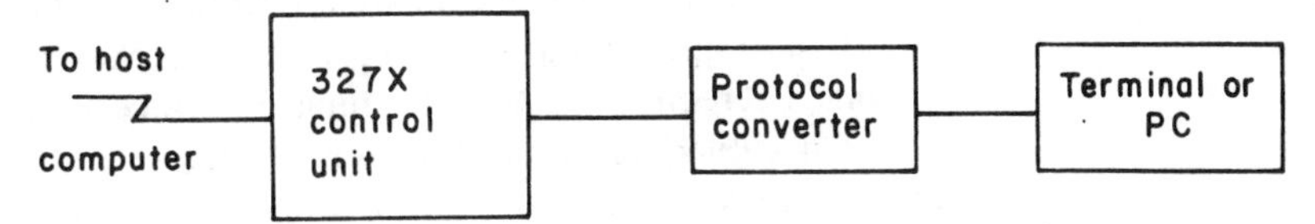

Control unit terminal emulator protocol converter

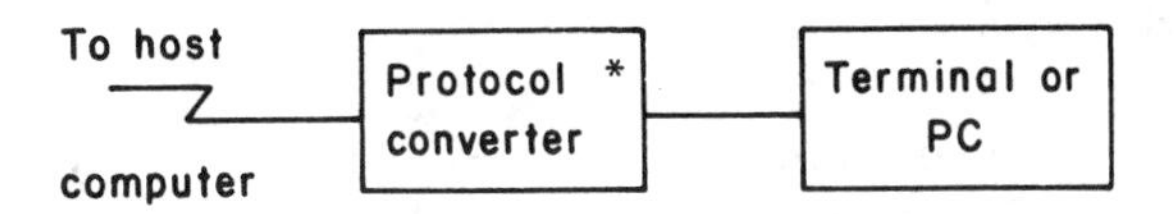

* Emulates both a control unit and terminal.

Figure 3.68 Protocol converters open the closed 3270 system. Through the use of protocol converters a variety of asynchronous terminals and personal computers can be connected to a 3270 network.

access asynchronous resources to include public databases and computer systems that support asynchronous transmission.

In an IBM 3270 network, terminal devices are connected directly to a control unit, precluding the use of the terminal from accessing other communications facilities. The terminal interface unit breaks this restriction, since it converts a coaxial interface used to connect many 3270 terminals to control units into an RS-232 interface at the flick of the switch on the unit. The use of this device is illustrated in Figure 3.69. When in a power-off state, the terminal interface unit is transparent to the data flow and the terminal operates directly

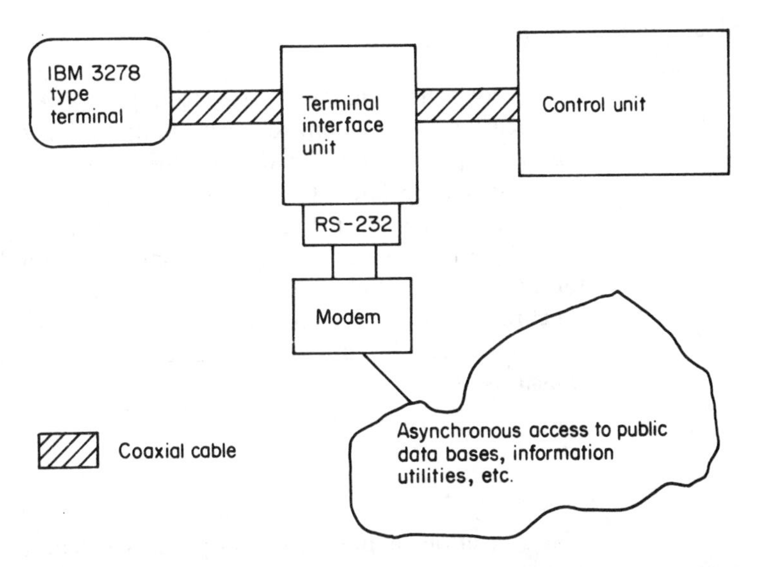

Figure 3.69. The terminal interface unit makes a coaxial cabled terminal multifunctional.

connected to the control unit. When in an operational mode, the terminal interface unit converts the keyboard-entered data into ASCII characters for transmission and similarly converts received characters for display on the IBM terminal's screen. Thus, the terminal interface unit can be employed to make a coaxial cabled terminal multifunctional, permitting it to access the coaxial cable port on a control unit as well as an RS-232/CCITT V.24 device, such as an asynchronous modem.

3.9 PORT SELECTORS

A traditional method used to provide service to new terminals as they are added to a network is through the expansion of the number of front-end processor ports. A variety of approaches can be used to accommodate such terminals. The number of dial-in lines at the computer center can be expanded, or additional dedicated or direct connect lines can be installed to service terminals added at the computer facility. For remote locations, the addition or upgrading of multiplexers and the installation of additional leased lines may be required to provide new channels to enable the new terminals to connect and transmit to the computer.

Even when such a network expansion is completed, it is unlikely that all ports will be busy at the same time. Thus, while extra lines and additional communications components may be required to provide additional transmission paths to the computer, some front-end processor ports may be operating only a fraction of the day. Port selectors, through their ability to cross-connect incoming transmission to available ports, permit more efficient front-end processor utilization.

Types of devices

Many equipment manufacturers sell port selection devices. Most of these products are stand-alone devices, built to function as an interface between computer ports and lines which may emanate from multiplexers, direct dial lines, or dial-up lines. Other manufacturers, however, offer port selection as an optional or built-in feature in their statistical and intelligent multiplexers.

Some port selectors are designed only to contend for asynchronous, teletype terminal traffic; other devices can contend for both asynchronous and synchronous traffic within the same unit, with each type of traffic being contended for by one or more computer ports servicing that mode of transmission.

Operation

The utilization of port selectors permits terminals to be added to a network without a corresponding increase in the number of computer ports. In addition, the utilization of this device may permit a system contraction whereby a number of computer ports become unnecessary and can be returned to the manufacturer.

The basic function performed by a port selector can be viewed as a dynamic data switch similar to telephone rotaries (stepping switches that sequentially search for available telephone lines), except that the selector provides appropriate interfaces between computers and terminals to route a large calling terminal population to a lesser number of called computer ports. Some selectors have additional features specifically applicable to data networks. Although users tend to confuse port selectors with port-sharing units, their applications are for specific line environments that result from the utilization of different types of terminals for specific applications. Port-sharing units are used in polled networks where the computer controls the traffic flow, and terminals must have a buffer area to recognize polls to their address. Port selection units are used in contention networks, in which terminals transmit to the host on a random basis; and the access to any port is normally on a first-come, first-served basis.

Computer site operations

From a network viewpoint, a port selector is similar to a black box with N line side input connections and n port side output connections, with $N \geq n$ as shown in Figure 3.70. The port selector continuously scans all line side connections for incoming data from terminals connected to that side. At the same time, the selector maintains a status check of available ports so that when a terminal becomes operational and requests access, the selector connects the terminal to an available port. Some port selectors can be arranged to form subgroups of contending terminals and ports, so that certain terminals can only be connected to one or more of an assigned group of ports. This capability is discussed more fully under "Selector features" at the end of this section. If all of the designated ports are taken, the selector will continue to scan until a port becomes available or until the request for access ceases. Another option is a "busy out" feature whereby the port selector signals new callers that all available ports are in use.

Other port selectors have the capability to place users into a queue when all ports are in use. Some port selectors offering a queuing feature inform the terminal user of their position in the queue only when their terminal access attempt is first initiated, while other port selectors continue to transmit the user's position in the queue each time the position changes or on a periodic basis.

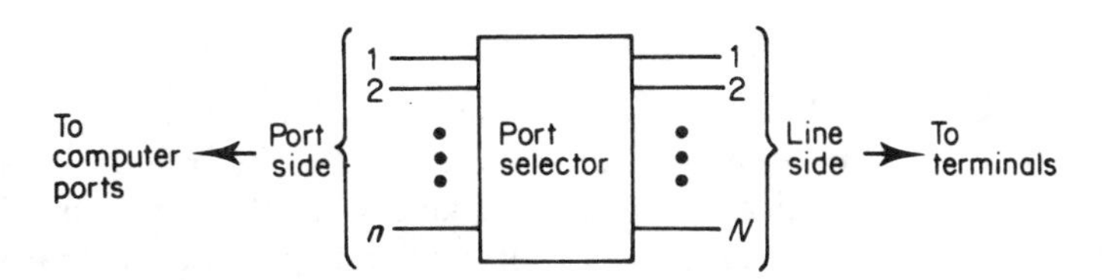

Figure 3.70 Port selector at computer site. At line side, port selectors can interface a variety of channels — from multiplexers, dial-in lines, leased lines, and direct connect lines. As terminal traffic becomes active, the port selector attaches the circuit to a predefined or randomly selected computer port for the duration of the transmission.

Remote operations

Port selectors can be used at remote locations to make input terminals contend for a lesser number of communications lines or multiplexer ports. Consider a remote location with 32 terminals, each communicating with the central computer at 300 bps, as illustrated in Figure 3.71. If all terminals at the remote location can be directly connected to the multiplexer to avoid telephone line and modem changes, one economical method to move the data is to multiplex the transmission from the 32 terminals over a single leased line, at an aggregate transmission speed of 9600 bps. This system requires two 32-port multiplexers and 32 computer ports. Assume that 16 additional 300 bps terminals must be added to the remote site after the system is installed. Since the leased line is already operating at its 9600-bps limit, one alternative is to upgrade the multiplexers by adding 16 ports to each and to replace the modems and voice-grade communications link with either higher-speed modems or wideband modems and a wideband circuit. This method requires 16 additional computer ports to service the additional terminal traffic. But, because wideband facilities are far more costly than voice-grade lines, expansion might be very expensive.

Another option would be to install a pair of 16-port multiplexers, one additional leased line between them, and two 4800-bps modems to transmit on the new line. But 16 additional computer ports would still be required to service the additional terminals. Increasing the number of terminals that can be handled by the central site by means of a port selection arrangement may indeed be less expensive than doing it by expanding the capacity of the multiplexing equipment. To find out if port selection is the practical answer to a particular situation, the user should first assess the demands the present terminals are making on the network and then estimate the additional demands that will arise because of the proposed new terminals.

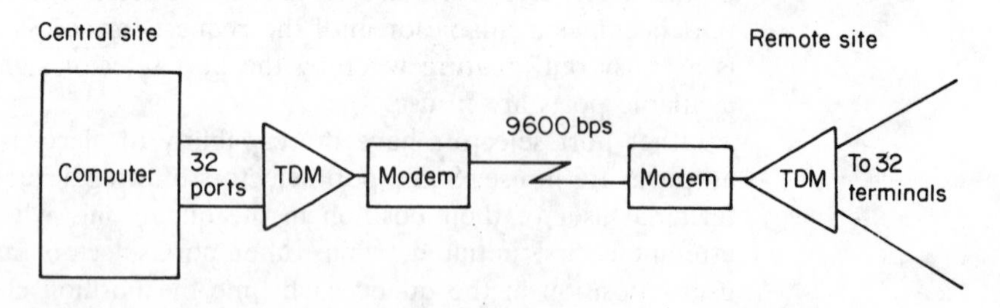

Figure 3.71 Traditional multiplexer utilization. When multiplexers are used for data concentration, one computer port is required for each multiplexer channel.

Let us assume a study was conducted and it was determined that although 48 terminals are necessary at the remote site, owing to time-dependent applications and terminal use habits (coffee breaks, different department shift schedules, etc.) a reasonable expectation is that with the exception of 2 hours per week, no more than 32 terminals will simultaneously use the computer. For the 2-hour high utilization period per week, the study concluded that at most 35 terminals will access the system. If we are willing to sacrifice three terminals from accessing the computer for 2 hours per week, then a port selector can be installed to service 48 lines and bring those lines to 32 ports on a contention basis, as illustrated in Figure 3.72. The port selector then offers a contention ratio of 48 : 32, or 3 : 2 (48 terminals connected to the line side of the selector and 32 outputs to the multiplexer ports). It eliminates the need to add 16 computer ports, to upgrade or add multiplexers, or to upgrade or add communications links. Although a telephone company rotary could be used for the port selection process, as shown in Figure 3.73, 80 modems would then be needed (48 for the terminals and 32 for the business lines connected to the multiplexer).

A port selector can thus cut back on the number of modems needed for port selection by rotaries, as well as the expense of 48 communications links. By directly connecting the terminals to the line side of the port selector it becomes possible to do away with the low-speed modems. A port selector that can make 48 lines contend for 32 ports might cost about $15,000. But a conservative cost for the rotary system of $1359 per month for the components enumerated in Table 3.20 would equal the cost of a port selector in a little less than one year.

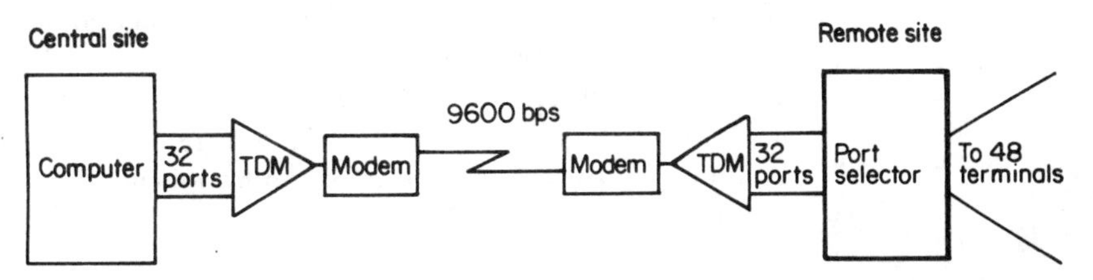

Figure 3.72 Servicing 48 terminals using a remotely located port selector. Port selector can be used to perform a remote concentration function by providing a 48 to 32 contention rate from the 48 terminals connected to the line side of the device to the 32 ports which are connected to the input side of the remotely located multiplexer.

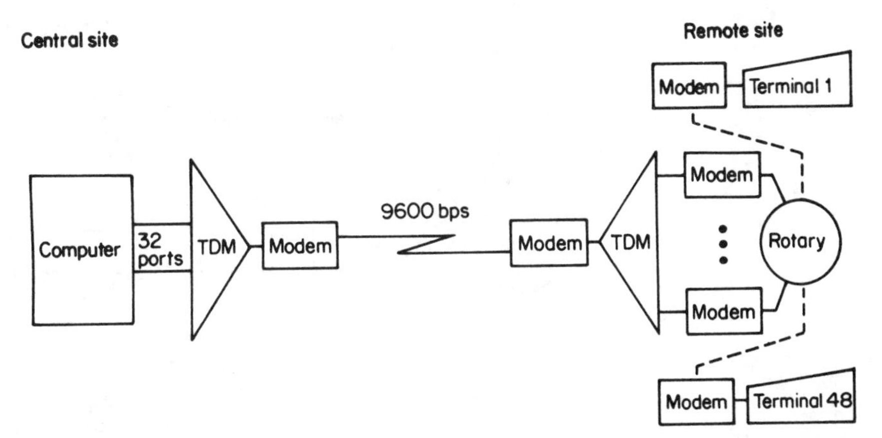

Figure 3.73 Using telco rotaries for the expansion. Although the installation of a telephone company rotary and lines will permit an infinite number of terminals equipped with modems to contend for the 32 multiplexer channels, the cost of such equipment may not prove to be the most economical method to employ.

Table 3.20 Monthly rotary equipment charges.

80 modems at $12.50	$1000
32 1FB business lines at $10.00	320
2 data cabinets at $4.50	9
1 804T autoanswer device at $30.00	30
Monthly cost	$1359

Another device which should be considered for this type of terminal expansion is a statistical or an intelligent multiplexer. This type of multiplexer strips the start and stop bits of each asynchronous transmission by only transmitting data as terminal activity occurs; and it permits the transmission of several terminals to occur on what would be one reserved time slot for an individual terminal when data is multiplexed on a traditional TDM. The use of this device could alleviate the necessity of installing additional line facilities. However, when terminals are directly connected to the multiplexer at the remote site a corresponding number of computer ports will still be required at the computer site, unless the composite high-speed data link was demultiplexed by specialized software in the front-end processor, in which case only one port would be required. Owing to the cost associated with developing specialized software as well as the memory requirements for the program, this technique is not normally employed unless the economics of scale can justify the effort and cost involved. The reader is referred to Section 3.1 of this book for additional information on the use and applications of statistical and intelligent multiplexers.

Usage decisions

A typical computer network, as in Figure 3.74, indicates what must be taken into account in coming to a decision about port selectors. In this example, 48

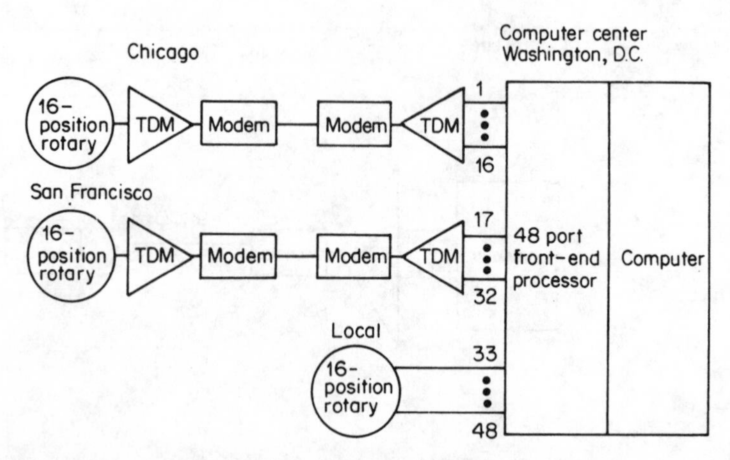

Figure 3.74 Traditional computer network. In a typical computer network a 48-port front-end processor would permit the 32 remote terminal and 16 local terminals to reach the computer simultaneously. This configuration insures that no terminal user ever meets a busy signal — but it is expensive.

computer ports handle messages from two multiplexers and a local rotary. Each multiplexer provides 16 terminal-to-computer connections, as well as 16 dial-in ports at the computer site. If the network is distributed over several time zones, peaks in utilization of the terminals will occur at different times in different places.

A typical profile of the number of users logged onto the network from each geographical area can be computed with statistical software packages provided by computer manufacturers. Profiles of the system's utilization are shown in Figure 3.75. While other networks may not reveal exactly the same patterns,

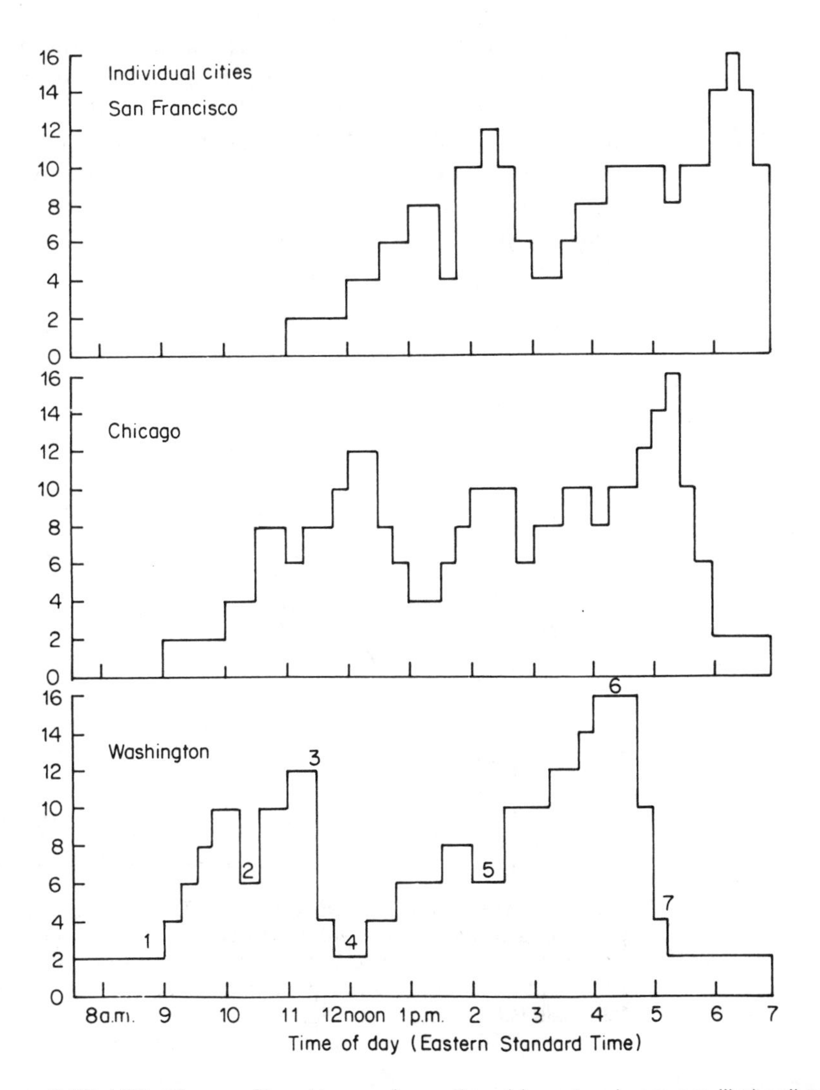

Figure 3.75 Utilization profiles. Users of a nationwide network are not likely all to be in their offices or doing the same amount of work at the same time. This situation works in favor of a port selection configuration; it will readily be seen that peak usage rarely coincides in the different, widely separated centers served by a large communications network.

they may be similar because normal working habits are a significant factor in the fluctuations of network utilization. At the start of working hours (Figure 3.75, point 1 on the third profile), use gradually builds as people arrive at the office and settle down to work. During the morning coffee break period (point 2), the number of users decreases temporarily, with the length of time and the degree of the drop-off varying from place to place. Morning peak use (point 3) is followed by a drop in activity during the lunch period (point 4). Use then builds up until the afternoon break (point 5) and peaks again as people rush to complete the day's work (point 6). As the close of business approaches (point 7), activity tapers off until only a few terminals remain on-line.

By combining the three local profiles, an overall network profile can be developed (Figure 3.76) which represents the number of terminals on-line as the day moves forward (Eastern Standard Time). The smooth curves above this profile plots activity in the 95th and 99th percentiles, and they indicate the maximum number of terminals that can be expected to be on-line for 95 and 99 out of every 100 observations. Put another way, 95 percent of the time there will be up to 37 terminals on-line or seeking access between 4:15 p.m. and 4:45 p.m., and 99 percent of the time there will be as many as 40 users (between 4 p.m. and 5:30 p.m.).

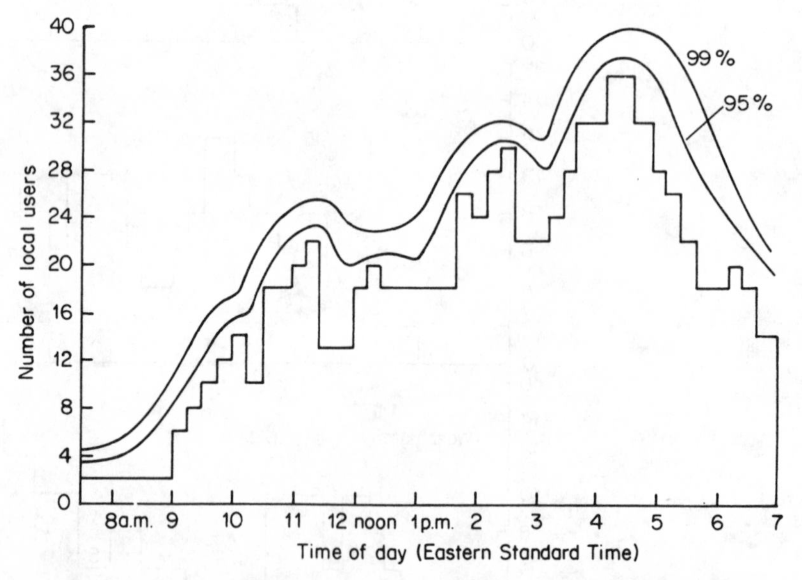

Figure 3.76 The total network profile. The curves represent the 95th and 99th percentiles of use for the entire network. They show that at peak periods only 37 or 40 terminals, respectively, will be either on-line or seeking access. The curves also show that during these same periods of peak use, a substantial number of the remote terminals will inevitably be idle.

Returning to the network in Figure 3.74, it is now apparent that 99 percent of the time, eight or more of the 48 computer ports are not in use, and that 95 percent of the time 11 or more ports will be idle. Thus, the use of a port selector becomes a question of economics against inconvenience. Is the cost of a number of mostly idle computer ports worth the seldom used advantage of being able to connect all terminals, simultaneously and without delay? A related question that the network designer must answer is whether the computer can process all messages rapidly enough when all terminals are on-line.

Let us assume that instead of installing the equipment illustrated in Figure 3.74, a port utilization study was conducted and that the 90th percentile of port usage was decided upon. This would mean that a 48-line by 32-port port selector would be required. After calculating the cost savings possible with such a port selector we can determine if the sacrifice of 16 continuously available ports is justifiable. Assuming such justification, the revised network which incorporates a port selector is illustrated in Figure 3.77. Other port selector configurations could result from an investigation into the savings possible with a 48-line by 37-port selector (95th percentile, Figure 3.76) or a 48-line by 40-port selector (99th percentile).

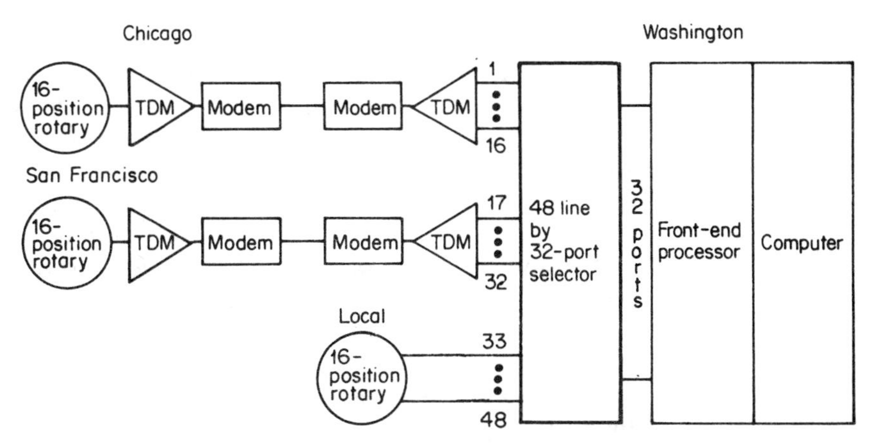

Figure 3.77 Network revision to incorporate a port selector. This network provides almost the same performance as the network shown in Figure 3.65. However, through the use of the port selector, only 32 computer ports are required. Because of the different time zones, it is unlikely that a remote terminal will get a busy signal.

Port costs

Besides a savings in the circuit boards that make up the ports themselves, other computer costs are reduced because of the fewer ports. Let us assume that the user has a Honeywell Series 60 computer and a Honeywell DATANET 6000 front-end network processor (FNP) which provides the network communications power for the Honeywell Series 60 computer system. The basic configuration of the DATANET 6000 FNP is shown in Figure 3.78.

Each FNP can contain up to two general-purpose communications bases (GPCB), each of which can accommodate up to 16 communication dual channel boards (up to 32 concurrently operating ports) at speeds of 50 to 50,000 bps and up to six asynchronous communications bases (ACB), each of which can service up to 17 ports at 30 characters per second through the installation of nine communication dual channel boards. Thus, the 32 ports from the port

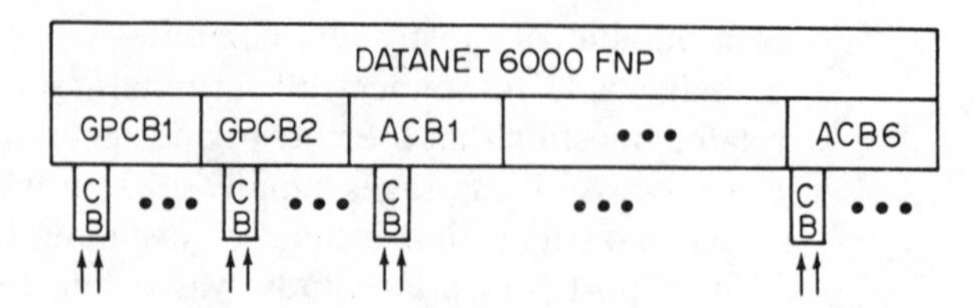

Figure 3.78 DATANET 6000 FNP basic configuration. The Honeywell 6000 front-end network processor contains one or more communications bases with each base containing one or more communication boards.

selector can be interfaced to one GPCB containing 16 communication channel boards or two ACBs containing eight communication channel boards apiece. Without a port selector to accommodate all 48 ports, an additional ACB or GPCB and eight communication channel boards would be required.

Since the installation of a 48-line by 32-port selector would save an additional ACB or GPCB and eight communication channel boards, the user must determine if the cost of the port selector functioning as a contention device outweighs the value of the extra ports. Later in this section, we will examine other functions which a port selector can perform that can result in its utilization from other than an economic perspective.

If the network serves an in-house computer system, productivity lost when terminals are denied access to the computer must be considered. If the network is used by outside customers, a possible loss of revenue should also be taken into account.

In some instances, using port selectors can result in additional savings for installations that have redundant computers, because every excess computer port that can be eliminated on one front-end processor can also be eliminated on the other. In addition, reductions may also be possible in the capital outlay for devices that switch between central processing units. One such redundant processor configuration is shown in Figure 3.79.

The installation of a port selector to front end a line transfer device is shown in Figure 3.80. Note that not only are the number of front-end processor ports reduced by 32 (16 per processor) but also that the size of the line transfer device can be reduced, since a 32 by 64 switch is now needed instead of a 48 by 96 switch if a selector was not used.

Load balancing

A key advantage obtained from the utilization of port selectors is the ability to balance the communications load when an installation has two or more front-end processors. This load balancing can be accomplished simply by wiring the cables from the port side of the selector in an alternate manner to each front-end processor. Thus, an installation with dual front-end processors would cable the leads from ports 1, 3, 5, 7, 9, . . ., $N - 1$ from the port selector to the first front-end processor and the leads from ports 2, 4, 6, 8, . . ., N to the second front-end processor.

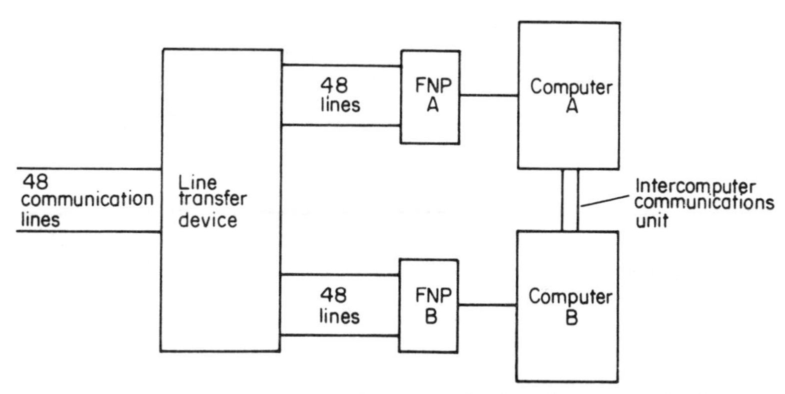

Figure 3.79 Redundant processor configuration. Savings in excess of twice the savings of a single processor may be obtained since a portion of the switching arrangement may be removed.

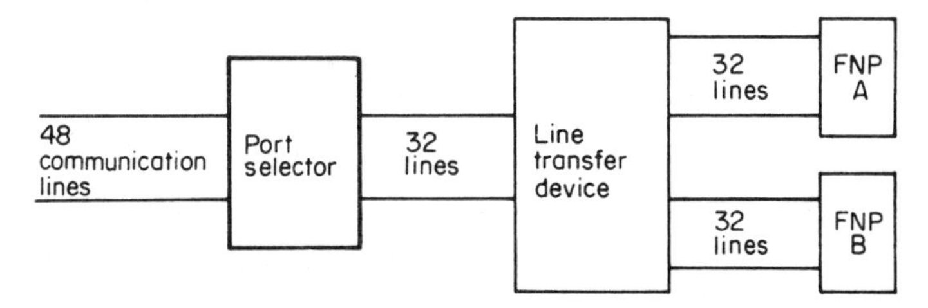

Figure 3.80 Port selector increases savings for redundant configurations. Using a port selector to front-end the line transfer device of a dual processor not only doubles the number of ports that can be eliminated but also reduces the size of the switch required.

Selector features

A wide range of standard and optional port selector features should be considered. First, the basic size or capacity of the port selector and its expansion capability, if any, should be determined. Selectors are offered as a base unit, with a predetermined number of calling channels and called ports available for interface. By the addition of line and port nests and adapters, the capacity of the selector can be increased.

Another feature to consider is the ability to partition the selector for interface to fixed speed front-end or multiplexer channels. This feature is illustrated in Figure 3.81, where port contention is accomplished within each of three speed/ code groupings for transmission to a lesser number of fixed speed front-end processor ports.

Even when using a speed-transparent selector where all incoming traffic transmits at the same rate, one may wish to consider assigning priorities for access to the computer by grouping certain incoming lines so that they contend for a different number of ports, thereby providing a higher access probability for certain users of the system. This is shown in Figure 3.82. Instead of being preconfigured to seek available ports within a predetermined group or groups

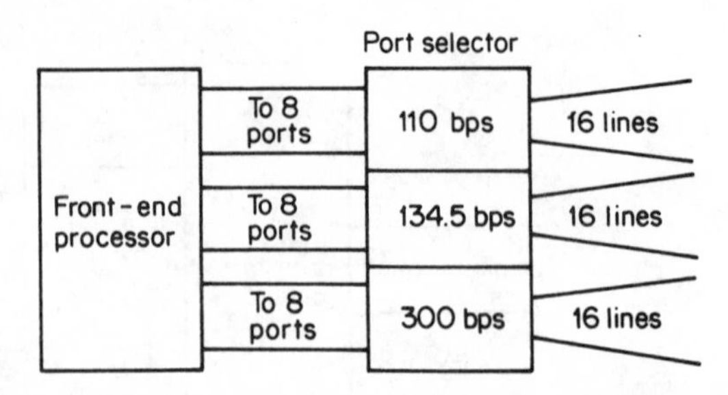

Figure 3.81 Speed-partitioned port selector. The 48 incoming lines are contended for in groups of 16 by 8 ports; each group can contend for eight predefined speeds.

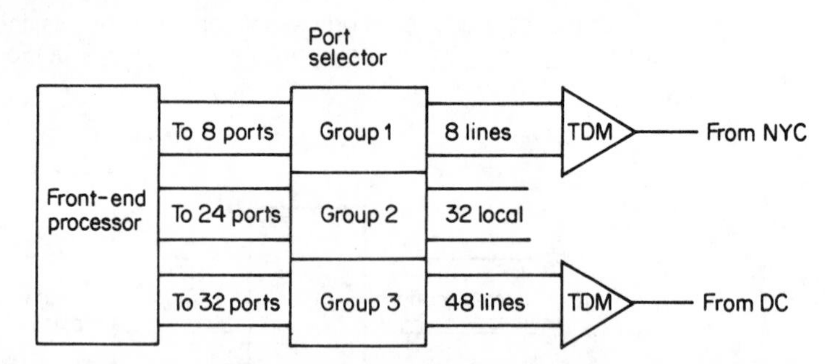

Figure 3.82 Grouping selector contention. Users from New York City always require immediate access so no contention is performed. Although 32 local terminals are connected to the selector, management has decided that at most 24 will be active at one time and has directed a 4:3 contention ratio be established for this group.

of ports, automatic group selection is an option which permits the selector to determine which of the port groups a given user requires based upon the transmission of a control character from the user's terminal.

Line-switching network

One of the most useful features of a port selector is its ability to serve as the foundation for the construction of a line- or circuit-switched network. This can be accomplished by obtaining a port selector that has an automatic group selection feature.

To understand how a port selector can function as the foundation for a line-switching network, let us assume our organization has two data-processing centers — one located in Macon, Georgia, and a second located in Miami, Florida. Suppose there are a large number of terminals located in Washington, D.C., and New York City that require access to both data centers. One method that might be employed to provide communications support to the terminal users would be the installation of multiplexer systems to link each city to each data center. This would require, as an example, two multiplexers and high-speed modems to be installed in New York City. Then two circuits would also be required, with one circuit routed to Macon and the second to Miami.

At each data center, the circuit would be terminated with a high-speed modem and a multiplexer. In addition to installing dual multiplexers and high-speed modems at the terminal locations, there would be significant distances where the circuits would be routed in parallel. Thus, the utilization of a common multiplexer system at each terminal location could result in significant cost savings as well as operational efficiencies if a method were obtained to allow terminal users to access one multiplexer and control the routing of their terminal traffic to either data center. One method that will permit this to occur is by the installation of a port selector with an automatic group selection feature as illustrated in Figure 3.83. A port selector is installed in Macon, Georgia, which is the initial termination point of the multiplexed data traffic from New York and Washington, D.C. The automatic group selection feature of the port selector is then configured to recognize two groups. Group 1 is assigned to the ports of the port selector that are cabled to the host computer system located in Macon. Group 2 is assigned to a group of ports that are routed from the port selector into a multiplexer that is then connected to a circuit that is routed to Miami.

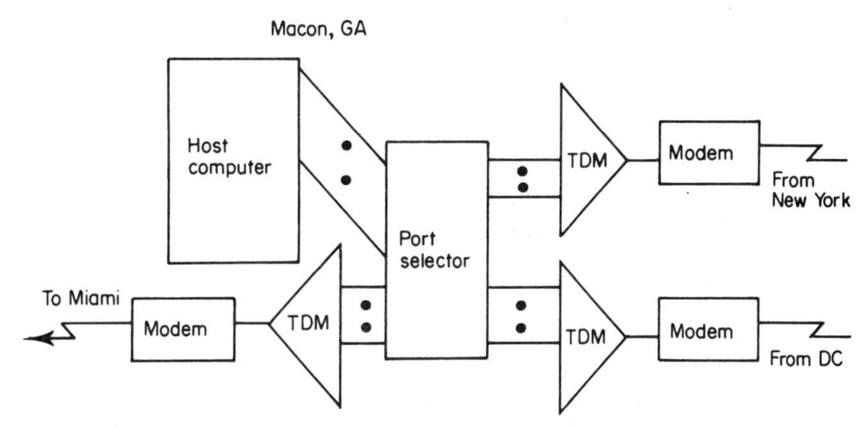

Figure 3.83 Constructing a circuit-switched network. A port selector with an automatic group select feature enables the utilization of a common network segment to access geographically separated computer systems.

When a terminal user in New York or Washington, D.C. reaches the port selector, he will be prompted by the port selector to enter the routing group number or similar parameter. Then, by entering a "1" the user would be routed to the computer in Macon while the entry of a "2" would cause him to be routed out of the port selector and into a multiplexer that would connect him to the computer in Miami. Thus, in this example, the port selector enables the utilization of a common network segment for obtaining access to geographically distant computer systems.

Additional features available on selectors include diagnostic modules and busy-out switches. Selector diagnostic modules are used to determine which line is connected to which port and to display key diagnostic information for that particular cross-connect while busy-out switches permit user flexibility in changing existing contention ratios.

REVIEW QUESTIONS

3.1 What are the major differences between time division multiplexing and frequency division multiplexing?

3.2 How many 300-bps data sources could be multiplexed by a frequency division multiplexer and by a traditional time division multiplexer?

3.3 Why does frequency division multiplexing inherently provide multidrop transmission capability without requiring addressable terminals and poll and select software?

3.4 Assume your organization has two computer systems, one located in city A and another located in city B. Suppose your organization has eight terminals in city C and 10 terminals in city B, with the following computer access requirements:

Terminal location	Terminal destination	
	City A computer	City B computer
City B	6	4
City A	4	4

Draw a network schematic diagram illustrating the use of time division hub-bypass multiplexing to connect the terminals to the appropriate computers.

3.5 Why is a front-end processor substitution for a multiplexer at a computer site the exception rather than the rule?

3.6 Assume the cost of communications equipment and facilities is as follows:
8-channel TDM at $2000
9600 modem at $3000
leased line city A to city B $1000/month
If eight 1200-bps terminals are located in city A and the computer is located in city B, draw a schematic of the multiplexing network required to permit the terminals to access the computer, assuming each terminal can be directly connected to the multiplexer. What is the cost of this network for one year of operation?

3.7 Assume the requirements of Question 3.6 are modified, resulting in all eight terminals being located beyond cabling distance of the multiplexer. Assume the following cost of communications equipment and facilities:

8 position rotary	$50/month
telephone lines at	$30/month
1200-bps modems at	$200

Draw a revised schematic of the multiplexing network. What is the cost of this network for one year of operation?

3.8 Why must a statistical multiplexer add addressing information to the data it multiplexes?

3.9 Why are most statistical multiplexers ill-suited for multiplexing multidrop circuits?

3.10 Assume the statistical multiplexer you are considering using has an efficiency 2.5 times that of a conventional TDM. If you anticipate connecting the multiplexer to a 9600-bps modem, how many 1200-bps data sources should the multiplexer multiplex?

3.11 Assume the multiplexer discussed in Question 3.10 services synchronous data by the use of a bandpass channel. What would be the effect upon the number of 1200-bps asynchronous data sources supported in Question 3.10 if you must service a 4800-bps synchronous data source by the use of a bandpass channel?

3.12 Assume you are considering the use of an intelligent multiplexer that has an efficiency three times that of a conventional multiplexer for asynchronous data traffic and 1.5 that of a conventional multiplexer for synchronous data traffic. If you have to multiplex 18 1200-bps asynchronous terminals and four 2400-bps synchronous terminals on a line operating at 9600 bps, could the multiplexer support your requirement? Why?

3.13 How could you use the statistical loading data available from an intelligent multiplexer to determine if the vendor's literature concerning its efficiency in comparison to a conventional TDM is reasonable?

3.14 Draw a network schematic showing how the port contention option of a multiplexer could be used in a hub-bypass network.

3.15 Why do T1 carriers operate at different data rates in the United States and Europe?

3.16 Why is the method a vendor uses in digitizing voice important to consider when evaluating a T1 multiplexer?

3.17 You are comparing the cost of inverse multiplexing to using wideband facilities to transmit data at 19,200 bps, and you determine the monthly cost of the following facilities and devices.

Facility/device	Monthly cost
Wideband line	$9000
Voice-grade line	$1250
Inverse multiplexer at	500
9600-bps modem at	300
19,200-bps modems at	700

Would it be economical with these figures to use inverse multiplexers?

3.18 What are the primary advantages of inverse multiplexing in comparison to wideband transmission?

3.19 Why is the utilization of most inverse multiplexers limited to two channel devices?

3.20 What are the major differences between concentrators and statistical and intelligent multiplexers? Under what circumstances might you consider using a concentrator instead of a multiplexer?

3.21 Compare the primary purpose of read-only memory in concentrators to the use of read-only memory in statistical and intelligent multiplexers.

3.22 Assume the cycle time of the processors used in a redundant store and forward message switching system is 500 ns. If the procedure for the failing system to inform the other system to take control requires 300 ms and there are three processor cycles per instruction, what is the maximum number of instructions one can code to initiate cutover?

3.23 Under what situations should you consider using a remote network processor?

3.24 Discuss the differences between concentrators and statistical multiplexers concerning the method of handling the situation where the aggregate data input into the device exceeds the data transmission rate of the high-speed line connected to the device.

3.25 Draw a network schematic diagram illustrating the relationship between a host computer system, its front-end processor, a contention concentrator, and a data concentrator. For devices that communicate with a front-end processor, show the relationship between the terminal connections to the device and the device's connection to the front-end processor.

3.26 Discuss the difference between a data concentrator and a statistical multiplexer with respect to the connection of the device to a front-end processor.

3.27 What is the major difference between a channel-attached and a link-attached communications controller?

3.28 On what type of circuits are modem- and line-sharing units used? What is the major difference between devices?

3.29 What is the primary constraint one should consider when employing port-sharing units in conjunction with line- or modem-sharing units?

3.30 Discuss the differences between sharing units and a control unit.

3.31 Why is it difficult to attach RS-232/CCITT V.24 devices to an IBM 3270 Information Display System network? Discuss some of the products one can obtain to connect RS-232/CCITT V.24 devices into a 3270 network to include their operational features.

3.32 What is the major benefit derived from the use of a port selector?

3.33 What is the advantage of obtaining a port selector with queuing and queuing position display capability?

3.34 When should one consider the utilization of a port selector at a remote site?

3.35 Assume your primary computer system is located in New York and your organization has a second computer system in Boston. If your organization has a 16-channel multiplexer connection between Chicago and New York and a 32-channel multiplexer connection between Philadelphia and New York, draw a network schematic diagram to illustrate the use of a port selector to provide access for up to 16 simultaneous users to the Boston computer system using a minimum number of leased lines.

CHAPTER 4

REDUNDANCY AND RELIABILITY AIDS

Although a variety of network configurations can be designed to attain specific levels of redundancy and reliability, basic to such designs will be two communications components – switches and line restoral units. Initially developed as devices to assist technical control center personnel in a wide range of operational tasks, data communications switches were recognized as an economical and simplistic series of devices that could be used to reconfigure network components and replace failing devices with alternate components simply by the turn of a switch.

As networks became more complex and additional emphasis was placed on the duration of component outages, a series of automatic and semiautomatic switches were developed to provide network users with a lower failed component replacement time. At the same time that component failures were addressed, manufacturers recognized the desirability of providing devices which could detect leased line failures and automatically provide an alternate communications path over the switched telephone network until the leased line outage was alleviated. This category of equipment is commonly referred to as line restoral units and can be used to provide alternate communications paths for a variety of components normally connected to a central computer via the installation of a leased line.

4.1 DATA COMMUNICATIONS SWITCHES

Communication switches are bringing a new freedom and economy to network design and operations. Until recently, they were found mainly in network technical control centers, where they help in on-line monitoring, fault diagnosis, and digital and analog testing. But now they are also being used to reroute data quickly and efficiently and to replace several dedicated backup units with just one, enabling a single terminal to act as standby for several on-line terminals.

The kinds of switches available and how they may be combined or chained to fulfill different functions are described in the first portion of this section. The second portion of this section will concentrate on applications, in particular on the use of switches to assure network uptime without a heavy investment in redundant equipment.

The four basic categories of switches are fallback, bypass, crossover, and matrix. Of these, two or more from the same or different categories may be chained to serve still other data communications requirements. Furthermore, within each category there are two types of switches: the so-called telco switches which transfer 4-wire leased or 2-wire dial-up telephone lines; and the Electronics Industry Association (EIA) or CCITT switches referred to in this section as EIA switches, which transfer all 24 leads of an EIA RS-232 or CCITT V.24 interface.

Fallback switches

The fallback switch is a rapid and reliable means of switching other network components from on-line to standby equipment. The EIA version selects either a pair of 24-pin-connected components, which, as shown in Figure 4.1 may be terminals, modems, or channels on a front-end processor.

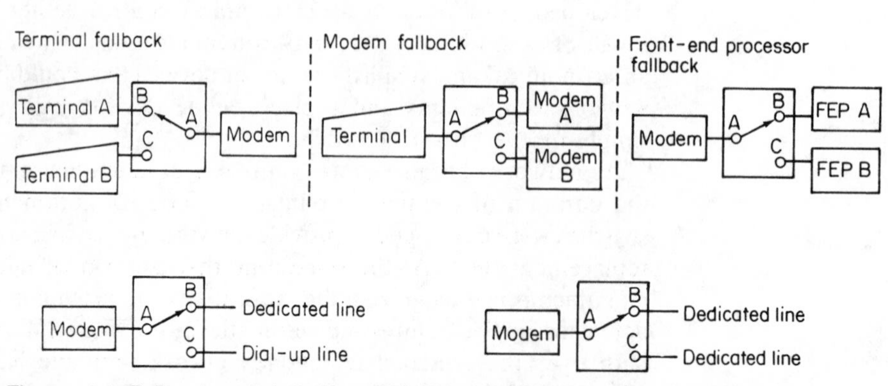

Figure 4.1 Fallback switches. Top: EIA (24-pin) fallback switch. Bottom: Telco fallback switch. The EIA fallback switch transfers 24 leads at a time, while the telco switch transfers the two or four leads associated with telephone lines.

In the first example, two terminals share a single modem. This configuration might be required, for example, when terminals have the same transmission speed but use different protocols, so that each communicates with a different group of remote terminals or computers.

In the second example, one terminal is provided with access to two modems, one of which is redundant but needed for uptime reliability. Alternately, the first modem might enable the terminals to transmit to another terminal at 2000 bps during one portion of the day, while the second lets it talk to a central computer at 9600 bps during other periods of the day. Then, depending on operational requirements, one terminal with a fallback switch for modem selection could be more practical than installing two terminals.

In a third application, an EIA fallback switch (Figure 4.1, top right) permits a modem to be transferred between front-end processors. Although called a line-transfer device by some manufacturers, in effect what one obtains is a device that selects which front-end processor will service the modem.

A telco fallback switch similarly allows the user to select one of two sets of telephone lines. As shown in Figure 4.1 (bottom), it can select one line from among various combinations of dedicated and dial-up lines that may have been installed to fit the needs of a particular application. Thus, for a large data transfer application which is of a critical nature, the telco fallback switch could be connected to a pair of leased lines, one of which is used as an alternative circuit in the event of an outage on the primary circuit.

Bypass switches

The EIA bypass switch connects several EIA interfaces of one type (for example, modems) to the same number plus a spare of another EIA interface type (for example, terminals) and can switch any member of the first group to the spare member of the second group. One application for bypass switches is at a computer installation (Figure 4.2, top left). Here, one front-end channel is reserved as a spare in case any of the existing channels, which normally

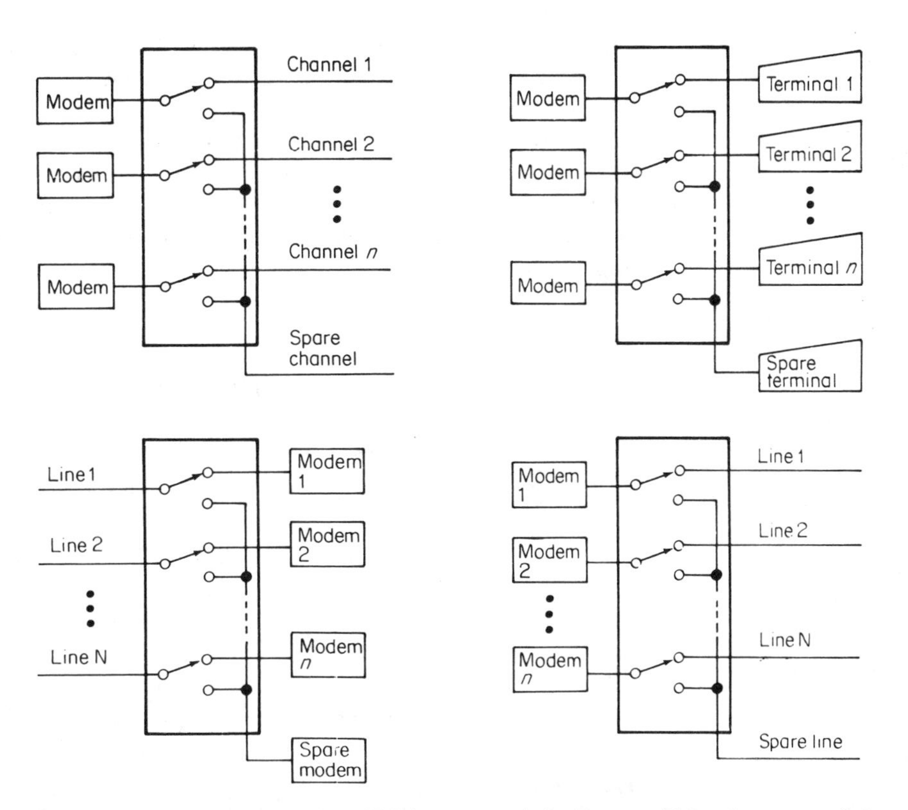

Figure 4.2 Bypass switches. Top: EIA bypass switch. Bottom: Telco bypass switch. Bypass switches transfer either EIA or telco interfaces to spare components with a similar interface on the other side of the switch.

service predetermined modems should need to be connected quickly to a spare channel.

In another application (Figure 4.2, top right), the EIA bypass switch can substitute a standby spare terminal for a failed on-line terminal and do away with the need for a spare modem. Although seldom used for multiple terminal access, a bypass switch can also enable many terminals to share a single modem and line.

A telco bypass switch transfers any one of a group of 2-wire or 4-wire telco lines to a spare communications component. For example, as shown in Figure 4.2 (bottom left), if modem 1 should fail, line 1 can be switched to the spare modem. Conversely, a telco switch may transfer a spare line to an operational communications component like a modem (Figure 4.2, bottom right). Telco bypass switches can be used to switch leased or dial-up lines to modems, automatic dialers, or acoustic couplers.

Crossover switch

Crossover switches provide the user with an easy method of interchanging the data flow between two pairs of communication components. Four connectors are associated with each switching module, one for each of the two pairs of communication components connected to that module.

As shown in Figure 4.3 (top), an EIA crossover switch permits the data flow to be reversed between two pairs of EIA interfaced components. In the example shown in Figure 4.3 (top), modem A, which is normally connected to the front-end processor channel A, and modem B, which is connected to the front-end processor channel B, are reversed when the switch is moved from the normal to the crossover mode of operation. Thus, modem A then becomes connected to channel B and modem B is connected to channel A upon crossover.

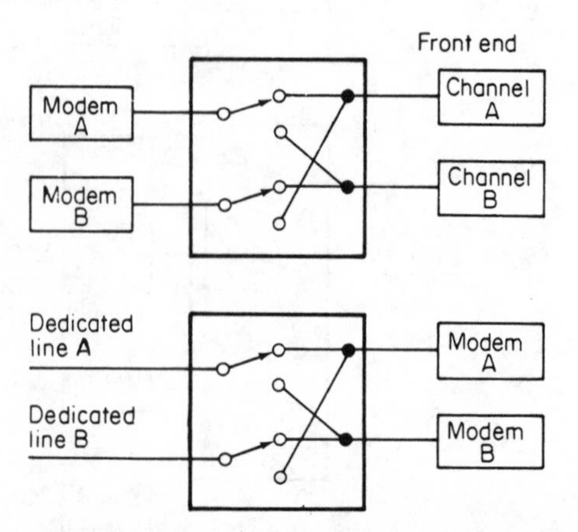

Figure 4.3 Crossover switches. Top: EIA crossover switch. Bottom: Telco crossover switch. Crossover switches, either EIA or telco, make it easy for the operator to reroute the flow of information between pairs of identical components.

Similarly, a telco crossover switch permits the user to interchange the data flow between two telco lines and two modems. Although two dedicated lines are shown connected to the crossover switch in Figure 4.3 (bottom), one can also connect one dedicated line and one dial-up line or two dial-up lines to the switch. Here, upon crossover, line A, which is normally connected to modem A, becomes connected to modem B, and vice versa.

Matrix switch

With a matrix switch the user can interconnect any combination of a group of incoming interfaces to any combination of a group of outgoing interfaces. Matrix switches are manufactured as an n by n matrix, with 4 by 4, 8 by 8, 16 by 16 combinations typically available. The user of a manual matrix switch makes an interconnection by depressing two pushbuttons on the switch simultaneously, one representing the incoming interface and the other representing the outgoing interface.

As shown in Figure 4.4 (top), an EIA 4 by 4 matrix switch is a quick and efficient way of connecting any combination of four modems to any combination of four front-end processor channels. The circles represent the depressed switch combinations, so that, in this case, modem 1 serves front-end processor channel 1, modem 2 serves front-end processor channel 3, modem 3 serves front-end processor channel 2, and modem 4 serves front-end processor channel 4. Further, with this configuration the user is free to designate one or more modem or front-end processor channels as spares or a combination of modems and channels as spares.

The telco 4 by 4 matrix shown in Figure 4.4 (bottom) similarly permits the transfer of any combination of four incoming lines to any combination of four outgoing lines.

A type of application warranting investigation of telco matrix switches arises when remote terminals require access to two or more adjacent computers. If the terminals are used heavily enough to justify installing leased lines from the remote sites to the central computers, the telco matrix switch enables the user to switch the incoming leased lines to outgoing cables which, via modems, are connected to different computers.

Additional derivations

A number of additional switching functions have been developed from the four categories of switches previously discussed. For instance, a spare component backup switch is basically a pair of fallback switches contained in one housing. As shown in Figure 4.5 (top left), this switch permits a normal and a backup mode of operation. The normal mode permits data to be transferred through the primary component, whereas the backup mode switches the data flow through the spare components.

In another configuration (Figure 4.5, top right), a pair of modems are the primary and spare components connected to one terminal, and the switch

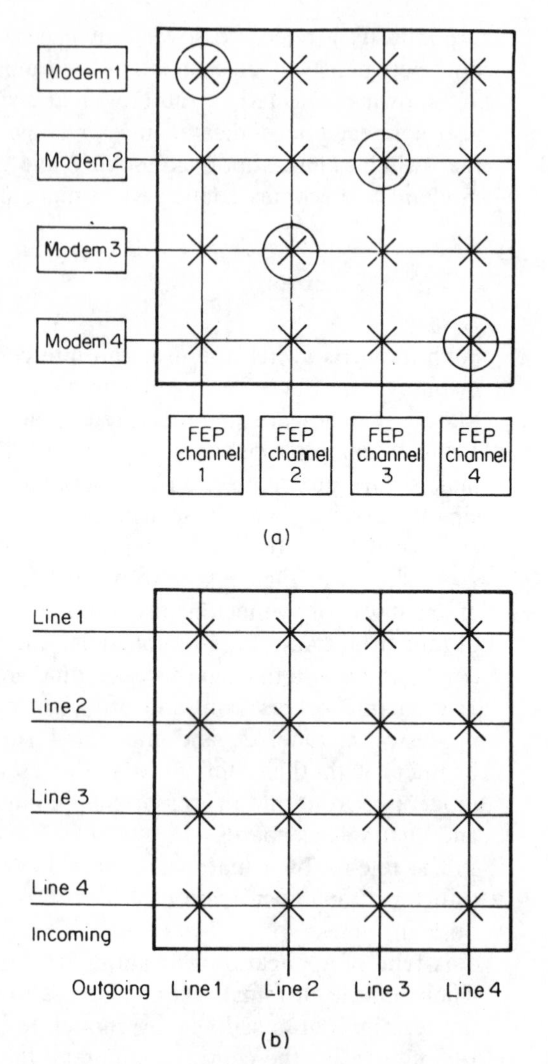

Figure 4.4 Matrix switches. Top: 4 by 4 EIA matrix switch. Bottom: 4 by 4 telco matrix switch. Matrix switches permit any combinations of a group of incoming interfaces to be rapidly connected to any combination of outgoing interfaces.

selects the modem to be used in transferring data between the terminal and the telco line. Because three EIA interfaces are involved, this configuration is called a 3 of 4 EIA interface bypass switch. In a 4 of 4 EIA interface (Figure 4.5, bottom), four interface devices are connected to the switch. In this configuration, the switch selects one of two encoders to encode terminal data for transmission through an attached modem.

A second common switch derivation is a multiple fallback switch (Figure 4.6). Besides the EIA and telco versions, this switch is manufactured in a 1 of *n* version, with *n* being the number of possible selections. Figure 4.6 (top) shows two possible configurations for a 1 of 4 EIA fallback switch. At the

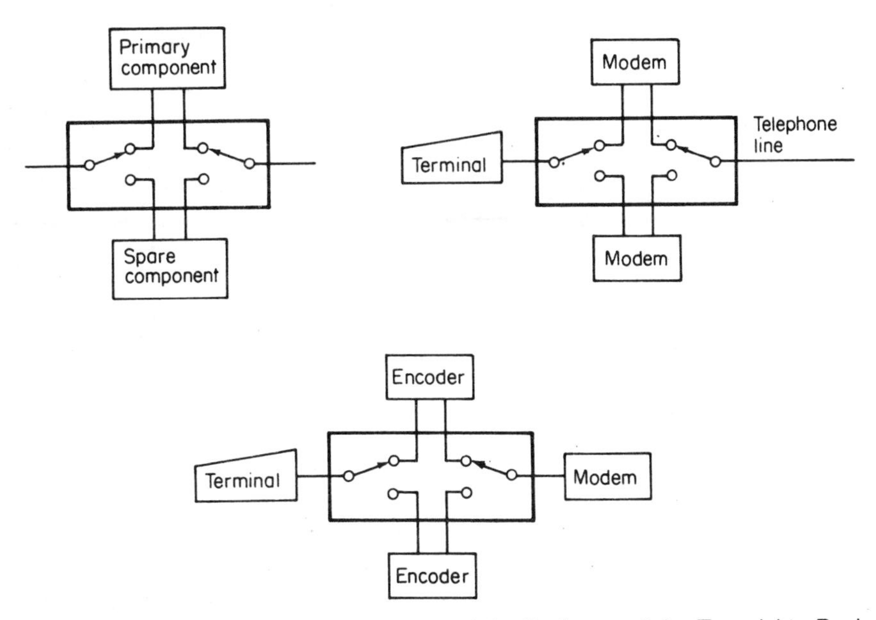

Figure 4.5 Backup switch variations. Top left: Backup switch. Top right: Backup switch , 3 of 4 EIA interface. Bottom: Backup switch, 4 of 4 EIA interface. Paired EIA and telco fallback switches, which often come in one package, provide both a backup and a normal mode of operation.

left, the switch allows the terminal to be connected to any one of four modems, while at the right, any one of four terminals may be connected to a single modem. Similarly, Figure 4.6 (bottom) shows how the 1 of 4 telco fallback switch allows either four modems to share a single line or four lines to share a single modem.

Chaining switches

No manufacturer produces a complete line of readymade switches, but it is often more convenient to deal with and install switches from a single maker. The user can do so by developing the switching functions required from combinations of one vendor's switches. In Figure 4.7 (left), for instance, four fallback switches are chained together to perform the function of a bypass switch. A single backup terminal can be used to replace any one of four primary terminals. In Figure 4.7 (right), four fallback switches are chained so that a single backup modem may be used by any terminal if its primary modem fails.

Other switches can be similarly chained to develop additional switching functions or to increase the capacity of existing network devices. Even more usefully, different categories of switches and different types of switches within the same category can be chained. Figure 4.8 shows a 4 by 4 telco matrix switch chained to a 4 by 4 EIA matrix switch so that the user may interconnect

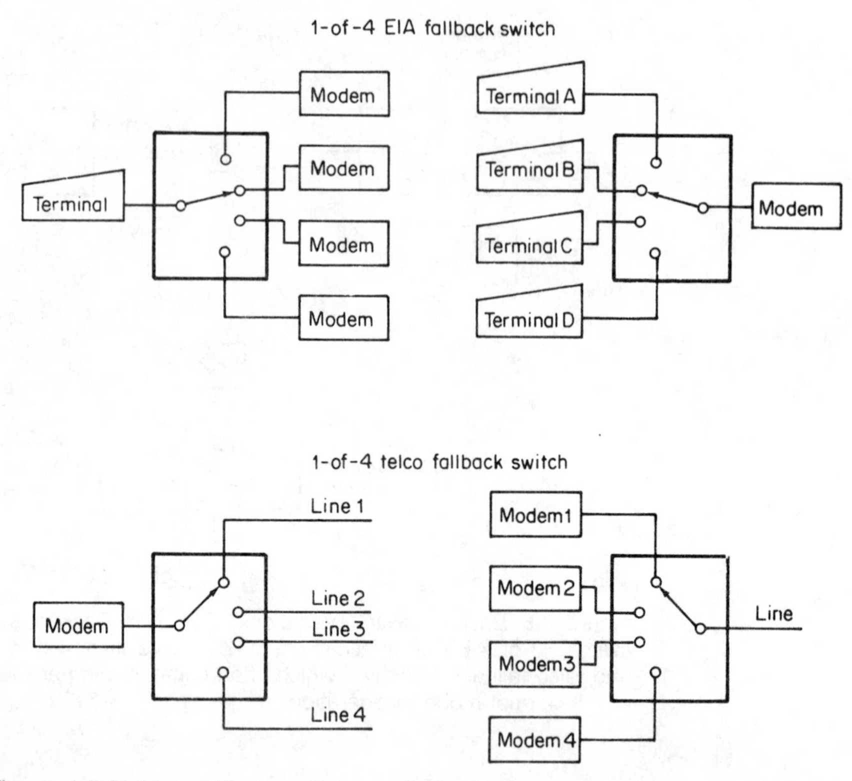

Figure 4.6 Multiple fallback switches. Multiple fallback switches, either EIA or telco, allow many terminals to share one modem or many modems to share one telephone line.

any combination of lines, modems, and front-end processor channels to arrange the information path desired. For this example, the number of possible configurations is increased to n^3 from the normal n^2 combinations available with a single n by n matrix switch.

Switch control

The four most common methods of transferring a switch are local and remote manual, American Standard Code for Information Interchange (ASCII) unattended remote, and via a business machine or central host computer. A local manual switch usually has a toggle or toggles, but many are also manufactured with pushbuttons and corresponding indicator lights. For a remote manual switch, one manufacturer produced a remote-control panel equipped with a pushbutton and a cable connecting it to the remote switch. This setup also has the advantage that shorter cable lengths can be run from communications components to the switch. Although toggle-operated units can be rack-mounted, they are normally available only in single-channel modules. Remote-control units, on the other hand, are normally manufactured in 4, 8

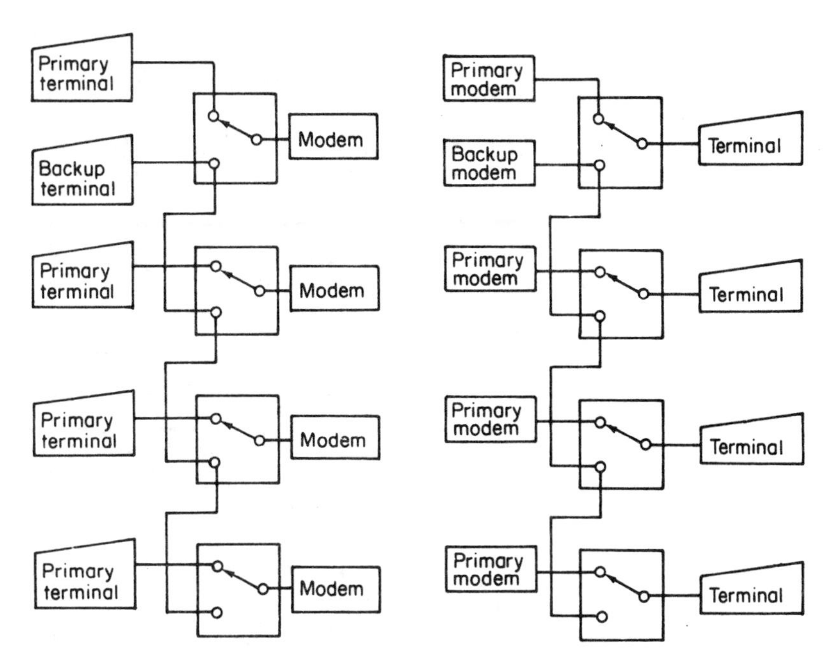

Figure 4.7 Chaining fallback switches. Chaining of simple switches yields a variety of functions, for instance, permitting any of several terminal locations to select a single spare modem.

or 16 unit configurations, and all units are master--switched from the remote control simultaneously, rather than one at a time.

The ASCII unattended remote control permits a switch to be controlled or monitored at any remote site at which a telephone line can be installed. An adapter interfaces the switch (or switches) to the telephone line and turns it on and off upon receiving a coded message consisting of the switch number and the state to which to transfer. The adapter then reports back the switch's new status. Also available is a query mode which allows the operator to check out a remote switch's position.

When a business machine (computer) is involved, switching is controlled directly by the machine, normally through a 5-V transistor–transistor logic (TTL) circuit. Let us now examine how such switches can be a cheaper alternative to obtaining overall network availability than the installation of typical redundant equipment. Furthermore, the cost of switches themselves can be kept low if the application allows a reasonable time interval for an operator to perform manual switching compared with the higher cost of master-controlled remotely operated switches.

Switching applications

When network equipment fails, a variety of communications switches can quickly get the network back into operation. They rapidly bring redundant

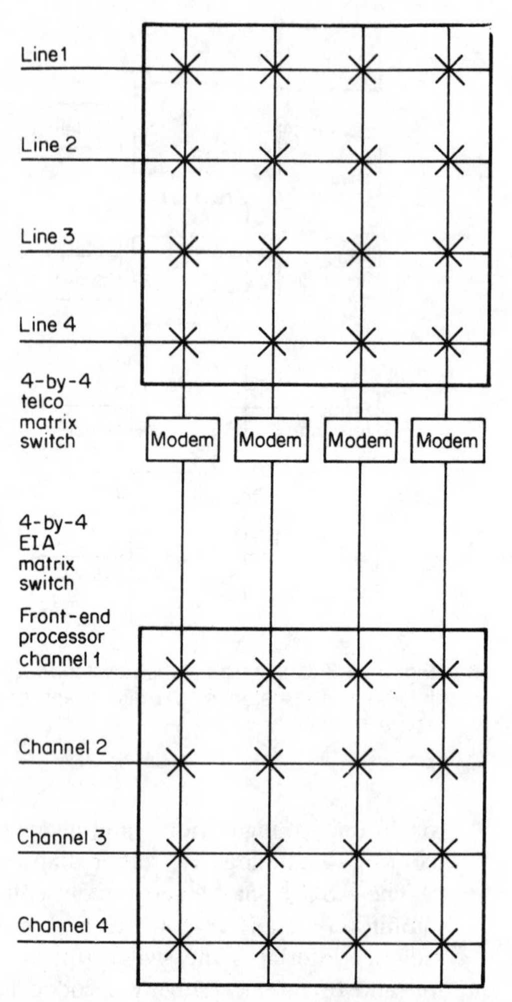

Figure 4.8 Chaining telco and EIA matrix switches. Here, an EIA matrix switch and a telco matrix switch are chained so the user may interconnect any combination of lines, modems, and front-end processor channels.

equipment into place to meet established requirements for overall network availability. The cost of providing the switches can range from less than a few hundred dollars to well over $50,000. What makes the price vary so much rests on the answers to such questions as:

Which devices are most likely to fail?

What tangible and intangible effects will a failed network device, such as a concentrator, have on the organization's operation?

Would the operational loss be so great that it warrants the cost of including backup equipment and transmission lines?

When a network component goes down, how much downtime, if any, is allowable to activate backup devices and get the network back into full-scale operation?

To obtain speedy network recovery, what are the best types of switches for the application, and where should they be placed in the network?

The significance of these and similar questions, and their answers, will become apparent during the discussions of typical redundancy/switching configurations that follow.

The four basic types of communications switches – fallback, bypass, crossover, and matrix—come in two versions: EIA for switching digital signals and telco for switching 2-wire and 4-wire telephone lines. Chaining these switches provides a variety of extended switching functions. Furthermore, the switches can be activated or controlled in local or remote manual modes, in an unattended remote mode in which the switch is activated by a specified ASCII-character code, and in a computer-controlled remote mode. Switches become more expensive in going from local manual mode up to computer-controlled mode. But changing a network from primary to backup mode manually may take 10 or 15 minutes, while a computer-controlled switch can activate all switch connections essentially instantaneously and automatically from a remote location.

In the first portion of this section, the use of switches to substitute spares for such devices as modems and terminals was discussed. In the second portion, the discussion will center around the ramifications of switching between dual-collocated concentrators. Here, one concentrator may be assigned completely to back up the other unit, or each concentrator may be servicing its own terminals during normal operation. In either case, on failure of one concentrator, the other takes over all duties if it has enough capacity to do so. In the latter case, if the reserve capacity is not available, then a secondary job such as driving a line printer, may be suspended as long as concentrator downtime continues. Although this discussion involves the use of concentrators, its points can be directly applied to most networking devices the user may wish to obtain a level of redundancy for to include statistical and intelligent multiplexers.

In the basic setup of the 10 following applications, each concentrator location services a number of relatively local low- and medium-speed terminals, so that each has a number of terminal-to-concentrator links. Each concentrator merges all traffic from its terminals and sends it on a high-speed line to a remote host computer.

As will be seen, the applications tend to become more complex and more expensive. The actual choice depends to some extent on network application and to some extent on the severity of the consequences of a device failure.

Hot-start configuration

The two main methods of integrating collocated concentrators to service remote terminals are commonly called 'hot-start' and 'cold-start'. The hot-start approach, Figure 4.9, means that a backup computer is energized, fully programmed with a duplicate of the software in the primary concentrator, and may be continuously tracking the traffic in and out of the primary concentrator. When the primary computer fails, a computer-controlled switch can put the backup concentrator in control instantaneously.

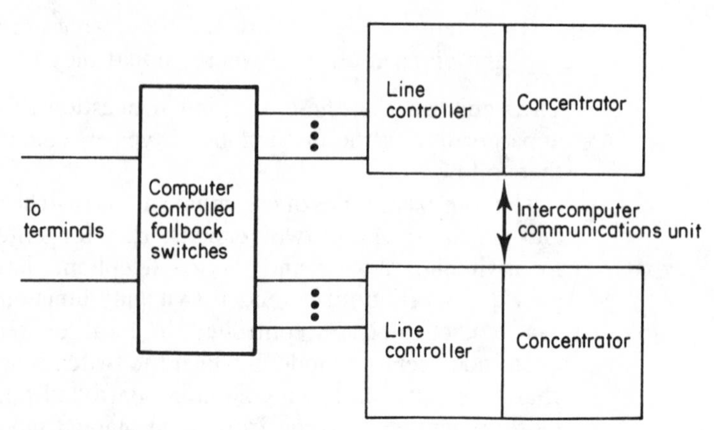

Figure 4.9 Hot-start configuration. In this hot-start configuration, when the primary concentrator fails, a computer-controlled switch puts the backup concentrator in control.

Full effectiveness of such a hot-start arrangement requires the installation of an intercomputer (that is, interconcentrator) communications unit. When a failure such as memory-parity errors or power loss occurs, the concentrator experiencing difficulty sends appropriate software commands through the communications unit. Additionally, an automatic command to a bank of computer-controlled telco fallback switches provides instantaneous transfer of the line from each terminal to the line controller of the operating concentrator.

The near-instantaneous switching and the minimization of the loss of data are the important advantages of the hot-start configuration. However, there are significant hardware costs associated with the computer-controlled switches and the intercomputer communications unit. In addition, the necessary software modifications to permit the desired switching are complex, involving experienced personnel, much patience, and large amounts of machine time for testing the developed software. Overall, the cost for a hot-start configuration may well reach over $100,000, not counting the cost of the concentrator itself. But it may be well worth the money to assure that the network remains continuously operational and available.

The remainder of this section will focus on variations of the cold-start redundancy configuration, with different methods of switchover available from manual (local toggles) and remote-control (pushbutton) switches into a communications network. However, computer-controlled switches can be used in any of these configurations with a consequent increase in complexity and cost combined with a salutary improvement in network uptime.

Cold-start configuration

Telco fallback switches represent one method of providing an alternate path between the remote terminals and the two concentrators (Figure 4.10). Here, the occurrence of a concentrator failure or a concentrator-to-host link failure will require manual intervention. When one or both failures occur, it becomes necessary to switch the telco units to insure that the remote terminals are connected to the operating concentrator. Furthermore, the standby concentrator

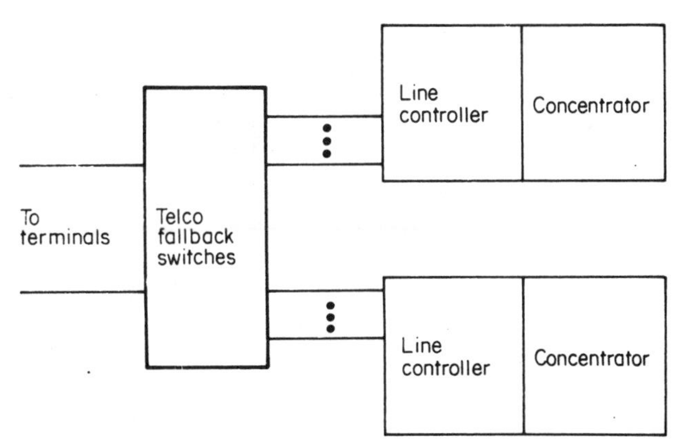

Figure 4.10 Cold-start configuration. In the cold-start configuration, failure of a concentrator or a concentrator-to-host link requires manual intervention by the operator.

must have its programs bootstrapped from a high-speed storage unit such as a disk.

Thus, if the concentrators are initially sharing the terminal workload, the failure of one concentrator may require the other concentrator's software to be reconfigured to service the entire workload. This configuration can be completed in a few minutes by manually throwing the switches and reading the backup programs from the disk into the operational concentrator's memory.

Some data being transmitted through the concentrator may be lost during the reconfiguration time. But the low cost of the cold-start configuration may justify the extra time associated with satisfying retransmission requests for lost messages.

Sharing a backup concentrator

The availability requirements of the network may be such that neither operating concentrator has the reserve to serve as backup for the other. But it may be possible to service both devices with a single backup concentrator, as shown in Figure 4.11. Here, telco fallback switches allow the terminals in building 1 or building 3 to be connected to the backup concentrator in building 2. The number of modems interfaced to the telco switching units in building 2 only need equal of maximum of the number of such devices in either building 1 or building 3. Thus, if the possibility of two concentrators failing at the same time is disregarded, the cost of the fallback switches is more than offset by the savings due to the lesser number of modems necessary at the backup concentrator.

Backup with EIA switches

An alternative approach to servicing the terminals in buildings 1 and 3 by the backup concentrator in building 2 can be obtained through the use of EIA fallback switches (Figure 4.12). Instead of installation between the modems as

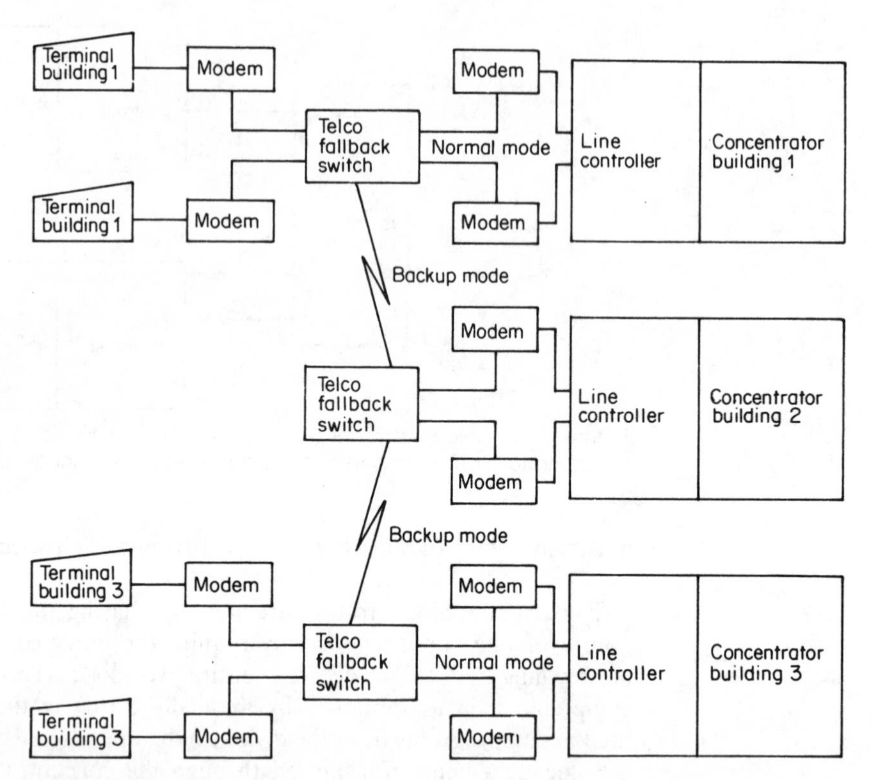

Figure 4.11 Sharing a backup concentrator. Here, both primary concentrators, neither of which can completely take over for the other, is serviced by a third backup concentrator.

with the telco switches in the preceding application, the EIA switches are between the modem and the line controller of the concentrator. Depending on the distance between either primary concentrator and the backup concentrator, line drivers or modems become necessary to permit an undistorted output signal to reach the backup concentrator. Assuming relatively short distances that permit the use of lower-cost line drivers, rather than modems, a telco fallback switch will suffice in building 2 for each pair of terminal-to-concentrator links in the other buildings.

In the normal mode of operation, the terminals in building 1 or 3 communicate with their respective concentrator via a pair of modems and an EIA fallback switch. Should either concentrator fail, the operator must position the fallback switch into its backup mode and position the telco switch in building 2. Doing this provides a new set of circuits from the affected terminals to the concentrator in building 2.

This and the previous application have a concentrator added to the basic configurations of two such devices. In either case, the user may set up a network of three primary concentrators that share the backup duties. As well as connecting terminals directly to the concentrator of building 2, the user would have to install EIA or telco fallback switches to transfer the new data paths to either of the other two concentrators when backup service is needed.

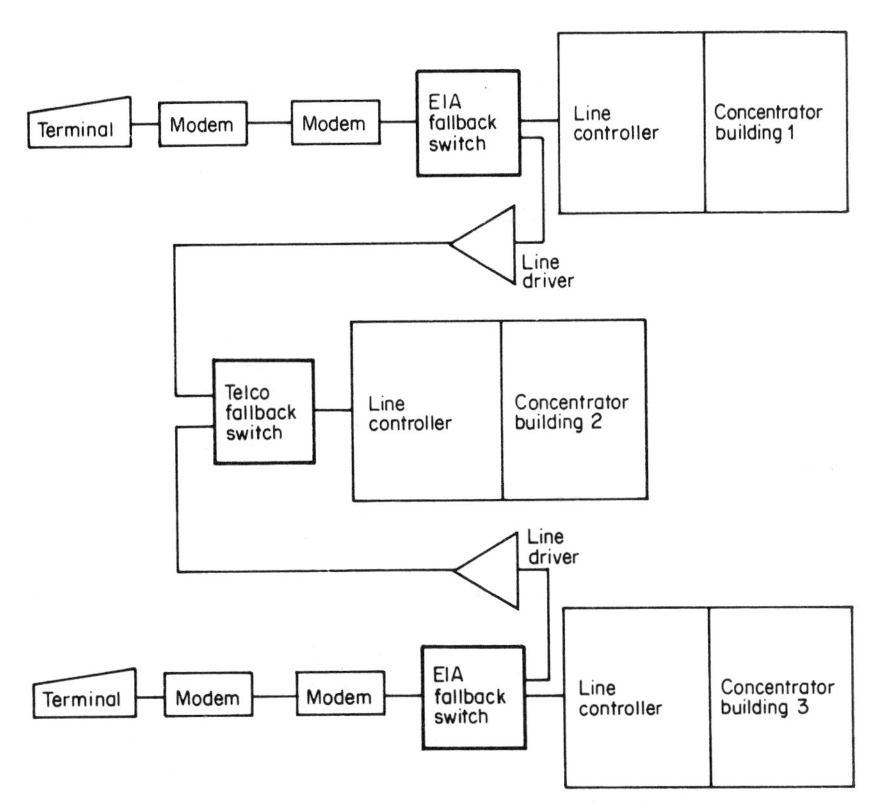

Figure 4.12 Backup with EIA switches. Another way to back up a failed concentrator is by using EIA fallback switches between the modem and the line controller of the concentrator.

Concentrator to central computer

If data transfer from each concentrator to the central computer is via a few high-speed lines, EIA fallback switches permit the transfer of modems and lines between the concentrators. In Figure 4.13, two switches permit each concentrator to communicate over its own dedicated link to the central computer complex. This type of configuration compensates for a concentrator failure by permitting the remaining concentrator to communicate with the host computer over its line and the line of the other device. However, the failure of either one of the dedicated lines or of a modem would require selection of one of the concentrators to use the remaining data communications link.

Adding a third EIA fallback switch

If the user wants to overcome the shortcomings of the preceding configuration, the inclusion of a third EIA fallback switch and another modem interfaced to the dual concentrator configuration can either prevent or minimize the failure

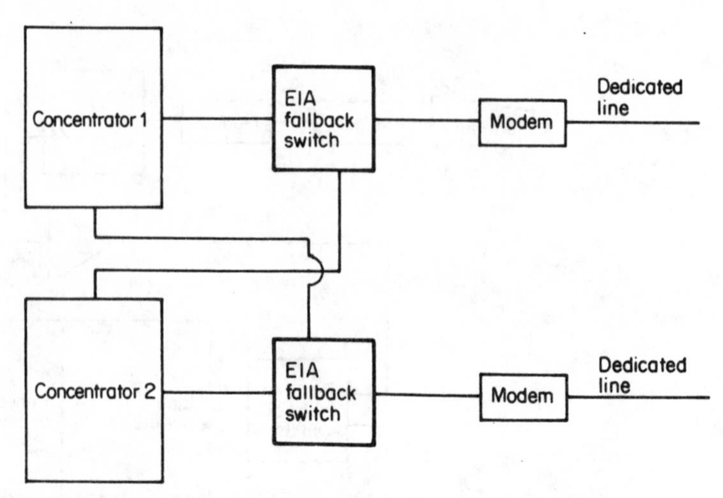

Figure 4.13 Concentrator-to-computer backup. Two EIA fallback switches permit each concentrator to communicate over its own dedicated link going to the central computer complex in the primary mode of operation. When a fallback switch is activated, the concentrator can be connected to the other concentrator's modem and line.

of a modem or of a dedicated line, as shown in Figure 4.14. In the normal mode of operation, each concentrator communicates with the central computer via its own dedicated line. If a modem or dedicated line should fail, the proper positioning of two of the switches allows the concentrator to communicate with the central computer via the middle modem over an alternative path, either a dial-up network or another dedicated line. A disadvantage of this configuration is that each concentrator has access to only one line at a time, unlike the configuration in the previous application.

Adding more switchable lines

Access to more than one dedicated line at a time may be obtainable by adding lines for each concentrator and reconfiguring the EIA fallback switches, as

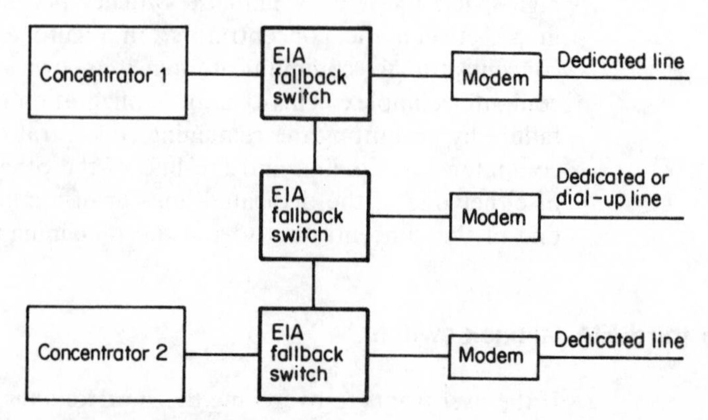

Figure 4.14 Adding a third EIA fallback switch. A third EIA fallback switch and another modem interfaced to the dual concentrators can prevent or minimize modem or line failure effects.

shown in Figure 4.15. If one concentrator should fail, the other can communicate over both dedicated lines, and it still has access to the backup line. In this manner, throughput degradation should be minimized.

Chaining adds options

Chaining two EIA fallback switches results in another way of providing an alternative central computer link for a dual-concentrator installation (Figure 4.16). Only one channel is required for each concentrator. In normal operation,

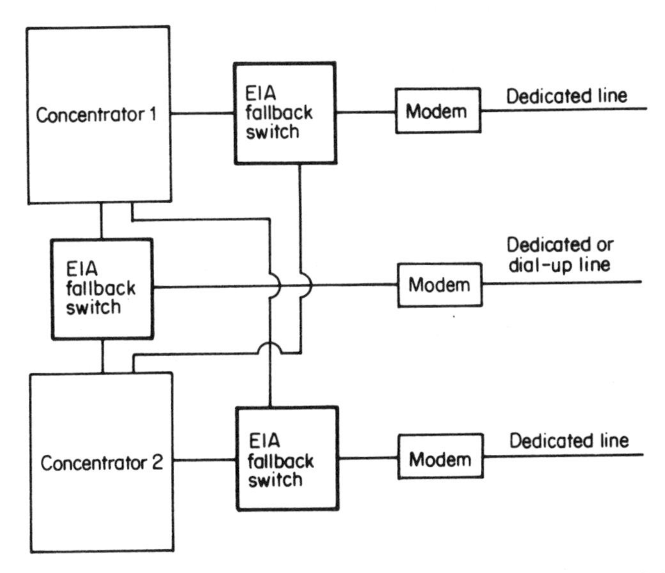

Figure 4.15 Adding more switchable lines. If one concentrator should fail, the other concentrator can communicate over both dedicated lines and still have access to the backup line.

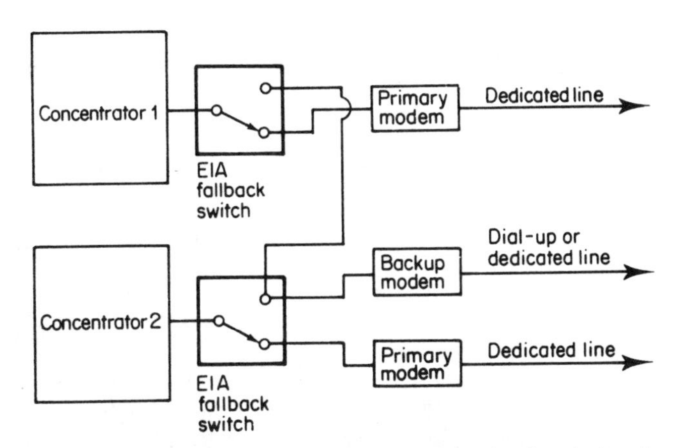

Figure 4.16 Chaining adds options. Chaining two EIA fallback switches is another way of providing an alternative central computer link for a dual-concentrator installation.

each switch interfaced to each concentrator channel remains in the primary modem position. If the dedicated line or the primary modem of either concentrator should fail, the associated switch is positioned so that a path is provided to the backup modem. As in the other application, this backup modem can use a dial-up network or a dedicated line to communicate with the central computer.

This configuration requires only one concentrator channel to provide a link in the event of modem or dedicated line failure. However, should a concentrator fail, the other one is not provided with access to the failing device's line. Therefore, if terminals from the failing concentrator are switched to the operational concentrator, the operational link to the computer may not be sufficient to satisfy the increased terminal traffic. Redundancy for this link through the use of EIA fallback switches can become rather complicated when more than a few lines require multiple access.

Access to other lines

Use of one or more EIA matrix switches, as shown in Figure 4.17, can alleviate switching complexity as well as provide each concentrator access to the other dedicated line. For example, with a single 4 by 4 switch, each concentrator can have easy access to the spare modem and to any modem and line connected to the other concentrator. Although only one spare modem is shown here, a

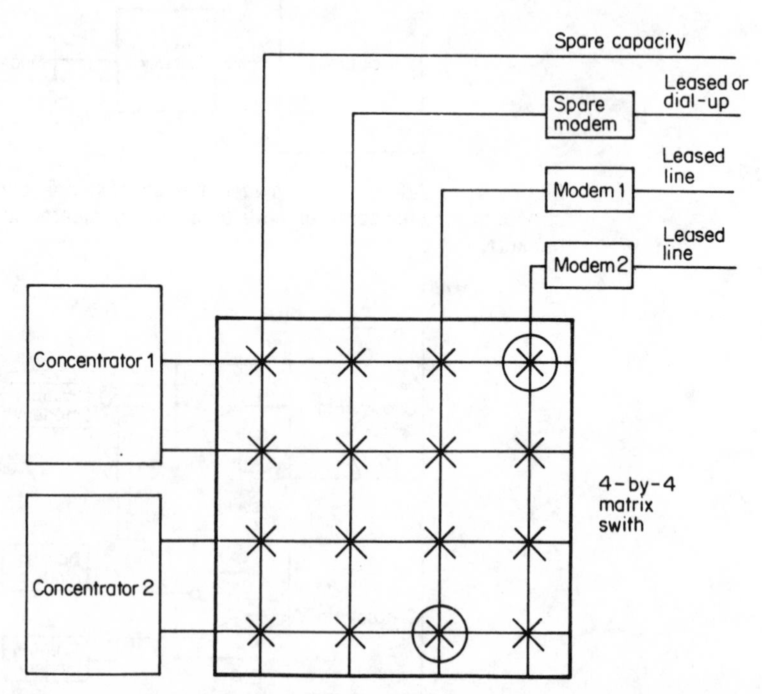

Figure 4.17 Access to other lines. One or more EIA matrix switches can alleviate switching complexity as well as provide each concentrator access to the other dedicated line.

second modem and its associated line facilities could be added, since the output side of the 4 by 4 switch can interface one additional device.

As shown by the circles, concentrator 1 normally transmits data through modem 2 to the central computer, and concentrator 2 via modem 1. Either concentrator can be connected to the spare modem and associated line, should its primary modem or line fail. If a concentrator fails, the other one can be connected to the failing device's primary modem and line, thus insuring the continuation of full throughput to the central computer. If each concentrator communicates with the central computer through more than one link, the use of an 8 by 8 or a 16 by 16 switch or the chaining of more than one matrix switch should be explored.

Integrating switches into both links

Although the number of concentrators and of communications lines from each concentrator to the host computer depends upon such factors as the number of terminals serviced by each concentrator in both primary and backup modes of operation, terminal traffic patterns, and line protocol overhead, the configuration of Figure 4.18 represents one possible way of integrating switches

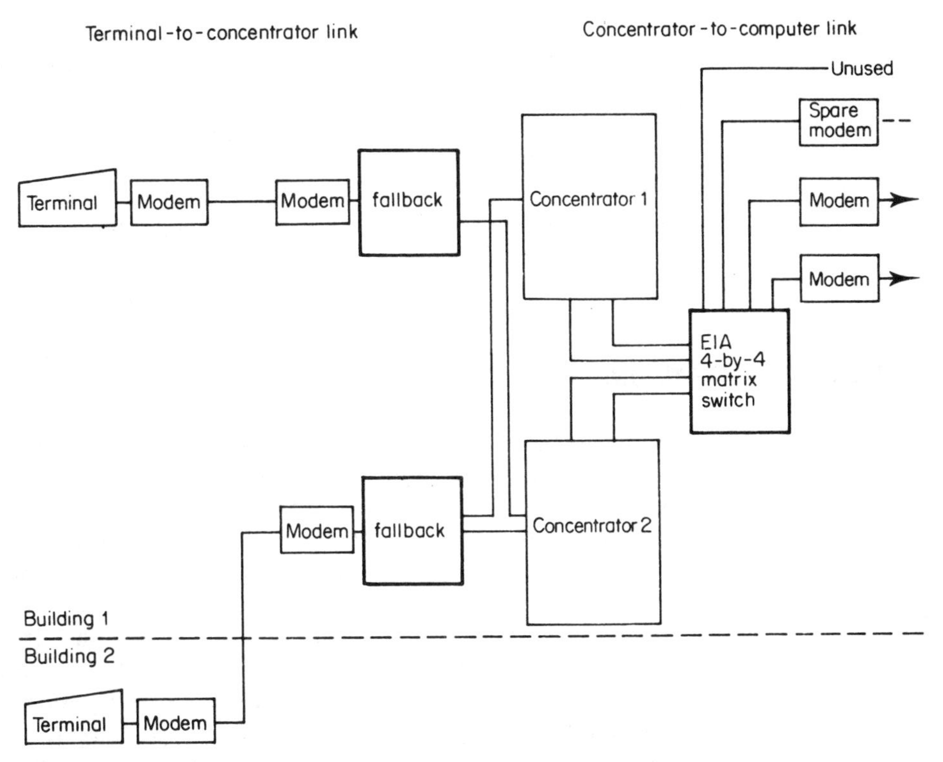

Figure 4.18 Integrating switches into both links. Integrating switches into both the terminal-to-concentrator and concentrator-to-host links provides an alternate path for both links.

into both the terminal-to-concentrator and the concentrator-to-host computer links. It provides an alternate path for both links when dual concentrators are within about 50 feet of each other. Here, it is assumed that the remote terminals are in two buildings. However, because the distances between each terminal and the concentrators preclude direct attachment or the use of line drivers, modems are necessary. An equipment study established each terminal's need for access to a second concentrator in order to maintain the desired level of backup. At the same time, to maintain throughput at the full transmission speed after the failure of one concentrator, it was found necessary to have the capability to switch the links of the failing device to the other one. Furthermore, should any modem or line of the concentrator link to the computer become inoperative, an easy switch to a spare modem communicating with the host computer via the dial-up network was desirable.

If the equipment study shows that each concentrator requires one channel for communicating with the host computer, then two channels become necessary on each device in order for each to use the other's link as well as its own at the same time. Thus, as shown here, the failure of concentrator 1 can be compensated for by positioning the EIA fallback switches to the terminal-to-concentrator link so that the terminals in building 1 connect to the second concentrator. In addition, each concentrator link to the computer is connected via a 4 by 4 EIA matrix switch to the other concentrator. The same procedure applies to the failure of the second concentrator.

Should a modem or line from either concentrator link to the computer become inoperative, the 4 by 4 matrix switch permits an easy reconfiguration to the spare modem and the dial-up network.

The procedures discussed here apply to a network with any number of terminal and host links. From the preceding examples, the utilization of switches provides the system designer with a low-cost option when designating a teleprocessing network that requires a level of redundancy which can range from a single redundant component requirement through multilayers of redundancy.

4.2 LINE RESTORAL UNITS

A recurring problem facing network designers is the method to employ in providing backup facilities when transmission is over a leased line. Consider a typical time-sharing network where numerous remote locations are linked to a central computer facility by multiplexers which combine the low-speed transmission of numerous terminals for retransmission over leased lines.

If a leased line outage occurs during normal working hours, several possibilities exist for reestablishing communications. If a number of dial-in lines are available at the computer center, remote users could use the switched network to continue. However, long-distance telephone charges may become prohibitive. Even if we disregard telephone charges, the number of dial-in lines may not be sufficient to provide service to a group of remote users whose primary communications path has failed. If a technique is available to reestablish the high-speed link, the time from when the outage is reported until someone physically effects a transfer to an alternate communications path for the multiplexed data can frustrate the remote users to the point where they may consider the use of another computer facility.

For other types of applications, such as a remote minicomputer monitoring telemetry signals and transmitting those signals to a distant computer, the loss of a leased line could result in the loss of data. Although switches can be used to provide backup for many applications, as discussed in Section 4.1, several problems are encountered if they are used for the applications previously discussed.

First, manually operated switches would result in the loss of a period of transmission time which is proportional to the time the line failure occurs until someone effects the switchover. Even if two leased lines are installed for critical applications, the switchover to the second line may not have the desired effect if they are similarly routed by the telephone company. Thus, a main cable failure or a fire such as the one at the American Telephone and Telegraph (AT&T) building in New York City could render both lines inoperative. One method to alleviate the problems associated with leased line outages is through the installation of line restoral units.

Operation

A line restoral unit is a device which monitors the signal strength on a leased line. If the signal strength should drop below a set level, the unit will automatically initiate a call over the switched network and route data via this new transmission path until the signal level and jitter on the leased line return to normal. Several modes of operation can be considered by line restoral unit users. Instead of internal control, where the device initiates switchover, the transfer to the switched network can occur under company control with the restoral unit then serving as an automatic dialer. If the leased line is a 4-wire circuit used for full-duplex transmission, the restoral unit can be configured to initiate two calls over the switched (2-wire circuit) network to reestablish a 4-wire connection. At the distant end a similar unit connects to the dialed calls, as shown in Figure 4.19.

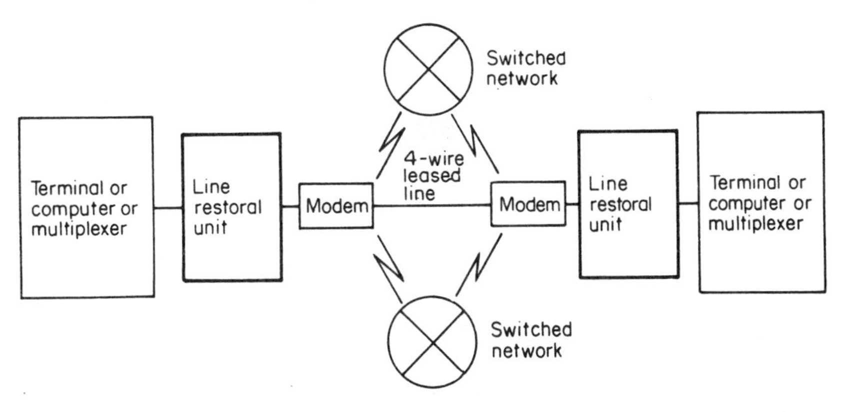

Figure 4.19 Line restoral unit operation. To reestablish communications when a 4-wire leased line fails, two calls are placed over the switched 2-wire network.

Prior to the initiation of transmission to the dialed path, the restoral unit will sample transmission quality and continue redialing over the switched network until a good link is established. Once a good connection is established, the modems are switched from the failed to the newly dialed connection. In addition, restoral units continuously monitor pilot tones put on the non-operating line, checking for phase jitter and signal strength until the line recovers. As soon as the leased line is restored to normal operations, the backup connections are either automatically or manually terminated and transmission can proceed once again over the leased line.

An audio alarm is built into these units to notify personnel in the vicinity that a switchover is occurring. In addition, delay circuits are built into these units to prevent constant switching between the switched network and the leased line. A strap option is available on some units which prevents the automatic return of transmission to the leased line. This option is most useful when one wants to delay the system from returning transmission to the leased line until that line has been fully diagnosed and repaired by the telephone company.

When placed into a network, one unit acts as a master while the second unit is strapped to operate as a slave. Although both master and slave units monitor the leased line, only the master unit can initiate the dial-up procedure. When the master unit senses a degradation in its received signal, it initiates the dial-up procedure. At the other end, if the slave unit finds a degradation of its received data it will stop its transmission of carrier to the master, which will then serve as a signal for the master unit to start the dial-up procedure.

Problems of utilization

Although the integration of restoral units into a network can alleviate a large percentage of leased line backup problems, for certain network configurations new problems can arise through the use of this equipment. A common problem associated with restoral units occurs when the transmission rate used over the leased line is greater than the transmission rate that can be established over the switched network. For the remote site shown in Figure 4.20, the 9600-bps composite speed developed by the multiplexer can represent a problem if the

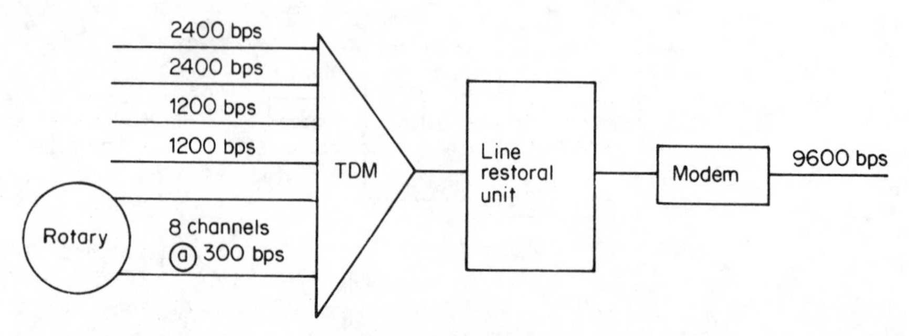

Figure 4.20 Alternate communications problems. If the switched network cannot accommodate 9600-bps transmission, the dial-backup may not be effective.

restoral unit cannot reestablish communications on the switched network at that speed.

If 9600-bps transmission cannot be established on the switched network, even restoral units that have the capability to switch speed-selectable modems to a lower data rate may prove ineffective. This is because conventional multiplexers require the physical strap out of one or more channels at both locations to produce a lower composite speed. The utilization of statistical and intelligent multiplexers that generate flow control signaling have been designed with these problems taken into account and can be integrated with restoral units to permit both dial backup and automatic transmission rate fallback in the event that transmission cannot proceed at the rate previously used on the leased line. If conventional TDMs are used, to effect the speed fallback a predetermined number of channels must be automatically 'busied out' as shown in Figure 4.21, whereas, the flow control feature built into statistical and intelligent multiplexers alleviates this requirement.

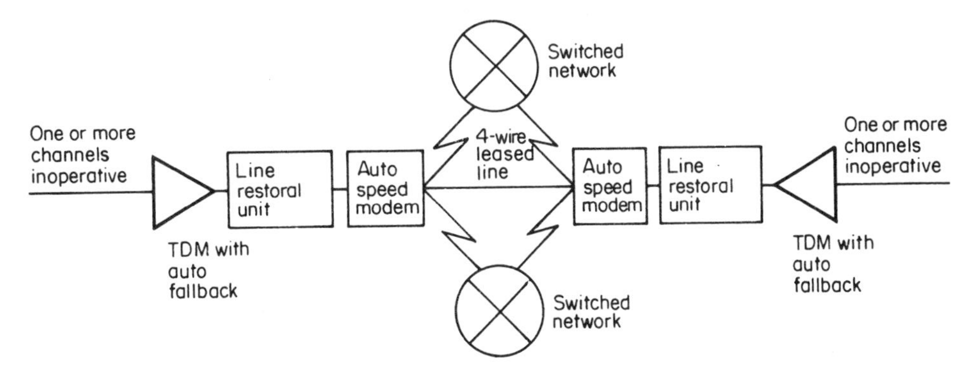

Figure 4.21 Automatic dial-backup and speed fallback. To accomplish automatic dial-backup and a reduction in the modem's transmitting speed, a number of multiplexer channels must be 'busied' out.

REVIEW QUESTIONS

4.1 What two types of switches are commonly used in networks?

4.2 Draw a network schematic illustrating the utilization of a fallback switch at a remote site and at a central computer site.

4.3 Discuss the utilization of a bypass switch at a central computer site.

4.4 What is the advantage of using a matrix switch over fallback or bypass switches?

4.5 How could you use a crossover switch at a remote site?

4.6 Draw a diagram illustrating how four fallback switches could be used to obtain a bypass switch capability linking m terminals to $m + 1$ modems.

4.7 What are the major differences between manual and automatic switches and when might you consider using the latter device?

4.8 What are the main differences between the 'hot-start' and 'cold-start' methods of integrating collocated concentrators?

4.9 Assume that the terminals to be installed at a remote location perform critical applications and you desire to install dual multiplexer systems to provide access to the central computer site. Draw two network schematic diagrams illustrating the use of two different switches to provide the terminal's access to each multiplexer via a direct connect cable.

4.10 Assume two collocated concentrators at a remote site are connected via individual leased lines to a common computer center. Draw a diagram illustrating the use of switches at the remote and central sites which would permit either concentrator to use a common backup circuit without requiring any additional ports at the central computer site.

4.11 Why are line restoral units essentially 'paired' devices?

4.12 Why does a line restoral unit monitor the signal strength on a leased line?

4.13 What is the function of delay circuits in a line restoral unit?

4.14 Why is it necessary, upon occasion, for a line restoral unit to make two calls over the switched telephone network?

4.15 Why is it more advisable to use an intelligent or statistical multiplexer instead of a conventional TDM with a line restoral unit?

CHAPTER 5

AUTOMATIC ASSISTANCE DEVICES

To reduce the necessity of operator intervention, two automatic assistance devices have been developed: automatic answering units and automatic calling units. In addition to reducing the requirement for operator intervention, these devices permit network components to operate at a high level of utilization since such devices can be used to effect data transmission after normal office hours when all employees have left and when calls over the public switched telephone network can be made at reduced rates.

Automatic answering units are normally integrated into modems which are interfaced to a computer. The answering unit detects the incoming telephone ring and automatically provides a connection to the calling party so that data transmission between a remote terminal's user and the computer can proceed without the intervention of the computer operator or other members of the computer room staff. Conversely, automatic calling units permit a business machine or computer to automatically dial the telephone number of other computers or business machines. Both the telephone numbers and time of call can be programmed in the computer, as well as such additional data as the number and sequence of redials if one or more busy numbers are encountered.

Automatic answering units are normally used in modems at computer time-sharing installations where a large calling population calls a number of computer dial-in lines on a random basis. Automatic calling units are usually employed for applications where it is desired to have equipment at a central location poll a number of remote locations during certain times of the day, but where the cost of installing leased lines is uneconomical since the transmission from each remote site is of short duration.

5.1 AUTOMATIC ANSWERING UNITS

Although some manufacturers produce automatic answering units on a card, such devices are known as an original equipment manufactured (OEM) component which will be integrated into another device at a later date. When integrated into a modem, the automatic answering unit provides automatic answer capability for the modem. The use of modems equipped with an automatic answering unit permits data transmission to occur between remote terminals; and a computer or another terminal may be interfaced to an automatic answering modem without operator intervention at the called end of the link. A typical employment of automatic answer modems is illustrated in Figure 5.1.

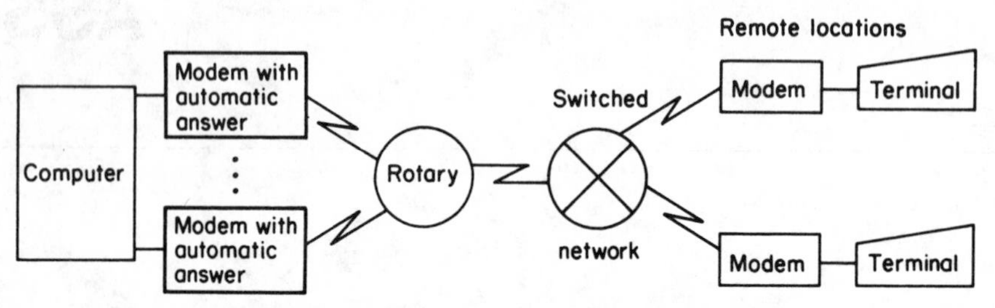

Figure 5.1 Automatic answer modem employment. Modems equipped with automatic answering units enable remote terminals to communicate with time-sharing computers using the telephone dial-up network without requiring operator intervention at the computer site.

Operation

When integrated into a modem, the automatic answering unit will place the modem in the answer mode upon the receipt of a ring indicator signal. Depending upon the device selected, the absence of an incoming carrier signal or its delay for a predetermined amount of time after the ring indicator signal is received can be used to disconnect the calling party. After the call is answered and data transmission is in effect the automatic answering unit will send a disconnect immediately; this is in response to a data terminal ready signal going off or when the incoming carrier signal is lost for greater than a predetermined amount of time, if the discount options on the device are strapped for those events. Some of the typical options which are strap-selectable for automatic answering units are listed in Table 5.1.

Alleviating problems

To prevent a tie-up of expensive computer facilities through the use of computer ports and lines due to such problems as wrong number calls, failure of the distant party to disconnect from a time-sharing system, or line failures, a number of techniques can be employed. The timer abort option permits the device to disconnect if a carrier signal is not received within a predetermined amount of time after the call is answered. This option can be effectively employed to resolve the situation where the modem answers a wrong number call. While the devices can be interfaced to a computer so that they will

Table 5.1 Automatic answering unit options.

Answer mode indicator
- Ring indicator OFF: ring indicator circuit follows ringing only
- Ring indicator ON: ring indicator circuit follows ringing and remains on for the duration of the incoming call

Time abort
- YES: automatic answering unit disconnects if incoming carrier is not received within a predetermined time after call is answered
- NO: automatic answering unit does not disconnect with absence of incoming carrier

Loss of carrier disconnect
- YES: automatic answering unit disconnects when incoming carrier is lost for longer than a predetermined amount of time
- NO: automatic answering unit does not disconnect on carrier loss

Disconnect immediately
- YES: automatic answering unit disconnects immediately in response to data terminal ready OFF
- NO: automatic answering unit transmits a spacing signal prior to disconnecting in response to data terminal ready OFF

disconnect on command when the computer receives a special terminal signal, such as a sign-off command from the terminal user, quite often such users forget to do so and merely lift their telephone handset out of the acoustic coupler or turn their modem off when they complete their terminal session. The loss of carrier disconnect option is valuable for this situation, since the answering unit disconnects when the incoming carrier is lost for a period longer than a predetermined amount of time.

5.2 AUTOMATIC CALLING UNITS

In trying to obtain a telecommunications link between a computer or business machine and another computer or business machine, the establishment of the link can be made manually or automatically, as shown in Figure 5.2. To initiate

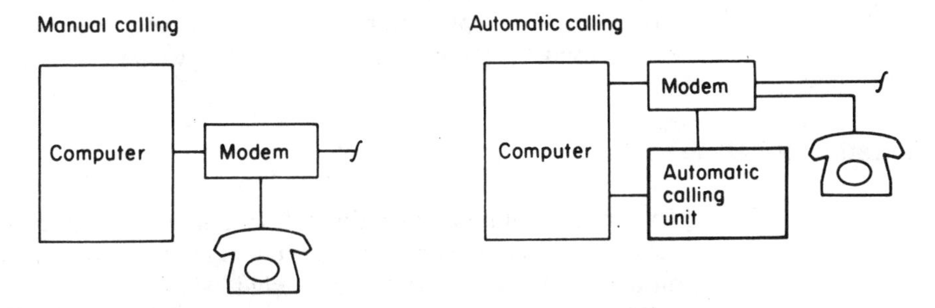

Figure 5.2 Call origination. Call origination can occur normally by having an operator dial a switched network number, or it can be performed automatically through the use of an automatic calling unit.

automatic calling, the business machine or computer directs a connected device known as an automatic calling unit to initiate and execute the call.

Although originally offered only by telephone companies, the growth of data communications has led a number of independent firms to offer equivalent-type devices, which in some cases exceed the capabilities offered by telephone company equipment. In addition, devices produced by independent firms can be purchased, whereas telephone company equipment is normally only available on a monthly lease basis.

The basic function of an automatic calling unit is to perform electronically for a business machine, such as a computer, what a human operator does by hand when he or she places a telephone call. In examining the functions performed by an automatic calling unit (ACU), these functions are electronically equivalent to operator functions when making a call; including lifting the telephone headset (referred to as going off-hook), waiting for the correct dial tone or tones, dialing each digit of the telephone number desired in its proper sequence, waiting while the phone rings until the dialed number answers, hanging up and possibly redialing at a later date if the number is busy or if the call is not answered after a reasonable length of time, and hanging up the headset at the end of the completed call. When properly configured into a data comunications network, automatic calling units dramatically increase line utilization while correspondingly reducing communications charges.

Type of calling units

There are two basic types of calling units which correspond to the type of dialing unit to which the equipment is connected. The dc dial pulse type of calling unit is commonly referred to as a Data Auxiliary Set 801A Automatic Calling Unit by the telephone company and permits dialing through rotary dialing connections. The second type of calling unit, the Data Auxiliary Set 801C Automatic Calling Unit, permits calls to be made over a Touch-Tone® telephone in about one-tenth of the time required when using the older rotary dial. For those sections of the country where Touch-Tone® dialing is available, the use of an 801C-type calling unit can shorten the required dialing time by about 10 seconds per call in comparison to using an 801A-type of calling unit. Calling units manufactured by independent firms are commonly referred to as 801A or 801C replacement devices, with the general designation of an 801 device used to designate some type of automatic calling unit.

Utilization of calling units

The most common utilization of automatic calling units is in a distributed network, where one or more computers are used to poll many remote terminals on a periodic basis, as shown in Figure 5.3.

The data access arrangement illustrated in Figure 5.3 is a separate unit, required to be installed as an interface between third-party products not 'certified' for use with the switched telephone network and that network.

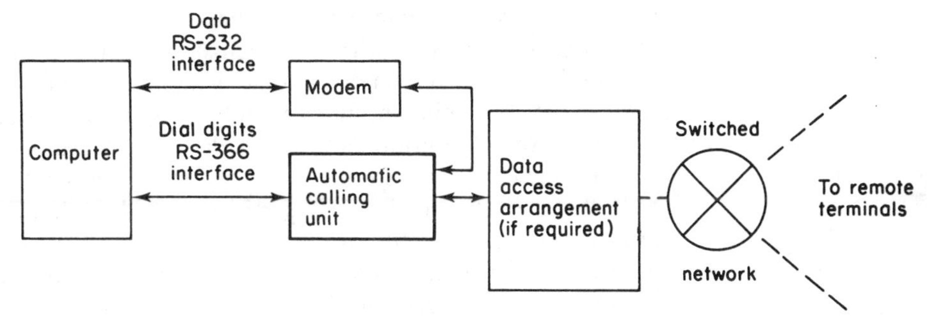

Figure 5.3 Polling of remote terminals. Calling units can be effectively employed to poll remote terminals at predetermined times to reduce communication toll charges. If the calling unit is not certified, a data access arrangement which serves as a protective device between the customer equipment and the telephone company line may be required.

Today, most third-party products designed for operation with the switched telephone network are built to comply with the Federal Communications Commission's Equipment Registration Program. Such devices include a built-in data access arrangement and eliminate the requirement for users to obtain a separate data access device. The use of data access arrangements is covered in more detail in Section 6.4.

One of the many users of the configuration shown in Figure 5.3 is the insurance industry. Most major insurance firms have offices scattered throughout the country. To assist their policy illustration effort a number of firms have their sales personnel fill out a policy illustration worksheet that specifies the type of policy under consideration, the amount of coverage desired, the name, age, sex, and occupation of the potential insuree, as well as insurance options, such as waiver of premium, that may be desired. This information is then entered by a secretary or by the sales personnel into a terminal's storage device, such as a cassette or diskette, at the insurance office.

In the evening, when the office has probably closed for business and also when telephone rates are lower, a computer which may be located at the firm's headquarters or a regional office polls the terminals located in the remote offices via the switched (dial-up) telephone network. If the office where the terminal is located did not have any transaction that day, the data terminal at that location is set so it will not respond to the computer's call, and therefore no toll charge is incurred. For those terminals which have been set to respond to the computer's call, the automatic answering device attached to the remote terminal accepts the call; and the data that has been stored on the terminal's storage device is now transmitted to the computer. This process is repeated until all of the preselected terminals have been called.

Later in the night, after the data from the terminals has been processed and the policy illustrations prepared in machine form, the results are returned to the requesting terminals, again by calls originating from the central or regional computer site and occurring over the switched telephone network. The returned data can either be stored on the terminal's storage device for later printout or printed on-line by the terminal's printer as transmission occurs. In either event, the sales personnel entering the office the next morning should find policy illustrations ready to deliver to clients. Since these calls are made without human intervention and at a time when telephone company charges are normally at a minimum, the monthly cost of the automatic calling unit is usually recovered during its first few days of operation during the month.

Depending upon the locations of the remote terminals, the duration of each call, the average number of terminals called each night, and the existing telephone rates for calls over the switched network, either ordinary dialing over the switched network or outward wide-area telephone service (WATS) lines may be used. Ordinary dialing over the switched network is most practical when the terminals to be called are geographically dispersed and the connect time required to accept data from each terminal and return the processed information at a later time is of a relatively short duration. For those applications where the connect time is long, WATS lines can be used to keep toll charges down, with a WATS line giving the user a block of hours per month of telephone usage to a specific geographical area at a fixed monthly cost. WATS line areas can be selected from several bands, with one band covering the entire United States with the exception of the state in which the user is located, while another band might consist of access to several adjacent states.

Interfacing automatic calling units

Two types of interfaces can be utilized by a system which employs an automatic calling unit. As shown in Figure 5.3, the more common RS-232 interface is used to connect the automatic calling unit to its associated data set and the data set to the business machine or computer. While some independently manufactured calling units can also be interfaced to the computer according to the RS-232 standard, telephone company provided calling units are interfaced to the computer according to the RS-366 (CCITT V.25 in Europe) standard which is used to define the interface between a business machine and an automatic calling unit. Within the RS-366 interface standard, five interface classes of automatic calling equipment have been defined, as shown in Table 5.2.

Although five classes of automatic equipment have been defined according to the RS-366 standard, basically the variations can be broken down into versions of type I or type II. Type I defines an automatic calling unit in which the numbers to be dialed are stored internally (stored number dialer), while type II defines one in which the numbers to be dialed are stored in the memory of the associated business machine or computer. Thus, types III and IV can be considered variations of the first two types.

Interface lead description

The interface leads required between an 801 ACU-type device and a business machine according to RS-366 are shown in Figure 5.4. Here, 13 of the 25

Table 5.2 Classes of RS-366 standard automatic calling equipment interfaces.

Class	Data terminal equipment (business machine, computer)	Data communications equipment (automatic calling unit)
Type I	Call request	ACU stores single or multiple telephone numbers which are automatically dialed in sequence; modem separate
Type II	Select number stored in ACU by single-digit control	ACU stores single or multiple telephone numbers: however, number to be called selected by the data terminal equipment; modem separate
Type III	Number to be dialed stored in the data terminal equipment and passed one at a time to the ACU	ACU receives number from the data terminal equipment a digit at a time via a parallel binary-coded decimal (BCD) interface; modem separate
Type IV	Call request or select numbers stored in the ACU by single-digit control	Combination ACU with built-in modem in one unit; stored number dialer can be type I or type II operation
Type V	Multiline automatic calling unit	Undefined at present time

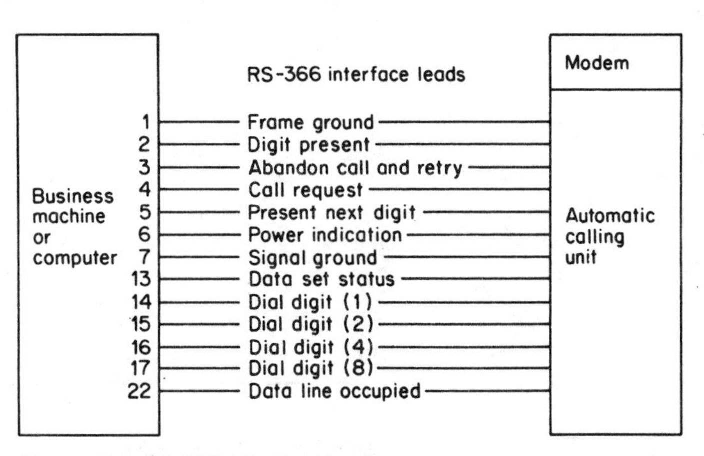

Figure 5.4 RS-366 interface leads.

available interface connections are used. Of the 13 interface leads, six are operated by the computer to signal the automatic calling unit. These leads include CALL REQUEST, four digit leads, and the DIGIT PRESENT lead.

Signals on the CALL REQUEST lead are generated by the business machine to request the calling unit to originate a call. The presence of an ON condition on this lead indicates a request to originate a call, while the OFF condition is used to indicate that the business machine has completed its use of the automatic calling equipment. During the call origination period, this lead must be maintained in the ON condition in order to hold the communication channel 'off-hook'. This lead is turned OFF between calls or call attempts and should not be turned ON unless the DATA LINE OCCUPIED lead is in the OFF condition.

By presenting parallel binary signals on the four digit leads, the business machine presents to the calling unit a 4-bit binary representation of the digit to be dialed. The information presented on these four leads may either be transmitted as the called number or used locally as a control signal, as shown in Table 5.3. As indicated in this table, to request the calling unit to dial a common number such as WE61212, the signals presented by the business machine on the four parallel leads to the ACU would be as shown in Table 5.4.

Signals on the DIGIT PRESENT lead are generated by the business machine to indicate that the calling unit may now read the code combinations presented on the four digit leads. The presence of an ON condition indicates that the business machine has set the states of the digit leads for the next digit. When the calling unit signals the business machine that it is ready to accept the next

Table 5.3 Digit leading coding.

Digit value	NB8 2^3	NB6 2^2	NB4 2^1	NB2 2^0
0	0	0	0	0
1	0	0	0	1
2	0	0	1	0
3	0	0	1	1
4	0	1	0	0
5	0	1	0	1
6	0	1	1	0
7	0	1	1	1
8	1	0	0	0
9	1	0	0	1
End of number (EON)	1	1	0	0

Table 5.4 Dialing WE61212.

Digit	NB8	NB4	NB2	NB1
9	1	0	0	1
3	0	0	1	1
6	0	1	1	0
1	0	0	0	1
2	0	0	1	0
1	0	0	0	1
2	0	0	1	0

digit by turning ON the PRESENT NEXT DIGIT (PND) lead, the business machine must turn ON, DIGIT PRESENT and keep it in the ON condition until the PND goes OFF. Then DIGIT PRESENT must be turned OFF and held in that state until PND comes ON again. When DIGIT PRESENT is ON, the states of the four digit leads are held constant and are changed during the DIGIT PRESENT transitions.

Of the thirteen interface connections which are used, five are operated by the calling unit to signal the associated business machine. As mentioned previously, the PND lead is used by the calling unit to control the presentation of digits on the four digit leads. During dialing, the ON condition indicates that the calling unit is ready to accept the next digit, which will be transmitted by the business machine on leads digit 8, digit 4, digit 2, and digit 1. By placing the PND signal lead in the OFF condition, the calling unit informs the business machine that it must turn OFF the DIGIT PRESENT lead and set the states of the digit leads for the presentation of the next digit. PND cannot revert to the ON condition as long as the DIGIT PRESENT signal is ON; however, it will come ON and remain in the ON condition after the business machine turns DIGIT PRESENT OFF following presentation of the last code combination on the digit leads. If the call is placed by the calling unit, throughout the data transmission interval PND will be in the ON condition. If the call is placed manually or if the business machine is receiving an incoming call, PND will be placed in the OFF condition.

To indicate that power is available within the calling unit, signals are generated on the POWER INDICATION lead by the calling unit. The presence of an ON condition indicates that power is available in the calling unit, while the OFF condition on this signal lead indicates that the calling unit is inoperative because of loss of power.

The DATA LINE OCCUPIED lead is used by the calling unit to indicate the status of the desired communications channel. The presence of an ON condition indicates that the communications channel is in use, while the OFF condition indicates that the business machine may originate a call provided that the POWER INDICATION lead is in the ON condition.

Signals on the DATA SET STATUS lead are generated by the calling unit to indicate the status of the local data communications equipment (i.e., attached data set). The presence of an ON condition indicates that the telephone line is connected to the data set to be used for data communications and that the data set is in the data mode.

The ABANDON CALL AND RETRY signal lead is used to indicate to the business machine that a preset time has elapsed since the last change of state of PND. The timer starts whenever CALL REQUEST is turned ON. The time-out interval on telephone company calling units can be set by a screwdriver-operated switch for a period of 7, 10, 15, 25, or 40 seconds, with a period of 25 to 40 seconds commonly used to allow sufficient time for the call to go to completion. Any time the business machine, the calling unit, or the telephone network takes more time than the preset timing interval is set to, from the last PND OFF, the ABANDON CALL AND RETRY lead is turned to the ON state. The ON state is a suggestion to the associated business machine to abandon the call and try again at a later time if the connection has not yet been completed. The response to the ON condition of the ABANDON CALL AND RETRY lead is left to the business machine, which can either act upon receipt of the signal or ignore it.

In addition to the previously mentioned signal leads, SIGNAL GROUND and FRAME GROUND leads are part of the 13-lead group utilized in the RS-366 standard. The SIGNAL GROUND provides a common ground reference for the interface circuits and is connected to the frame of the calling unit, while the FRAME GROUND lead provides an electronic bond between the frames of the calling unit and the business machine.

Although telephone company supplied calling units are only interfaced to business machines using the RS-366 interface standard, some independently manufactured calling units permit either RS-232 or RS-366 interfacing. Using a single RS-232 asynchronous data port as shown in Figure 5.5, data is transmitted serially from the business machine to the calling unit. In addition to common ground leads, nine other leads of the 25-pin connector are utilized in this type of interface.

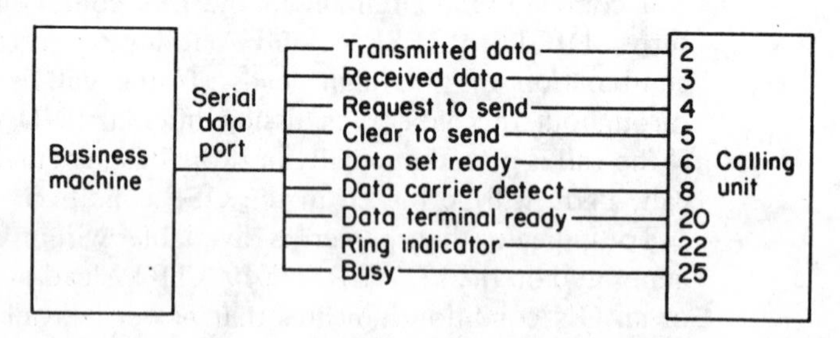

Figure 5.5 Serial data port RS-232 interface.

When dial digit data is transmitted via a serial data port, any four of the first six bits after the start bit can be used to define the dial digit, with the remaining bits being ignored by the calling unit. The four bit positions selected must be coded in the same manner as that shown in Table 5.3. An example of an American Standard Code for Information Interchange (ASCII) character for dialing a seven is shown below, where S is the START BIT and R is RESET (stop) BIT.

S 1 1 1 0 0 0 0 0 R R

RS-232 calling unit operation

If the interface is according to the RS-232 standard, the business machine turns the DATA TERMINAL READY (DTR) lead in the ON condition to initiate a call. The calling unit in response goes 'off-hook' and sets a dial tone timer for a few seconds to insure that the dial tone is present before proceeding with dialing. Next, the calling unit turns the CLEAR-TO-SEND (CTS) lead to the ON condition to inform the business machine of a request by the calling unit for a serial dial digit.

Upon receipt of the CTS request from the calling unit, the business machine transfers a serial character to the calling unit via the TRANSMIT DATA lead. The calling unit stores the serial dial digit and turns the CTS lead to the OFF condition once the entire character has been received. After receiving the full character, the calling unit outputs dial pulses or tones, times out for a second or for a predetermined number of dial tones if so programmed, and then turns on the CTS lead to request the next dial digit. This process is repeated for all the digits of the number to be dialed.

Depending upon the manufacturer of the calling unit, the transfer of a serial end of number character to the calling unit or the non-transfer of this character for a preset time interval of from 1 to 10 seconds will perform the end of the number function.

If the ABANDON CALL AND RETRY timer expires before the answer tone or carrier is detected, the calling unit turns OFF the DATA SET READY (DSR) lead to signal the business machine to abandon the call. To abandon or conclude a call, the business machine turns OFF the DTR lead, and the calling unit then turns ON the DSR lead and resets to await the next CALL REQUEST from the business machine.

Calling unit operation RS-366 interface

The business machine originates a call by turning the CALL REQUEST lead to the ON condition. In response to this signal, the calling unit goes 'off-hook' and 'holds' the telephone line in much the same manner as a telephone which is in use. Since the telephone line is now in use, the calling unit turns the DATA LINE OCCUPIED lead to the ON condition. If the DATA LINE OCCUPIED lead was on prior to the CALL REQUEST being transmitted, the calling unit will not respond since the line would already be in use. Next, the calling unit generates a PND signal which is used to tell the business machine that the calling unit is ready to receive the first digit to be dialed.

Upon detection of the PND condition, the business machine transmits the first digit to be dialed to the calling unit via the four digit leads. When the digit leads have been set, the business machine then places the DIGIT PRESENT lead in the ON condition which tells the calling unit to dial the first digit. After the first digit has been dialed, the calling unit places the PND lead in the OFF condition. This process is repeated until all of the digits have been dialed, after which either of two actions can occur.

If the calling unit receives a binary 12 (end of number) digit to signify that the preceding digit was the last digit to be dialed, the calling unit will give the telephone line back to the data set and signal the data set to enter the data mode, thus making it ready to send or receive data. If instead of transmitting a binary 12 the business machine fails to output another digit when requested, the calling unit will time out and assume that the last digit has been dialed.

Although the preceding example is rather straightforward, numerous variations in the calling unit's method of operation can occur depending upon the calling unit's selected options as well as upon the manufacturer of the device. As shown previously, some calling units interface not only with business machines that output dial digit information in parallel according to the RS-366 standard, but also with devices that output the more common RS-232 control signals and asynchronous start–stop data.

Data sets that can be called by the calling unit are not necessarily the same as those data sets which can be connected to a calling unit. One example of this is a network where a calling unit is used to call remote terminals where each terminal is connected to a Bell System 113B (answer only) type modem. Here, the data set associated with the calling unit could be a 113A (originate mode) data set but obviously would not be another 113B type modem. Thus, manufacturer's specifications concerning both data sets that can be called as well as data sets that can be connected to the calling unit should be carefully checked. While data sets provided by the telephone company are separate devices, some non-telephone company calling units may be ordered with built-in data sets, making possible savings in space, cabling, and power. If existing modems are to be used with the calling unit and they do not comply with the FCC's equipment registration program, their use over the switched network will require the use of a data access arrangement to serve as a protective device between the modem and the switched network. The reader is referred to Chapter 6, Section 6.4, for additional information about this device.

Additional calling unit applications

As previously mentioned, the most common type of calling unit application is for the polling of remote terminals, as shown in Figure 5.3. For this type of application, the business machine conducts automatic polling of remote terminals by utilizing the switched telephone network. This switched network utilization can consist of the ordinary direct distance dialing network or it can be through one or more WATS lines which provide access to the switched telephone network, with the selection based upon such economic factors as the location of the polled terminals, the time the polling occurs, and the duration of each call.

For the situation where only a few cities have to be called but each city contains numerous terminals, an automatic calling unit which can be interfaced to a business machine according to RS-232 standards and thus be remotely located from the business machine. When equipment is used in this manner, long-distance dialing over a leased or foreign exchange line to a remotely located calling unit can be accomplished as shown in Figure 5.6.

The configuration shown in Figure 5.6 may permit long-distance dialing to be made more economical than when using WATS lines or the switched network; this is done through direct distance dialing when a large number of terminals to be polled are concentrated in only a few cities and the polling time required for each terminal is of relatively long duration. When compared with ordinary long-distance dialing, this configuration would also eliminate the possibility of long-distance trunk busy conditions, since private leased lines are now used to connect the business machine to each remotely located calling unit.

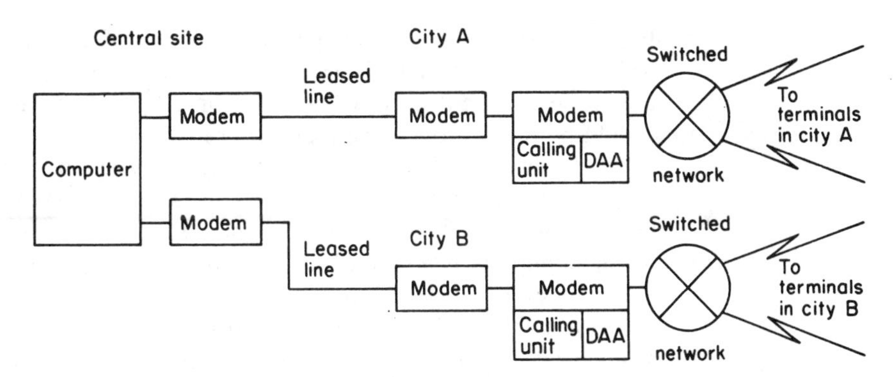

Figure 5.6 Using leased lines to connect remotely located calling units. When a large number of terminals to be polled are located in only a few distant cities, it may be more economical to place calling units in each city and connect them to the computer by installing leased lines between the computer and each calling unit.

If the number of terminals located within a city or geographical area has a total transmission time that exceeds the computer time available for remote polling owing to transmission speed limitations when one calling unit is used, multiple calling units can be used as shown in Figure 5.6. If instead of two calling units several additional units are required, the leased line charges for the configuration shown in Figure 5.6 could become prohibitive or result in unnecessary telephone line charges which can be reduced through the utilization of multiplexers which are illustrated in Figure 5.7. The configuration shown in

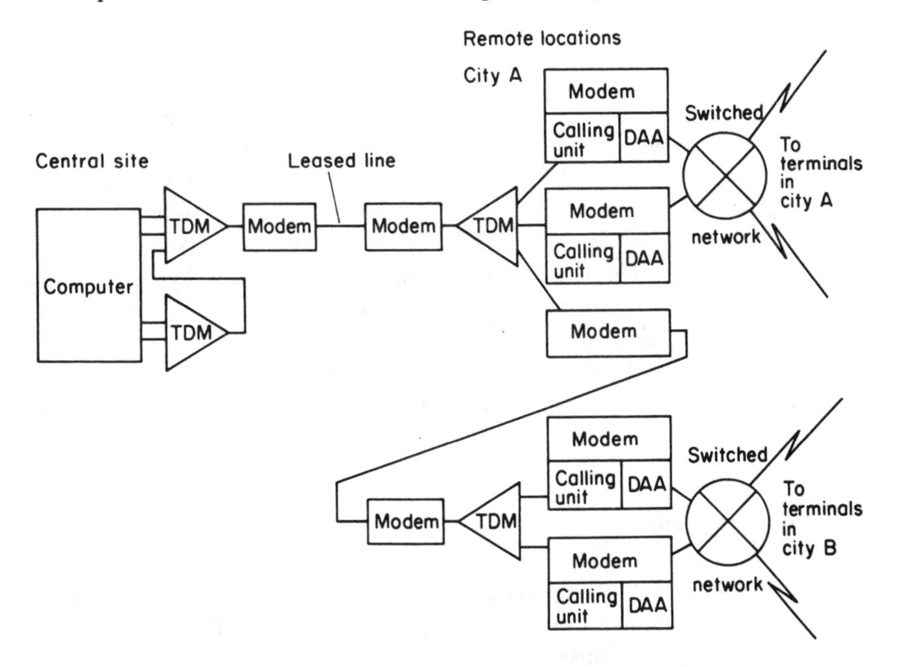

Figure 5.7 Multiple long-distance dialing using multiplexers. Multiple remotely located calling units may have their transmissions multiplexed to reduce communications costs.

Figure 5.7 should also be compared to the cost of multiple leased line charges as well as of multiple WATS line charges. When many channels of low-speed data are to be transmitted to a few cities or geographical areas, this arrangement should be more economical than using multiple WATS lines or multiple leased lines. However, the failure of the leased line between multiplexers would terminate a portion of all remote polling activity and should be taken into consideration by users whose requirements include a method of backup servicing of terminals.

For situations where the duration of each call will be appreciably longer than the time necessary to dial and establish the call, it may be feasible to utilize a single calling unit to dial on several lines. This multiple calling with a single calling unit can be effected through the installation of a device referred to as a calling line selector, and it can be used to switch a single calling unit to a number of lines with the call being switched to any free line by the calling line selector, as shown in Figure 5.8.

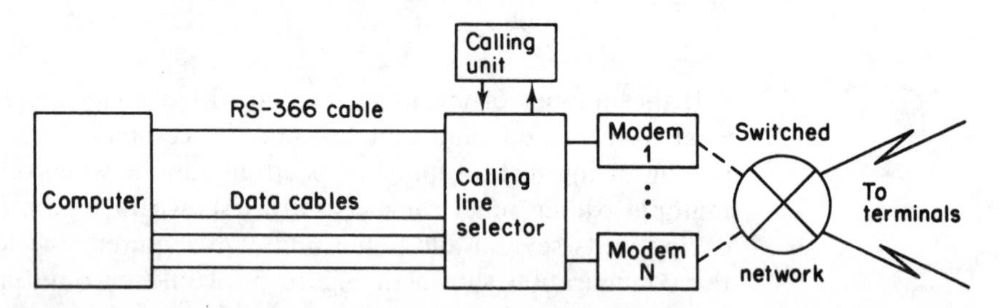

Figure 5.8 Multiple calling using calling line selector and a single calling unit. One calling unit can establish a number of concurrent calls through the use of a calling line selector.

As the figure also shows, the calling line selector can be effectively used when the terminals to be polled are located within the same city as the business machine; or it can be used where a number of foreign exchange lines are used to give the business machine located in one city local telephone access to terminals located in different telephone exchange areas. When the calling line selector is used, the business machine may be programmed to select the next line to be dialed by transmitting a parallel 4-bit digit ahead of the first dial digit; or, the calling line selector will automatically go to the next free line in sequence if the business machine selection mode is not used. In addition to selecting the next line to be dialed, the calling line selector provides for the automatic answering of incoming calls.

REVIEW QUESTIONS

5.1 What control signal normally activates an automatic answering unit?

5.2 Why should the timer abort option normally to be set ON on an automatic answering unit?

5.3 Why should the loss of carrier disconnect option normally be set ON on an automatic answering unit?

5.4 What is the major difference between an 801A and an 801C automatic calling unit?

5.5 Discuss an application where the use of automatic calling units could be economically viable.

5.6 What is the major difference between RS-232 and RS-366 with respect to transmitted data?

5.7 Construct a table showing the digit lead coding required to dial the number BOYGIRL via an RS-366 interface.

5.8 What is the main advantage in using an RS-232 interfaced automatic calling unit?

5.9 When would you consider using leased lines to connect remotely located automatic calling units instead of having them directly connected to the switched telephone network?

5.10 What is the advantage obtained in using a calling line selector and under what circumstances should you consider its use?

CHAPTER 6

SPECIALIZED DEVICES

The devices explored in this section can be used in a variety of situations. To provide a level of protection to transmitted data, security devices can be installed to encode such data in order to reduce the possibility of the transmission being intercepted and understood by an unauthorized person. Other security devices to include special types of modems and switches can now be obtained with password verification capability, permitting access to the desired facility only upon the entry of an appropriate code.

Both speed and code converters and voice adapters can be employed to reduce communications costs. Speed and code converters may permit data communications users both to standardize transmission speeds and reduce the number of lines required for the transmission of data from remote terminals to a central computer. Voice adapters permit a line used for data communications to be alternating or simultaneously used for voice communications, permitting voice contact between remote terminal operators and computer center personnel over common communication facilities.

Another device similar to, but far more sophisticated than, a speed and code converter is a protocol converter. By the utilization of an appropriate protocol converter incompatible physical interfaces, device characteristics, and/or transmission protocols can become compatible.

Although much controversy and many court rulings do provide a colorful history to data access arrangements, such devices can basically be viewed as a protective component to shield the switched telephone network from the possibility of malfunctioning customer-provided equipment when such equipment has not been registered for usage by the Federal Communications Commission (FCC). While basically all devices currently manufactured for use on the switched telephone network are registered and do not require the use of a data access arrangement, devices manufactured prior to the FCC equipment registration program coming into effect require this device. So this device is also examined in this chapter.

Since Pulse Code Modulation (PCM) which is the primary method of voice digitization results in a 64-kbps data stream it is unsuitable for use in networks where lines operate at data rates below that obtainable on wideband facilities. Due to this, vendors developed two devices to permit voice and data to be transmitted over conventional analog facilities. In this chapter we will examine both Speech plus Data Units and Voice Digitizers, denoting their similarities and differences as well as their network employment.

One method to achieve greater transmission efficiency is to be able to transmit less data. This can be accomplished by data compression and we will examine several methods that can be used to perform compression as well as several devices that perform this function.

Concluding this chapter, we will examine the use of fiber optic transmission systems to include both fiber optic modems and multiplexers. Based upon the transmission of light energy instead of electrical pulses, these devices use the large bandwidth of light to obtain extremely high data transmission rates in comparison to conventional modems and multiplexers that transmit and receive data in the form of electrical pulses. Since energy travels in the form of light in fiber optic transmission systems we will also examine the advantages and disadvantages of the use of this transmission medium in this chapter.

6.1 SECURITY DEVICES

As networks expand and proliferate, more and more people have access to them, and it becomes harder to guard against unauthorized access to the network and entry into data files on computers which normally may be restricted to only a few users. A door that is locked and patrolled is, of course, some measure of security against sabotage or theft; but it offers little protection against the wily, white-collar evil-doer for whom entry is no serious problem. Hence, in the absence of adequate operational controls, someone using a time-sharing system may be billed for someone else's transactions; or, more seriously, a company transmitting confidential bid information could have its message intercepted and read by a competitor.

To reduce the possibilities of these events occurring, a number of devices have been introduced by several manufacturers to encrypt both data files and transmitted data. Prior to discussing how these devices operate and where they can be employed, a review of sign-on and database security features will be undertaken to provide the reader with an insight into the problems associated with unauthorized access as well as some of the methods that can reduce the probability of such access with computer software.

Password shortcomings

The most frequently employed methods of preventing unauthorized access to networks and databases, that of identification codes, has a number of shortcomings. The code can be glimpsed over a person's shoulder, found on a discarded printout in a trash basket, or on the ribbon of a printer terminal. A remedy, then, is to use characters that are non-printable. (The term 'character' in this context refers to any distinct electrical signal initiated at a terminal and does not necessarily imply a visible figure in the usual sense of the word.) This is the so-called password approach. It is a precaution that is gaining increasing attention among data communications users, and it is also fairly simple, being essentially an extension of the use of terminal identification numbers.

Whether printed or non-printed, however, careful consideration is required in putting together the elements of an adequate access control system. These include immunity to access by repeated random tries and the ability of the network to report repeated attempts at access.

Practically any element in a network can be the target of unauthorized access, but the most likely point of entry is a remote terminal because it is out of sight of the central office. The sign-on procedure, involving a code that must be keyed in to the terminal to gain clearance, is the standard method of assuring access. Figure 6.1(a) depicts the normal response of a computer to a sign-on request. If an illegal sign-on is attempted, the response might be as shown in Figure 6.1(b).

For example, in a time-sharing system, the procedure starts with the user establishing the telephone connection and identifying the type of terminal (code and transmission speed), usually by sending a carriage return. Typically, after getting a go-ahead from the computer, the user then transmits his or her own personal identification number. If the number is invalid, as might happen if a wrong key is accidentally struck, the user receives a message, such as 'illegal sign-on'. The computer then allows the user a fixed amount of time, usually several seconds, to send another sign-on. If the time is up or the user exceeds a certain number of allowable retries, the computer automatically disconnects. In this case the user must redial and try again.

This procedure is susceptible to sophisticated attack, however, unless additionally protected; but more on that later. A flow chart outlining a typical sign-on procedure in a time-sharing system is shown in Figure 6.2.

```
PLEASE SIGN-ON
ID UcUcUcVcVcVcVc
GENERAL TIME SHARING COMPANY
ON AT DATE, TIME
```

(a)

```
PLEASE SIGN-ON
ID 127 ABCD
ILLEGAL SIGN-ON THRESHOLD EXCEEDED
CONTACT OPERATOR
OFF AT XX/YY
```

(b)

Figure 6.1 Identification messages. The format of the message requesting the user's identification and password (underlined) is similar to that shown in diagram (a). U_C is the user's identification number, while V_C represents a verification code, typically non-printable control characters. Notification that the maximum number of retries has been exceeded is shown in part (b). Operator intervention may follow.

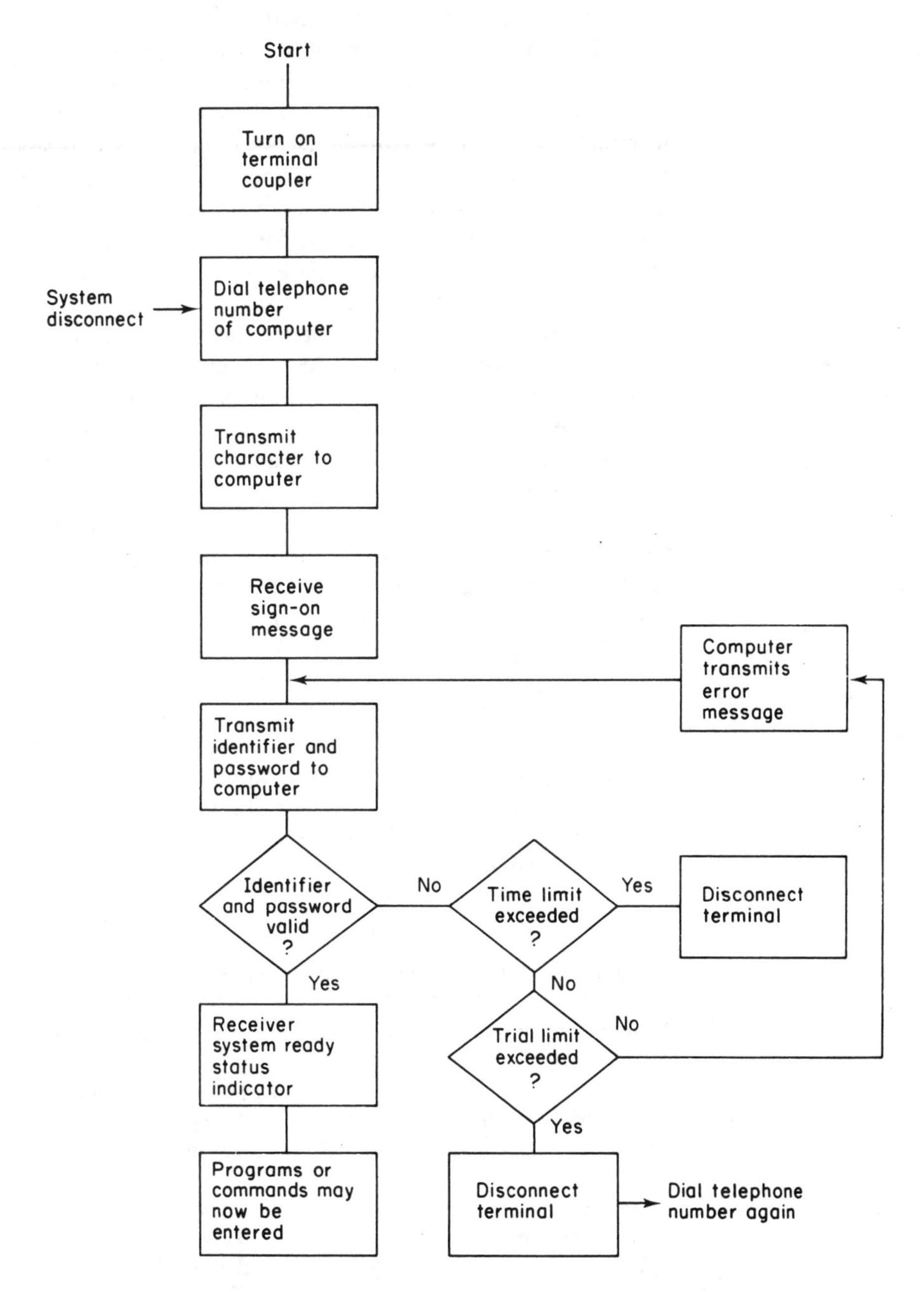

Figure 6.2 Sign-on procedure. Typical in sign-on procedures for time-sharing systems, the response to a user attempting to gain access is a request for identification and password.

Legal, but unprintable

A password approach incorporated into the sign-on procedure following the identification number has the obvious advantage of being less liable to illicit discovery. However, there are limitations. The most obvious limit is the number of characters on the terminal keyboard. For instance, the most common terminal, the Teletype Corporation Model 33-ASR, limits the user to a subset of the American Standard Code for Information Interchange (ASCII). And not all control characters on a keyboard can be used in the password. Some systems reserve certain characters for terminal function control, system control, and communications control. Examples are a carriage return to indicate the end of a line and a backspace to cause deletion of the last character entered. Other characters may turn a paper tape reader on or off. A list of frequently unavailable non-printable characters is given in Table 6.1. For some terminals there may be few characters left. In such a case, a long password is needed to provide enough combinations. Since the list is extensive, it can easily be seen that not many characters in an average system may really be available for a password.

Experience has shown, however, that a three- to six-character password is the optimum length to provide enough characters in most systems to take care of the job. The optimum length, incidentally, is taken to mean one that does not defy an average user's ability to remember the password and one that does not add substantially to computer overhead in the processing of it.

It may appear at first that the number of combinations possible in a password of a given length would simply be C^i, where C is the size of the available character repertoire and i the number of character positions to be filled. Actually, additional combinations are possible by using fewer than the maximum number of positions that one has decided upon. This is accomplished by typing a carriage return immediately following the shortened password to indicate that it is terminated.

Table 6.1 Frequently unavailable characters.

Character meaning	Possible reversed usage
Break	System interrupt
End of transmission	Communications
Bell	Terminal
Line feed	Terminal
Form feed	Terminal
Carriage return	Terminal, system
X-On	Terminal
Tape	Terminal
Cancel	System or communications
Space (field separation)	System
Delete line	System
Delete character	System
Rub-out	System or communications

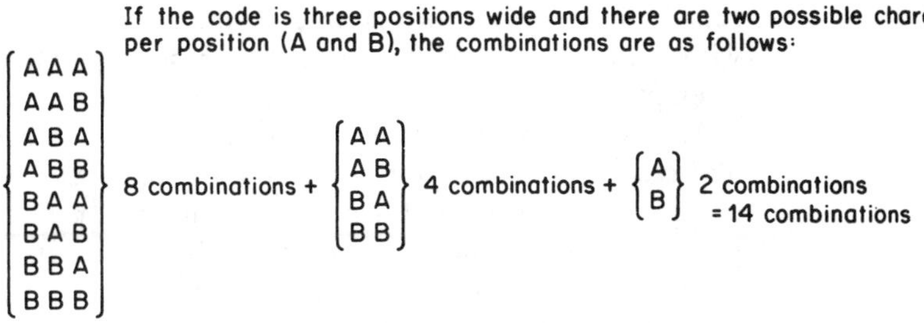

If the code is three positions wide and there are two possible characters per position (A and B), the combinations are as follows:

$$\left\{\begin{matrix} AAA \\ AAB \\ ABA \\ ABB \\ BAA \\ BAB \\ BBA \\ BBB \end{matrix}\right\} \text{8 combinations} + \left\{\begin{matrix} AA \\ AB \\ BA \\ BB \end{matrix}\right\} \text{4 combinations} + \left\{\begin{matrix} A \\ B \end{matrix}\right\} \text{2 combinations} = \text{14 combinations}$$

Figure 6.3 Code combinations. By permitting the password to be smaller than the maximum number of chosen positions, extra combinations are possible. A and B denote non-printable characters, making up to 14 combinations.

Consider, for example, a three-position password in which each position can take either of two non-printable characters, which are designated A and B in Figure 6.3. By using all three positions there are eight possible combinations. But by using two positions and leaving the third position blank, four more combinations are possible. If only one position is filled, two more still are possible, for a total of 14 possibilities.

In this routine, the user's personal identification number, which is usually keyed in at the start of the request for access, tells the computer what to expect. Identification numbers and passwords must correlate. (From a practical standpoint, however, one- or two-position passwords might be inadvisable because it might make it easier for an intruder to come upon the correct password by simple random selection.)

In any event, the total number of realistic combinations, T, can be calculated as:

$$T = \sum_{i=1}^{i=w} C^i$$

where i represents individual positions (first, second, third, etc.), w is the total number of positions in the password, and C is the number of characters available. For example, a four-position password with a character set of two provides 30 combinations. Increasing the character set to eight raises the combinations to 4680.

The unwitting accomplice

A potent tool available to the illicit network user is another computer. Here the human malefactor harnesses the machine, which need be little more than a microcomputer and automatic dialer, to repeatedly dial, try a password, and retry if the password does not work.

If, for example, such a system were used to access a computer that permits five sign-on attempts before automatically disconnecting and the terminal transmits at 30 characters per second, then the five tries could be performed in less than 10 seconds. Upon sensing a disconnect, the microcomputer could redial and try five more code combinations. Over a 3-day holiday weekend, with little traffic and the central computer unattended, this scheme could attempt over 40,000 passwords with nobody being the wiser.

To counter this brute force approach, provision should be made to monitor and record the repetition of sign-on attempts. Once a predetermined number of tries has been made for a given identification number, the system could be programmed to inhibit additional sign-on efforts until a system operator intervenes. The maximum number of tries can be established daily or at more frequent intervals. The threshold for a given identification number could be 10 attempts in a given hour or 20 attempts a day, depending on the individuality of the system. When the threshold is exceeded, the user is locked out of the network and receives a message requesting that he or she contact the system operator for return of service.

Automatic intervention does not eliminate the possible success of a concerted effort to gain access by numerous retries, but it does sharply reduce the probability of success. The only way in which access might be obtained is to try the maximum number of sign-on attempts short of intervention, disconnecting for the balance of the timing interval, and trying again. If this were to succeed at all, it could take months or years in a system with a small threshold, a long password, and a long time interval.

As shown in Figure 6.4, with a three-position password drawing on 15 available characters, the probability of successful entry (a probability of 1) rises sharply if there is no intervention and is assured by the 3615th trial. With intervention after 20 repeated tries, however, the probability does not exceed 0.005. The diagram also shows that if the character repertoire is reduced to 10 and there is no intervention mechanism, the intruder is assured of access by the 1110th try, two-thirds sooner than with a 15-character repertoire. Another factor that can have a major bearing on access security is the number of characters reserved to perform communications functions. While a large number of such characters are standardized, installations that remove characters from the character set available for passwords run the risk of increasing the access vulnerability of the system.

While the preceding discussion was primarily concerned with sign-on access, similar problems and solutions can be effected with respect to the passwords employed to protect data files.

Transmission security

Although a number of methods have been developed to promote access and database security further; unless the transmission medium utilized is secure, the user may become vulnerable to having data transmission compromised by such means as line tapping or line monitoring. In addition, if transmission is over such store and forward message switching systems as Telex or TWX, or by courier or telegram, the message may be read by unauthorized personnel or obtained through active penetration by insiders as well as outsiders. In certain US Department of Defense installations, terminal transmission is over secure transmission lines to a central facility or another terminal which makes

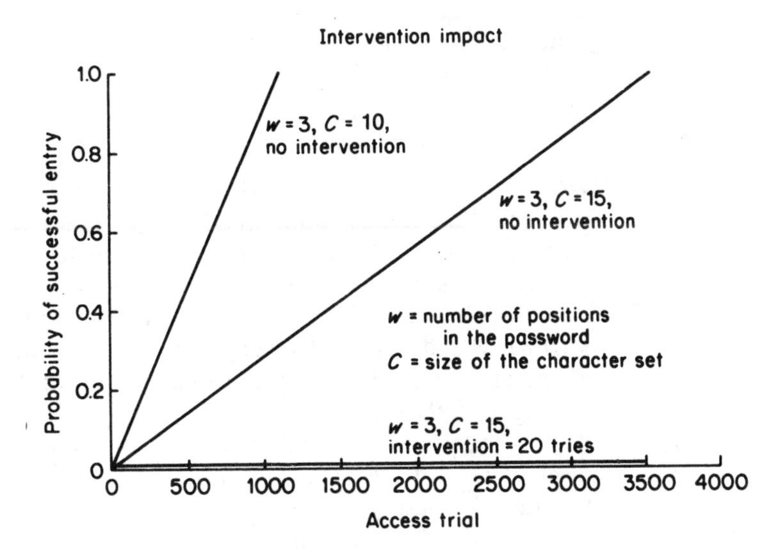

Figure 6.4 Intervention impact. With intervention software, the probability of successful entry by random password selection becomes extremely small.

access security their major concern. This is due to the fact that, although all transmission lines on the installation are secure, personnel with different security level clearances have access to the same computer and a method of differentiating who obtains what computer resources is primarily determined by the terminal's location and the identification code of the user.

For the situation where data transmission occurs over public, non-secure facilities, a method of making the transmitted data unintelligible to unauthorized parties becomes as important as having a good access security procedure. Fortunately, numerous techniques are available for the user to make transmitted data unintelligible to unauthorised parties; the oldest and most widely used method is the various types of manual coding processes where, through the use of code books and pads, the original text is encoded before transmission and then decoded by the recipient of the message. The foundations for the various coding schemes go back thousands of years and have been used to protect a wide range of messages ranging from diplomatic and military communications from before the time of the Roman Empire through commercial messages of industrial companies today.

Manual techniques

In spite of the fact that most manual coding techniques can be broken by trained cryptanalysts, they do offer a measure of protection because of the time element: there may elapse a period of considerable length before the message is decoded by an unauthorized party. Thus, the information that company A will bid 2 million dollars on a contract whose bid is to be opened on 1 February is worthless to a competitor that decodes the message from the home office to company A's field agent after that date. In addition, manual coding schemes can also be used as a backup in the event of the failure of a coding device that the user may have installed.

Caesar cipher

One of the earliest coding techniques was the so-called Caesar Cipher, which was probably known and used long before Caesar was born. Using this technique, Julius Caesar enciphered his dispatches by displacing each letter by a fixed amount. If the displacement was two, then the message INVADE ENGLAND TONIGHT would be transformed into KPXCFG GPINCPF VQPKIJV and sent by messenger to the appropriate recipient. Upon receipt of the message, a reverse transformation would develop the plain text of the message. Although this scheme may appear primitive, some encryption devices today use electronic circuits to perform a continuous and alternating displacement of the plain text to develop an encrypted message which could frustrate the best cryptanalyst.

Checkerboard technique

A group of encryption techniques have been developed based upon what is known as the basic checkerboard. Here, the alphabet and numerals are written into a six by six block with coordinates for row and column used to specify the cells. This technique can involve the use of different indices as well as a rearrangement of cell data, as shown in Table 6.2. Again, with the advent of modern electronics it is possible to construct a device to continuously change their indices after a certain period of transmission or to change the cells, or both.

Table 6.2 Basic checkerboard.

	Block 1							Block 2					
	A	E	I	O	U	V		G	H	I	J	K	L
A	A	B	C	D	E	F	A	Z	Y	X	W	V	0
E	G	H	I	J	K	L	B	U	T	S	R	Q	1
I	M	N	O	P	Q	R	C	P	O	N	M	L	2
O	S	T	U	V	W	X	D	K	J	I	H	G	3
U	Y	Z	0	1	2	3	E	F	W	D	C	B	4
V	4	5	6	7	8	9	F	A	9	8	7	6	5

Using the above blocks, the word CODE becomes: from Block 1: AI II AO AU; from Block 2: EJ CH EI EH.

A variation of the checkerboard technique is accomplished through the utilization of a keyword which is commonly referred to as a Vigenere cipher technique. In its simplest form the Vigenere cipher consists of a table of alphabets, as shown in Figure 6.5. For ease of remembrance, a meaningful phrase or mnemonic is selected as the key, although this can be a major weakness, and incoherent keys which reduce the number of clues are preferable for use.

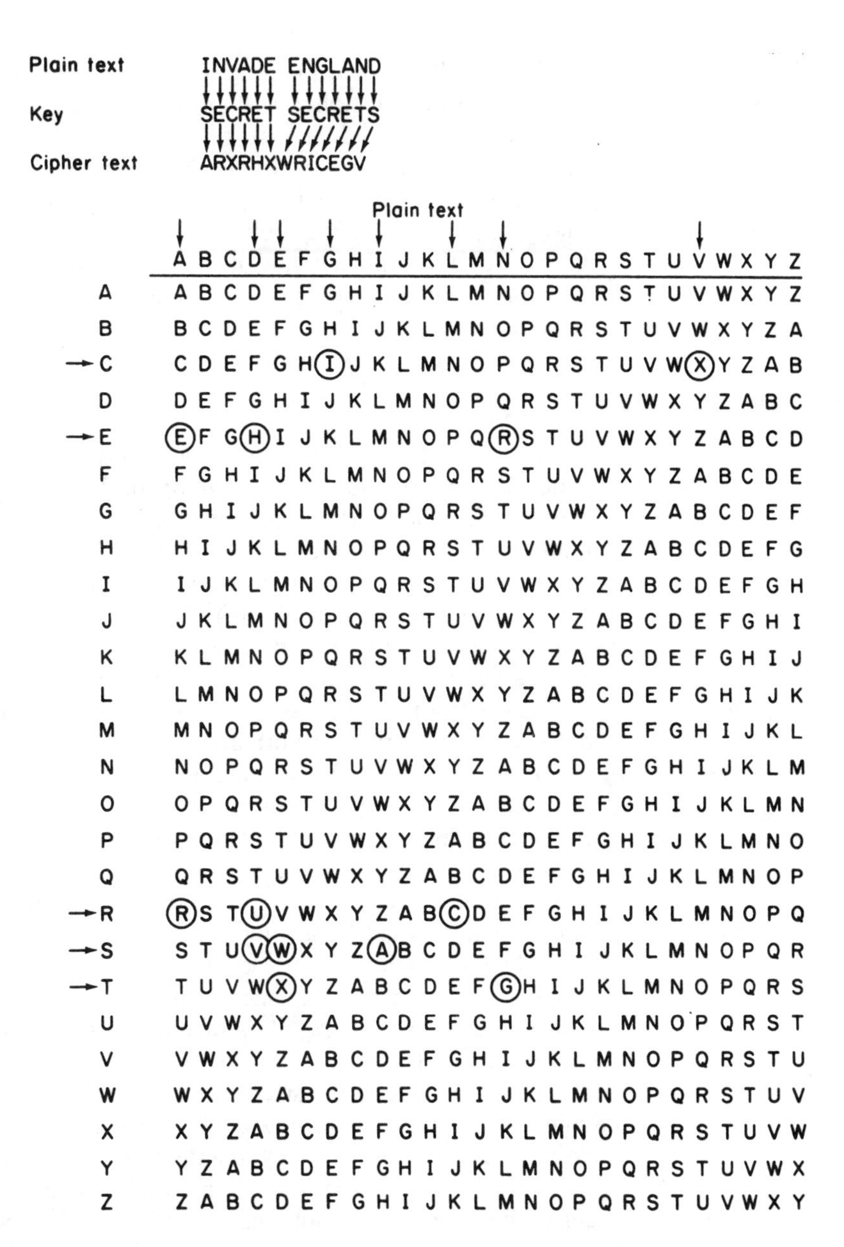

Figure 6.5 Vigenere cipher. A one-to-one correspondence between the plain text and the characters of a key is used to develop the cipher text. Here, the intersection of the plain text character I with the key character S produces a cipher text character A.

Here, encipherment begins by establishing a one-to-one correspondence between the characters of the plain text and the characters of the keys, with the key partially or completely repeated if shorter than the plain text. The cipher character is then obtained from the intersection of the appropriate keyletter row and the plain text column. For example, if SECRET SECRETS is used as the key to encipher the plain text message INVADE ENGLAND, the table provides a cipher text of ARXRMX WRICEGV. As shown, the cipher text is developed character by character, with the first character of the cipher text obtained from the intersection of the S character of the key SECRET with the first character (I) of the plain text, and so on. Since 676 memory locations are required for 26 characters of the alphabet or 1296 locations for the alphabet and digits, some devices encode data by using a fixed memory but generating pseudorandom numbers which are used to develop a pseudorandom key.

Automated techniques

Concurrent with the development of electromechanical devices, several methods of encoding information were developed. In 1917, Gilbert Vernam of the American Telephone and Telegraph Company (AT&T) developed a method for insuring that the information contained in a punched paper tape would remain unintelligible to unauthorized users. In Vernam's technique, each text letter was enciphered with its own cipher letter. If the key tape was as long as the message, and its key perfectly random, the text was then theoretically unbreakable. However, the inconvenience of preparing thousands of feet for high-volume traffic as well as the security problems inherent in guarding tape supplies and accounting for both active and cancelled tape rolls prevented most users from considering this technique.

A practical compromise between convenience and security was the development of pseudorandom events which appear to be as unpredictable as those generated by white noise, sunspots, and other physical phenomena, but in reality are developed from a reproducible mathematical relationship. An example of this would be a program to manipulate two 18-bit registers where the product of each register's contents through a predetermined process is extracted into two numbers which return to the registers where the process is repeated again. Thus, the 36 bits of the two registers could produce a period length of $2^{36} - 1$ before returning to a repeatable pattern. Owing to the continuing shrinkage of the size as well as a reduction in the costs of integrated circuits, pseudorandom keying devices became available to the commercial user at realistic prices. The key lengths of some of these devices have become so long that it appears that the communications equipment whose transmission security they are safeguarding may become obsolete before the end of the first key period.

In addition to a number of firms which manufacture encoding devices solely for use by intelligence agencies, the US Armed Forces cryptologic agencies, and other government users, several industrial firms have actively entered the market and manufacture a family of security devices for commercial users. These manufacturers provide a family of devices designed to operate with a wide variety of terminals to include facsimile devices. These encryption devices utilize a built-in key generator which is similar to a multilevel register device.

Using proprietary encryption techniques, the user selects the code family by either turning appropriate rotary switches inside the device to a specific setting or entering the code through a built-in keypad. Then, the specific code in each family is user-selected by thumbwheel switches on the device's front panel, access to which is obtained through the unlocking of a steel door or by the entry of the code through the keypad at the same time as a security key is inserted into the device and positioned into the activate position. Depending upon manufacturer, up to 32 trillion key settings or codes are available to the user. One device that offers 32 trillion code settings has those settings arranged as 16 million code families, with 2 million codes in each family. Using this arrangement the user's security officer can develop various code administration techniques in hierarchical arrangements.

Security devices available for the commercial market operate either off-line or on-line. The off-line devices preceded the development of the on-line devices and are primarily used with punched paper tape transmission over the Telex and TWX networks. The off-line systems are used to prepare enciphered message tapes prior to transmission over teleprinter circuits with the device connected to an auxiliary off-line teleprinter which provides the paper tape and keyboard/printer functions with the operator typing the clear message into the keyboard of the teleprinter and the encryption device simultaneously punching the encoded tape.

Some encryption devices also permit a two-pass operation to be used where a clear tape is punched by the operator and then read by the device at high speed to punch the encoded tape. The encoded tape is then transmitted by the subscriber over the network to the recipient as any punched paper tape normally would be transmitted. At the receiving end the punched paper tape of the receiving terminal is decoded off-line by reading it into the encryption device at that end and having the plain text printed at the connected teleprinter which is now turned to the off-line mode of operation. This is shown in Figure 6.6. Since the header of the message contains routing and destination information, it obviously cannot be encoded. Thus, the off-line device has either switches or buttons which allow the user to start and stop the encoding of information at the appropriate points within the message.

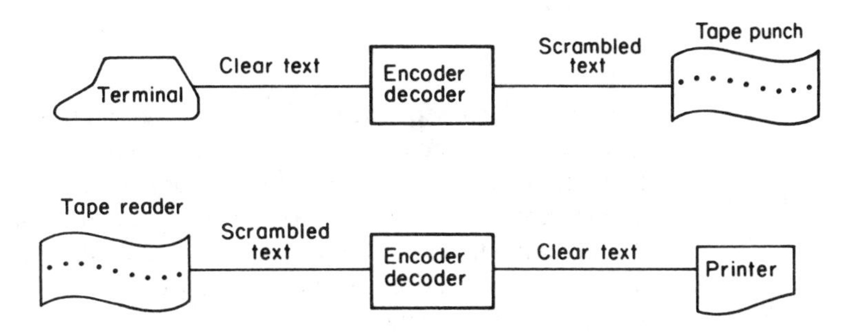

Figure 6.6 Off-line encoding (top) and decoding (bottom).

Modern developments

Advances in electronics to include large-scale integrated circuits and microprocessors have produced a technology base for the development of a family of data security devices. These devices normally employ the National Bureau of Standards (NBS) data encryption standard algorithm which provides a set of rules for performing the encryption and decryption of data which reduces the threat of code-breaking by illegal personnel to virtually an insignificant possibility. This is due to the fact that if the data is intercepted, the time required to decipher the encoded information would require many tens or hundreds of years of machine time, probably resulting in the information being of no value by the time it is decoded.

New encryption devices use a feedback-shift register and associated circuits or a microprocessor to generate a pseudorandom bit sequence following the NBS algorithm. To make the transmitted data appear to be a random stream of ones and zeros, modulo 2 addition is employed to add the data to an apparently orderless bit stream developed by a pseudorandom number generator.

Modulo, or modulus, is a capacity or unit of measurement, and a modulo sum is a sum with respect to a modulus while the carry is ignored. When a two-digit decimal counter is used, it is a modulo 100 counter. Such a counter cannot distinguish the numbers 99 and 199. When 9 is added to 8, the sum is 17, but assuming that the modulus is 10 in this case, the modulo sum is 7. In modulo 2 addition, one and zero and zero and one result in one, and both one and one and zero and zero make zero. Modulo 2 addition and subtraction are illustrated in Table 6.3.

Table 6.3 Modulo 2 addition and subtraction.

Modulo 2 addition				Modulo 2 subtraction			
0	0	1	1	0	0	1	1
+0	+1	+0	+1	−0	−1	−0	−1
—	—	—	—	—	—	—	—
0	1	1	0	0	1	1	0

When the bit sequence or key text generated by the pseudorandom number generator is added by modulo 2 addition to the original information or clear text, the result is an apparently random stream of ones and zeros referred to as the cipher or encoded text. This encoded data is then transmitted to its destination where another encoder/decoder using an identical key text performs modulo 2 subtraction on the encoded data to develop the original clear text, as illustrated in Table 6.4.

Another method employed to effect data security which eliminates the requirement of hardware devices is through the utilization of software packages that can be installed on many computers and which utilize the NBS algorithm. These software packages can be utilized with most programming languages, including such languages as Fortran and Cobol through the use of CALL statements and result in a data file being encrypted. Once encrypted, the file can be transmitted to another location where another software package is

Table 6.4 Modulo 2 addition keeps data secure.

Encoding	
Source (clear) text	1 0 1 1 1 1 0 1 1 1 1 0 0 0 1
Pseudorandom (key) generator	1 0 0 0 1 0 1 0 1 1 0 0 1 0 0
Modulo 2 addition (encoded) text	0 0 1 1 0 1 1 1 0 0 1 0 1 0 1
Decoding	
Encoded text	0 0 1 1 0 1 1 1 0 0 1 0 1 0 1
Pseudorandom (key) generator	1 0 0 0 1 0 1 0 1 1 0 0 1 0 0
Modulo 2 subtraction (clear) text	1 0 1 1 1 1 0 1 1 1 1 0 0 0 1

available on a computer to enable the user at that location to decode the file. While this method eliminates hardware, users remotely located from the computer still face the threat of having their data intercepted if they are building a data file, since the file is not protected until it is stored and encrypted.

A problem associated with using software to encode and decode data according to the NBS data encryption algorithm is the overhead associated with the required processing. To alleviate this overhead most methods of implementing the algorithm have been through the utilization of stand-alone devices incorporating microprocessors to encode and decode data.

On-line applications

With the commercial development of automatic cipher synchronization techniques, continuous on-line encryption devices are no longer the exclusive preserve of government agencies and the military and intelligence communities. These on-line encryption devices are capable of operating with both asynchronous and synchronous data streams at various data rates. In the on-line asynchronous mode of operation the encryption device uses the start–stop pulses of the individual characters to develop the synchronization between the sending and the receiving units.

Plain text, which can be typed on the terminal's keyboard or read from a paper tape reader, is automatically encrypted on a character-by-character basis; the encrypted data is transmitted via the communications channel to the receiving unit where it is automatically deciphered and furnished to the receiving terminal. For synchronous transmission, the encryption devices are automatically synchronized in time by a short character sequence typed by the user or through the depression of an INITIATE button on the front panel of one of the devices. Once synchronization has occurred, the two units step under the control of crystal-controlled clocks, which keep the units in synchronization. Figure 6.7 shows typical on-line encryption device applications.

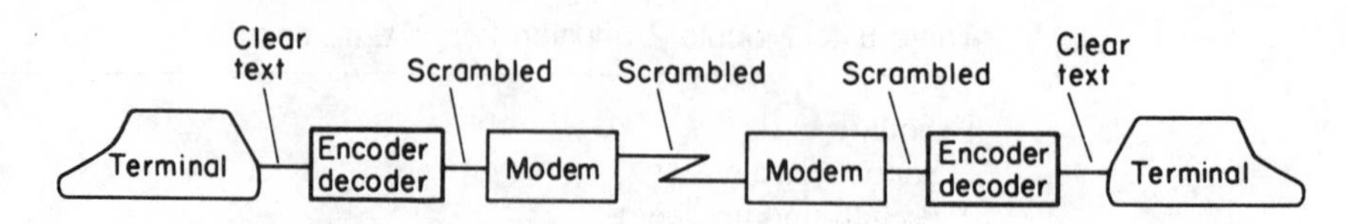

Figure 6.7 On-line encryption device applications. Terminal-to-terminal transmission via message-switching systems such as Telex or TWX can be secured by interfacing security devices between the modem and each terminal.

As shown in Figure 6.7, on-line encryption for terminal-to-terminal transmission over such common carrier facilities as TWX or Telex can be accomplished by the installation of an encoder/decoder between the terminal and the terminal's associated modem. Thus, clear text originating at the terminal is scrambled by the encoder and passed to the modem which transmits the data over the common carrier's facility. At the receiving end, scrambled data passes through the modem to a decoder which now produces clear text that is passed to the terminal at this end.

In Figure 6.8, a typical time-sharing application for an encryption device is shown. In this example, the terminal which required the security measures associated with encryption devices is connected via a leased line to a time division multiplexer (TDM). Since this is a dedicated port connection, there is no problem in determining what port of the demultiplexer to which a similar encoder/decoder should be interfaced.

In Figure 6.9, the terminal user now dials a rotary over the switched telephone network in order to connect to the multiplexer. For this configuration, the terminal may be connected to any port on the multiplexer, which means that at the other end a method is necessary to determine which port of the demultiplexer has the scrambled data. The encoder/decoder selector performs

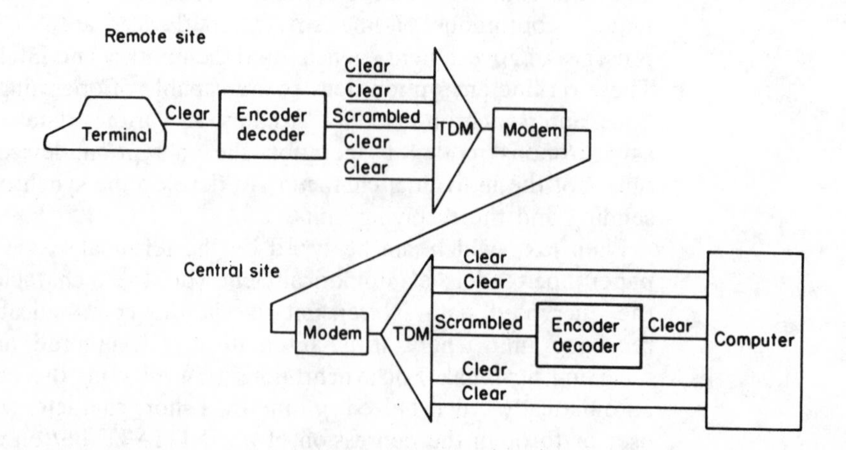

Figure 6.8 Time-sharing encryption application. Terminals connected to a time-sharing system through multiplexers must have a dedicated port to effect secure communications.

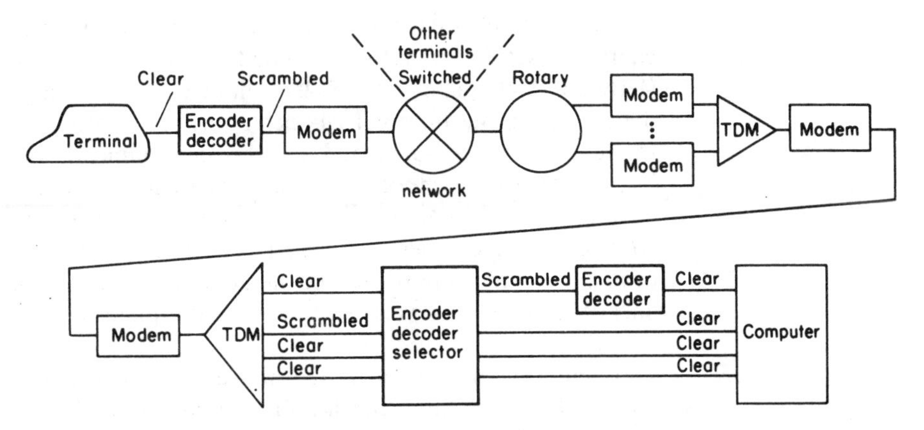

Figure 6.9 Data security for time-sharing systems. To transmit encoded information via the switched network through a multiplexer, a selector must be employed at the computer site to switch the scrambled output of a TDM channel into a decoder to produce clear text.

this function by sampling the output of each of the demultiplexer's ports and switching the port that has the scrambled data to the encoder/decoder, which then decodes the scrambled text and passes it to the computer. This selector in effect performs limited switching and acts as a transparent device, passing the data from the other ports directly to the computer.

Before selecting an encryption device the user should examine in detail the application requirements and develop a set of specifications which will aid in the selection process. Table 6.5 contains a checklist of some common encryption device specifications which should be considered. Although some devices are code transparent, many are not. Sine some devices are manufactured for specific applications such as for use on the Telex or TWX network, the character sets to include the crypto and plain text character sets should be examined.

Similarly, the transmission speed supported should also be examined since some devices only support teletype operating speeds while other devices can operate at speeds up to 1 million bps.

The operational mode of encryption devices falls into three categories: off-line, on-line asynchronous and on-line synchronous, with the mode selected dependent upon the user's requirements.

Table 6.5 Encryption device specification.

Character set supported	Message key
Transmission rate supported	Key code change
Operational mode	Power supply
Crypto character set	Terminal interface
Plain text character set	Alarm circuitry
Carrier compatibility	

The crypto character set and plain text character set can vary from the Telex character set through the extended binary-coded decimal interchange code (EBCDIC) character set of 256 characters. The character set selected will depend upon the terminal character set, the computer's character set, and/or the character set supported by the communications facility used. Carrier compatibility is an important feature when the user wishes to transmit encrypted messages over a commercial network such as Telex or TWX. Since certain control characters perform unique functions, carrier compatibility alleviates problems by the suppression of the generation of control characters during the encoding process as well as by sending the plain text control characters in the clear.

As mentioned previously, the message key and key code change determine the number of coding variations as well as the total codes available. Like most devices used in data communications, encryption devices have a wide range of interface options which the user must properly select from to match the terminal's requirements.

To prevent transmission from being compromised if the device should fail, some encryption devices have built-in alarm circuitry which monitors the key generator output and will inhibit data transmission upon detection of encoding failure.

Access control devices

Another group of security devices that warrants attention is designed to control access to a facility instead of protecting the transmitted data. Two devices that fall into this category are security modems and switches.

Security modem

A security modem incorporates a microprocessor, random access memory, and an RS-232 (CCITT V.24) port for programming the device. Two common features incorporated into most security modems are password verification and dialback capability.

Upon installation of the security modem, the user can temporarily cable a terminal to the device and enter a password. Then, users dialing that modem will be prompted for the appropriate password prior to the modem passing the remote user's data traffic onto the device attached to the security modem. Normally this feature of the security modem is used to protect access to a computer system whose operating system does not provide password access security. One example of this might be a personal computer in an organization that is used as a bulletin board for access by company employees to post and retrieve messages and other information.

A second feature incorporated into many security modems is a dialback capability. When this feature is implemented, at installation time a list of telephone numbers will be entered into the modem's memory as well as a password that will be linked to each telephone number. After the password/

telephone number table is completed, remote users accessing the modem will be queried for their authorization code or password.

If successfully entered, the user will then be informed to place his or her device in the answer mode of operation and the modem will drop carrier to break the connection. Then the security modem will dialback the call originator, using the telephone number associated with the password. Although this feature insures that access over the switched telephone network only occurs from authorized telephone numbers, it does require a second call for each transmission session. In addition, it may be impractical for organizations that have traveling salespersonnel or whose employees use portable terminals and personal computers from many different locations. For additional information concerning security modems the reader is referred to Chapter 2.

Security switch

A security switch is similar in functionality to a security modem employing a password verification function, with the major difference between devices being the number of access routes permitted to users whose password entry is acceptable to the security switch. Security switches can be obtained in varying configurations, ranging from one to seven output ports.

A one-port switch in essence permits an existing modem without a security feature to be used to provide access by verification of password entry to a device that does not have password access protection. The use of this type of device is illustrated in the top part of Figure 6.10. A multiport security switch can be viewed as a single line port selector, enabling the user to be routed to

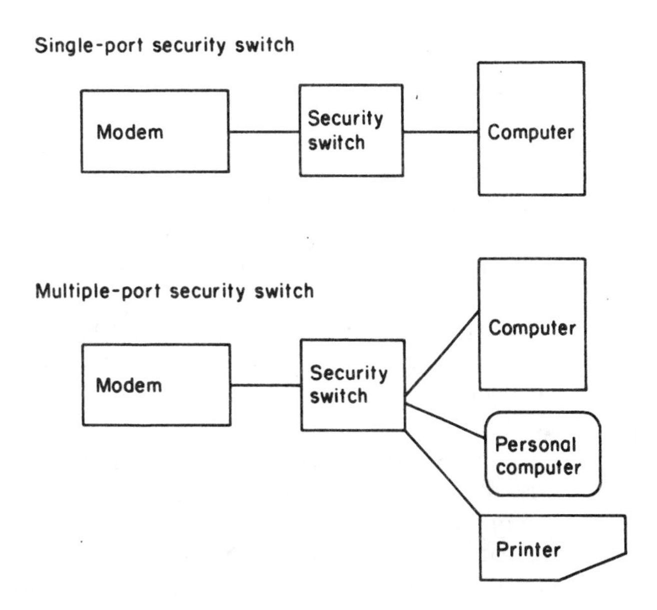

Figure 6.10 Both single- and multiple-port security switches can be used to provide controlled access.

one of many devices upon the successful entry of a password and port route. In the lower portion of Figure 6.10, the use of a security switch illustrates how this type of device can be employed to control remote access to a computer, personal computer, and printer.

6.2 SPEED AND CODE CONVERTERS

Owing to the rapid growth in the utilization of terminals for remote computing and message transmission applications, new terminal products have outpaced developments in many other areas of the computer communications field. Users who previously purchased what was state-of-the-art equipment in many cases may face costly network redesign problems if they desire to take advantage of new terminal developments. Other users who previously purchased terminals for specific applications may now have to obtain additional terminals as new applications materialize. A large portion of terminal problems are a result of the speed or code limitations of the user's existing equipment.

To reduce the time and cost involved in network redesign as well as to permit terminals procured for specific applications to become multifunctional, several conversion devices have been designed and manufactured with the problems of the terminal user in mind. These devices not only increase the flexibility and extend the life of existing terminals but can also be utilized during a network design or redesign effort to economize costs by reducing the number of computer ports and multidrop lines which may be required by a user's network without the incorporation of such a device. Although intelligent concentrators and front-end processors can also be programmed to perform these functions, quite often the memory and software processing overhead required warrants the consideration of a stand-alone device.

Operation

Basically, speed and code converters are built around a hardware data regenerator or microprocessor. For the former, the addition of plug-in assemblies expands the device functions to include speed and code conversion. For the latter, its adaptability is achieved by software and firmware programming for each application. Most speed converters are designed to operate with asynchronous, slow-speed terminals. When data is transmitted from a computer to a terminal at a higher data rate than the terminal can accept, the speed converter buffers the data in its memory and retransmits it at the slower rate to the terminal.

Since asynchronous transmission in a human–machine interaction environment usually consists of short messages to the computer and medium-length responses to the terminal, memory for such devices is normally 4000 words or less. Some speed converters designed for synchronous operations have storage capabilities up to 64,000 characters of information. When transmission occurs from the terminal to the computer, the data rate from the terminal is 'stepped up' to the computer's data rate by the speed converter.

Like speed conversion, code conversion permits terminals designed to operate with one type of code to communicate with other terminals or a computer that operates in a different code. As data is transmitted from the terminal to the computer in one code format, the code converter translates each character to the code acceptable to the computer. Conversely, data transmitted from the computer in one code is converted by the device on a character-by-character basis into the terminal's code.

By combining the functions of speed and code conversion, a speed and code converter both translates the code of a transmitted character and changes its transmission speed according to the plug-in assemblies used in the device or the software and firmware programming developed for the particular application.

General use of converters

Consider the situation where an inventory of 110- and 150-bps terminals exists. Suppose a user wishes to obtain a 300-bps terminal because of existing requirements. A method of economizing the number of computer ports and multidrop lines to service the new terminal and maintain support for existing terminals can be obtained through the use of speed converters.

As illustrated in Figure 6.11, top, the user who has a mixture of 110- and 150-bps terminals connected to the computer and who uses multidrop lines for communications will require a minimum of two computer ports and two multidrop lines to service the remote terminals. Two different types of modems – one to support 110-bps and the other to support 150-bps transmission – are required at the computer site as well as the remote sites, depending upon the speed of the terminal at each site. The two different transmission rates could also cause a large investment in spare modems if the user purchased modems and desired to stock spares at each site.

If the user now desires to install 300-bps terminals at a few remote locations, he or she would normally have to add an additional multidrop line, computer port, and enough 300-bps modems to service this new requirement. Through the installation of speed converters for each of the 110- and 150-bps terminals, all terminals may now be serviced on a common 300-bps multidrop line. In addition to reducing the number of lines required, only one modem and one computer port are now required at the central site. Since all modems in the network would now be speed standardized, a reduction in modem spare parts may now be practical. The network after the installation of speed converters is illustrated in the bottom portion of Figure 6.11.

Like speed converters, code converters permit different types of terminals to be mixed on a multidrop line. Code converters like speed converters can be utilized in a variety of ways. Code converters can be very useful during the transition period of system upgrading, such as when a user gradually replaces a large number of older Baudot terminals with more modern ASCII or EBCDIC terminals. In this case, depending upon the number of existing terminals and the conversion sequence, the user can either use code converters to make the 5-level Baudot-coded terminals interface an 8-level (7 bits plus a parity bit) ASCII code transmission line; or the user might let newly installed

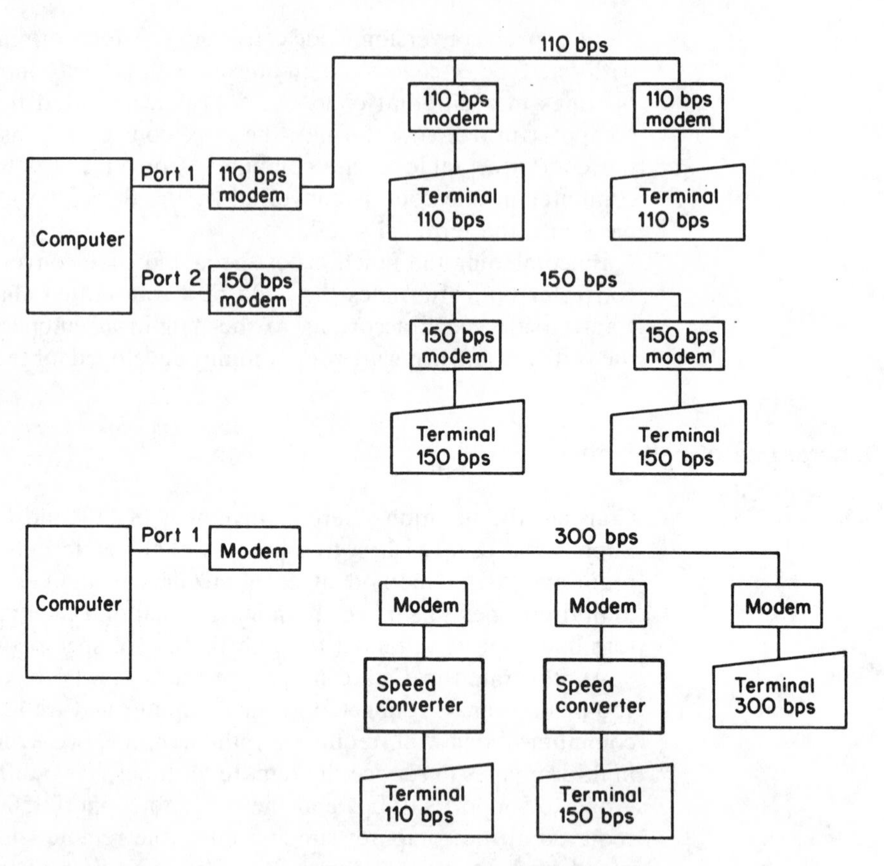

Figure 6.11 Adding higher-speed terminals. The network in the top portion has a mixture of 110- and 150-bps terminals and requires at least one computer port and multidrop line to service each type of remote terminal. Furthermore, two different modems are required at the computer site. With the installation of code converters, a standard speed can be obtained and only one multidrop line and one central site modem may be required, as shown in the lower portion.

8-level ASCII-coded terminals interface the existing 5-level Baudot-coded transmission line until the time is right for switchover to ASCII transmission on the line. This is shown in Figure 6.12. In Figure 6.12(a), the existing system operates by the computer transmitting 5-level Baudot code on the multidrop line to the 5-level Baudot code terminals interfaced to that line. As shown in Figure 6.12(b), the user can convert transmission to 8-level ASCII code by installing ASCII to Baudot code converters between each 5-level Baudot code terminal and the transmission line. Another option the user can consider is shown in Figure 6.12(c). Here the user can continue to transmit in 5-level Baudot code and still service the new ASCII terminal by the installation of a Baudot to ASCII converter between the new terminal and the transmission line. When all terminals have been upgraded to ASCII terminals or the user decides that it is time to switch transmission to ASCII, the appropriate action can then be taken to change the transmission code. It should be noted that

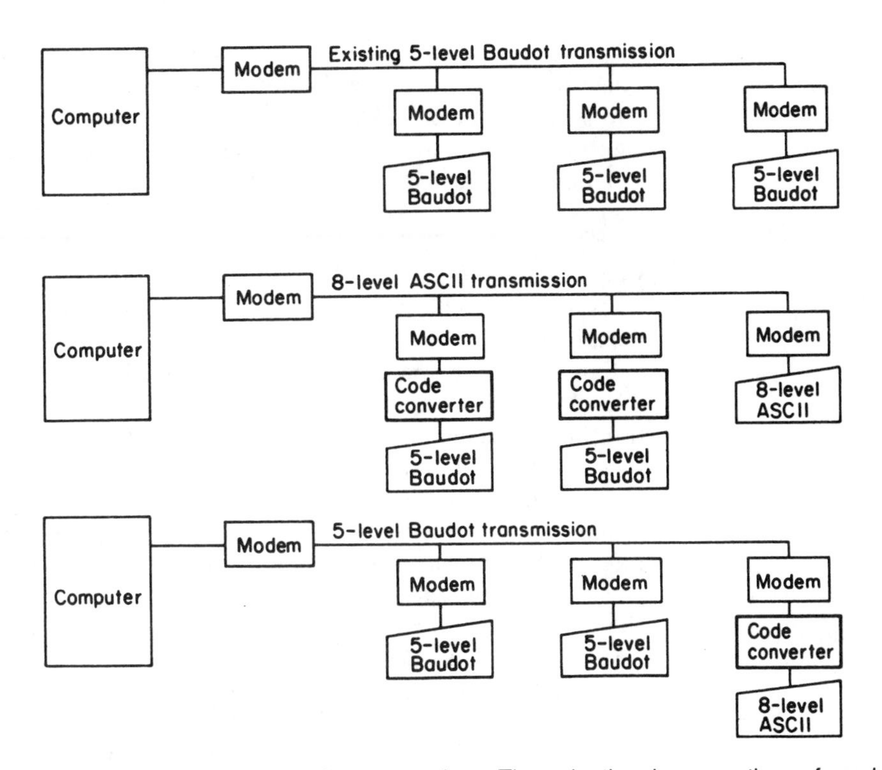

Figure 6.12 Implementing code conversion. Through the incorporation of code converters, terminals with different codes can share the use of common multipoint lines and a common modem at the computer site.

although this example leaves the reader with the impression that code converters are useful during transmission switchover or during the period when all terminals on a multidrop circuit are being upgraded, the user can also elect to retain existing terminals through the permanent utilization of code converters.

Specific application example

By combining the functions of speed and code conversion, users may be able to satisfy two or more discrete applications on one terminal by the use of a speed and code converter. One particular area where different types of terminals may be employed is for access to the various message-switching systems in operation today as well as the utilization of terminals to communicate with time-sharing computer systems.

Both TWX and Telex are switched network service offerings provided by Western Union. TWX was developed by AT&T but was sold to Western Union in 1971. Of the two classes of service offered on TWX, one at 45 bps uses the 5-level Baudot code and the other operating at 110 bps uses the 8-level ASCII code. Through the use of speed and code conversion equipment located at the switching office of Western Union, terminals operating at the two different speeds can communicate with each other. In 1956, Western Union introduced Telex service in the United States. This service originated in Germany in 1934 and was expanded into an international communications service. For Telex, transmission is limited to 50 bps using the 5-level Baudot code.

Through the use of a speed and code converter it becomes possible for the user of the Telex network to obtain the more readily available Teletype models 33 and 35 or equivalent devices for use on the Telex network. In addition, since these terminals can be used on the TWX network as well as for data transmission to most time-sharing computers, more efficiency and greater flexibility is obtained.

Functionally, a speed and code converter provides the means by which the 8-level ASCII code of Teletype models 33 and 35 terminals can have their transmission speed and code converted to the 5-level Baudot code and 50-bps speed of the Telex network. The ASCII even parity serial code of 11 bits (1 stop, 8 character bits, and 2 stop bits) of the Teletype models 33 and 35 terminals are converted by the speed and code converter from a transmission rate of 110 bps into a Baudot serial code of 7.5 bits (1 start, 5 character bits, and 1.5 stop bits) at a speed of 50 bps, and vice versa. The ASCII control '0' character is translated to perform the Baudot figures (FIGS) function, and the ASCII control 'N' character is used to perform the Baudot letters (LTRS) function. Since most any ASCII terminal can be interfaced through a converter, even cathode ray tubes (CRT) can be used to transmit messages to the Telex network as well as being used on the TWX network or for time-sharing operations. This is illustrated in Figure 6.13.

6.3 PROTOCOL CONVERTER

Although similar to speed and code converters since they may also perform this function, protocol converters are usually much more sophisticated devices.

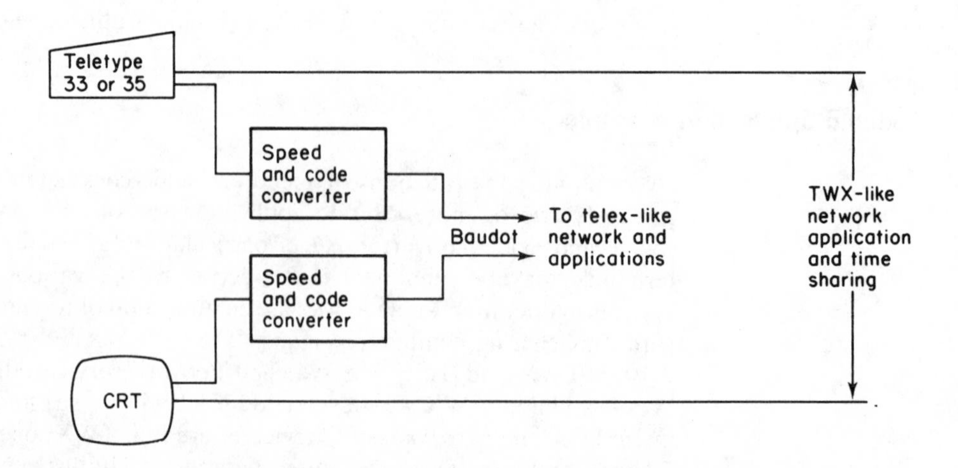

Figure 6.13 Combining speed and code conversion. Speed and code conversion may enable a mixture of terminals to access various communication systems.

Table 6.6 Operational conversion performed by protocol converters.

Operational conversion	Level
Device functionality	1
Device operation	2
Protocol	3
Data code/speed	4
Physical and electrical interface	5

A wide range of protocol converters are marketed, ranging in scope from devices that convert the physical and electrical interface from one standard to another, to devices that may perform five levels of conversion to include the physical and electrical interface. Table 6.6 lists the different levels of operational conversion that may be performed by a protocol converter.

The reader should note that to properly connect incompatible devices, a protocol converter may need to provide more than one level of operational conversion, with the number of levels of conversion based upon the incompatibility of the devices.

Physical/electrical conversion

At the lowest level of conversion, it may be necessary to convert the link-level connection from one interface standard to another. Thus, some protocol converters can be obtained which perform, as an example, RS-232 to RS-449 conversion or RS-232 to X.21 conversion.

Data code/speed conversion

Owing to the variation in operating codes and speeds of operation among devices, often code and speed conversion is required to obtain compatibility between equipment. At this level of conversion, the protocol converter performs as a speed and/or code converter. Although in many instances code conversion is obtained by the simple mapping of a bit string that represents a character in one code into a bit string that represents the same character in a different code, upon occasion code conversion can become quite complex.

One example of this would be a protocol converter employed to convert an asynchronous terminal's transmission into a bisynchronous communications format for connection to a host computer's BISYNC port. In this type of operation, the computer's port will probably be transmitting EBCDIC data in blocks, with each block containing a block check character at the end of the block. Similarly, the computer port will be expecting to receive EBCDIC data grouped into blocks, with a block check character appended to each block. If the asynchronous terminal device transmits ASCII data, not only must that data code be converted to EBCDIC, but in addition a computational operation must occur to generate the block check character.

Protocol conversion

Although the previous illustration of code conversion involved developing the appropriate block check character from ASCII data, in actuality a protocol conversion is also being conducted.

Protocol conversion can fall into three primary categories since most communications protocols fall into the categories of asynchronous, byte-oriented, and bit-oriented protocols. Common examples of protocols in each category include teletype (TTY) asynchronous transmission, IBM's various types of byte-oriented bisynchronous protocols, and the International Standards Organization Higher-level Data Link Control (HDLC) bit-oriented protocol.

Converting a data stream from one protocol into another requires a considerable amount of work. As an example, in TTY transmission each character is framed by a start bit and one or more stop bits, with either parity used for simple error detection or no error detection and correction scheme employed. In bisynchronous transmission data is grouped into blocks, with a number of synchronization characters prefixed to the block to provide synchronization between the transmitting and receiving devices. In addition, special control characters consisting of unique patterns of bit configurations inform the receiver of such information as when the message starts, when the actual text starts, if a message consists of more than one block; it also informs the receiver about the end of the block containing the block check character, which is formed by a cyclic redundancy checking process that enables the receiver to ascertain if the block was received without a transmission error affecting the block.

Thus, a protocol converter, converting asynchronous data to bisynchronous data, must eliminate the start and stop bits from each asynchronous character (character stripping), block the characters into a bisynchronous block format, generate and appropriately place the transmission control characters within the data block, and compute a cyclic redundancy check character and append it to the block as the block check character.

Device operation conversion

Since the physical characteristics of many devices vary, one of the functions a protocol converter may have to perform is device operation conversion. Again returning to the previous asynchronous to bisynchronous example, one device operation conversion would involve the translation of bisynchronous printer spacing information into an equivalent number of carriage returns, line feed sequences recognized by the asynchronous device.

Device functionality conversion

At the highest level of protocol operation, device functionality conversion ensures that the data transmitted by one device is correctly interpreted by the other device. Thus, as an example, a computer's cursor positioning code sequences transmitted in one protocol would be mapped by the protocol converter into the equivalent cursor control codes of the device the converter supports. Other examples of device functionality conversion include the translation of PF keys on 3270 terminals to equivalent codes on other devices, translation of field attributes to include underline, blinking and high intensity display, and other terminal functions.

Applications

The number of applications to which protocol converters can be employed is only limited by one's imagination such is the diversity of the devices. Figure 6.14 illustrates two popular application areas where protocol converters are typically employed.

In the top part of Figure 6.14, a protocol converter functioning as an asynchronous/synchronous converter is employed to permit an asynchronous terminal to be interfaced to a multiport modem. Since the limited function multiplexer in the multiport modem only multiplexes synchronous data, the asynchronous data stream must be converted into a synchronous data stream prior to the modem multiplexing the data. Thus, the use of an asynchronous

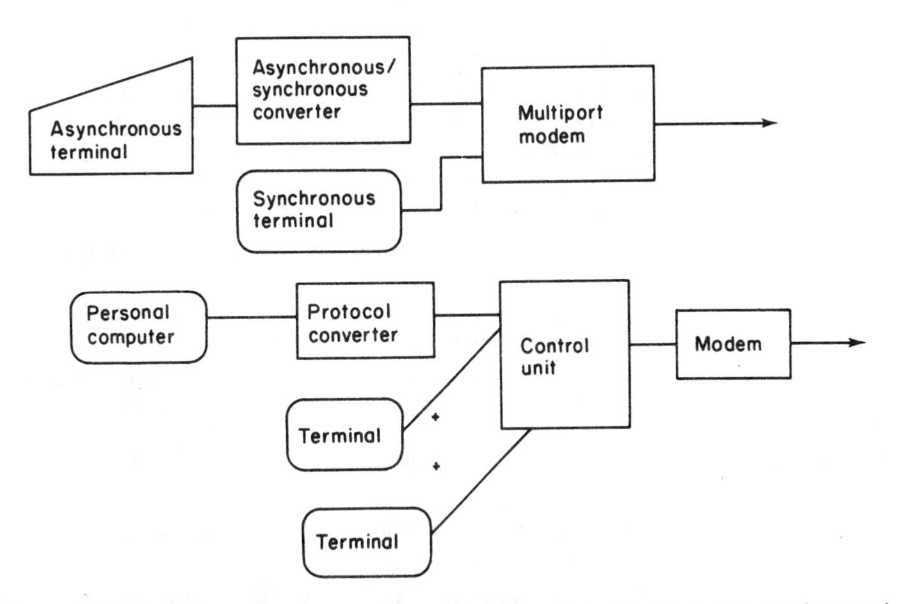

Figure 6.14 Typical protocol converter applications. Protocol converters can be used for asynchronous to synchronous conversion (top) or to make asynchronous personal computers that transmit data on a line by line format compatible with a screen-oriented display protocol (bottom).

to synchronous protocol converter allows both synchronous and asynchronous devices to share the use of a common multiport modem.

A protocol converter functioning as a terminal emulator is illustrated in the lower part of Figure 6.14. In this mode of operation, the protocol converter changes the line by line transmission sequences of the personal computer into a screen-oriented format that matches the protocol supported by the control unit. In addition, the protocol converter will most likely perform data code, speed, device operation, and functionality conversion.

6.4 DATA ACCESS ARRANGEMENTS

Acting as a protective device between customer-provided equipment and the switched telephone network, data access arrangements (DAA) or data couplers have a long history and evolution as a result of numerous court decisions.

In what is now referred to as the Carterphone decision, the FCC ruled that non-Bell System transmitting and receiving equipment could be attached to the switched telephone network. Contained in the ruling was a requirement that a protective device be furnished, installed, and maintained by the telephone company. This device was to act as an interface between the non-Bell System equipment and the network to provide protection for telephone company personnel and equipment from the possibility of hazardous voltages applied from the customer-furnished equipment, to limit abnormally high frequency and signal levels, and to prevent line seizure except in response to ringing. While data communications users were now permitted to interconnect the products of any vendor to the telephone network, users were also responsible for compliance with the data access arrangement requirements.

A subsequent series of court cases resulted in an interconnect equipment registration program which modifies the requirements for data access arrangements. Under the FCC interconnect equipment registration program, customer-provided equipment that is registered as a result of meeting a series of operational characteristics can be interfaced via a plug to a telephone company jack for connection to the switched network. In effect, such equipment contains a built-in data access arrangement, avoiding the requirement to obtain a separate device.

Unregistered equipment can be used providing that it is interfaced with the telephone company jack through a registered data coupler provided either by the telephone company or by an independent manufacturer.

Today the use of data access arrangements are primarily directed toward enabling equipment manufactured prior to the FCC registration program to be connected to the switched network. As vendors continue to manufacture new equipment according to the FCC program and as older equipment becomes obsolete, the requirement for DAAs will dramatically decrease. Since several million devices required DAAs for use with the switched telephone network, it may be another decade until their utilization is no longer required. Figure 6.15 illustrates the effect of the FCC registration program when various types of transmitting and receiving equipment are interfaced to telephone company facilities.

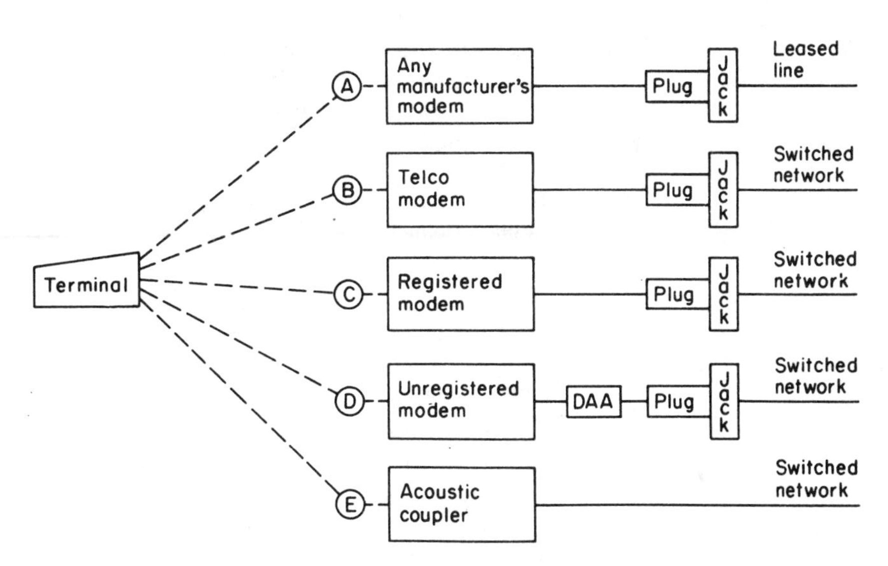

Figure 6.15 FCC registration program requirements. Data couplers may be required based upon the type of communications equipment installed and the line to which it is connected.

When the user desires to transmit data over a leased line (Figure 6.15(A)), no data access arrangement is necessary regardless of the type of modem used. This is due to the fact that the leased line is for the exclusive use of that user, and any interference caused by non-telephone company equipment on that line should only affect that user. For the user who obtains telephone company equipment (Figure 6.15(B)), the interface is via a plug, and no DAA is required since such telephone company equipment is registered. A similar interface arrangement occurs when any vendor-registered modem (Figure 6.15(C)) is connected to the switched network. When an unregistered modem (Figure 6.15(D)) is connected, a data coupler must be provided to interface the telephone company jack. For users who install terminals with built-in acoustic couplers or who use a stand-alone coupler as their transmission device (Figure 6.15(E)), neither a DAA nor a plug–jack connection is required since the interface to the telephone line is via the telephone headset and both the terminal's and coupler's power supplies are segregated from the line.

Types of DAA

There are three types of data access arrangements available. The DAAs or data couplers should be identified by the telephone company ordering codes when the user requests installation from the telephone company or from an independent vendor since such equipment is standardized by those codes. The coupler ordering codes are CDT, CBS, and CBT, with the basic differences between each model depending upon the mode of operation (manual or unattended), the presence or absence of a built-in power supply, and the type

Table 6.7 Data access arrangement comparison.

DAA type	Mode	Principal use	Line control	Power source
CDT	Originate only or manual originate/ answer terminals only	AC analog devices	Telephone set controls line	By telephone company lines
CBT	Manual originate/ automatic answer terminals only	Devices employing contact closures for control leads	Coupler controls line	Optional
CBS	Manual originate/ automatic answer terminals only	Devices employing EIA RS-232 interface	Coupler controls line	Customer three-pronged wall outlet

of line control. The basic differences between each type of data access arrangement are listed in Table 6.7.

The CDT data access arrangement is designed for use with devices which can only originate or can originate and answer calls but only through manual operation. When using a CDT coupler, the operator dials the desired number and the receiving operator answers. Upon verification that the equipment at both ends is ready for data transmission, the operators depress the data keys on their telephone sets, and the modems which are connected to the CDT DAA may then transmit and receive data. Once transmission is completed, both operators hang up their telephones and the connection is broken, as illustrated in Figure 6.16. Since control is manual, the CDT coupler is wired so that the telephone set controls the line. Likewise, being powered by the telephone company line obviates the necessity of the user supplying a power source.

CDT couplers are also required when manual dial-backup for leased lines are employed using unregistered modems, as shown in Figure 6.17. To effect switching between the leased line and the dial-up or switched network will

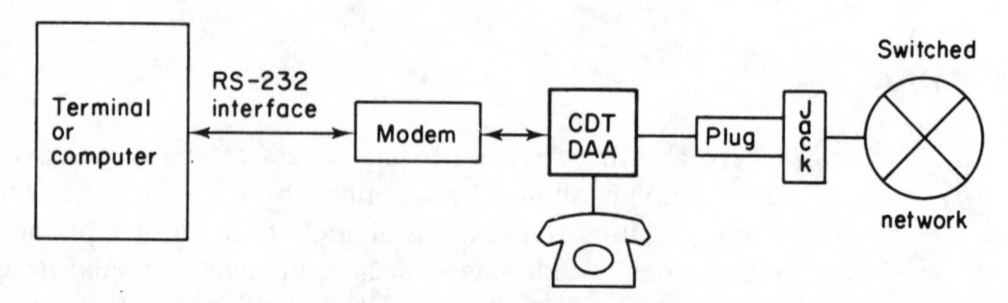

Figure 6.16 Manually operated CDT utilization. Data transmitted manually through a customer-provided unregistered modem must pass through a CDT coupler prior to accessing the switched network.

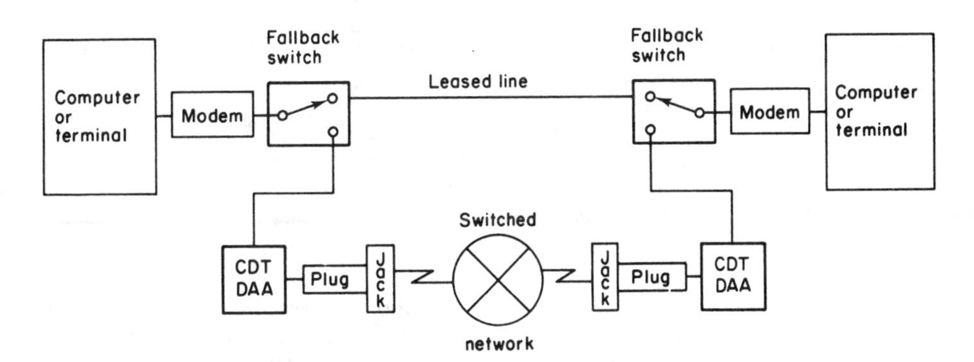

Figure 6.17 Manual dial-backup operation. By turning both fallback switches to their secondary position, data may be transmitted through the CDT couplers when backup operations are required.

require the installation of a fallback switch at each end of the link. If the leased line should become inoperative, the fallback switches are turned to the secondary position which permits access to the switched network through the CDT coupler.

The CBS and CBT couplers are designed to operate with devices that in addition to having manual originate capability can automatically control the originating and answering of data calls. For the automatic originating of calls a separate automatic calling unit will be required to provide this capability. The two major differences between the CBS and CBT data access arrangements are the type of control leads furnished and the source of power. The CBT coupler is designed to provide contact-type control leads, while the CBS coupler provides EIA RS-232 voltage-type control leads. Users requesting the CBS coupler receive a power transformer external to the coupler's housing and must supply a three-pronged wall outlet power source while the CBT coupler can operate with the telephone company line's power source, the dc voltage from the modem, or from an external power supply operated by conventional ac power. The utilization of CBS and CBT couplers are illustrated in Figures 6.18 and 6.19.

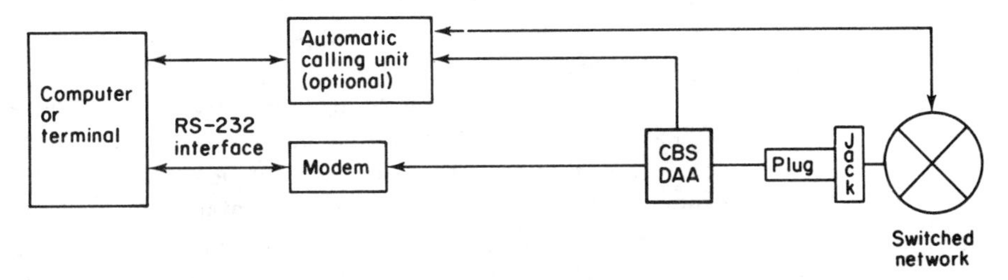

Figure 6.18 CBS coupler permits automatic operations. By exchanging RS-232-type signals, a CBS coupler permits automatic dialing, answering and disconnecting so that the terminal or computer may operate unattended. The site at the opposite end of the transmission path must be similarly equipped with a CBS coupler. In addition, automatic answering capability must be built into the modem.

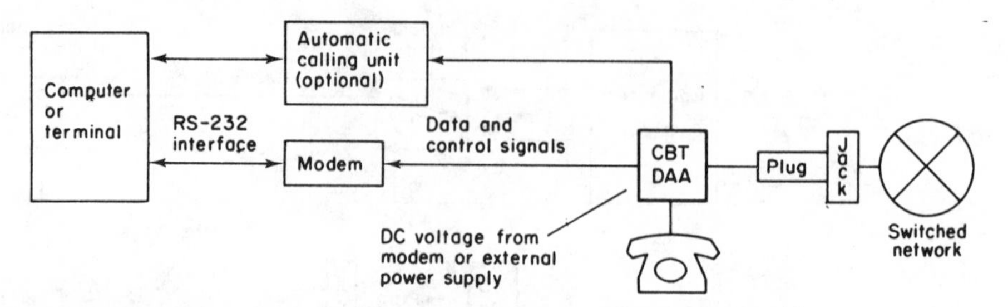

Figure 6.19 CBT coupler requires external power source. The CBT coupler requires an external dc power source which can usually be obtained from the modem. If the connection is impractical or the capability is not available, either the telephone company can furnish the power supply or the user can provide the external power supply.

When data access arrangements are obtained from the telephone company, charges are typically 50 to 100 dollars for installation, plus 10 to 25 dollars per month for the lease of such equipment, since the DAAs are tariffed on a state-by-state basis, and pricing varies considerably. Although only available from telephone companies on a lease basis, couplers are available for purchase from a number of communications companies. Telephone company tariffs should be checked in order to determine which method of procurement is most advantageous.

6.5 SPEECH PLUS DATA UNITS

When the operator of a computer, remote batch terminal, or some other type of data transmission equipment experiences difficulties in the initiation or transmission of data, a common procedure is to telephone the other site and attempt to alleviate the problem. If the equipment at each site is connected by a leased line to dedicated modems, the telephone call must be made over the switched network. For this situation, not only is the leased line left unutilized until the problem is resolved, but the company may face the prospect of a costly long-distance telephone bill. Similarly, those sites that rearrange equipment configurations to differentiate between daytime and night-time operations may require one or more calls each day between sites to clarify the configuration used, the service expected, or the initiation times and procedures to be followed.

In addition to being able to arrange wideband channels so that the remaining capacity not used for data transmission can be used for simultaneous individual voice communications, users of a single voice-grade leased line circuit can configure their equipment to take advantage of alternate or simultaneous voice data transmission. Using proper equipment, this alternate or simultaneous data transmission can occur over the same leased line which is normally used for data transmission.

Supplementing alternate voice data transmission services which can be obtained from the telephone company for leased line facilities is a range of devices manufactured by commercial communications companies known as voice adapters. These voice adapters provide the user with alternate or simultaneous voice data transmission capability and can be used in a wide range of applications.

Types of unit

A device known as a voice adapter can be obtained as a base unit which supports a standard telephone headset. Several different models are available which can be interfaced to different speed modems to provide alternate or simultaneous voice and data contact between points connected by a leased line. By incorporating all summing and switching circuits within the adapter and its associated data set, the necessity of obtaining auxiliary switching equipment which is normally used to perform this function is eliminated. Voice adapters must be interfaced to modems at each end of the leased line, as shown in Figure 6.20.

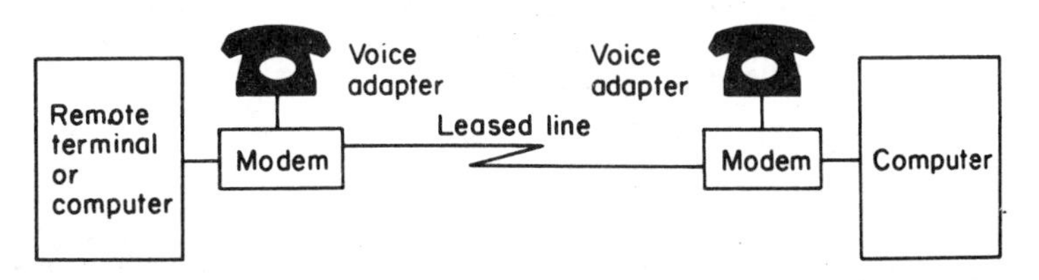

Figure 6.20 Installing voice adapters on leased lines. By interfacing voice adapters to the modems at each end of the leased line, alternate or simultaneous voice and data transmission may be accomplished.

Voice adapters are equipped with a pushbutton switch and, depending upon the model, a selector switch. The pushbutton switch permits the user to signal the distant end of the leased line when voice communications are desired. In addition to causing a tone to be produced and transmitted to the opposite end of the leased line, a connection is available on some models which can be used to operate a remote indicating device, such as a lamp or bell, which, upon receipt of the tone, alerts the operator at the other end of the line that voice coordination is requested. The selector switch, depending upon operating mode selected, permits full voice communications, simultaneous voice with 2000-bps or 2400-bps data transmission, or simultaneous teletype (up to seven 75-bps channels) and either 2000- or 2400-bps data transmission. Two typical applications where voice adapters could be employed are shown in Figure 6.21.

In Figure 6.21, if data transmission occurs at 9600 bps, the voice adapter would be used to provide an alternate voice link between the two sites. Since 2000- or 2400-bps transmission would probably be inefficient for this type of network configuration, alternate voice with a return to full 9600-bps data transmission via a voice adapter is selected. In Figure 6.21(b), if the CRT normally communicates with the CPU at 2000 or 2400-bps, simultaneous voice data transmission can be used. In this example, the bandwidth required by the

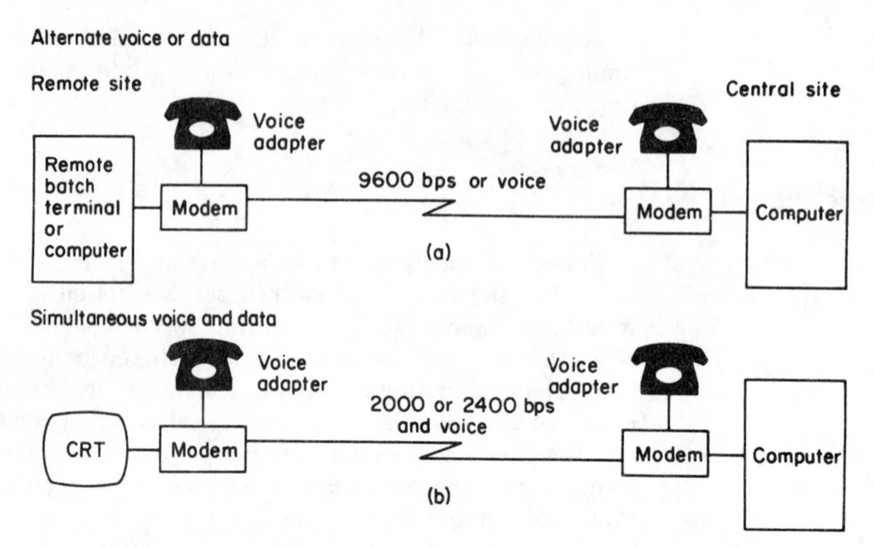

Figure 6.21 Alternate and simultaneous voice and data. Top: Alternate voice or data. Bottom: Simultaneous voice and data. Depending upon model selected and data transmission requirements, voice adapters may be used for alternate or simultaneous voice communications.

voice communication does not degrade the transmission speed requirements of the CRT.

Users considering voice adapters will find that they are simple to install and no special power is required. By attaching the connector of the voice adapter cord to a connection on the associated modem, power is furnished for the adapter from the modem's power supply, and the voice adapter becomes ready for use.

A second type of device is known as a speech plus data terminal which is used to provide an audio channel in conjunction with a number of telegraph channels that are transmitting at any bit rate up to 600 bps or a mixture of bit rates up to 2400 bps. This type of terminal uses amplifiers to combine and then separate voice signals from the data signals. Currently, up to eight teletype channels can be operated in addition to a voice channel on such devices.

6.6 VOICE DIGITIZERS

Through the employment of voice digitizers, it is now possible to integrate voice transmission into low-speed data networks. In addition, this device may be employed to drive multiple voice channels on a single analog line facility in an economical manner dependent upon the distance of the line and the cost of the digitization equipment.

To obtain an understanding of the networking capabilities afforded through the use of voice digitizers, we will first review the conventional pulse code modulation (PCM) of analog signals. Using PCM as a comparative reference, linear predictive coding (LPC), which is the digitization technique used in voice digitizers, will be examined next. This will permit the reader to understand how the low bit rate of a digitally encoded analog signal from a voice digitizer can be used to develop a variety of networking strategies. These strategies can include the derivation of multiple voice logical channels on one physical analog channel and the integration of voice onto a low-speed data network.

Pulse code modulation

Over the last 30 years, a variety of techniques has been developed to digitize voice signals. Among the earliest techniques and the one most often employed by telephone companies to convert analog voice conversations into a digital data stream is pulse code modulation, commonly referred to as PCM.

Three major steps are involved in the PCM process. In the first step, the analog signal is sampled at a high enough rate to insure that all of the necessary information that will be required to reconstruct the signal is obtained. In the second step, a digital value of each sample is obtained by the process known as signal quantization. During the quantization process a binary value representing the amplitude of the signal is assigned to the sample period. This sample value is then transmitted as a binary number. Figure 6.22 illustrates the steps involved in the application of pulse code modulation to an analog signal.

Nyquist's Theorem states that all of the information of a sampled message can be obtained if the samples are taken at a rate which is at least twice that of the value of the highest significant signaling frequency in the bandwidth. Since a voice-grade telephone call has its highest frequency at 4 kHz, even though actual speech is transmitted from 300 to 3300 Hz, then 8000 samples must be taken each second. With 8 bits used to encode the amplitude of each sample, the resulting data rate required to transmit digitized voice is 8000 samples times 8 bits per sample per second or 64 kbps.

The 64 kbps data rate associated with PCM makes this voice digitization technique unsuitable for low-speed data networks which are based upon the utilization of analog 3002-type leased lines or Digital Dataphone Service. This unsuitability is based upon the maximum data rates obtainable on the previously mentioned facilities. On 3002-type leased lines, a maximum data rate of 19.2 kbps has only recently been obtained through the use of Trellis coded modulation modems. On Digital Dataphone Service, the data rate is limited to a maximum of 56 kbps. Since PCM results in a data rate of 64 kbps, this technique cannot be used to transmit a digitized voice signal on analog leased lines nor can it be used for transmitting a digitized voice signal on a Digital Dataphone Service facility. Due to its high data rate requirement, PCM is used to encode multiple voice conversations on T1 carriers which operate at 1.544 Mbps. At this data rate, 24 voice conversations or a mixture of digitized voice and data can be multiplexed onto a T1 facility. However, due to the cost of T1 multiplexers and T1 lines only a small percentage of network users can economically integrate voice and data on T1 circuits. For other network users, an alternative to PCM must be considered if they desire to integrate voice

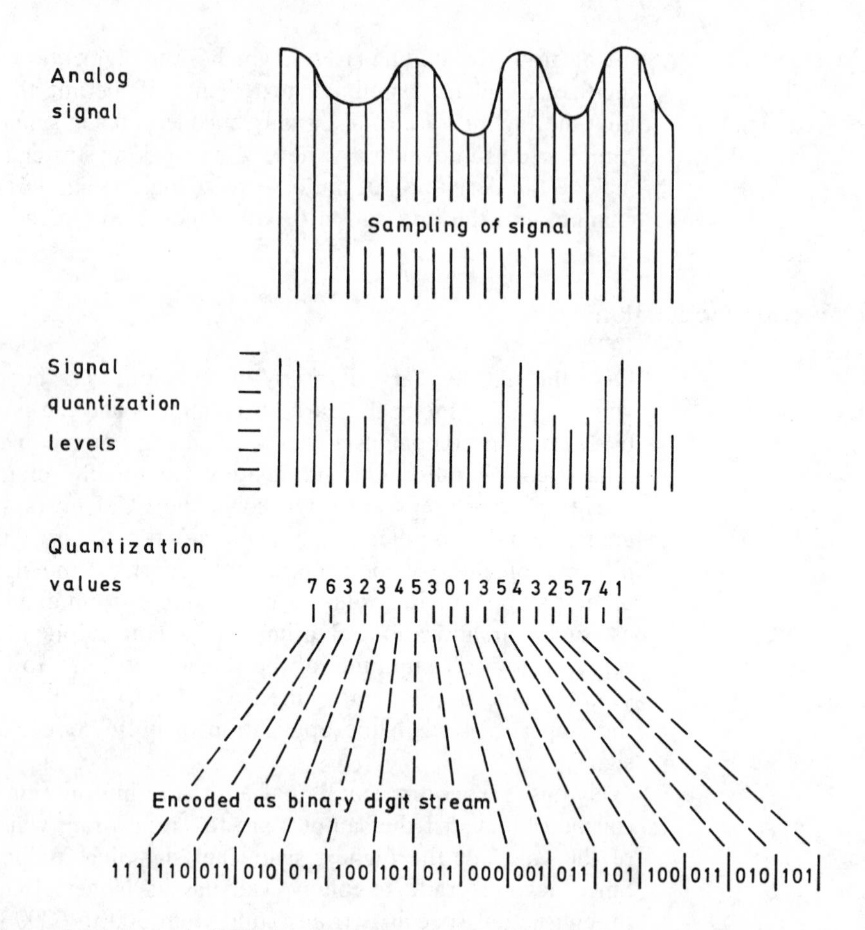

Figure 6.22 Applying PCM to an analog signal. (Reprinted with permission from *Data Communications Management*, © 1988 Auerbach Publishers, New York.)

with data on a common network or derive multiple logical voice conversations on one physical line facility. This alternative is obtained through the use of voice digitizers which are based upon the use of linear predictive coding to convert voice into a digital data stream.

Linear predictive coding

In linear predictive coding, the analog voice input is analyzed and then converted into a set of digital parameters that represent the voice input. To reconstruct a digitized voice signal a synthesizer is used to develop an analog voice output based upon the received set of digital parameters. The key to the utility of linear predictive coding is the limiting of the analysis of voice signals to four sets of voice parameters, permitting a very low data rate to be obtained for transmitting analog data in digital form.

LPC operations

In a linear predictive coding system, the voice signal is first sampled by a 12-bit analog to digital converter. Next, the output of the converter is used as input to four parametric detectors. These parametric detectors include a pitch detector, a voice/unvoiced detector, a power detector and a spectral data decoder.

To appreciate the employment of parametric detectors, consider Figure 6.23 which illustrates the speech-producing elements of the human vocal tract. The pitch detector analyzes the data to obtain the fundamental pitch frequency at which human vocal cords vibrate. Next, the voice/unvoiced detector senses whether sound is caused by the vibration of vocal cords (voice) or by sounds such as 'shhh' (unvoiced), which do not vibrate. The power detector determines the amplitude or loudness of the sound, while the spectral data detector models the resonant cavity formed by the throat and the mouth.

Through the use of parametric detectors, a parametric model of human speech is constructed, resulting in the generation of a series of speech coefficients which represent the human voice. These speech coefficients developed through the use of linear predictive coding are then transmitted by the voice digitizer instead of the actual amplitude of the signal which is used by the PCM technique. As a result of the voice digitizer transmitting a model of human speech instead of actual speech, a very low data rate is obtained to represent a voice conversation. Today, voice digitizers can operate at data rates as low as 2400 or 4800 bps, enabling the communications manager to consider a variety of strategies to integrate voice into low-speed data networks or to obtain multiple voice channels on one physical line. Since the costs associated with the employment of voice digitizers are the primary decision criteria for their use, an overview of voice digitizer economics is warranted.

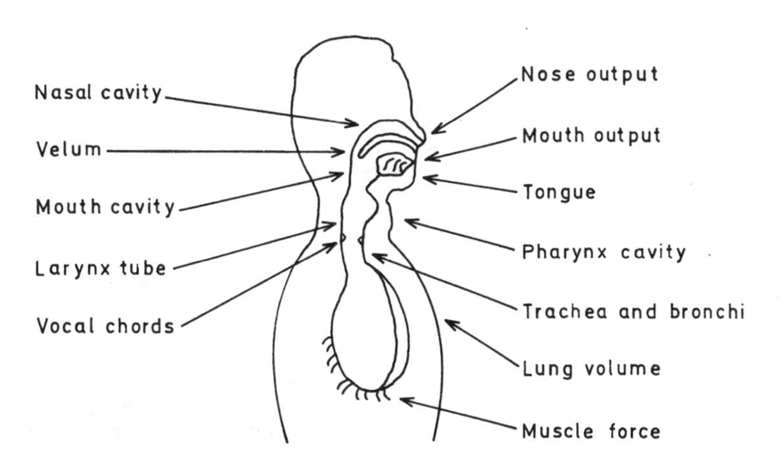

Figure 6.23 Speech producing elements of the human vocal tract. (Reprinted with permission from *Data Communications Management*, © 1988 Auerbach Publishers, New York.)

Voice digitizer economics

Voice digitizers are relatively expensive devices in comparison to such network components as statistical multiplexers and modems. Costing approximately $10,000 per unit, two voice digitizers are required to encode and decode a voice conversation, resulting in a system cost of approximately $20,000 for each analog channel one wishes to digitize.

Although a discussion of exact analog line tariffs is beyond the scope of this book due to the frequently changing rates of tariffs, the reader should note that the cost of an analog 3002 leased line from coast to coast is approximately $1400 per month. On an annual basis, this represents an expenditure of $16,800 for a voice-grade circuit which your organization may be able to replace by the integration of two voice digitizers into an existing data network. If this integration of a voice circuit into a data network is accomplished, the one-time expenditure of $20,000 for two voice digitizers would result in an 84 percent payback of the equipment cost during its first year of operation. Since the cost of a leased line decreases as the distance between the points to be connected decreases, the payback percentage will also decrease. Figure 6.24 illustrates the relationship between the annual percentage of equipment payback and the distance between locations where voice and data have been merged through the use of voice digitizers. At a maximum distance of approximately 2000 miles a payback exceeding 80 percent can be obtained, while for distances under 250 miles between locations the annual payback will be under 5 percent. Since communications equipment normally has a life expectancy between 3 and 5 years prior to obsolescence, most organizations would look for a minimum 20 to 33 percent payback prior to considering integrating voice onto a data network with voice digitizers. This payback period would then preclude the use of voice digitizers on all circuits under 1000 miles in length. In fact, due

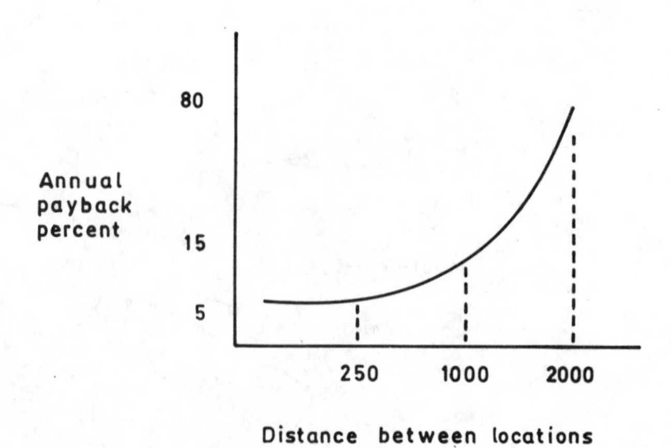

Figure 6.24 Integrating voice into data networks through the use of voice digitizers. Due to most communications equipment having a 3–5 year life, communications managers normally require a 20–33 percent return on their investment when adding equipment based upon potential economic savings. (Reprinted with permission from *Data Communications Management*, © 1988 Auerbach Publishers, New York.)

to international circuits having the highest cost of all categories of telephone company line facilities, voice digitizers are primarily used on such facilities. This is due to the use of voice digitizers generating cost savings through their ability to eliminate lines which can equal or exceed the cost of the digitizers in just a few months.

Networking options to consider

Figure 6.25 illustrates a typical network where two locations interconnect Private Branch Exchanges (PBX's) through the use of tie lines for voice communications and operate a separate data communications network between similar locations. In the upper part of Figure 6.25, it was assumed that three tie lines were used to provide voice communications capability between locations A and B. In the lower portion of Figure 6.25, it was assumed that a number of terminals at location A required access to a computer at location B, resulting in the installation of a leased line and a pair of modems and multiplexers to support the data communications transmission requirements between locations. As indicated, separate voice and data transmission facilities are installed to support the voice and data transmission requirements between the two locations.

If the data communications network is operating at 9.6 or 14.4 kbps, it becomes possible to upgrade the modems at each end of the analog line to obtain additional bandwidth. This additional bandwidth could then be used to service a digital data stream generated by voice digitizers. Assuming the

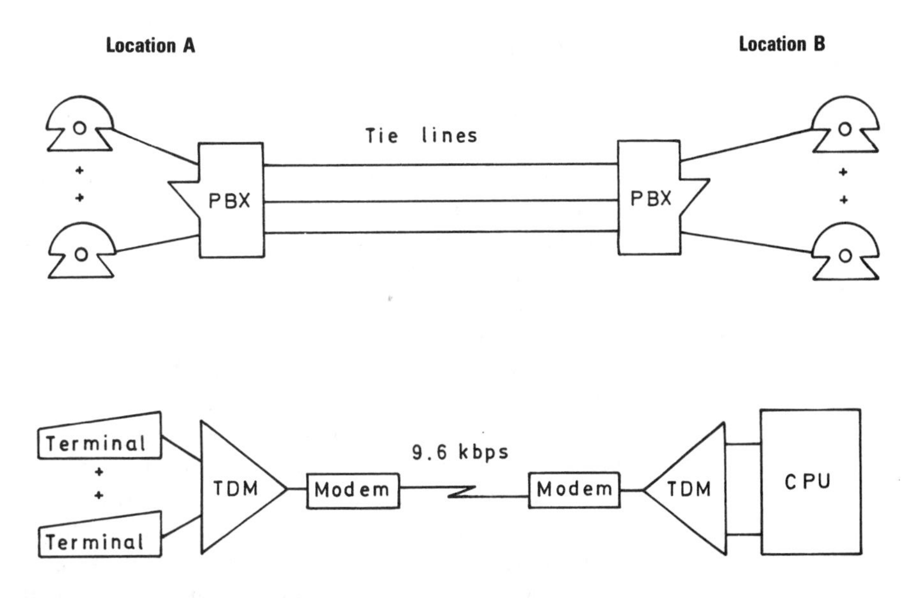

Figure 6.25 Separate voice and data networks. Many organizations maintain separate voice and data networks between common locations where they have offices or other facilities. (Reprinted with permission from *Data Communications Management*, © 1988 Auerbach Publishers, New York.)

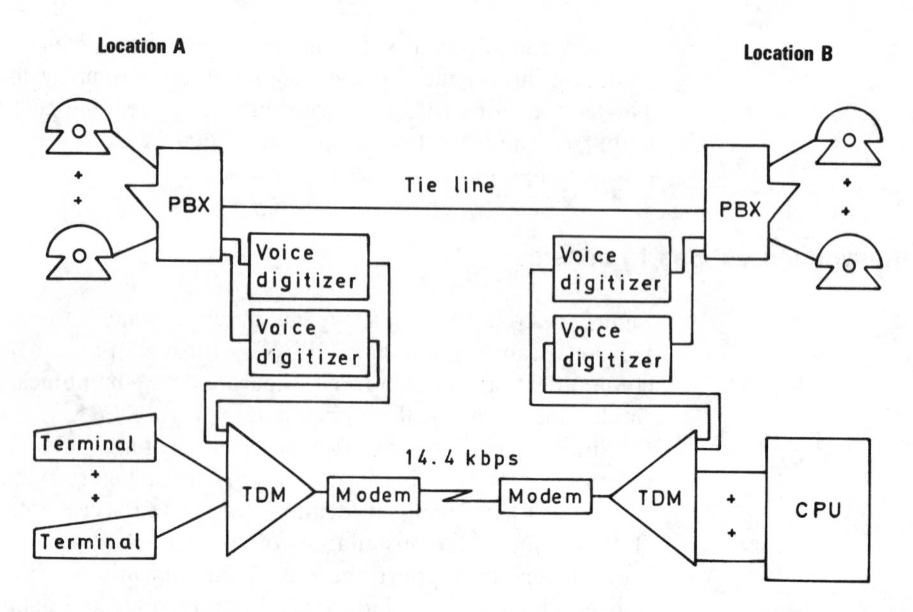

Figure 6.26 Employing voice digitizers. Through the utilization of voice digitizers, the 2400 or 4800-bps digital data stream can be integrated into an existing digital network. In this example, two tie lines were eliminated through the utilization of two pairs of voice digitizers whose outputs were multiplexed on to a line originally used to support the data transmission requirements between the two corporate locations. (Reprinted with permission from *Data Communications Management*, © 1988 Auerbach Publishers, New York.)

modems of the data network were operating at 9.6 kbps, Figure 6.26 illustrates the effect of upgrading the modem's data rate to support the use of voice digitizers. In this example, it was assumed that the voice digitizers operate at 2.4 kbps, enabling two of the three tie lines between locations to be eliminated by the employment of two pairs of voice digitizers.

Although the voice digitizers are connected to time division multiplexers in Figure 6.26, they could also be connected to individual ports on multiport modems. In fact, if statistical multiplexers are employed in a network you should attempt to avoid using this type of equipment unless a bandpass channel is available. This is due to the fact that when the statistical multiplexer's buffer fills to a predefined point the device will generate XOFF's, drop a Clear-to-Send (CTS) signal or reduce clocking to synchronous inputs to enable the buffer to empty to another predefined point prior to enabling transmission to resume. Although these techniques are used to reduce or inhibit data input to prevent the multiplexer's buffers from overflowing when too many terminals become active, they do not affect the results of data transmission. Unfortunately, they will result in random gaps of time occurring in a voice conversation.

Since a bandpass channel in a statistical multiplexer bypasses statistical multiplexing, you can alleviate time gaps by using this type of facility. If a vendor does not provide a bandpass channel for their statistical multiplexers then you may wish to examine the use of multiport modems, which are modems containing a limited-function, synchronous, time division multiplexer in one housing. The multiport modem uses a conventional time division multiplexer which reserves a time slot on the composite bandwidth for each input, thus no statistical multiplexing occurs and the potential of delays due to buffer fills is avoided, permitting voice digitizers to be connected to these devices without worrying about potential time gaps in conversations occurring due to the traffic load on the network.

Multiple voice logical channels

Another strategy that can be considered is the derivation of multiple voice logical channels on one physical circuit. Figure 6.27 illustrates how two logical voice channels can be derived from one physical circuit through the use of two pairs of voice digitizers. Here, the annual payback becomes:

$$\text{annual payback} = \frac{\text{line savings} \times 12}{\text{cost of 4 modems} + \text{cost of 4 voice digitizers}}$$

where the line savings is the monthly cost of one voice-grade tie line which is eliminated if two logical voice channels were derived from one physical channel.

Since the integration of voice onto a low-speed data network may permit existing modems or multiplexers to be used the annual payback can be expected to increase under such circumstances. In fact, under the ideal situation where no additional communications equipment is required to integrate voice onto a low-speed data network the annual payback becomes:

$$\text{annual payback} = \frac{\text{line savings} \times 12}{\text{cost of 2 voice digitizers}}$$

Based upon the preceding, the annual payback associated with deriving multiple voice channels on one physical channel will be significantly less than the potential payback when a line used for voice transmission is replaced through the integration of voice onto an existing data network. Due to this, most organizations limit the use of voice digitizers for deriving multiple voice

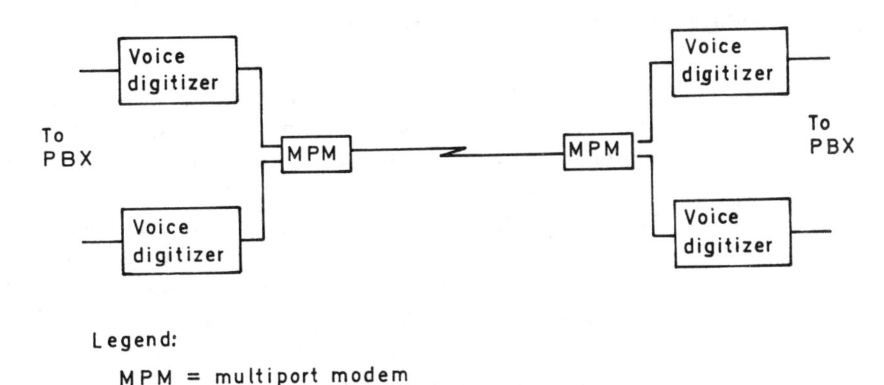

Figure 6.27 Deriving two voice channels on one physical circuit. Two pairs of voice digitizers can be used to derive two logical voice channels on one physical circuit. Similarly, three pairs of voice digitizers would enable three logical voice channels to be derived from one physical channel. (Reprinted with permission from *Data Communications Management*, © 1988 Auerbach Publishers, New York.)

channels to international circuits. This is due to the high cost of such circuits in comparison to the cost of domestic circuits, permitting significant savings to be obtained from the elimination of each voice circuit. One example of this might be a bank which has two tie lines between branches in London and New York. Since each circuit would cost approximately $5000 per month, eliminating one circuit would result in an annual line savings of $60,000 which is significantly greater than any savings that can occur through the elimination of a domestic circuit.

Utilization problems

Although voice digitizers can be a viable solution for integrating voice onto low-speed data networks or deriving multiple voice channels on one physical circuit their utilization is not problem free. Foremost among the problems associated with the use of voice digitizers are the environment of the person using the digitizer and backup upon device failure.

The environment affects the use of voice digitizers since linear predictive coding will consider background noise when it models speech parameters. Thus, voice conversations through a voice digitizer should not originate from a factory floor or similar noisy environment. This means that for most office environments the synthesizer in a voice digitizer should be able to faithfully reproduce a voice conversation first digitized into a series of speech parameters.

Concerning backup, in the event a voice digitizer fails users can simply use the direct distance dial network until repairs or a replacement of the failed device occurs. Although numerous long-distance calls within a 24 or 48 hour period until repairs or replacement occurs may result in extra operating expenses, such expenses are normally considerably less than stocking a backup pair of voice digitizers to be used in the event of device failure.

6.7 DATA COMPRESSION PERFORMING DEVICES

Although data compression is correctly considered as a technique used to reduce the duration of a transmission session, other benefits may be derived from the use of data compression performing devices. Since compression results in the transmission of a lessor number of characters in comparison to sending uncompressed data, the probability of a transmission error occurring is reduced. Similarly, a reduction in the number of characters can reduce one's transmission cost. Examples of cost reductions can range from the ability to connect additional terminals to a multidrop line to the use of a packet network where one of the cost elements associated with the use of the network is the number of characters transmitted and received by a terminal device.

From the preceding, it is obvious that the use of compression performing devices should be considered whenever possible since their use can boost network efficiency as well as potentially reduce the cost of operating a network. In this section, we will first review a few of the methods that can be employed to compress data. This will be followed by an examination of the benefits associated with the utilization of data compression performing devices. Using the previous information, we will then examine several compression performing data communication devices currently marketed to investigate how they can be effectively used, focusing our effort upon the benefits that can be derived by their use.

Compression techniques

While the number of algorithms that have been developed over the years to compress data may exceed one hundred, in essence compression techniques can be subdivided into two classes—character and statistical encoding.

Character oriented

In character oriented compression techniques data is examined on a character by character basis, with special characters or a small group of characters employed to represent a larger string of characters. In this manner, redundancy can be reduced or eliminated prior to the transmission of data. At the receiver, a check of the incoming data occurs for the special characters or group of characters used to represent compressed data. Upon detection, the compressed data is decompressed according to a predefined algorithm, resulting in the duplication of the original data. Examples of character oriented compression techniques include null compression, run length encoding and pattern substitution.

Null compression

Null or space compression is probably the first data compression technique to be applied to data transmission. As illustrated in Figure 6.28, a special compression indicating character is employed to indicate the occurrence of null compression. This character is then followed by the count of the number of null or blank

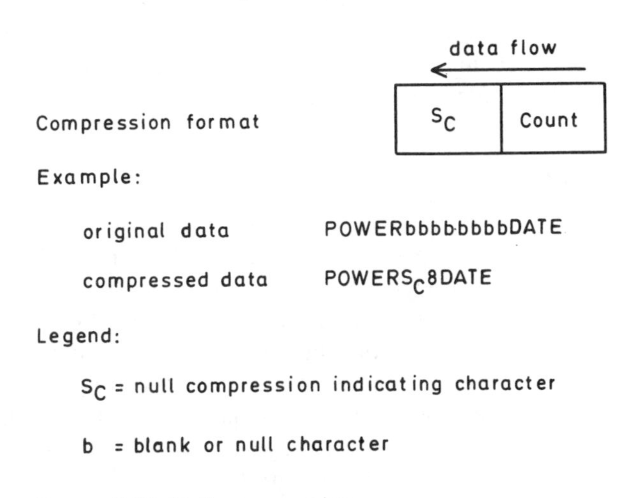

Figure 6.28 Null compression.

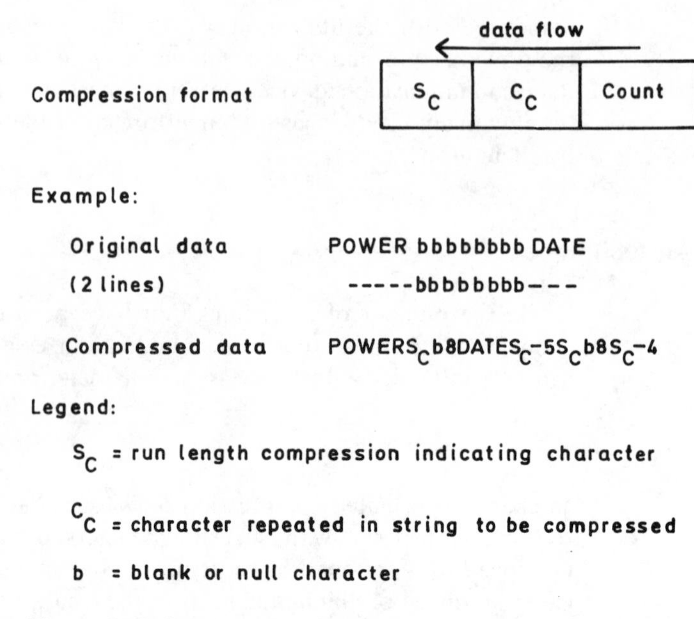

Figure 6.29 Run length compression.

characters that occurred in the string. One of the earliest examples of the employment of null compression for data transmission is the IBM 3780 protocol which incorporates this technique.

Run length compression

Run length compression can be viewed as a superset of null compression. This technique is applicable to any repeating character string, whereas, null compression is limited to a repeating sequence of null characters. Figure 6.29 illustrates the basic format and operation of run length compression.

As indicated in Figure 6.29, run length compression can be used to compress any string of repeating characters to include nulls. Since three characters in this compression format are substituted for the repeating string, one must encounter four or more repeating characters for this technique to compress data.

Pattern substitution

Pattern substitution requires data to be matched against entries in a pattern table. If the pattern is encountered in the table, the data that forms the pattern is replaced by a compression indicator which can range in scope from a single character to a group of characters. One version of pattern substitution that is frequently implemented in communications devices is diatomic encoding. In this compression technique, the most frequently encountered pairs of characters in the English language supplemented by common pairs of characters used in programming languages are used as entries in the pattern table.

An example of pattern substitution using diatomic compression is illustrated in Figure 6.30. Normally, the compression indicating substitution characters

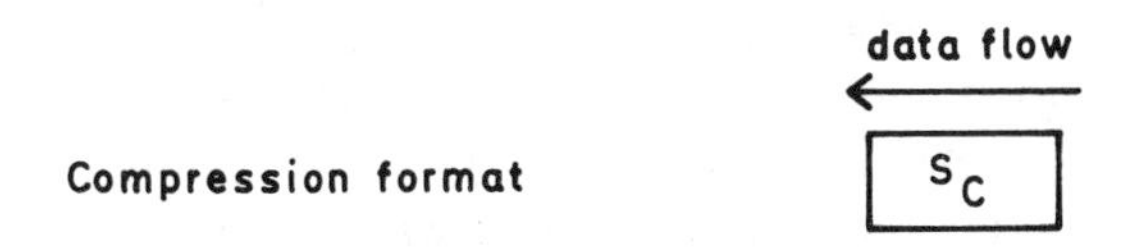

Example:

assumed pattern table

Entry	Substitution
th	S_{C1}
ig	S_{C2}

Original data the night of the iguana

Compressed data S_{C1}e S_{C2}ht of S_{C1}e S_{C2}uana

Figure 6.30 Pattern substitution using diatomic compression.

are obtained from undefined or unused characters in the character set being used. An example of the former might be the large number of undefined characters in the EBCDIC data code, while an example of the latter could be those characters from ASCII 127 through ASCII 255 which are known as extended ASCII.

Statistical encoding

In statistical encoding the probabilities of occurrence of single characters and patterns are employed to develop short codes to represent frequently occurring characters and patterns while longer codes are used for data that occurs less frequently. To better understand the utilization of statistical encoding assume the probability of occurrence of each character in a 5-character set is as indicated in Table 6.8.

Table 6.8 Five-character set probability of occurrence.

Character	Probability of occurrence
A	0.250
B	0.250
C	0.250
D	0.125
E	0.125

Based upon the data contained in Table 6.8, we can develop a coding scheme to minimize the average codeword length of each character by employing their probability of occurrence in our codeword construction scheme. For illustrative purposes, we will develop what is known as a Huffman code for the character set presented in Table 6.8.

The Huffman code not only minimizes the average codeword length but, in addition, results in the generation of a code that is instantaneously decodable. That is, in the Huffman code every codeword is uniquely decipherable and no codeword is a prefix of another longer codeword. Figure 6.31 illustrates the development of a code set based upon the Huffman coding algorithm. First, the character set is arranged in a column in order of decreasing frequency of occurrence. Starting at the smallest probability of occurrence, the lines with the two smallest probabilities are merged, with their probabilities added to obtain a composite probability on a new line. This combining is continued until all lines have been merged. Then, for each pair of branches attached to a node we arbitrarily assign a 0 bit to one branch and a 1 bit to the other. The bit sequence assigned to each character is then determined by tracing the route from the 1.0 probability node backwards to the character, using the assigned bits in each branch that provides the route to the character.

Figure 6.32 illustrates the compression of character data into a binary string based upon the utilization of the previously developed Huffman code set. The reader should note that the original data consisting of the characters ABADABD would contain 56 bits if encoded 8 bits per character. In comparison, through the utilization of Huffman encoding only 16 bits are required to encode the data.

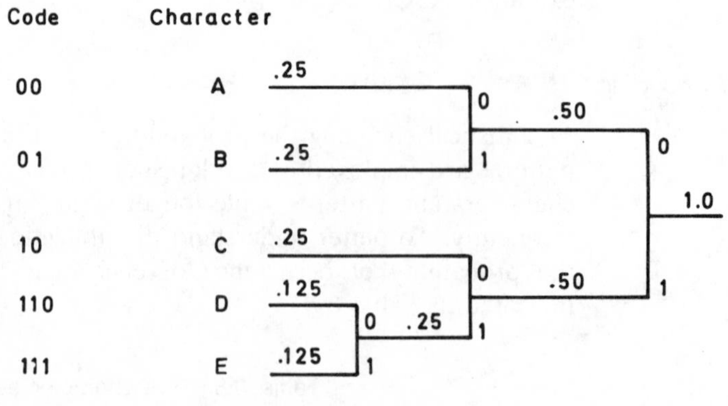

Code	Character	Probability
00	A	.25
01	B	.25
10	C	.25
110	D	.125
111	E	.125

Figure 6.31 Huffman code set construction.

Original data	ABADABD
Compressed data	00|01|00|110|00|01|110

Figure 6.32 Huffman encoding.

Benefits of compression

Table 6.9 lists some of the primary benefits that may result from the use of data compression performing devices. Since few of these benefits are self-explanatory, we will examine each in detail.

Transmitting less data equates to lowering the duration of a transmission session. Although we normally associate some cost per unit of transmission time leading to a reduction in communications costs when compressing data, there are other benefits that may result from compression. First, if we transmit less data while the probability of a transmitted bit being in error remains constant the overall probability of a transmission error occurring will decrease. Thus, data compression will reduce the probability of transmission errors occurring.

Table 6.9 Potential benefits of data compression.

Reduces probability of transmission errors
Distorts clarity of transmitted data
Can reduce or eliminate RBT shifts
Provides ability to restructure a network
Provides capability to reduce costs

By converting text that is in a conventional code (such as standard ASCII) into a different code, compression algorithms may offer some security against illicit monitoring. As previously indicated in Figure 6.32, the original data stream 'ABADABD' was converted into the compressed binary string '0001001100001110.' Here, the 16-bit compressed string might be interpreted as two 8-bit characters by a person illicitly monitoring the transmission. Thus, compression can be considered as a technique that reduces the clarity of transmitted data to observers of the data traffic.

In many remote batch processing environments, data compression has been successfully used to reduce or eliminate workshifts. Typically, the primary function of second- and third-shift remote batch terminal (RBT) operations is the retrieval of output from deferred batch processing jobs run on a central computer in the evening. Since the limiting factors for this type of shift operation are the processing power of the central computer and the data transfer rate on the communications link, increasing the data transfer rate through compression may result in the reduction or elimination of an RBT shift.

One of the most promising features of data compression is its ability to provide a mechanism to alter the structure of a communications network. How such a change might occur can be seen from a brief examination of a typical network application.

Assume an existing multidrop line services eight terminal locations. Perhaps an expansion to add two additional terminal locations to the network is planned. Assume that the network planner estimated the activity on the multidrop line based upon its expansion to ten drops and determined that due to the additional data traffic the response time of the terminals would be unacceptable. As a result of this analysis, a separate line consisting of two drops might be installed. If data compression is implemented there would be less total traffic, reducing the time it takes a computer to poll and service each terminal on a multidrop line. Thus, by implementing data compression one might be able to service the two additional locations by extending the multidrop line without increasing the response time of the terminals.

Using compression performing devices

Often, one can obtain the benefits of data compression while avoiding the efforts required to analyze actual or potential data traffic and develop software to perform compression. This can be accomplished by leasing or purchasing data compression performing devices that are specifically designed to be used in a particular networking environment. Thus, we will examine several hardware products to obtain a better understanding of the use of compression performing devices. The products covered in this section were selected for illustrative purposes only and should not be construed as an endorsement of the device.

Asynchronous data compressors

Since asynchronous terminals outnumber synchronous operating terminals by a factor of 10 or more, it should be of no surprise that many vendors have developed products for use with asynchronous transmission. Such products can normally be used for transmission occurring on both leased lines and the switched telephone network, however, their primary use is for transmission on the public switched telephone network. When used on this transmission facility the major advantage of the compression device is its ability to reduce the duration of the transmission session. Since the cost of a long-distance call is approximately proportional to its duration, decreasing the length of the transmission session reduces the cost of the call. One product specifically designed to compress asynchronous data that will be examined for illustrative purposes is RAD Computers Inc.'s CompressoRAD-1.

CompressoRAD-1

The CompressoRAD-1 is a stand-alone compression unit that also performs asynchronous-to-synchronous/synchronous-to-asynchronous conversion and provides error detection and correction capability to one's data flow. Figure 6.33 illustrates how the CompressoRAD-1 could be utilized for transmitting data via the switched telephone network.

The CompressoRAD-1 accepts asynchronous data at 1200, 2400, 4800 or 9600 bps. The asynchronous data can consist of 7 or 8 bits per character, with either 1 or 2 stop bits and odd, even, mark, space or no parity. Utilizing an automatic adaptive compression algorithm, data compression ratios from 2:1 to 4:1 are obtainable according to the company.

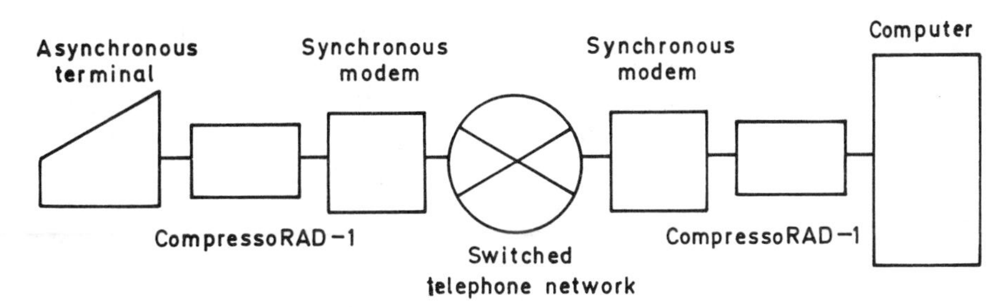

Figure 6.33 Using the CompressoRAD-1.

In addition to compressing data, the device converts the asynchronous data flow into a synchronous modified higher-level data link control (HDLC) protocol. By adding a cyclic redundancy check character to each transmitted data block, end-to-end error detection and correction capability is added to the data transmission.

The output data rate of the CompressoRAD-1 is determined by the clock of the attached modem, with the device capable of operating between 600 and 4800 bps. A buffer in the device is used to compensate for the differences in the compressibility of input data. Thus, when data input into the CompressoRAD is compressed and represents a data flow greater than the rate at which data is output to the attached synchronous modem the buffer fills, serving as a temporary storage area. Since the buffer is finite in size, the CompressoRAD employs two methods to inhibit additional data input once the buffer is filled to a predefined level. Known as flow control, data is regulated into the device by the transmission of XOFF and XON characters and the raising and lowering of the Clear-To-Send RS-232 control signals. When buffer storage fills to a predefined level, the CompressoRAD will either transmit an XOFF character or drop the CTS signal on the RS-232 interface. The choice of which method one selects is based upon the operational specifications of the terminal or computer port attached to the CompressoRAD. Some terminals and computer ports recognize XON and XOFF as flow control character signaling, enabling the CompressoRAD to be configured to use this method of flow control. For terminal and computer ports that do not recognize this method of flow control, the CompressoRAD can be configured to enable and inhibit transmission to the device via the CTS control signal. Thus, after data is inhibited from transmission into the device by the CompressoRAD dropping CTS, it will raise this control signal once the buffer is emptied to another predefined level. The issuance of flow control signals based upon the quantity of data in the device's buffer is illustrated in Figure 6.34.

To understand the economics associated with the utilization of the CompressoRAD and similar compression performing products, assume that the daily cost of transmission on the switched telephone network between a remote terminal and a central computer facility is $5, a cost typically representing a long-distance call duration of under 30 minutes. Assuming one call per day and 22 working days per month, the cost of using the switched telephone network is $110 per month or $1320 on an annual basis. Now, let us assume that the use

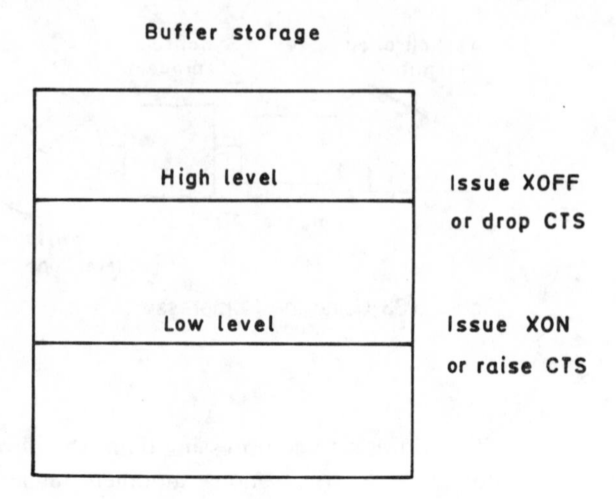

Figure 6.34 Buffer control.

of two asynchronous data compression devices reduces the transmission session duration by one-half, resulting in the ability to reduce the cost of communications by $660 during 1 year of operation. To what should one then compare this potential cost savings?

In using the CompressoRAD illustrated in Figure 6.33 or a similar compression performing product one must obtain two synchronous modems as well as two compression devices. In comparison, one would use two lower cost asynchronous modems when transmitting data without the use of asynchronous compression devices. Thus, from an economic perspective:

$$(\Delta M) \times 2 + 2 \times C \leq 660 \times E$$

where ΔM = cost difference between synchronous and asynchronous modems, C = unit cost of each compression performing device, and E = expected life or use of the compression performing devices in years. If, for example, 1200-bps 212A modems are used and each compression performing device has an expected life of 3 years and costs $500, applying the formula results in the equation:

$$(0) \times 2 + 2 \times \$500 \leq 660 \times 3$$
$$\text{or } \$1000 \leq \$1980$$

This shows that the cost is less than the expected savings and the use of compression performing devices is economically warranted.

Multifunctional compression devices

One of the more popular uses of compression performing devices is to reduce the quantity of data to be multiplexed, increasing the servicing capacity of the multiplexer. A natural evolution of the development of compression devices was to include both statistical multiplexing and data compression in one

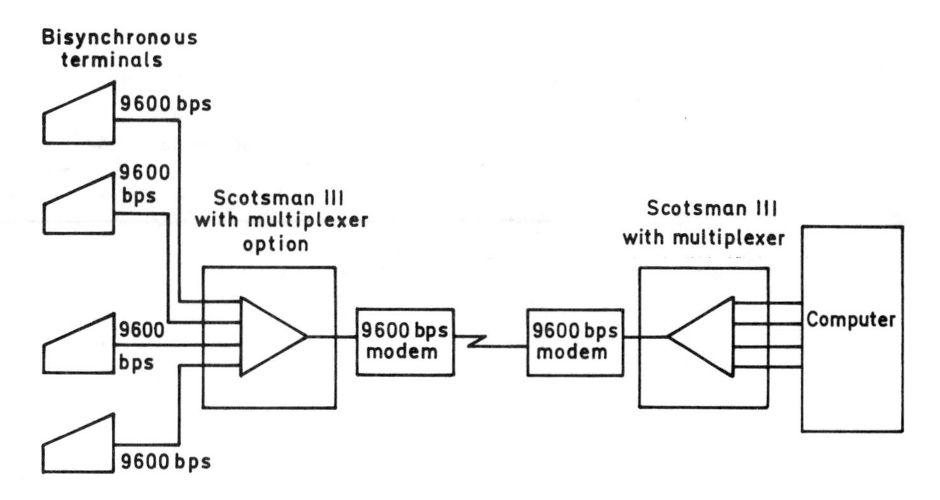

Figure 6.35 Using the Racal–Vadic Scotsman III as a multiplexer.

hardware device. Two products representing this multifunctional capability are the Racal-Vadic Scotsman III and the Datagram Corporation Streamer product line.

Scotsman III

The Racal-Vadic Scotsman III is a stand-alone data compression device that can be configured in a variety of ways to include the addition of an optional four-channel multiplexer. Figure 6.35 should be compared to the employment of a statistical multiplexer. Since most statistical multiplexers do not perform data compression, significant time delays would occur in attempting to multiplex four 9600-bps bisynchronous data sources onto one 9600-bps transmission line. In other situations, statistical multiplexers may service bisynchronous data through the use of bandpass channels. When this occurs, only asynchronous data is actually statistically multiplexed, while bisynchronous data is time division multiplexed onto a predefined time slot on the composite multiplexed link as illustrated in Figure 6.36. When bisynchronous data is multiplexed via a bandpass channel, the data rate of the bisynchronous input data reduces the portion of the composite channel's bandwidth available for multiplexing other data sources. As an example, one bisynchronous data source operating at 4800 bps would reduce the bandwidth of the composite channel for multiplexing other data sources by 4800 bps when the bisynchronous data is multiplexed via a bandpass channel. Thus, if the composite channel operates at 9600 bps, only 4800 bps would be available for the statistical multiplexing of other data sources.

Due to the limitations of servicing bisynchronous data with bandpass channels, most statistical multiplexers using this technique are limited to supporting only one bisynchronous data source. Thus, the Scotsman III can be effectively employed for situations where several high-speed bisynchronous data sources transmit to a common location.

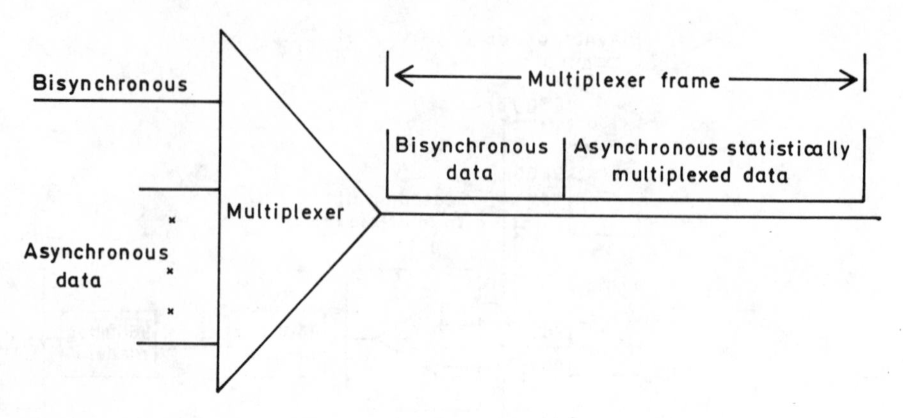

Figure 6.36 Band pass multiplexing.

Streamer

Datagram Corporation's Streamer series of devices are statistical multiplexers that incorporate data compression, resulting in an overall statistical efficiency of up to 8 to 1 according to the company. In comparison to many similar products, the Streamer series of devices supports a wide variety of bit-oriented protocols to include synchronous data link control (SDLC), X.25, HDLC, Univac data link control (UDLC) and character oriented protocols to include asynchronous and bisynchronous transmission.

Due to the ability of the Streamer series of devices to support numerous data link protocols, this compression performing communications device can be used in a large variety of network applications. Figure 6.37 illustrates one potential application that shows the versatility of the Streamer. The Streamer first performs typical statistical multiplexer functions to include stripping start and stop bits from asynchronous transmission, and passing data only from active terminals. After the data is statistically multiplexed, the device's compressor uses an adaptive compression table to encode the data, resulting in overall statistical and compression efficiencies of up to 8 to 1. Due to the efficient multiplexing and compression of data, more data sources can be multiplexed onto one high-speed line in comparison to the use of non-compression performing multiplexers. This may eliminate the necessity of installing dual multiplexer systems and multiple leased lines, resulting in a considerable economic benefit to the user.

6.8 FIBER OPTIC TRANSMISSION SYSTEMS

While the majority of attention focused upon fiber optic transmission relates to its use by communication carriers, this technology can be directly applicable to many corporate networks. Due to the properties of light transmission, fiber optic systems are well suited for many specialized applications to include the high-speed transmission of data between terminals and a computer or between computers located in the same building.

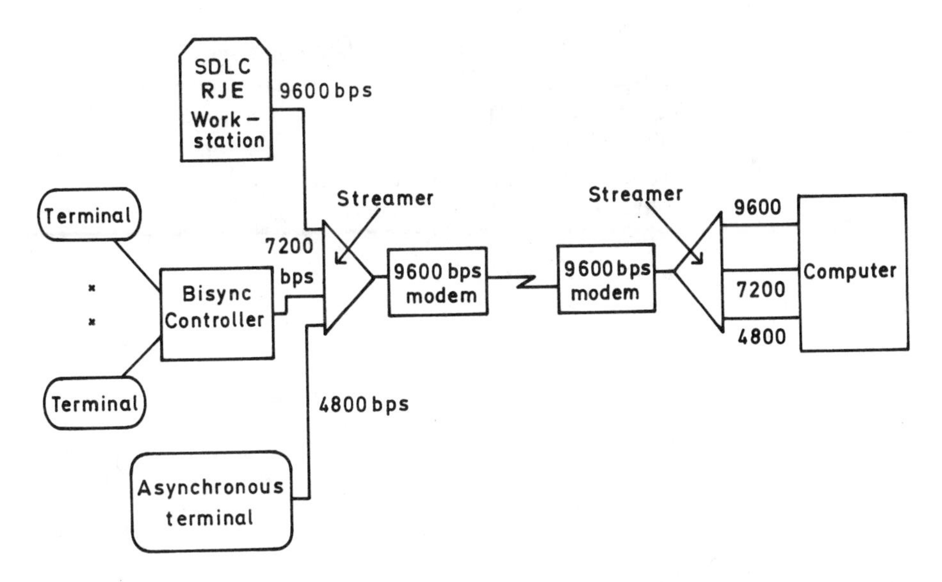

Figure 6.37 Using the datagram streamer. Datagram's streamer series of compression performing multiplexers supports a large number of bit and character oriented protocols.

In this section we will first focus our attention upon the components of a fiber optic transmission system. After examining the advantages and limitations associated with such systems we will use the previous information presented as a foundation to analyze economics associated with cabling terminals to a computer, illustrating how this technology can be directly applicable to many computer installations. In doing so, we will examine the use of fiber optic modems and multiplexers to denote the advantages and limitations associated with their use in a communications network.

Fiber optics is a rapidly evolving technology that has moved from a laboratory and consumer-product curiosity into a viable mechanism for low-cost, high data rate communications. Permitting the transmission of information by light, fiber optics is most familiar to individuals by their incorporation into contemporary desk lamps that were first marketed in the 1970s. In such lamps, bundles of individual fibers were attached to a common light source at the base of the lamp's neck. Radiating outward in a geometric pattern, the fibers were shaped to form a contemporary design. By polishing the end of each fiber, the common light source was reflected at the tip of each fiber, resulting in a very impressive visual display. Since the early 1970s a significant amount of fiber research has resulted in such fibers becoming available for practical data communications applications.

System components

Similar to conventional transmission systems the major components of a fiber optic system include a transmitter, transmission medium and receiver. The transmitter employed in a fiber optic system is an electrical to optical (E/O) converter. The E/O converter receives electronic signals and converts those signals into a series of light pulses. The transmission medium is an optical fiber cable which can be constructed out of plastic or glass material. The receiver used in fiber optic systems is an optical to electric (O/E) converter. The O/E converter changes the received light signals into their equivalent electrical signals. The relationship of the three major fiber optic system components is shown in Figure 6.38.

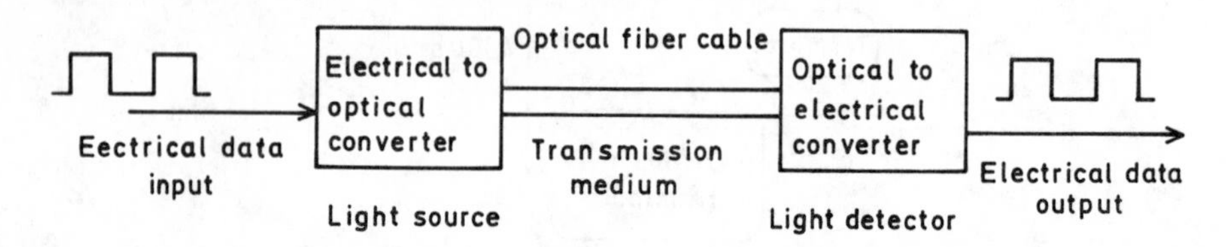

Figure 6.38 Fiber optic system component relationship. The electrical to optical converter produces a light source which is transmitted over the optical fiber cable. At the opposite end of the cable an optical to electrical converter changes the light signal back into an electrical signal. (Reprinted with permission from *Data Communications Management*, © 1986 Auerbach Publishers, New York.)

The light source

Currently two types of light sources dominate the electronic to optical conversion market: the light emitting diode (LED) and the laser diode (LD). Although both devices provide a mechanism for the conversion of electrically encoded information into light encoded information, their utilization criteria depends upon many factors. These factors include response times, temperature sensitivity, power levels, system life, expected failure rate, and cost.

A LED is a PN junction semiconductor that emits light when biased in the forward direction. Typically a current between 25 to 100 mA is switched through the diode, with the wavelength of the emission a function of the material used in doping the diode.

The laser diode offers users a fast response time in converting electrically encoded information into light encoded information, thereby, they can be used for very high data transfer rate applications. The laser diode can couple a high level of optical power into an optical fiber, resulting in a greater transmission distance than obtainable with light emitting diodes. Although they can transmit further at higher data rates than LEDs, laser diodes are more susceptible to temperature changes and their complex circuitry makes such devices more costly.

Optical cables

Many types of optical cables are marketed, ranging from simple 1-fiber cables to complex 18-fiber, commonly jacketed, cables. In addition, a large variety of specially constructed cables can be obtained on a manufactured-to-order basis from vendors.

In its most common form an optical fiber cable consists of a core area, cladding, and a protective coating. This is illustrated in the upper part of Figure 6.39. As a light beam travels through the core material the ratio of its speed in the core to the speed of light in free space is defined as the refractive index of the core.

A physical transmission property of light is that while traveling in a medium of a certain refractive index, if it should strike another material of a lower refractive index, the light will bend towards the material containing the higher index. Since the core material of an optical fiber has a higher index of refraction than the cladding material, this index differential causes the transmitted light signal to reflect off the core-cladding junctions and propagate through the core. This is shown in the lower part of Figure 6.39.

The core material of fiber cables can be constructed with either plastic or glass. While plastic is more durable to bending, glass provides a lower attenuation of the transmitted signal. In addition, glass has a greater bandwidth, permitting higher data transfer rates when that material is used for fiber construction.

The capacity of a fiber optic data link of a given distance depends upon the numerical aperture (NA) of the cable as well as the core size, attenuation and pulse dispersion characteristics of the fiber. The NA value can be computed from the core and cladding refractive indices as follows:

$$NA = \sqrt{m_1^2 - m_2^2}$$

where: m_1 = core material refractive index, and m_2 = cladding material refractive index.

The numerical aperture value indicates the potential efficiency in the coupling of the light source to the fiber cable. Together with core material diameter,

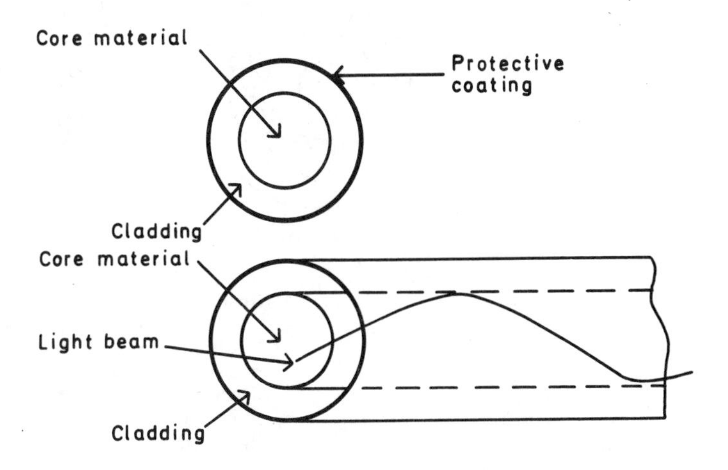

Figure 6.39 Optical fiber cable. (Reprinted with permission from *Data Communications Management*, © 1986 Auerbach Publishers, New York.)

the numerical aperture indicates the level of optical power that can be transmitted into a fiber.

A function of both the fiber material and core/cladding imperfections, attentuation determines how much power can reach the far end of an unspliced link. When the numerical aperture, core diameter and attenuation are considered together one can determine the probable transmission power loss ratio.

Pulse dispersion is a measure of the widening of a light pulse as it travels down an optical fiber. Dispersion is a function of the cable's refractive index. Fibers with an appropriate refractive index permit an identifiable signal to reach the light detector at the far end of the cable.

Types of fibers

Two types of optical fibers are available for use in cables: Step Index and Graded Index.

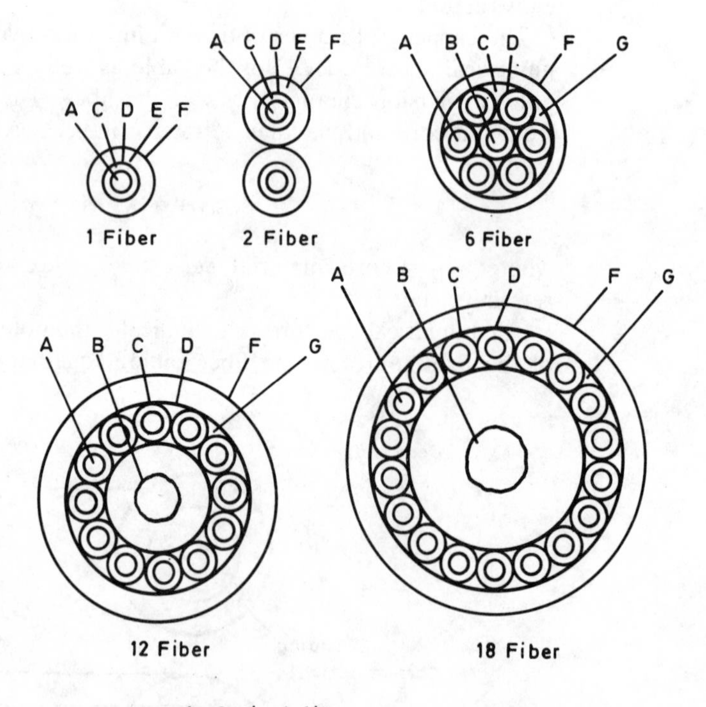

Figure 6.40 Common cable types. (Reprinted with permission from *Data Communications Management*, © 1986 Auerbach Publishers, New York.)

In a Step Index fiber an abrupt refractive index change exists between the core material and the cladding. In a Graded Index fiber the refractive index varies from the center of the core to the core-cladding junction. The gradual variation of the refractive index in this type of fiber serves to minimize the optical signal dispersion as light travels along the fiber core. The minimization of dispersion results from the light rays near the core-cladding junction traveling faster than those near the center of the core. The minimization of signal dispersion permits greater bandwidth, allowing transmission to occur over greater distances at higher data rates than available with a Step Index fiber. Thus, Graded Index fibers are commonly used for long haul, wide bandwidth communications applications. Conversely, Step Index fibers are used for communications applications requiring a narrow bandwidth and relatively low data rate requirement.

Common cable types

Five common types of optical cables are illustrated in Figure 6.40. Note that when a 1-fiber cable has a transmitter connected to one cable end and a receiver to the opposite end, it functions as a simplex transmission medium. As such, transmission can only occur in one direction. The 2-fiber cable can be considered a duplex transmission medium since each fiber permits transmission to occur in one direction. When transmission occurs on both fibers at the same time a full-duplex transmission sequence results. Thus, the 18-fiber cable is capable of providing nine duplex transmission paths.

The light detector

To convert the received light signal back into a corresponding electrical signal a photodetector and associated electronics are required. Photodetection devices currently available include a PIN photodiode, an avalanche photodiode, a phototransistor and a photomultiplier. Due to their efficiency, cost, and light signal reception capabilities at red and near infrared (IR) wavelengths, PIN and avalanche photodiodes are most commonly employed as light detectors.

In comparison to PIN detectors, avalanche photodiode detectors offer greater receiver sensitivity. This increased light sensitivity results from their high signal-to-noise ratio, especially at high bit rates.

Since avalanche photodiode detectors require an auxiliary power supply which introduces noise, circuitry to limit such noise results in the device having a higher overall cost than a PIN photodiode. In addition, they are temperature sensitive and their installation environment requires careful examination. A block diagram of the major components of a light detector are illustrated in Figure 6.41.

Other optical devices

Besides the optical devices previously discussed, two additional devices warrant coverage. The first device is an optical modem which performs 'optical' modulation and demodulation. The second device is an optical multiplexer which permits the transmission of many data sources over a single optical fiber.

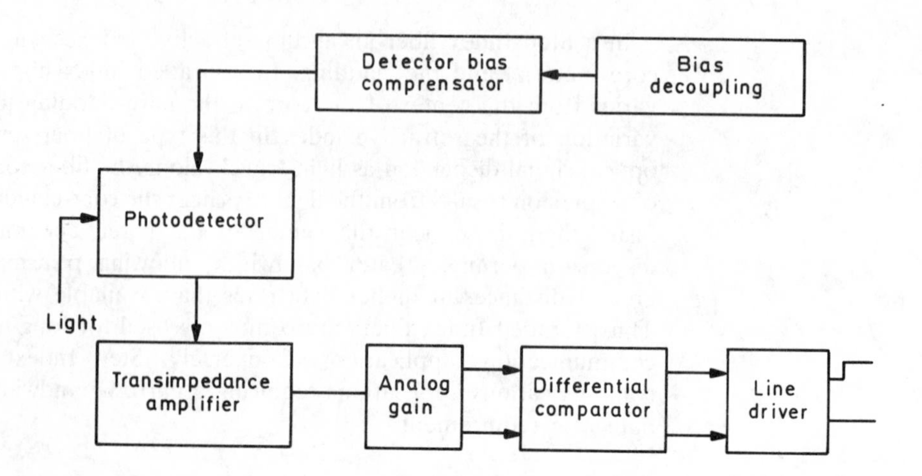

Figure 6.41 Light detector component. To filter DC input voltage, protect the photodiode and reduce the effect of electromagnetic interference, a light detector module has bias decoupling. Since an avalance photodiode is a temperature sensitive device, a detector-bias compensator is used to compensate for temperature changes that could affect the diode.

The photodetector converts the received optical signal into a low-level electrical signal. This detector can either be an avalance or PIN diode, depending upon the optical sensitivity requirements. The transimpedence amplifier is a low noise, current-to-voltage converter while the analog gain element increases the voltage gain from the preceding amplifier to the level required for the decision circuitry. The differential comparator converts the analog signal into digital form by interpreting signals below a certain threshold as a '0' and above that threshold as a '1'. The line driver regenerates and drives the squared signal from the comparator for transmission over metallic cable. (Reprinted with permission from *Data Communications Management*, © 1986 Auerbach Publishers, New York.)

Optical modem

An optical modem is a device housing both an optical transmitter and an optical detector as shown in the top portion of Figure 6.42. Similar to conventional modem development, a variety of optical modems have evolved. These variations range from single-channel stand-alone devices to multiport optical modems, the latter capable of functioning as a synchronous multiplexer and optical modem. The multiport optical modem permits conventional electronic bit streams from up to four data sources to be multiplexed and converted into one stream of light pulses for transmission on one optical fiber. In the lower part of Figure 6.42, a multiport optical modem servicing four data sources is shown.

Optical multiplexers

Functioning similar to a conventional time division multiplexer, an optical multiplexer accepts the RS-232 electrical input of many data sources and multiplexes such signals for transmission over a single optical fiber. Included in the optical modem are a light source generator and detector. Thus, such devices are equivalent to a multiplexer with a built-in modem since they convert

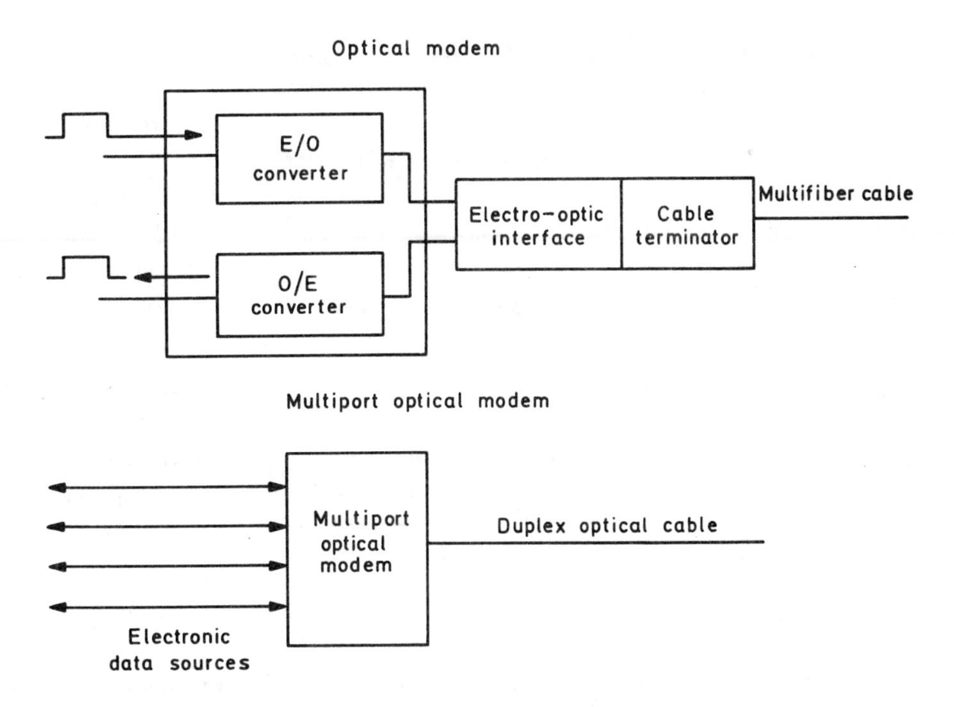

Figure 6.42 Optical modem. An optical modem can transmit and receive data over one multifiber cable, converting the electronic data source to light and the received light back into its corresponding electronic signal. Serving as a synchronous multiplexer, the multiport optical modem transmits data from up to four electronic sources as one optical signal. (Reprinted with permission from *Data Communications Management*, © 1986 Auerbach Publishers, New York.)

electronic multiplexed data into an optical signal and detect and convert light signals to their equivalent electronic signals prior to demultiplexing.

Transmission advantages and limitations

When used for data transmission, fiber optic cables offer many advantages over cables with metallic electrical conductors. These advantages result from several distinct properties of the optical cables. The more common advantages associated with the utilization of a fiber optic transmission system are listed in Table 6.10.

Bandwidth

One of the advantages of fiber optic cable in comparison to metallic conductors is the wide bandwidth of optical fibers. With potential information capacity directly proportional to transmission frequency, light transmission on fiber cable provides a transmission potential for very high data rates. Currently, data rates of up to 10^{14} bps have been achieved on fiber optic links. When compared to the 9.6 to 19.2 kbps limitation of telephone wire pairs, fiber cable

Table 6.10 Fiber optic system advantages.

Large bandwidth	Mixed voice, video, and data on one line
No electromagnetic interference (EMI)	No specially shielded conduits required Cable routing simplified Bit error rate improved
Low attentuation	Permits extended cable distances
No shock, hazard, or short circuits	Can be used in dangerous atmospheres Common ground eliminated
High security	Transmission TEMPEST acceptable Tapping noticeable
Light-weight and small-size cable	Facilitates installation
Cable ruggard and durable	Long cable life

makes possible the merging of voice, video, and data transmission on one conductor. In addition, the wide bandwidth of optical fiber provides an opportunity for the multiplexing of many channels of lower speed, but which are still significantly higher data rate channels than are transmittable on telephone systems.

Electromagnetic non-susceptibility

Since optical energy is not affected by electromagnetic radiation, optical fiber cables can operate in a noisy electrical environment. This means special conduits formerly required to shield metallic cables from radio interference produced by such sources as electronic motors and relays are not necessary.

Similarly, cable routing is easier since the rerouting necessary for metallic cables around fluorescent ballasts does not cause concern when routing fiber cables. Due to its electromagnetic interference immunity, fiber optic transmission systems can be expected to have a lower bit error rate than corresponding metallic cable systems. By not generating electromagnetic radiation, fiber optic cables do not generate cross talk. This property permits multiple fibers to be routed in one common cable, simplifying the system design process.

Signal attenuation

The signal attenuation of optical fibers are relatively independent of frequency. In comparison, the signal attenuation of metallic cables increases with frequency. The lack of signal loss at frequencies up to 1 GHz permits fiber optic systems to be expanded as equipment is moved to new locations. In comparison, conventional metallic cable systems may require the insertion of line drivers or other equipment to regenerate signals at various locations along the cable.

Electrical hazard

On fiber optic systems light energy in place of electrical voltage or current is used for the transfer of information. The light energy alleviates the potential of a shock hazard or short circuit condition. The absence of a potential spark makes fiber optic transmission particularly well suited for such potentially dangerous industrial environment uses as petrochemical operations as well as refineries, chemical plants, and even grain elevators. A more practical benefit of optical fibers for most corporate networks is the complete electrical isolation they afford between the transmitter and receiver. This results in the elimination of a common ground which is a requirement of metallic conductors. In addition, since no electrical energy is transmitted over the fiber, most building codes permit this type of cable to be installed without running the cable through a conduit. This can result in considerable savings when compared to the cost of installing a conduit required for conventional cables, whose cost can exceed $2500 for a 300-foot metal pipe.

Security

Concerning security, the absence of radiated signals makes the optical fiber transmission TEMPEST acceptable. In comparison, metallic cables must often be shielded to obtain an acceptable TEMPEST level. Although fibers can be tapped like metallic cable, doing so would produce a light signal loss. Such a loss could be used to indicate to users a potential fiber tap condition.

Weight and size

Optical fibers are smaller and lighter than metallic cables of the same transmission capacity. As an example of the significant differences that can occur, consider an optical cable of 144 fibers with a capacity to carry approximately 100,000 telephone conversations. The cable would be approximately one inch in diameter and weigh about 6 ounces per foot. In comparison, the equivalent capacity copper coaxial cable would be about three inches in diameter and weigh about 10 pounds per foot. Thus, fiber optic cables are normally easier to install than their equivalent metallic conductor cables.

Durability

Although commonly perceived as being weak, glass fibers have the same tensile strength as steel wire of the same diameter. In addition, cables containing optical fibers are reinforced with both a strengthening member inside and a protective jacket placed around the outside of the cable. This permits optical cables to be pulled through openings in walls, floors, and the like without fear of damage to the cable.

With better corrosion resistance than that of copper wire, transmission loss at splice locations has a low probability of occurring when optical fiber cable is used.

Limitations of use

As discussed, optical fiber cables offer many distinct advantages over conventional metallic cables. Unfortunately, they also have some distinct usage disadvantages. Two of the main limiting factors of fiber optic systems are cable splicing and system cost.

Cable splicing

When cable lengths of extended distances become necessary, optical fibers must be spliced together. To permit the transmission of a maximum amount of light between spliced fibers, precision alignment of each fiber end is required. This alignment is time consuming and depending upon the method used to splice the fibers, one's installation cost can rapidly escalate.

Fibers may be spliced by welding, gluing, or through the utilization of mechanical connectors. All three methods result in some degree of signal loss between spliced fibers.

Welding or the fusion of fibers results in the least loss of transmission between splice elements. Due to the time required to clean each fiber end and then align and fuse the ends with an electric arc does not make splicing easily suitable for field operations. An epoxy, or gluing method of splicing requires the utilization of a bonding material that matches the refractive index of the core of the glass fiber. This method typically results in a higher loss than obtained with the welding process. While mechanical connectors have the highest data loss among the three methods, it is by far the easiest method to employ. Although mechanical connectors considerably reduce splicing time requirements, the cost of good quality connectors can be relatively expensive. Currently, typical connectors cost approximately $30.00.

System cost

Good quality, low-loss single-channel fiber optic cable costs between $1.50 and $2.00 per meter. A typical fiber optic modem having a transmission range of 1 km costs between $200.00 and $600.00. In comparison, conventional metallic cable for synchronous data transmission costs approximately 30 cents per foot while a line driver capable of regenerating digital pulses at data rates up to 19.2 kbps cost approximately $200.00.

Based upon the preceding, the cost of a limited-distance fiber optic system at data rates up to 19.2 kbps will generally exceed an equivalent conventional metallic based, transmission system. Only when high data rates are required or transmission distances expand beyond the capability of line drivers and to some extent limited distance modems are fiber optic systems economically viable.

Utilization economics

One of the most commonly used duplex fiber optic cables cost $2.50 per meter which is equivalent to 76 cents per foot. An optical modem containing an electric to optic and an optic to electric converter capable of transferring data at rates up to 10 Mbps costs approximately $600.00 per unit. The system cost of a pair of optical modems as a function of distance is illustrated in Figure 6.43. Note that as long as no splicing is required, costs are a linear function of distance.

Suppose a requirement materializes which calls for the communications linkage of four high-speed digital devices located in one building of an industrial complex to a computer center located in a different building 10,000 feet away. What fiber optic systems can support this requirement and what are the economic ramifications of each configuration?

Dedicated cable system

Equivalent to individually connecting devices on metallic cables, four individual fiber optic systems and four cables could be installed. Here eight optical modems would be required, resulting in the cost of the modems being $4800.00.

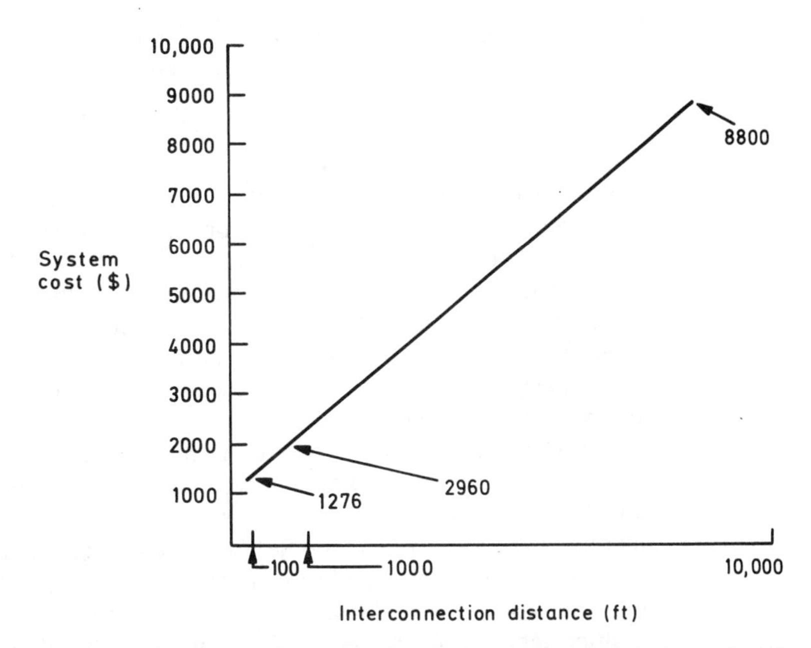

Figure 6.43 System cost varies with distance. Cost of a typical duplex optical fiber cable and a pair of optical modems capable of transmitting up to 10 Mbps. (Reprinted with permission from *Data Communications Management*, © 1986 Auerbach Publishers, New York.)

With four cables being required, 40,000 feet of cable would cost $30,400.00, resulting in a total cost of $35,200.00 for this type of network configuration. In addition, a substantial amount of personnel time may be required to install four individual ten-thousand-foot cables.

Multichannel cable

A second method that can be employed to link multiple devices at one location is by the employment of multichannel cable. We can examine the economics associated with multichannel cable by considering the cost of an eight-channel cable.

A typical eight-channel cable capable of supporting four duplex transmissions costs approximately $10.00 per meter or $3.05 per foot. On a duplex channel basis, this represents a cost of 76 cents per foot per channel. This is the same cost as an individual duplex cable. The use of most multichannel cable does not offer any appreciable savings over individual cable until ten or more channels are packaged together. Prior to excluding the use of a multichannel cable, one should carefully consider cable installation costs since the time required to install one multichannel cable can be significantly less than the time required to install individual cables.

Optical multiplexers

Similar to conventional metallic cables, the potential installation of parallel opical cables indicates that multiplexing should be considered. Prior to deciding upon the use of an optical multiport modem or optical multiplexer, one should examine the type of data to be transmitted as well as the data transfer rates required. If each data type is synchronous and no more than four data sources exist, the utilization of an optical multiport modem can be considered. If a mixture of asynchronous and synchronous data must be supported or, if more than four data sources exist, one should consider an optical multiplexer.

Currently, four-channel synchronous optical multiport modems cost approximately $1000 per unit. Returning to our system requirement, a pair of optical multiport modems and one cable could support the four data terminals. Here the total system cost would be reduced to $9600, of which $2000 would be for the pair of optical multiport modems and the remainder for the cable.

Suppose one or more of the data terminals was an asynchronous device, or more than four terminals required communications support. For such situations an optical multiplexer should be considered. One optical multiplexer currently marketed costs $3300 and supports up to eight data channels at data rates up to 64 kbps. Since an optical receiver/transmitter is included in the multiplexer, only the cost of one duplex cable must be added to the cost of a pair of multiplexers. Doing so, we will obtain the cost of an optical transmission system capable of supporting up to eight data sources at data rates up to 64 kbps. For such a system the cost of ten-thousand feet of cable and the two optical multiplexers would be $14,400.

In Table 6.11 the reader will find a comparison of the four previously discussed network configurations. As indicated, both cost and expansion

Table 6.11 Network configuration comparisons of four data sources and 10,000-foot transmission distance.

	Individual cable	Multichannel cable	Optical multiport modem	Optical multiplexer
System cost	$35,200	$35,200	$9600	$14,400
Expansion capability	Extra cable and transmitter/ receiver per data source	Requires more expensive cable and additional transmitters/ receivers	Cannot support more than four channels	No cost to add support for up to four additional channels

capacity varies widely between configurations. Although fiber optic systems represent new technology, the financial aspects and expansion capability must be considered similar to the process involved with conventional metallic cable based transmission systems.

REVIEW QUESTIONS

6.1 What are some of the limitations of security access implemented through passwords?

6.2 What is the effect of reserving characters from a character set for additional system or communication functions upon access security implemented via passwords?

6.3 Assuming you required a measure of privacy in transmitting a message discuss two techniques you could implement using a personal computer to scramble your message.

6.4 What is the effect of encoding both your header and text message for transmission on a Telex or TWX network?

6.5 If the data to be transmitted is 1 0 1 1 0 0 1 0 and the encoding key has the composition 0 1 1 0 1 0 1 0, what is the resulting encoded transmission if modulo 2 addition is used to develop the encoded text?

6.6 If the received data consists of the binary string 1 1 0 0 1 0 0 1 and your key is 0 1 0 1 0 1 1 0, what is the original encoded data if modulo 2 addition was used to form the encoded binary string?

6.7 What precaution should be taken when connecting encryption devices to a multiplexer?

6.8 What common features are incorporated into most security modems? What is a potential economic problem resulting from one of these features?

6.9 Discuss the use of a security switch in a data communications network.

6.10 What is the primary use for speed and code converters?

6.11 What are the different levels of operational conversion that may be performed by protocol converters and how do they compare to conversion performed by speed and code converters?

6.12 Draw a network schematic diagram illustrating how you could connect an asynchronous terminal to a multiport modem.

6.13 Why is the market for data access arrangements a contracting market?

6.14 What is the difference between a speech plus data unit and a voice digitizer? How could you use each device in a network?

6.15 Draw a schematic diagram illustrating how you could interconnect the switchboards at two offices using voice digitizers. Assume you will use 4-port multiport modems operating at 9600 bps to obtain three simultaneous voice channels on one leased line connecting the two offices.

CHAPTER 7

INTEGRATING COMPONENTS

As shown from the examination of components presented in this book, numerous factors govern their employment. Even after such devices are selected and installed, a review of the network and its components should be accomplished on a periodic basis.

In addition to changes in user requirements, new communications tariffs must be examined to determine their impact on the network. While at one particular time the utilization of specialized components may not be economical, a revision in a tariff may warrant the use of such components at a later date. Similarly, a requirement to support additional time-sharing users on a computer system where all ports are used would warrant the investigation of using such devices as modem-sharing units, line-sharing units, or port selectors, as an example.

While the combinations of communications components that can be used present an almost unlimited number of possibilities, careful analysis of user requirements and tariff structures will enable the system designer to reduce the number of such devices for consideration. Although this chapter is primarily concerned with integrating inverse multiplexers and multiport modems, the systems designer can similarly integrate other components to respond to user requirements in a cost-effective manner.

7.1 INVERSE MULTIPLEXING WITH MULTIPORT MODEMS

With a little imagination, the user of a small- to medium-size data communications system can fit several data streams into a voice-grade line, even if some of the signals already exceed the 19,200-bps range limit of voice-grade facilities. Thus the user can avoid the high cost of wideband service while enjoying similar advantages. The trick is to combine an inverse time division multiplexer (TDM) with a multiport modem for remultiplexing data.

Simply put, the combination of multiport modem and inverse TDM is a means of splitting high-speed data streams into two lines and multiplexing them with other data streams to circumvent the limits of voice-grade lines. At the receiving end, another multiport modem/inverse multiplier system reunites the various signal components and routes them to their destinations. An inverse TDM provides the advantage of wideband transmissions to any point where two voice-grade lines are available, at a fraction of the wideband cost (Chapter 3, Section 3.3) when the distances between transmission points exceed a range from 80 to 100 miles. In addition, an inverse multiplexer permits a user to reduce the number of computer ports required for data transmission, since it accepts data from a single computer port at speeds up to 38,400 bps for a two-subchannel device and 115,200 bps for a six-subchannel device.

An inverse multiplexer splits an incoming data stream into two or more paths for transmission over two to six voice-grade lines and reassembles the data stream at the original composite speed at the receiving end, as shown in Figure 7.1.

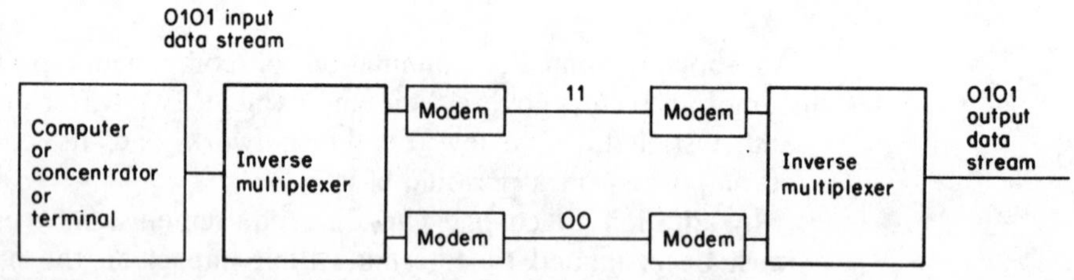

Figure 7.1 Inverse multiplexing. Basic inverse time division multiplexing splits data streams into odd and even bit streams at the transmitting station and recombines them at the receiving station.

The advantages obtainable with multiport modems were examined in Chapter 2, Section 2.5. Basically, a multiport modem consists of a high-speed synchronous modem with a built-in, limited-function TDM. Although multiport modem capabilities vary by manufacturer as well as by the modem's aggregate speed, all are similar in that they can service two to six synchronous data streams, multiplexing these streams onto a single line, as shown in Figure 7.2.

For many network applications, the system designer can take advantage of both the inverse multiplexer and the multiport modem to configure a network.

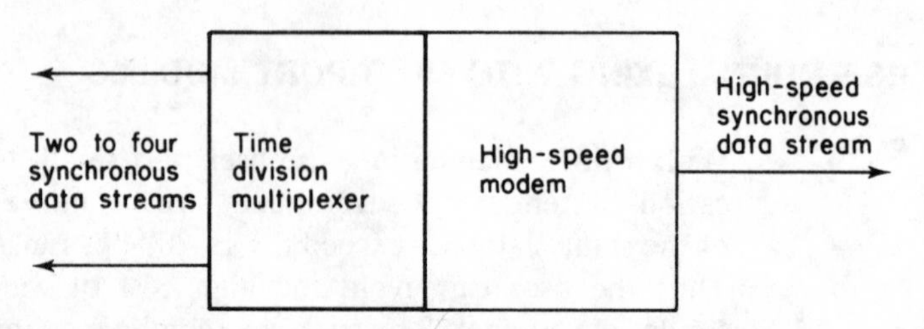

Figure 7.2 Multiport modem. Multiport modems have the ability to multiplex from two to six synchronous data streams into a single line for transmission over facilities.

The combination of both devices can provide flexibility and an economy of operation that may not be available when using more conventional communications devices, and the system designer can take the advantage of both the inverse multiplexer and the multiport modem to configure a network. In certain cases, a mixture of these devices can be used to reduce computer loading and still maintain cost-effective line utilization. At the very least, the remultiplexing configurations made possible by a combination of these devices ought to be looked at by the system designer as a possible alternative solution to his or her network design problem. A typical data communications application where inverse multiplexers and multiport modem components could be used is shown in Figure 7.3.

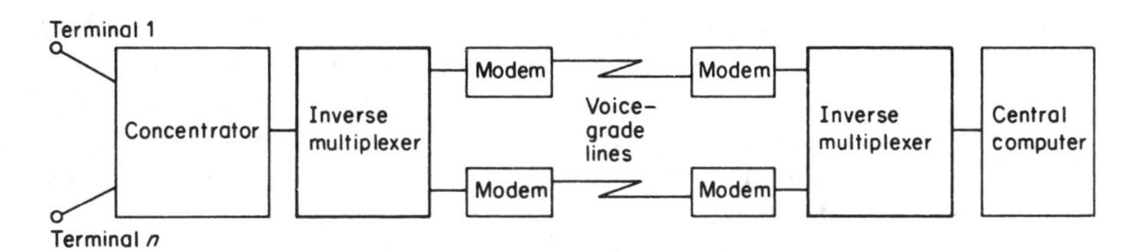

Figure 7.3 Typical remultiplexing operation. The combination of multiport modems with inverse multiplexers (remultiplexing) offers the user wideband services over two or more voice-grade lines.

Here, if the concentrator services 64 low-speed teletypewriter terminals transmitting at 300 bps, by allowing for overhead, the concentrator-to-computer transmission speed could become 19,200 bps. This normally exceeds the capacity of many single 3002-type voice-grade lines, so that if the user wishes to transmit from a single port of the data concentrator to a single port of the central-site computer, he or she is faced with installing wideband facilities. Alternately, the user may install two of the telephone company's voice-grade circuits and a pair of inverse multiplexers to split the transmitted data over the two lines. When inverse multiplexers are used, the breakeven point is reached after the transmission distance exceeds a range between 80 and 100 miles or so, based upon the leased line charges and the rental of four modems and two inverse multiplexers.

Post-installation economy

If such an installation must be expanded or modified, use of conventional equipment and methods could be costly. One example could be a growing demand that the concentrator serve additional terminals at the remote location. If the concentrator's CPU utilization factor is high, a second concentrator may have to be installed at the remote site at a cost of several thousand dollars per month. In addition, the installation of another line may be required between the remote location and the central site to provide a communications path for the second concentrator. When terminals must be added at a remote site, a configuration such as that shown in Figure 7.4 may be more appropriate. The solution would depend on such factors as the types of terminals already serviced by the concentrator, the types of remote terminals to be installed, the utilization factor of the existing lines, and the central-site computer's utilization factor.

For this type of configuration, the limiting factor is the remote concentrator utilization, rather than line utilization between the concentrator and the central computer. Thus, although additional terminals can be handled by existing lines, the installation of multiport modems makes it unnecessary to add a more expensive second concentrator. Installing multiport modems is a better solution than adding multidrop lines or installing multiport modems that only service additional terminals.

Through the installation of 14,400-bps multiport modems, rather than conventional ones, two new 4800-bps channels can be connected through modem-sharing units to serve as many as 24 additional polled terminals, which can be serviced at the remote location over the two type-3002 lines. In addition, line utilization is increased because the speed of each has been increased from 9600 bps to 14,400 bps, and the two lines provide a communications path for the 64 terminals shown in Figure 7.3, as well as the two dozen 4800-bps polled terminals that are shown in Figure 7.4.

If only 12 additional terminals are needed at the remote site, yet the concentrator is reaching its saturation point in servicing only 64 low-speed terminals, the user could also remove 12 terminals from the concentrator, upgrade these to 4800-bps cathode ray tubes (CRTs), and have them serviced through a second modem-sharing unit. In this way, the user could provide service capability to the 24 polled terminals by installing two additional ports at the central computer and by interfacing the output of the multiport modem's 4800-bps channels to these new computer ports.

Cost consideration

The configuration shown in Figure 7.4 entails the addition of two modem-sharing units at a typical lease price of $25 per month per unit and two synchronous computer ports at approximately $40 per month per port, plus the costs of upgrading the four 9600-bps modems to four 14,400-bps multiport modems which may add approximately $100 per month per modem. These costs should be compared to (1) the cost of installing two multidrop lines at the appropriate line-rate schedule, (2) adding four 4800-bps modems at $70 per modem per month, (3) adding two synchronous computer ports at $40 per port per month plus the two modem-sharing units, or (4) directly connecting the 24 polled terminals to eight 14,400-bps multiport modems.

To implement the latter choice, the user would need 16 multiport modems (eight at each end) at $320 per month per modem, as well as eight leased lines. At the main computer site, the user would also need 24 synchronous computer ports. The two alternate configurations that can be used to service the 24 4800-bps polled terminals are shown in Figure 7.5. Table 7.1 compares the monthly cost for all three configurations necessary to service systems

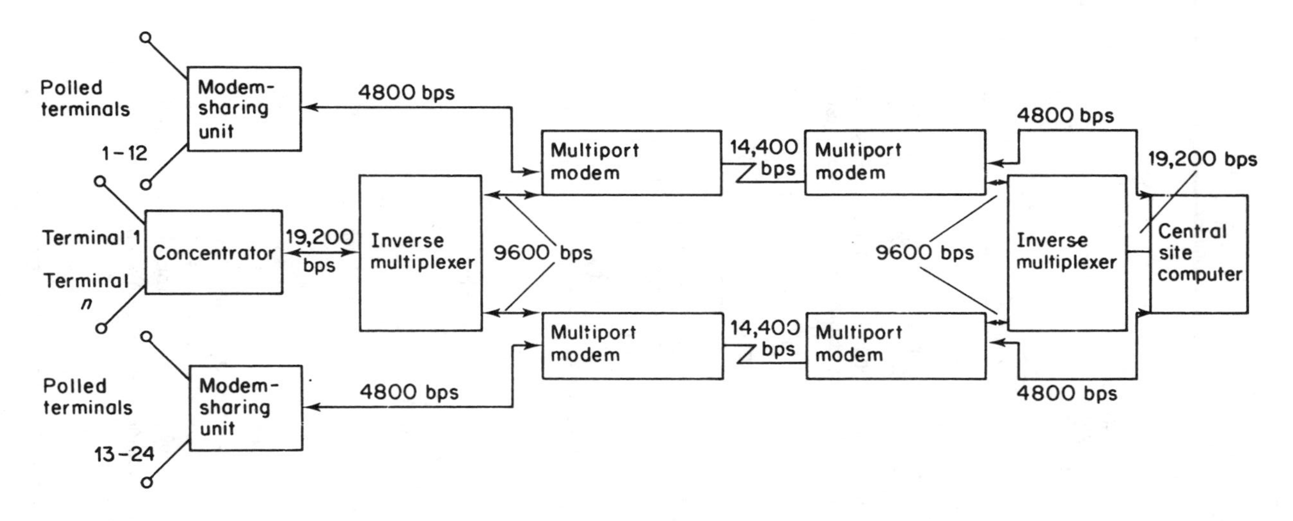

Figure 7.4 Remultiplexing with multiport modems. This example of an alternate remultiplexing configuration allows the servicing of as many as 12 polled terminals by means of just two modem-sharing units.

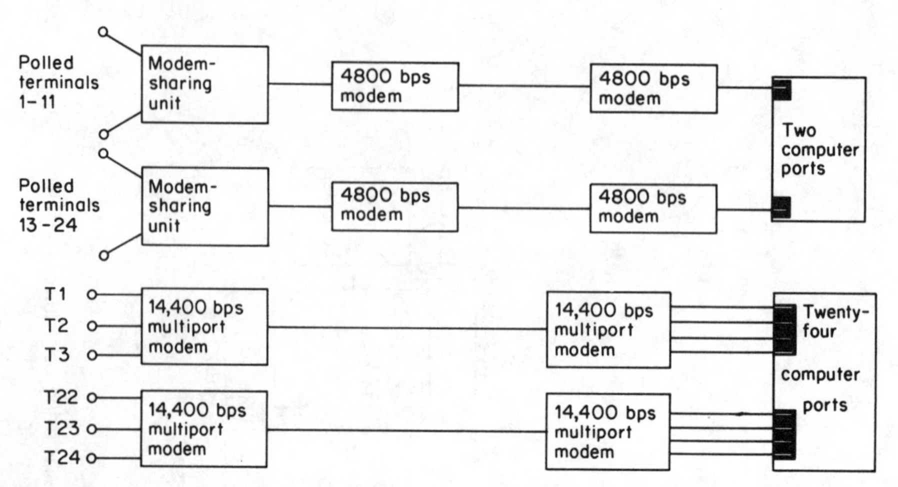

Figure 7.5 Alternate service methods. Top: Serviced by multidrop lines. Bottom: Serviced by multiport modems. Alternatives to remultiplexing shown here involve the expense of either two computer ports and two modem-sharing units or 24 ports and eight multiport modems. Both require extra leased lines.

Table 7.1 Additional costs for servicing 12 2400-bps terminals.

Reflexing with multiport modems (Figure 7.4)		
1.	Upgrade 4 9600-bps modems to 4 14,400-bps multiport modems $100/mo. × 4 modems	$ 400.00
2.	Add 2 modem-sharing units at $25/mo.	50.00
3.	Add 2 synchronous computer ports at $40/mo.	80.00
		$ 530.00
Servicing by multidrop lines (Figure 7.5(a))		
1.	Add 4 4800-bps modems at $70/mo.	$ 280.00
2.	Add 2 modem-sharing units at $25/mo.	50.00
3.	Add 2 synchronous computer ports at $40/mo.	80.00
4.	Add 2 leased lines, from Washington D.C. to Macon, Ga. (560 miles) at $800	1600.00
		$2010.00
Servicing by multiport modems only (Figure 7.5(b))		
1.	Add 16 14,400-bps modems at 320/mo.	$5120.00
2.	Add 24 synchronous computer ports at $40/mo.	960.00
3.	Add 3 leased lines from Washington D.C. to Macon, Ga. (560 miles) at $800	2400.00
		$8480.00

involving the use of 24 extra 4800-bps polled terminals. As shown, replexing with multiport modems can effectively support a network upgrade at up to one-fourth to one-tenth the cost of the two alternative methods considered in the table.

Remote terminal service

A variation of the configuration shown in Figure 7.4 entails the installation of multiport modems with a data communications equipment (DCE) option at the remote site. When additional terminals are required at locations distant from the primary remote site, through the use of the DCE option, one can interface two low-speed modems to the multiport modems, allowing distant interactive terminals at a secondary remote site to be added to the system as shown in Figure 7.6.

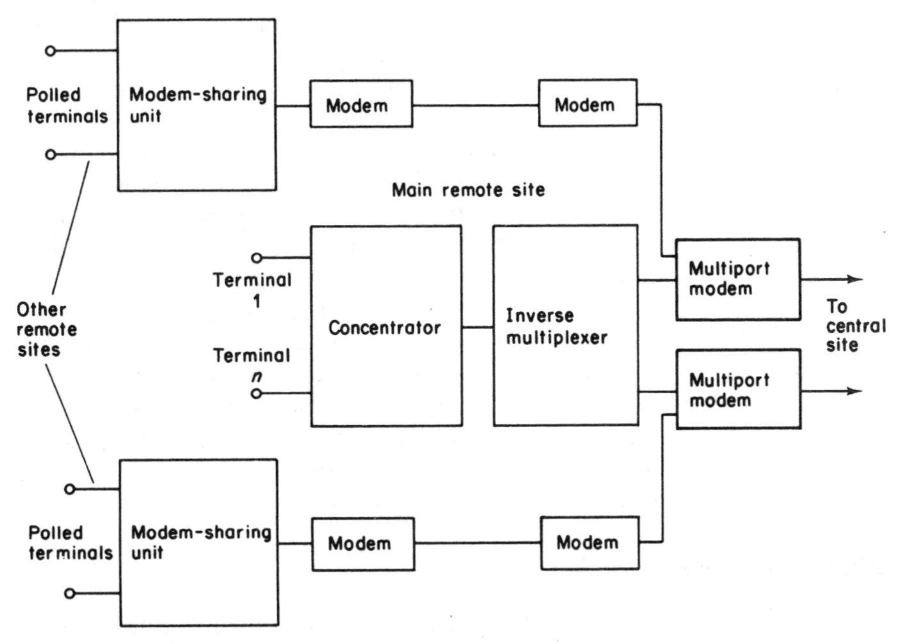

Figure 7.6 Multiport modems with DCE option. Remultiplexing with the addition of a data communications equipment (DCE) option, employing a separate, low-speed modem interface, allows the servicing of terminals distant from the main remote site.

Another interesting configuration could be developed through the installation of TDMs, multiport modems, and a pair of inverse multiplexers. For the configuration shown in Figure 7.7, the basic CPU-to-CPU wideband transmission is accomplished at 19,200-bps over two voice-grade channels by the utilization of inverse multiplexers. Up to eight 1200-bps interactive terminals can be serviced at the remote CPU location by installing two four-channel multiplexers. By transmitting the output of the multiplexers into 4800-bps ports of the multiport modems, no additional lines would be required between the remote site and the central computer. Again, line utilization would be increased.

If the conventional TDMs were replaced by statistical or intelligent multiplexers additional line utilization increases become possible, resulting in a corresponding increase in the number of terminals that could be supported without requiring the installation of additional lines. As an example, if the statistical multiplexers had a service ratio of 3 for asynchronous data traffic, then each multiplexer could support 12 1200-bps terminals and operate at 4800 bps into the multiport modem shown in Figure 7.7. This would increase the total number of remote terminals capable of being supported to 24 without requiring any additional leased lines.

Flexibility equals savings

As shown in the preceding examples, the use of multiport modems and remultiplexing data permits the system designer to consider numerous configurations to satisfy operational requirements. Since the objective in using an inverse multiplexer is to obtain wideband transmission speeds, only two 4800-bps ports of all the various multiport modem combinations available were considered. This configuration permits a remote concentrator or remote computer to achieve 19,200-bps transmission and also allows the remaining capacity of each of the two lines to be shared by different devices through different configurations. Although the transmission speed of the remote concentrator or remote CPU can be further reduced (with a resulting increase in the remaining transmission capacity through the multiport modems), this increase would not normally be attempted below 14,400 bps because it defeats the purpose of using inverse multiplexers. For those cases where the highest channel is 14,400 bps, the inverse multiplexer can be removed and the configuration developed by considering the addition of various communications components to multiport modems. For those who have installations that require wideband service capabilities only at certain times, the utilization of up to four ports on each multiport modem may be considered during those periods when the service of the inverse multiplexer is not required.

Network integration problems

The following problems are presented as illustrative examples of networking requirements that can be solved by the appropriate selection and integration of data communications devices.

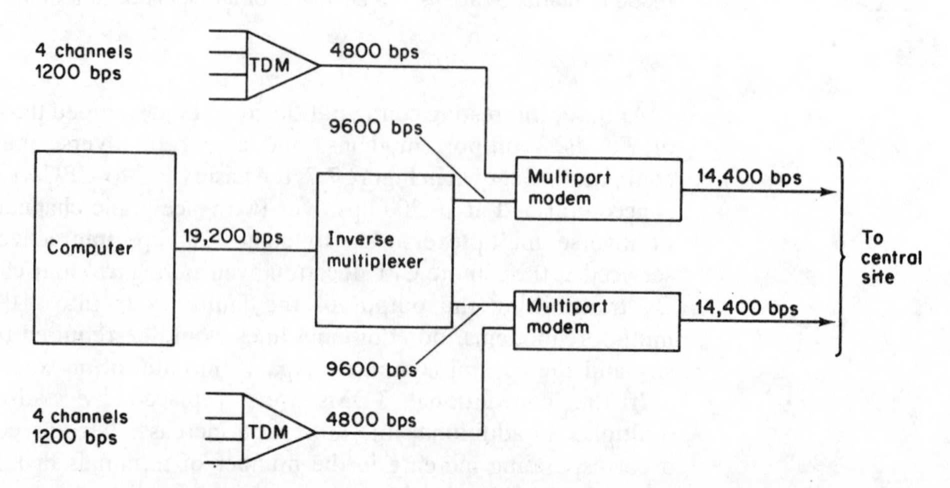

Figure 7.7 Adding TDMs to a remultiplexing configuration. Two 4-channel TDMs added to the multiport modems permit access by 8 low-speed terminals at no increase in line costs.

Network integration I

Assume eight terminals are to be installed in a terminal room 800 feet from the computer. Suppose cable costs $1 per foot and data pulses become distorted at a distance of 600 feet to the point where serious transmission errors occur. If you wish to connect the terminals to the computer, what configurations could be used and which one would you recommend based upon the following.

(a) Terminals operate at 2400 bps.
(b) 4-port multiport modems operating at 9600 bps cost $500.
(c) Synchronous line drivers operating at a selectable data rate from 2400 bps to 19,200 bps cost $100.
(d) 8-channel TDMs cost $1000 and the resulting composite high-speed output at 19,200 bps distorts after 450 feet.

Network integration II

The V.S. Cracker Company operates a large-scale computer system at its corporate headquarters in Greensboro, N.C. Until recently, all corporate accounts were in New England and as a result of this the company's network consisted of a number of leased lines from Greensboro to several branch offices in the northeast portion of the United States.

Owing to a high demand for crackers on the West Coast, a branch office is scheduled to open in San Francisco. Based upon a recent feasibility study, the San Francisco office will require the installation of one remote batch terminal (RBT) operating at 4800 bps and ten 1200-bps interactive teletype-compatible terminals. The prime time activity factor of the RBT is expected to be unity, meaning that the device will be connected to the system continuously during the prime shift although it may not necessarily be transmitting or receiving data during that period of time. For the interactive terminals, during prime time an average of four terminals are expected to be connected to the system and for 90 percent of the time six or less terminals will be connected to the system.

At San Francisco all interactive terminals will be located on the same floor of one building, geographically distributed into several work areas. A room is reserved for the RBT and any required communications equipment and the average distance from any interactive terminal to the inside of the RBT room is approximately 275 feet. In discussing terminal operations with San Francisco personnel, it was ascertained that all terminals will be in fixed locations. Since terminal operators also serve as order entry clerks, separate telephones must be installed for each terminal to insure that no customer encounters a period of busy signals when a terminal is in use.

At the central computer site, 30 slots are available on the front-end processor for network expansion. Each slot can service a dual capacity asynchronous/synchronous channel module which is available from the computer manufacturer.

Requirement Design an equipment configuration to service the terminals to be installed in San Francisco so that the remote and central-site costs are minimized. For the interactive terminals, assume that the servicing of four such terminals would result in a loss of productivity of $400 per month, servicing 6 terminals would not result in any measurable loss of productivity and the servicing of 10 terminals might actually result in a gain to the company of $100 per month. The following equipment should be considered at the denoted monthly lease rates:

Equipment	Monthly cost
Acoustic coupler	$ 30
Auto answer modem 1200 bps	40
Telco rotary	30
Telephone	15
Dial-in line	15
Line driver	10
TDM (4-channel)	90
TDM (8-channel)	120
Statistical TDM (12-channel)	275
9600-bps modem	200
9600-bps multiport modem	220
Front-end processor channel	35
Cable per foot	0.10
Leased line San Francisco – Greensboro	1400

Network integration III

Your organization operates a large-scale, multidimensional (batch, remote batch and time-sharing) computer system in City A. Currently, two dual functions (asynchronous/synchronous) ports are unused and available on your existing front-end processor. An additional front-end processor can be obtained with a minimum of 16 dual function ports and leases for $1500 per month.

Expansion requirements Management has decided to open a branch office in City B and a customer inquiry office in City C. At the branch office five CRTs and one remote batch terminal are required. At the office in City B, one additional CRT is required. If each terminal is connected to an individual port, asynchronous 1200-bps CRTs can be used. If CRTs are clustered, 4800-bps asynchronous devices must be obtained. The remote batch terminal must operate at 4800 bps.

Communication cost Voice-grade leased line costs per month are as follows:

From	To	Monthly cost
City B	City A	$200
City C	City A	$450
City C	City B	$250

Equipment cost The following equipment should be considered at the indicated monthly cost:

Equipment	Monthly cost
4800-bps modem	$ 120
9600-bps modem	200
9600-bps multiport modem	220
8-channel TDM	160
Modem-sharing unit	35
1200-bps terminal	100
4800-bps terminal	125
Computer port	35
Front-end processor	1500
Remote batch terminal	500
1200-bps modem	40

Assuming that existing facilities cannot be modified, what network configurations should be considered and what are the economic and operational implications of those configurations?

APPENDIX A

SIZING DATA COMMUNICATIONS NETWORKS COMPONENTS

A.1 DEVICE SIZING

Of many problems associated with the acquisition of data communications network components, one item often requiring resolution is the configuration or sizing of the device. The process of insuring that the configuration of the selected device will provide a desired level of service is the foundation upon which the availability level of a network is built.

The failure to provide a level access acceptable to network users can result in a multitude of problems. First, a user encountering a busy signal might become discouraged, take a break or do something other than redial a telephone number of a network access port. Such action will obviously result in a loss of user productivity. If the network usage is in response to customer inquiries, a failure to certify a customer purchase, return, reservation or other action in a timely manner could result in the loss of that customer to a competitor. With a little imagination, it becomes easy to visualize that the lifeline of the modern organization is its data communications network. An unacceptable level of access to the network can be considered akin to a blockage in the human circulatory system – harm will result and additional analysis and testing may become necessary to alleviate the problem.

Sizing problem similarities

There are many devices that can be employed in a data communications network whose configuration or sizing problems are similar. Examples of such devices include the number of dial-in lines required to be connected to a telephone company rotary and the number of channels on such devices as multiplexers, data concentrators, and port selectors.

Basically, two methods can be used to configure the size of communications network devices. The first method, commonly known as experimental modeling, involves the selection of the device configuration based upon a mixture of previous experience and gut intuition. Normally, the configuration selected is less than the base capacity plus expansion capacity of the device. This enables the size of the device to be adjusted or upgraded without a major equipment modification if the initial sizing proves inaccurate. An example of experimental modeling is shown in Figure A.1.

(a) Initial configuration

Channel adapters

Power supply | Central logic | 1 2 | 3 4 | 5 6 | 7 8 | 9 10 | Empty slots

(b) Adjusted configuration

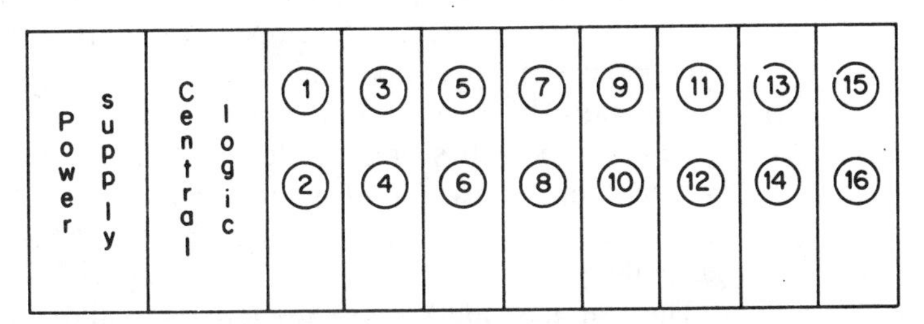

Figure A.1 Experimental modeling. Experimental modeling results in the adjustment of a network configuration based upon previous experience and gut intuition.

A rack-mounted time division multiplexer is shown in Figure A.1(a). Initially, the multiplexer was obtained with five dual channel adapters to support ten channels of simultaneous operation. Since the base unit can support eight dual channel adapters, if the network manager's previous experience or gut intuition proves wrong, the multiplexer can be upgraded easily. This is shown in Figure A.1(b). Here, the addition of three dual channel adapters permits the multiplexer to support sixteen channels in its adjusted configuration.

The second method which can be employed to size network components ignores experience and intuition. This method is based upon a knowledge of data traffic and the scientific application of mathematical formulas to traffic data, hence, it is known as the scientific approach or method of equipment sizing. Although some of the mathematics involved in determining equipment sizing can become quite complex, a series of tables can be employed to reduce many sizing problems to one of a simple table look-up process.

Although there are advantages and disadvantages to each method, the application of a scientific methodology to equipment sizing is a rigorously defined approach. This should provide a much higher degree of confidence and accuracy of the configuration selected when this method is used. On the negative side, the use of a scientific method requires a firm knowledge or an accurate estimate of the data traffic. Unfortunately, for some organizations, this may be difficult to obtain. In many cases a combination of two techniques will provide an optimum situation. For such situations sizing can be conducted using the scientific method with the understanding that the configuration selected may require adjustment under the experimental modeling concept. In the remainder of this appendix, we will focus our attention upon the application of the scientific methodology to equipment sizing problems.

Telephone terminology relationships

Most of the mathematics used for the sizing of data communications equipment evolved from work originally developed from the sizing problems of telephone networks. From a discussion of a few basic telephone network terms and concepts we will see the similarities between the sizing problems associated with data communications components and the structure of the telephone network. Building upon this foundation, we will learn how to apply the mathematical formulas developed for telephone network sizing to data communications network component configurations.

To study the relationships between the telephone network and communications component sizing problems let us examine a portion of the telephone network and study the structure and calling problems of a small segment formed by two cities, assuming that each city contains 1000 telephone subscribers.

The standard method of providing an interconnection between subscribers in a local area is to connect each subscriber's telephone to what is known as the local telephone company exchange. Other synonymous terms for the local telephone company exchange include the 'local exchange' and 'telephone company central office'. Whether the term central office or exchange is used, the subscriber's call is switched to the called party number at that location. If we assume each city has only one local exchange, then all calls originating in that city and with a destination located within that city will be routed through that exchange.

Since our network segment selected for analysis consists of two cities, we will have two telephone company exchanges, one located in each city. To provide a path between cities for intercity calling, a number of lines must be installed to link the exchanges in each city. The exchange in each city can then act as a switch, routing the local subscribers in each city desiring to contact parties in the other city.

As shown in the upper part of Figure A.2, a majority of the telephone traffic in the network segment consisting of the two cities will be among the subscribers of each city. Although there will be telephone traffic between the subscribers in each city, it will normally be considerably less than the amount of local traffic in each city. The path between the two cities connecting their telephone central offices is known as a trunk. One of the problems in designing the telephone network is determining how many trunks should be installed between telephone company exchanges. A similar sizing problem occurs many times in each city at locations where private organizations desire to install switchboards. An example of the sizing problem with this type of equipment is illustrated in the lower portion of Figure A.2. In effect, the switchboard functions as a small telephone exchange, routing calls carried over a number of trunks installed between the switchboard and the telephone company exchange to a larger number of the subscriber lines connected to the switchboard. The determination of the number of trunks required to be installed between the telephone exchange and the switchboard is called dimensioning and is critical for the efficient operation of the facility. If insufficient trunks are available, company personnel will encounter an unacceptable number of busy signals when trying to place an outside telephone call. Once again, this will obviously affect productivity.

Returning to the intercity calling problem, consider some of the problems one faces in dimensioning the number of trunks between the two cities. Let us assume that, based upon a previously conducted study, it was determined that no more than 50 people would want to have simultaneous telephone conversations where the calling party was in one city and the called party in the other city. If 50 trunks were installed between cities and the number of intercity callers never exceeded 50, at any moment the probability of a

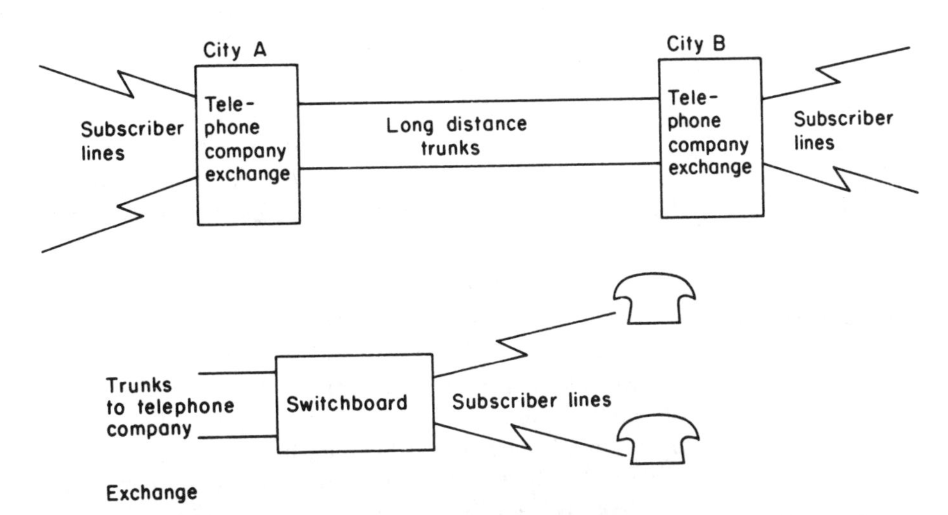

Figure A.2 Telephone traffic sizing problems. Although most subscriber calls are routed locally through the local telephone company exchange or local switchboard to parties in the immediate area, some calls require access to trunks. The determination of the number of trunks required to provide an acceptable grade of service is known as line dimensioning and is critical for the effective operation of the facility.

subscriber completing a call to the distant city would always be unity, guaranteeing success. Although the service cost of providing 50 trunks is obviously more than providing a lesser number of trunks, no subscriber would encounter a busy signal.

Since some subscribers might postpone or choose not to place a long-distance call at a later time if a busy signal is encountered, a maximum level of service will produce a minimum level of lost revenue. If more than 50 subscribers tried to simultaneously call parties in the opposite city, some callers would encounter busy signals once all 50 trunks were in use. Under such circumstances, the level of service would be such that not all subscribers are guaranteed access to the long-distance trunks and the probability of making a long-distance call would be less than unity. Likewise, since the level of service is less than that required to provide all callers with access to the long-distance trunks, the service cost is less than the service cost associated with providing users with a probability of unity in accessing trunks. Similarly, as the probability of successfully accessing the long-distance trunks decrease, the amount of lost revenue or customer waiting costs will increase. Based upon the preceding, a decision model factoring into consideration the level of service versus expected cost can be constructed as shown in Figure A.3.

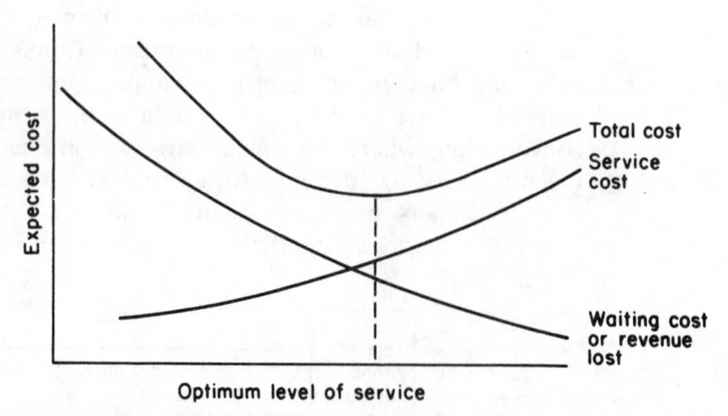

Figure A.3 Using a decision model to determine the optimum level of service. Where the total cost is minimal represents the optimum level of service one should provide.

The decision model

For the decision model illustrated in Figure A.3, suppose the optimum number of trunks required to link the two cities is 40. The subscriber line to trunk ratio for this case would be 1000 lines to 40 trunks, for a 25:1 ratio.

To correctly dimension the optimum number of trunks linking the two cities requires both an understanding of economics as well as subscriber traffic. In dimensioning the number of trunks, a certain tradeoff will result that relates the number of trunks or level of service to the cost of providing that service and the revenue lost by not having enough trunks to satisfy the condition when a maximum number of subscribers dial subscribers in the other city. To determine the appropriate level of service, a decision model as illustrated in Figure A.3 is required. Here, the probability of a subscriber successfully accessing a trunk corresponds to the level of service provided. As more trunks are added, the probability of access increases as well as the cost of providing such access. Correspondingly, the waiting cost of the subscriber or the revenue loss to the telephone company decreases as the level of service increases. Where the total cost representing the combination of service cost and waiting cost is minimal represents the optimal number of trunks or level of service that should be provided to link the two cities.

Decision models can be easily adapted to data communications problems. As an example, consider a network designed to provide access to computational facilities for time-sharing customers. Here lost revenues could be equated to lost business at some fixed hourly rate while the service cost can be computed by determining the cost of extra equipment required to increase the level of service for customers.

Traffic measurements

Telephone activity can be defined by the calling rate and the holding time (duration of the call). The calling rate is the number of times a particular route or path is used per unit time period while the holding time is the duration of the call on the route or path. Two other terms that warrant attention are the offered traffic and the carried traffic. The offered traffic is the volume of traffic routed to a particular telephone exchange during a predetermined time period while the carried traffic is the volume of traffic actually transmitted through the exchange to its destination during a predetermined period of time.

The key factor required to dimension a traffic path is knowledge of the traffic intensity during the time period known as the busy hour. Although traffic varies by day and time of day, traffic is generally random but follows a certain consistency one can identify. In general, traffic peaks prior to lunchtime and then rebuilds to a second daily peak in the afternoon. The busiest period of the day is known as the busy hour (BH). It is the busy hour traffic level which is employed in dimensioning telephone exchanges and transmission routes since one wants to size the exchange or route with respect to its busiest period.

Telephone traffic can be defined as the product of the calling rate per hour and the average holding time per call. This measurement can be expressed mathematically as:

$$T = C^*D$$

where C = calling rate per hour and D = average duration per call.

Using the above formula, traffic can be expressed in call-minutes (CM) or call-hours (CH) where a call-hour is the quantity represented by one or more calls having an aggregate duration of one hour.

If the calling rate during the busy hour of a particular day is 500 and the average duration of each call was 10 minutes, the traffic flow or intensity would be 500* 10 or 5000 CM which would be equivalent to 5000/60 or approximately 83.3 CH.

Erlangs and call-seconds

The preferred unit of measurement in telephone traffic analysis is the erlang, named after A.K. Erlang, who was a Danish mathematician. The erlang is a dimensionless unit in comparison to the previously discussed call-minutes and call-hours. It represents the occupancy of a circuit where one erlang of traffic intensity on one traffic circuit represents a continuous occupancy of that circuit.

A second term often used to represent traffic intensity is the call-second (CS). The quantity represented by one hundred call-seconds is known as 1 CCS. Here the first C represents the quantity 100 and comes from the French term cent. Assuming a one-hour unit interval, the previously discussed terms can be related to the erlang as follows:

$$1 \text{ erlang} = 60 \text{ call-minutes} = 36 \text{ CCS} = 3600 \text{ CS}$$

If a group of twenty trunks were measured and a call intensity of 10 erlangs determined over the group, then we would expect one-half of all trunks to be busy at the time of the measurement. Similarly, a traffic intensity of 600 CM or 360 CCS offered to the 20 trunks would warrant the same conclusion. A traffic conversion table is located in Table A.1 which will facilitate the conversion of erlangs to CCS and vice versa. Since the use of many dimensioning tables are based upon traffic intensity in erlangs or CCS, the conversion of such terms is frequently required in the process of sizing facilities.

Grade of service

One important concept in the dimensioning process is what is known as the grade of service. To understand this concept, let us return to our intercity calling example illustrated in example A.1, again assuming 50 trunks are used to connect the telephone exchanges in each city. If a subscriber attempts to originate a call from one city to the other when all trunks are in use, that call is said to be blocked. Based upon mathematical formulas, the probability of a call being blocked can be computed given the traffic intensity and number of available trunks. The concept of determining the probability of a call being blocked can be computed given the traffic intensity and number of available trunks. The concept of determining the probability of blockage can be easily adapted to the sizing of data communications equipment.

From a logical analysis of traffic intensity, it follows that if a call will be blocked, such blockage will occur during the busy hour since that is the period when the largest amount of activity occurs. Thus, telephone exchange capacity is engineered to service a portion of the busy hour traffic, the exact amount of service being dependent upon economics as well as the political process of determining the level of service one desires to provide to customers. One could over-dimension the route between cities and provide a trunk for every subscriber. This would ensure that a lost call could never occur and would be equivalent to connecting every terminal in a network directly to a front-end processor port. Since a 1:1 subscriber to trunk ratio is uneconomical and will result in most trunks being idle a large portion of the day, we can expect a

Table A.1 Traffic conversion table.

Dimension		Erlangs (intensity)	
Minutes	Hours	Call-hours (quantity)	CCS (quantity)
12	0.2	0.2	6
24	0.4	0.4	12
36	0.6	0.6	18
48	0.8	0.8	24
60	1.0	1.0	36
120	2.0	2.0	72
180	3.0	3.0	108
240	4.0	4.0	144
300	5.0	5.0	180
360	6.0	6.0	210
420	7.0	7.0	252
480	8.0	8.0	288
540	9.0	9.0	324
600	10.0	10.0	360
900	15.0	15.0	540
1200	20.0	20.0	720
1500	25.0	25.0	900
1800	30.0	30.0	1080
2100	35.0	35.0	1260
2400	40.0	40.0	1440
2700	45.0	45.0	1620
3000	50.0	50.0	1800
6000	100.0	100.0	3600

lesser number of trunks between cities than subscribers. As the number of trunks decreases and the subscriber to trunk ratio correspondingly increases, we can intuitively expect some sizings to result in some call blockage. We can specify the number of calls we are willing to have blocked during the busy hour. This specification is known as the grade of service and represents the probability (P) of having a call blocked. If we specify a grade of service of 0.05 between the cities, we require a sufficient number of trunks so that only one call in every twenty or five calls in every one hundred will be blocked during the busy hour.

Route dimensioning parameters

To determine the number of trunks required to service a particular route one can consider the use of several formulas. Each formula's utilization depends upon the call arrival and holding time distribution, the number of traffic sources, and the handling of lost or blocked calls. Regardless of the formula employed, the resulting computation will provide one with the probability of call blockage or grade of service based upon a given number of trunks and level of traffic intensity.

Concerning the number of traffic sources, one can consider the calling population as infinite or finite. If calls occur from a large subscriber population and subscribers tend to redial if blockage is encountered, the calling population can be considered as infinite. The consideration of an infinite traffic source results in the probability of call arrival becoming constant and does not make the call dependent upon the state of traffic in the system. The two most commonly employed traffic dimensioning equations are both based upon an infinite calling population.

Concerning the handling of lost calls, such calls can be considered cleared, delayed, or held. When such calls are considered held, it is assumed that the telephone subscriber immediately redials the desired party upon encountering a busy signal. The lost call delayed concept assumes each subscriber is placed into a waiting mechanism for service and forms the basis for queuing analysis. Since we can assume a service or non-service condition, we can disregard the lost call delayed concept.

Traffic dimensioning formulas

The principal traffic dimensioning formula used in North America is based upon the lost call concept and is commonly known as the Poisson formula. In Europe, traffic formulas are based upon the assumption that a subscriber upon encountering a busy signal will hang up the telephone and wait a certain amount of time prior to redialing. The Erlang formula is based upon this lost call cleared concept.

A.2 THE ERLANG TRAFFIC FORMULA

The most commonly used telephone traffic dimensioning equation is the Erlang B formula. This formula is predominantly used outside the North American continent. In addition to assuming that data traffic originates from an infinite number of sources, this formula is based upon the lost calls cleared concept. This assumption is equivalent to stating that traffic offered to but not carried by one or more trunks vanishes and this is the key difference between this formula and the Poisson formula. The latter formula assumes that lost calls are held and it is used for telephone dimensioning mainly in North America. Since data communications system users can be characterized by either the lost calls cleared or lost calls held concept, both traffic formulas and their application to data networks will be covered in this appendix.

If E is used to denote the traffic intensity in erlangs and T represents the number of trunks designed to support the traffic, the probability $P(T,E)$ represents the probability that T trunks are busy when a traffic intensity of E erlangs is offered to those trunks. The probability is equivalent to specifying a grade of service and can be expressed by the erlang traffic formula as follows:

Table A.2 Factorial values.

N	Factorial N
1	1
2	2
3	6
4	24
5	120
6	720
7	5040
8	40320
9	362880
10	3628800
11	3.99168E 07
12	4.79002E 08
13	6.22702E 09
14	3.71788E 10
15	1.30767E 12
16	2.09228E 13
17	3.55687E 14
18	6.40237E 16
19	1.21645E 17
20	2.48290E 18
21	5.10909E 19
22	1.12400E 21
23	2.58520E 22
24	6.20448E 23
25	1.55112E 25
26	4.03291E 26
27	1.03889E 28
28	3.04888E 29
29	8.84176E 30
30	2.65258E 32
31	8.22284E 33
32	2.63131E 35
33	8.68832E 36

$$P(T,E) = \frac{E^T/T!}{1 + E + (E^2/2!) + (E^3/3!) + \ldots + (E^T/T!)} = \frac{E^T/T!}{\sum_{m=0}^{T} (E^M/m!)}$$

where $T! = T^*(T-1)^*(T-2) \ldots 3^*2^*1$.

In Table A.2 a list of factorials and their values are presented to assist the reader in computing specific grades of service based upon a given traffic intensity and trunk quantity.

To assist in the computation of a range of grades of service based upon specific traffic loads and channel or trunk capacity, a BASIC computer language program was developed. The BASIC program is listed in Appendix B and can be easily modified to compute the probability of all channels busy when a call is attempted for different levels of traffic loads by changing line 40 of the program. Currently, line 40 in conjunction with line 50 varies the traffic intensity from 0.5 through 40 erlangs in increments of 0.5 erlangs.

Extracts from the execution of the BASIC program are listed in Table A.3. While the use of the Erlang formula is normally employed for telephone dimensioning, it can be easily adapted to sizing data communications equipment. As an example of the use of Table A.3, consider the following situation. Suppose one desires to provide customers with a grade of service of 0.1 when the specific traffic intensity is 7.5 erlangs. From Table A.3, 10 channels or trunks would be required since the use of the table requires one to interpolate and round to the highest channel. Thus, if it was desired to offer a 0.01 grade of service when the traffic intensity was 7 erlangs, one could read down the 7.0 erlang column and determine that between 13 and 14 channels are required. Since one cannot install a fraction of a trunk or channel, 14 channels would be required as we round to the highest channel number.

Multiplexer sizing

In applying the Erlang B formula to multiplexer sizing, an analogy can be made between telephone network trunks and multiplexer channel adapters. Let us assume that a survey of terminal users in a geographical area indicated that during the busy hour normally 10 terminals would be active. This would represent a traffic intensity of 6 erlangs. Suppose we wish to size the multiplexer

Table A.3 Erlang B distribution extracts. Probability all channels busy when call attempted.

	(Load in erlangs)				
Channels	5.500	6.000	6.500	7.000	7.500
1	.846154	.857143	.866667	.875000	.882353
2	.699422	.720000	.737991	.753846	.767918
3	.561840	.590164	.615234	.637546	.657510
4	.435835	.469565	.499939	.527345	.552138
5	.324059	.360400	.393910	.424719	.453016
6	.229022	.264922	.299099	.331330	.361541
7	.152503	.185055	.217365	.248871	.279209
8	.094897	.121876	.150100	.178822	.207455
9	.054814	.075145	.097803	.122101	.147397
10	.029265	.043142	.059772	.078741	.099544
11	.014422	.022991	.034115	.047717	.063557
12	.006566	.011365	.018144	.027081	.038206
13	.002770	.005218	.008990	.014373	.021566
14	.001087	.002231	.004157	.007135	.011421
15	.000398	.000892	.001798	.003319	.005678

to ensure that at most only 1 out of every 100 calls to the device encounters a busy signal. Then our desired grade of service becomes 0.01. From Table A.3, the 6 erlang column indicates that to obtain a 0.011365 grade of service would require 12 channels while a 0.005218 grade of service would result if the multiplexer had 13 channels. Based upon the preceding data, the multiplexer would be configured for 13 channels as illustrated in Figure A.4.

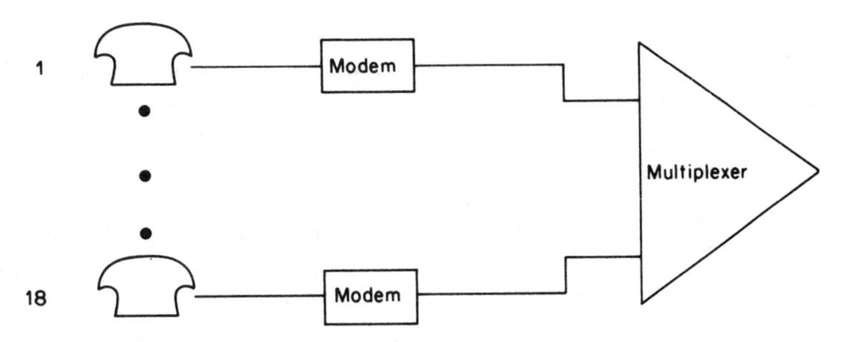

Figure A.4 Multiplexer channel sizing. Based upon a busy hour traffic intensity of 10 erlangs, 19 dial-in lines, modems and multiplexer ports would be required to provide a 0.01 grade of service.

From a practical consideration, the Erlang B formula assumption that lost calls are cleared and that traffic not carried vanishes can be interpreted as traffic overflowing one dial-in port being switched to the next port on the telephone company rotary as each dial-in port becomes busy. Thus, traffic overflowing dial-in port m is offered to port $m + 1$ and the traffic lost by the mth dial-in port, Lm, is the total traffic offered to the entire group of dial-in ports multiplied by the probability that all dial-in ports are busy. Thus:

$$Lm = E^{*}P(m,E)$$

where E = traffic intensity in erlangs and m = number of ports or channels.

For the first dial-in port, when m is one, the proportion of traffic blocked becomes:

$$P_i = \frac{E}{E + 1}$$

For the second dial-in port, the proportion of traffic lost by that port becomes:

$$P_2 = \frac{E^2/2!}{1 + E + (E^2/2!)}$$

In general, the proportion of traffic lost by the mth port can be expressed as:

$$P_m = \frac{E^m/m!}{\sum_{i=1}^{m}(E^i/i!)}$$

To reduce the complexity of calculation, let us analyze the data traffic carried by a group of 4 dial-in ports connected to a 4-channel multiplexer when a traffic intensity of 3 erlangs is offered to the group.

For the first dial-in port, the proportion of lost traffic becomes:

$$P_1 = \frac{3}{3+1} = 0.75$$

The proportion of lost traffic on the first port multiplied by the offered traffic provides the actual mount of lost traffic on port 1. Thus:

$$T_1 = P_1{*}E = \frac{E}{1+E}{*}E = \frac{E^2}{1+E} = 2.25 \text{ erlangs}$$

The total traffic carried on the first multiplexer port is the difference between the total traffic offered to that port and the traffic that overflows or is lost to the first port. Denoting the traffic carried by the first port as C_1 and the amount offered as A, we obtain:

$$C_1 = A - T_1 = 3 - 2.25 = 0.75 \text{ erlangs}$$

Since we consider the rotary as a device that will pass traffic lost from port one to the remaining ports, we can compute the traffic lost by the second port in a similar manner. Substituting in the formula to determine the proportion of traffic lost, we obtain for the second port:

$$P_2 = \frac{E^1/2!}{1 + E + (E^2/2!)} = 0.5294$$

The amount of traffic lost by the second port, T_2, becomes:

$$T_2 = P_2{*}E = 0.5294{*}3 = 1.588 \text{ erlangs}$$

The traffic carried by the second port, C_2, is the difference between the traffic lost by the first port and the traffic lost by the second port, thus:

$$C_2 = T_1 - T_2 = 2.25 - 1.588 = 0.662 \text{ erlangs}$$

In Table A.4, a summary of individual port traffic statistics are presented for the 4-port multiplexer based upon a traffic intensity of 3 erlangs offered to the device. From Table A.4, the traffic carried by all 4 ports totaled 2.3817 erlangs. Since 3 erlangs were offered to the multiplexer ports, then 0.6183 erlangs were lost. The proportion of traffic lost to the group of 4 ports is T_4/E or 0.6183/3 which is 0.2061. These calculations become extremely important from a financial standpoint if a table lookup results in a device dimensioning which causes an expansion nest to be obtained to service one or only a few channel adaptors. Under such circumstances, one may wish to analyze a few of the individual high order channels to see what the effect of the omission of one or more of those channels will have upon the system.

If data tables are available, the previous individual port calculations are greatly simplified. From such tables the grade of service for channels 1 through 4 with a traffic intensity of 3 erlangs is the proportion of traffic lost to each port. Thus, if tables are available, one only has to multiply the grade of service by the traffic intensity to determine the traffic lost to each port.

Table A.4 Individual port traffic statistics.

Port	Proportion of lost traffic	Amount of lost traffic	Traffic carried
1	.7500	2.25	75
2	.5294	1.588	.662
3	.3462	1.038	.550
4	.2061	6183	.4197

A.3 THE POISSON FORMULA

The number of arrivals per unit time at a service location can vary randomly according to one of many probability distributions. The Poisson distribution is a discrete probability distribution since it relates to the number of arrivals per unit time. The general model or formula for this probability distribution is given by the following equation:

$$P(r) = \frac{e^{-\lambda}(\lambda)^r}{r!}$$

where r = number of arrivals, $P(r)$ = probability of arrivals, λ = mean of arrival rate, e = base of natural logarithm (2.71828), and $r!$ = r factorial = $r^*(r-1)^*(r-2) \ldots 3^*2^*1$.

The Poisson distribution corresponds to the assumption of random arrivals since each arrival is assumed to be independent of other arrivals and also independent of the state of the system. One interesting characteristic of the Poisson distribution is that its mean is equal to its variance. This means that by specifying the mean of the distribution, the entire distribution is specified.

Multiplexer sizing

As an example of the application of the Poisson distribution let us consider a multiplexer location where user calls arrive at a rate of 2 per unit period of time. From the Poisson formula, we obtain:

$$P(r) = \frac{2.71828^{-2}2^r}{r!}$$

Substituting the values 0, 1, 2,. . ., 9 for r, we obtain the probability of arrivals listed in Table A.5, rounded to four decimal points. The probability of arrivals in excess of 9 per unit period of time can be computed but is a very small value and was thus eliminated from consideration.

The probability of the arrival rate being less than or equal to some specific number, m, is the sum of the probabilities of the arrival rate being 0, 1, 2, . . . to m. This can be expressed mathematically as follows:

$$P(r \leq m) = \sum_{r=0}^{m} \frac{e^{-\lambda}\lambda^r}{r!}$$

Table A.5 Poisson distribution arrival (rate of 2 per unit time).

Number of arrivals per period	Probability
0	.1358
1	.2707
2	.2707
3	.1805
4	.0902
5	.0361
6	.0120
7	.0034
8	.0009
9	.0002

To determine the probability of 4 or less arrivals per unit period of time we obtain:

$$P(r \leq 4) = \sum_{r=0}^{4} \frac{e^2 2^r}{r!}$$

$$P(r \leq 4) = 0.1358 + 0.2707 + 0.2707 + 0.1804 + 0.0902 = 0.9478$$

From the preceding, almost 95% of the time 4 or less calls will arrive at the multiplexer at the same time, given an arrival rate or traffic intensity of 2. The probability that a number of calls in excess of 4 arrives during the period is equal to one minus the probability of four or less calls arriving.

$$P(r \leq 4) = 1 - 0.9478 = 0.0522$$

If 4 calls arrive and are being processed, any additional calls are lost and cannot be handled by the multiplexer. The probability of this occurring is 0.0522 for a 4-channel multiplexer, given a traffic intensity of 2 erlangs. In general, when E erlangs of traffic are offered to a service area containing m channels, the probability that the service area will fail to handle the traffic is given by the equation:

$$P(r \geq m) = e^{-E} \sum_{r=m+1}^{\infty} \frac{E^r}{r!}$$

The key difference between the Poisson and Erlang formulas is that the Poisson formula assumes that lost calls are held or retired immediately after a busy signal is encountered. When the Erlang formula is used, it is assumed that lost calls are cleared.

In Appendix C the reader will find a BASIC program written to compute a table of grades of service for varying traffic intensities when lost calls are assumed to be held and thus follow the Poisson distribution. By changing the values of line 40 of the program in conjunction with line 50, different traffic

Table A.6 Poisson distribution program result extract A. Probability all channels busy when call attempted.

	Traffic in erlangs (grade of service)				
Channel	3.000	3.500	4.000	4.500	5.000
1	.950213	.969802	.981684	.988891	.993262
2	.800851	.864111	.908422	.938900	.959572
3	.576809	.679152	.761896	.826421	.875347
4	.352767	.463366	.566529	.657703	.734973
5	.184735	.274553	.371161	.467895	.559505
6	.083916	.142384	.214867	.297068	.384037
7	.033507	.065286	.110672	.168947	.237814
8	.011902	.026736	.051131	.086584	.133369
9	.003801	.009871	.021361	.040255	.068090
10	.001100	.003312	.008129	.017090	.031825
11	.000290	.001017	.002837	.006666	.013692
12	.000069	.000286	.000912	.002402	.005450
13	.000014	.000073	.000271	.000802	.002015
14	.000001	.000016	.000074	.000249	.000694
15	.000001	.000002	.000017	.000071	.000223
16	.000001	.000002	.000002	.000017	.000065
17	.000001	.000002	.000002	.000002	.000016

intensities can be easily computed. Currently, the program computes grades of service as the traffic intensity varies from 0.5 to 40 erlangs in increments of 0.5 erlangs.

Formula comparison

In order to contrast the difference between Erlang B and Poisson formulas, let us return to the multiplexer examples previously considered. When 7 erlangs of traffic are offered and it is desired that the grade of service should be 0.01, 14 multiplexer channels are required when the Erlang B formula is employed. If the Poisson formula is used and the program in Appendix C is executed, one of the tables from running that program would appear as indicated in Table A.7. By using this table a grade of service of 0.01 for a traffic intensity of 7 erlangs results in a required channel capacity somewhere between 10 and 11. Rounding to the next highest number results in a requirement for 11 multiplexer channels. Now let us compare what happens at a higher traffic intensity. For a traffic intensity of 10 erlangs and the same 0.01 grade of service, it was determined that 19 channels were required when the Erlang formula was used. If the Poisson formula is used a 0.01 grade of service based upon 10 erlangs of traffic requires between 18 and 19 channels. Rounding to the next highest channel results in the Poisson formula providing the same value as computed previously with the Erlang formula.

Table A.7 Poisson distribution program result extract B. Probability all channels busy when call attempted.

	Traffic in erlangs (grade of service)				
Channel	5.500	6.000	6.500	7.000	7.500
1	.995913	.997521	.998497	.999088	.999447
2	.973436	.982649	.988724	.992705	.995299
3	.911623	.938031	.956964	.970364	.979743
4	.798300	.848796	.888150	.918234	.940854
5	.642481	.714942	.776327	.827008	.867937
6	.471079	.554319	.630957	.699290	.758562
7	.313961	.393695	.473474	.550287	.621843
8	.190511	.256017	.327239	.401283	.475358
9	.105640	.152759	.208423	.270905	.338029
10	.053774	.083920	.122611	.169500	.223587
11	.025247	.042617	.066834	.098516	.137756
12	.010984	.020088	.033874	.053345	.079235
13	.004447	.008824	.016021	.026995	.042660
14	.001681	.003625	.007095	.012807	.021558
15	.000594	.001396	.002950	.005712	.010254
16	.000196	.000505	.001154	.002402	.004602
17	.000059	.000171	.000425	.000953	.001953
18	.000015	.000053	.000146	.000357	.000784
19	.000001	.000014	.000045	.000125	.000297
20	.000001	.000001	.000011	.000039	.000104

In general, the Poisson formula produces a more conservative sizing at lower traffic intensities than the Erlang formula. At higher traffic intensities the results are reversed. The selection of the appropriate formula depends upon how one visualizes the calling pattern of users of the communications network.

Economic constraints

In the previous dimensioning exercises the number of trunks or channels selected was based upon a defined level of grade of service. Although we want to size equipment to have a high efficiency and keep network users happy, we must also consider the economics of dimensioning. One method that can be used for economic analysis is the assignment of a dollar value to each erlang-hour of traffic.

For a company such as a time-sharing service bureau, the assignment of a dollar value to each erlang-hour of traffic may be a simple matter. Here the average revenue per one hour time-sharing session could be computed and used as the dollar value assigned to each erlang-hour of traffic. For other organizations, the average hourly usage of employees waiting service could be employed.

As an example of the economics involved in sizing, let us assume lost calls are held, resulting in traffic following a Poisson distribution and that 7 erlangs of traffic can be expected during the busy hour. Let us suppose we initially desire to offer a 0.001 grade of service. From the extract of the execution of the Poisson distribution program presented in Table A.7, between 16 and 17 channels would be required. Rounding to the highest number, 17 channels would be selected to provide the desired 0.001 grade of service.

Multiplexers normally consist of a base unit of a number of channels or ports and an expansion chassis into which dual port adapter cards are normally inserted to expand the capacity of the multiplexer. Many times one may desire to compare the potential revenue loss in comparison to expanding the multiplexer beyond a certain capacity. As an example of this consider the data in Table A.6 which indicates that when the traffic intensity is 5 erlangs a 14-channel multiplexer would provide an equivalent grade of service. This means that during the busy hour 2 erlang hours of traffic would be lost and the network designer could then compare the cost of three additional ports on the multiplexer and modems and dial-in lines if access to the multiplexer is over the switched network to the loss of revenue by not being able to service the busy hour traffic.

A.4 APPLYING THE EQUIPMENT SIZING PROCESS

Many methods are available for end-users to obtain data traffic statistics required for sizing communications equipment. Two of the most commonly used methods are based upon user surveys and computer accounting information.

Normally, end-user surveys require each terminal user to estimate the number of originated calls to the computer for average and peak traffic situations as well as the call duration in minutes or fractions of an hour, on a daily basis. By accumulating the terminal traffic data for a group of terminals in a particular geographical area one can then obtain the traffic that the multiplexer will be required to support.

Suppose a new application is under consideration at a geographical area currently not served by a firm's data communications network. For this application, 10 terminals with the anticipated data traffic denoted if Figure A.5

	Calls originated per day		Call duration (minutes)	
Terminal	Average	Peak	Average	Peak
A	3	6	15	30
B	2	3	30	60
C	5	5	10	15
D	2	3	15	15
E	2	4	15	30
F	2	4	15	30
G	3	3	15	35
H	4	6	30	30
I	2	3	20	25
J	2	2	15	60

Figure A.5 Terminal traffic survey.

are to be installed at five small offices in the greater metropolitan area of a city. If each terminal will dial a centrally located multiplexer, how many dial-in lines, auto-answer modems, and multiplexer ports are required to provide users with a 98 percent probability of accessing the computer upon dialing the multiplexer? What would happen if a 90 percent probability of access was acceptable?

For the 10 terminals listed in Figure A.5 the average daily and peak daily traffic is easily computed. These figures can be obtained by multiplying the number of calls originated each day by the call duration and summing the values for the appropriate average and peak periods. Doing so, one obtains 480 minutes of average daily traffic and 1200 minutes of peak traffic. Dividing those numbers by 60 results in 8 erlangs average daily traffic and 20 erlangs of peak daily traffic.

Prior to sizing, some additional knowledge and assumptions concerning the terminal traffic will be necessary. First, from the data contained in most survey forms, information containing busy hour traffic is non-existent although such information is critical for equipment sizing. Although survey forms can be tailored to obtain the number of calls and call duration by specific time intervals, for most users the completion of such precise estimates is a guess at best.

Normally, busy hour traffic can be estimated fairly accurately from historical data or from computer billing and accounting tape data. Suppose that either source shows a busy hour traffic equal to twice the average daily traffic based upon an 8-hour normal operational shift. Then the traffic would be 8/8*2 or 2 erlangs while the busy hour peak traffic would be 20/8*2 or 5 erlangs.

The next process in the sizing procedure is to determine the appropriate sizing formula to apply to the problem. If we assume that users encountering a busy signal will tend to redial the telephone numbers associated with the multiplexer, the Poisson formula will be applicable. From Table A.8 the 2.0 erlang traffic column shows a 0.0165 probability (1.65 percent) of all channels busy for a device containing six channels, 0.0527 for five channels and 0.1428 for four channels. Thus, to obtain a 98 percent probability of access based upon the daily average traffic would require six channels while a 90 percent probability of access would require five channels.

If we want to size the equipment based upon the daily peak traffic load, how would sizing differ? We would now use the 5 erlang traffic column of Table A.6 from the table, 11 channels would provide a 0.0137 probability (1.37 percent) of encountering a busy signal while 10 channels would provide a 0.0318 probability. To obtain a 98 percent probability of access statistically would require 11 channels. Since there are only 10 terminals, logic would override statistics and 10 channels or one channel per terminal would suffice. It should be noted that the statistical approach is based upon a level of traffic which can be generated from an infinite number of terminals. Thus, one must also use logic and recognize the limits of the statistical approach when sizing equipment. Since a 0.0681 probability of encountering a busy signal is associated with 9 channels and a 0.1334 probability with 8 channels, 9 channels would be required to obtain a 90 percent probability of access.

Table A.8 Poisson distribution program result extract C. Probability all channels busy when call attempted.

	Traffic in erlangs (grade of service)				
Channel	.500	1.000	1.500	2.000	2.500
1	.393469	.632120	.776870	.864665	.917915
2	.090204	.264241	.442174	.593994	.712702
3	.014387	.080301	.191152	.323323	.456186
4	.001751	.018988	.065642	.142875	.242422
5	.000172	.003659	.018575	.052652	.108820
6	.000014	.000594	.004455	.016562	.042019
7	.000001	.000083	.000925	.004532	.014185
8	.000000	.000010	.000169	.001095	.004245
9	.000000	.000000	.000027	.000236	.001138
10	.000000	.000000	.000003	.000045	.000275
11	.000000	.000000	.000000	.000007	.000059
12	.000000	.000000	.000000	.000000	.000010

Probability of access (%)	Daily average	Traffic peak
90	5	9
98	6	10

Figure A.6 Channel requirements.

In Figure A.6 the sizing required for average peak daily traffic is listed for both 90 and 98 percent probability of obtaining access. Note that the difference between supporting the average and peak traffic loads is four channels for both the 90 and 98 percent probability of access scenarios, even though peak traffic is $2\frac{1}{2}$ times average traffic.

The last process in the sizing procedure is to determine the number of channels and associated equipment to install. Whether to support the average or peak load will depend upon the critical nature of the application, funds availability, how often peak daily traffic can be expected, and perhaps organizational politics. If peak traffic only occurs once per month, we could normally size equipment for the average daily traffic expected. If peak traffic was expected to occur twice each day, we would normally size equipment based upon peak traffic. Traffic between these extremes may require the final step in the sizing procedure to one of human judgment, incorporating knowledge of economics, and the application into the decision process.

APPENDIX **B**

ERLANG DISTRIBUTION PROGRAM

```
  5 FOR Z=1 TO 5
  6   PRINT
  7 NEXT Z
 10 REM A IS THE OFFERED LOAD IN ERLANGS
 20 REM S IS THE NUMBER OF PORTS, DIAL IN LINES OR TRUNKS
 30 DIM [[80],B[48,30]
 35 C=0
 40 FOR I=5 TO 400 STEP 5
 45   C=C+1
 50   A=I/10
 60   A[C]=A
 70   FOR S=1 TO 48
 80     X=S
 90     GOSUB 1000
100     N=(A**S)/F
110     D=1
120     FOR D1=1 TO S
130       X=D1
140       GOSUB 1000
150       D=D+(A**D1)/F
160     NEXT D1
170     B[S,C]=N/D
180   NEXT S
190 NEXT I
200 FOR I=1 TO 76 STEP 5
201   FOR Z=1 TO 5
202     PRINT
203   NEXT Z
204   PRINT "              ERLANG B DISTRIBUTION"
205   PRINT &
  "PROBABILITY ALL CHANNELS BUSY WHEN CALL ATTEMPTED (GRADE OF
SERVICE)"
```

```
206     PRINT"CHANNEL         TRAFFIC IN ERLANGS"
210     PRINT USING 220;A[I],A[I+1],A[I+2],A[I+3],A[I+4]
220     IMAGE 5XDD.DDD
225     PRINT
230     FOR S=1 TO 48
235       IF B[S,I+4]<=1E-7 THEN 260
240       PRINT USING 250;S,B[S,I],B[S,I+1],B[S,I+2],B[S,I+3],B[S,I+4]
250       IMAGE DDD,5(2XD.DDDDDD)
260    NEXT S
270  NEXT I
800  STOP
990  REM SUBROUTINE TO COMPARE FACTORIAL S VALUES
1000 F=1
1005 IF X=0 THEN 1045
1010 FOR F1=X TO 1 STEP -1
1020 LET F=F*F1
1030 NEXT F1
1040 RETURN
1045 F=1
1050 RETURN
```

APPENDIX C

POISSON DISTRIBUTION PROGRAM

```
  5 FOR Z=1 TO 5
  6    PRINT
  7 NEXT Z
 10 REM A IS THE OFFERED LOAD IN ERLANGS
 20 REM S IS THE NUMBER OF PORTS, DIAL IN LINES OR TRUNKS
 30 DIM [[80],B[48,80]
 35 C=0
 40 FOR I=5 TO 400 STEP 5
 45    C=C+1
 50    A=I/10
 60    A[C]=A
 65    K=0
 70    FOR S=1 TO 47
 75       K=K+1
 80       X1=0
 90       FOR X=0 TO S
100          GOSUB 1000
110          X1=X1+(A**X)/(F*2.71828**A)
120       NEXT X
130       B[K,C]=1-X1
140       B[K,C]=ABS9B[K,C]0
150    NEXT S
160 NEXT J
170       B[S,C]=N/D
180    NEXT S
190 NEXT I
200 FOR I=1 TO 76 STEP 5
201    FOR Z=1 TO 5
202       PRINT
203    NEXT Z
204    PRINT "              POISSON DISTRIBUTION"
```

```
 205     PRINT &
  "PROBABILITY ALL CHANNELS BUSY WHEN CALL ATTEMPTED (GRADE OF
SERVICE)"
 206     PRINT"CHANNEL         TRAFFIC IN ERLANGS"
 210     PRINT USING 220;A[I],A[I+1],A[I+2],A[I+3],A[I+4]
 220    IMAGE 5XDD.DDD
 225    PRINT
 230    FOR S=1 TO 48
 235       IF B[S,I+4]<=1E-7 THEN 260
 240       PRINT USING 250;S,B[S,I],B[S,I+1],B[S,I+2],B[S,I+3],B[S,I+4]
 250       IMAGE DDD,5(2XD.DDDDDD)
 260   NEXT S
 270 NEXT I
 800 STOP
 990 REM SUBROUTINE TO COMPARE FACTORIAL S VALUES
1000 F=1
1005 IF X=0 THEN 1045
1010 FOR F1=X TO 1 STEP -1
1020    LET F=F*F1
1030 NEXT F1
1040 RETURN
1045 F=1
1050 RETURN
```

APPENDIX D

MULTIDROP LINE ROUTING ANALYSIS

One of the most frequent problems encountered by organizations is to determine an economical route for the path of a multidrop circuit. This type of circuit is used to interconnect two or more locations that must be serviced by a common mainframe computer port. Although there are several commercial services that the reader can subscribe to as well as a free service offered by AT&T to obtain a routing analysis, in many situations this analysis can be conducted internally within the organization. Doing so not only saves time but may also eliminate some potential problems that can occur if one relies upon programs that do not consider whether the resulting number of drops on a circuit can support the data traffic while providing a desired level of performance.

In this appendix, the use of a simple algorithm that can be employed to minimize the routing distance and resulting cost of a multidrop circuit will be discussed. Since there is a finite limit to the number of drops a multidrop circuit can support, we will also investigate a method that will enable users to estimate the worst case and average terminal response times as the number of drops increase. Then if the response time exceeds the design goal of the organization, the networks manager can consider removing one or more drops and placing them on a different multidrop circuit.

The minimum-spanning-tree technique

When the total number of drops to be serviced does not exceed the capacity of the front-end processor software, the minimum-spanning technique can be used. This technique results in the most efficient routing of a multidrop line by the use of a tree architecture which is used to connect all nodes with as few branches as possible. When applied to a data communications network, the minimum-spanning-tree technique results in the selection of a multidrop line whose drops are interconnected by branches or line segments which minimize the total distance of the line connecting all drops. Since the distance of a circuit is normally proportional to its cost, this technique results in a multidrop line which is also cost optimized. To better understand the procedure used in applying the minimum-spanning-tree technique, let us examine an example of its use.

Figure D.1 illustrates the location of a mainframe computer with respect to four remote locations that require a data communications connection to the computer. Assuming that remote terminal usage requires a dedicated connection to the computer, such as busy travel agency offices might require to their corporate computer, an initial network configuration might require the direct connection of each location to the computer by separate leased lines. Figure D.2 illustrates this network approach.

When separate leased lines are used to connect each terminal location to the mainframe, a portion of many line segments can be seen to run in parallel. Thus, from a visual perspective, it is apparent that the overall distance of one circuit linking all locations to the computer will be less than the total distance of individual circuits. Other factors that can reduce the cost of a composite multidrop circuit in comparison to separate leased lines include differences in the number of computer ports and modems required between the use of a multidrop line and individual leased lines.

A multidrop line requires the use of one computer port with a common modem servicing each of the drops connected to the port. Thus, a total of $n + 1$ modems are required to service n drops on a multidrop circuit. In comparison, separate point-to-point leased lines would require one computer port per line as well as $n \times 2$ modems, where n equals the number of required point-to-point lines.

The minimum-spanning-tree algorithm

In the minimum-spanning-tree algorithm, the farthest point from the computer is first selected. In our example illustrated in Figure D.1 this would be location E. Next, the cost for connecting location E to each of the remaining locations is calculated. Since the cost of a line segment normally corresponds to the length of a line, users can simply consider the distance from location E to each of the remaining locations.

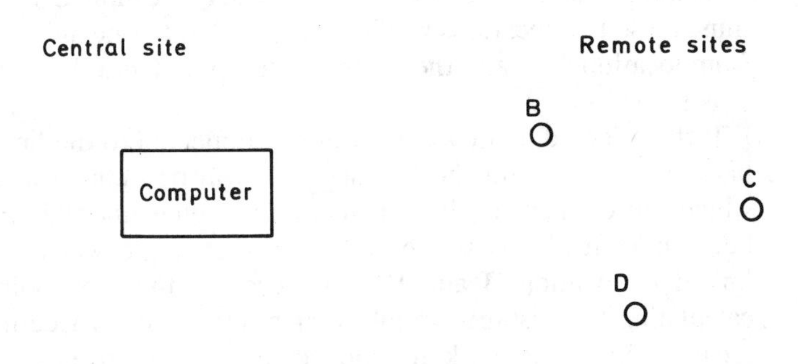

Figure D.1 Terminal locations to be services.

When a computer program is used to perform this operation, the user is normally required to enter the area code and telephone exchange of each location. The program uses this information to assign what is known as a (VH) coordinate pair to each entered location, where V and H refer to a vertical and horizontal coordinate system AT&T developed which subdivides the United States into grid squares. Then the computer uses the following equation to calculate the distance between pairs of (V,H) coordinate locations.

$$D = \text{INT}\left(\sqrt{\frac{(V_1 - V_2)^2 + (H_1 - H_2)^2}{10}} + 0.5\right)$$

Where D = distance between locations, V_1 = V coordinate of first location, V_2 = V coordinate of second location, H_1 = H coordinate of first location, and H_2 = H coordinate of second location. In this equation, 0.5 is added to the result and the integer taken since the telephone company is permitted by tariff to round the answer to the next higher mile in performing its cost calculations.

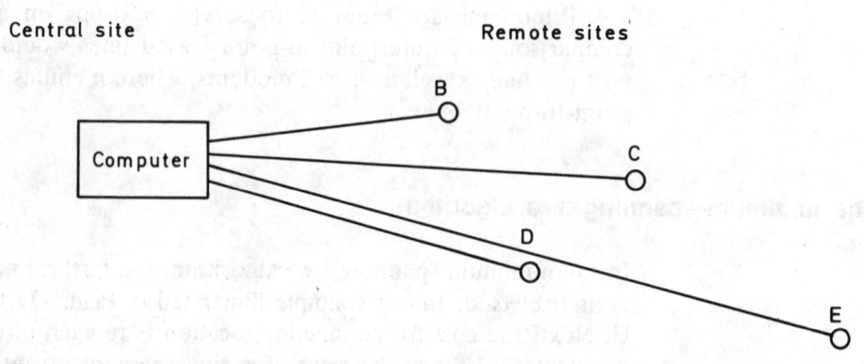

Figure D.2 Using separate leased lines.

Returning to the example in Figure D.1, location C is closest to location E, thus, these two locations would be connected to one another. Next, the distance from location C to all other points is computed and the closest point to location C is selected.

If the selected point was previously connected to the line, the line has looped back upon itself and the line segment will be defined as a cluster location to which more than one line segment can home upon. If the line did not loop back upon itself, as is the case in the example we are using, location C is linked to location B and B then becomes the next point of origination for calculating the distance to all other points. This procedure continues until all locations in the network are connected, with any resulting line clusters merged together by calculating the distance between the homing point of one cluster to the homing point of any other cluster using the minimum-distance path to connect clusters together. Figure D.3 illustrates the application of the minimum-

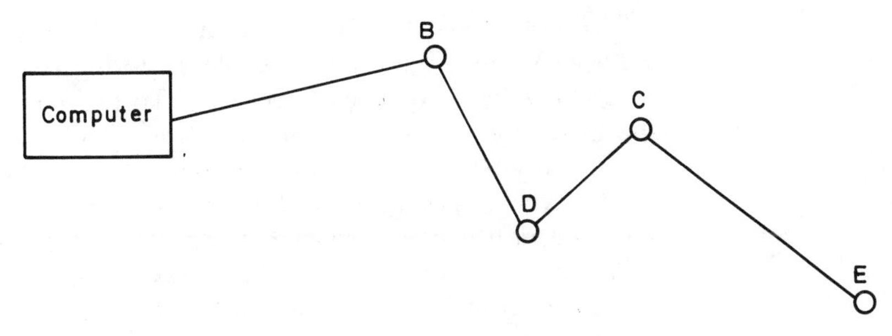

Figure D.3 Applying the minimum spanning-tree algorithm.

spanning-tree algorithm to the four network locations previously illustrated in Figure D.1.

Minimum-spanning-tree problems

The minimum spanning-tree algorithm, while economically accurate, does not consider two key variables which could make its implementation impractical – the terminal response time of the locations interconnected and the capacity of the front-end processor to service the total number of locations connected on one multidrop line.

Terminal response times

Normally, full-screen display terminals are used on a multidrop circuit. The terminal response time is defined as the time from the operator pressing the enter key to the first character appearing on the terminal's screen in response to the data sent to the computer. This response time depends upon a large set of factors, of which the major ones are listed in Table D.1.

Table D.1 Terminal response time factors.

Line speed
Type of transmission line
Modem turnaround time
Number of characters serviced per polling used
Computer processing time
Polling service time

The line speed refers to the transmission data rate which determines how fast data can be transported between the terminal and the computer once the terminal is polled or selected. The type of transmission line, full- or half-duplex, determines whether or not an extra delay will be incurred to turn the line around after each poll. If a half-duplex transmission protocol is used, then the modem turnaround time will effect the terminal response time.

The number of characters serviced per poll refers to how the communications software services a terminal capable of storing 1920 characters on a full screen of 25 lines by 80 characters per line. To prevent one terminal from hogging the line, most communications software divide the screen into segments and services a portion of the screen during each poll sequence.

This type of polling can occur "round robin" where each terminal receives servicing in a defined order or it can occur based upon a predefined priority. Although the computer processing time can greatly affect the terminal response time, it is normally beyond the control of the communications staff. The polling service time is the time it takes to poll a terminal so the communications software can service another segment of the screen when the data to be read or written to the terminal exceeds one segment. Finally, the probability of a transmission error occurring will affect the probability of transmitting the same data again, since detected errors are corrected by the retransmission of data.

Probability of transmission errors

To estimate the average terminal response time requires an estimate of the average number of users that are using the terminals on a multidrop circuit. Next, the average number of characters to be transmitted in response to each depression of the enter key must be estimated. This data can then be used to estimate the average terminal response time.

Suppose there are 10 terminals on the multidrop circuit and at any one time four are active, with approximately 10 lines of 30 characters on the display when the enter key is pressed. If the communications software services segments of 240 characters, two polls will be required to service each terminal. Assuming a transmission rate of 4.8 kbps, which is equivalent to approximately 600 characters per second, it requires a minimum of 240/600 or 0.4 s to service the first segment on each terminal, excluding communications protocol overhead. Using a 25% overhead factor which is normally reasonable, the time to service the first segment becomes 0.5 s, resulting in the last terminal having its first segment serviced at a time of 2.0 s if all users requested servicing at the same time. Since 60 characters remain on each screen, the second poll requires 60/600 or 0.1 s per terminal plus 25% for overhead, or a total of 0.5 s until the fourth terminal is again serviced. Adding the time required to service each segment results in a total time of 2.5 s transpiring in the completion of the data transfer from the fourth terminal to the computer.

Assuming the average response is 300 characters, the transmission of two screen segments is also required in the opposite direction. The first segment would then require 2.0 s for displaying on each terminal while the second segment requires 0.3 s plus 25% overhead or 0.375 s until the first character starts to appear on the fourth terminal, for a total response time (inbound and outbound) of 2.5 plus 2.375 or 4.875 s.

If a "round robin" polling sequence is used, the computer has an equal probability of polling any of the four terminals when the enter key is pressed. Thus, the computed 4.875 s response time is the worst case response time. The best case response time would be the response time required to service the first terminal sending data, which in the previous example would be 2.1 s inbound and 2.0 s outbound until the first character is received, or a total of 4.1 s. Thus, the average terminal response time would be (4.1 + 4.875)/2 or approximately 4.5 s.

Front-end processing limitations

A second limitation concerning the use of multidrop circuits is the capability of the front-end processor. In a large network that contains numerous terminal locations, the polling addressing capability of the front-end processor will limit the number of drops that can be serviced. Even if the processor could handle an infinite number of drops, polling delay times as well as the effect of a line segment impairment breaking access to many drops usually precludes most circuits to a maximum of 16 or 32 drops.

Large network design

When the number of drops in a network requires the use of multiple multidrop circuits, the network designer will normally consider the use of a more complex algorithm, such as the well-known Esau and Williams formula. While such algorithms are best applied to network design problems by the use of computer programs, the reader can consider a practical alternative to these complex algorithms. This alternative is the subdivision of the network's terminal locations into servicing areas, based upon defining a servicing area to include a number of terminals that will permit an average response time that is acceptable to the end-user. Then each segment can be analyzed using the minimum-spanning-tree algorithm to develop a minimum cost multidrop line to service all terminals within the servicing area.

SOLUTIONS

CHAPTER 1: FUNDAMENTAL CONCEPTS

1.1 The three major elements of a transmission system are a transmitter, receiver and a transmission medium. The transmitter translates information into a form suitable for transmission over the transmission medium while the medium can be considered the path used to convey information to the receiver. At the receiver, information is converted from its transmitted form back into its original form.

1.2 A dedicated or direct connect line normally has a one-time cost which is its installation cost. Thus, its transmission use has no effect upon its recurring cost and the line can be used for both short or long duration calls.

The cost of a switched (dial-up) call is a function of the distance between the calling and called party, the time of day when the call occurred, whether or not operator intervention was required and the duration of the call. Since the call duration is usually the major cost element in a switched call, such calls are usually limited to short duration transmission sessions.

The cost of a leased line is based upon the distance between connected locations and is a fixed fee per month. Thus, transmission activity has no bearing on the cost of the line and it can be used for both long duration or short duration calls.

1.3 To transmit digital data over the analog telephone network, such data must be converted into an analog format. This translation is accomplished by modems which modulate digital data into an analog format and demodulate analog data into its original digital format.

To transmit data on a digital network, the unipolar digital signals of terminals and computers must be converted into a bipolar signal. This signal translation is performed by a digital service unit (DSU).

1.4 Wide Area Telephone Service (WATS) can be used to provide access to a common location to terminals distributed over a geographical area. This is accomplished by the installation of one or more IN-WATS telephone numbers at a computer center that terminal users then dial and contend for their availability.

A foreign exchange (FX) line can be employed to provide access from a group of terminal devices located within a local telephone exchange area to a computer located distant from that area. Here each terminal operator contends for access to the telephone number assigned to the foreign exchange.

A leased line provides a permanent connection between two locations. This type of facility is normally used to connect terminal devices that have large data transfer requirements to a computer.

The switched telephone network provides a temporary connection between a computer terminal and computer system connected to the public switched telephone network.

1.5 Since many locations do not have access to a digital network, an analog extension must be employed to connect terminal devices at such locations to a digital network. This extension consists of a pair of modems and an analog leased line that connects the nearest digital network serving location to the user's location.

1.6 Simplex transmission is transmission that can only occur in one direction.

Half-duplex transmission can occur in both directions but only one way at a time.

Full-duplex transmission permits transmission to occur in both directions simultaneously.

1.7 When referring to terminal operations, the term half-duplex means that each time a character is pressed on the keyboard it is printed or displayed on the local terminal as well as transmitted. When a terminal operates in a full-duplex mode, each time a character is pressed on the device's keyboard it is transmitted but not printed or displayed locally.

When we refer to half and full-duplex with respect to computer systems, we reference whether or not they echo received characters back to the originator. A half-duplex computer system does not echo received characters while a full-duplex system echoes each character it receives.

Based upon the preceding, one of three responses can occur for each character entered at a terminal's keyboard as indicated by the following table.

Operating Mode		
Terminal device	Host computer	Character display
Half-duplex	Half-duplex	1 character
Half-duplex	Full-duplex	2 characters
Full-duplex	Half-duplex	No characters
Full-duplex	Full-duplex	1 character

1.8 In asynchronous transmission, the line is held in a marking condition between characters. Thus, the start bit which results in a transition from a mark to a space informs the receiver by the change in the line condition that a start bit has occurred.

1.9 Asynchronous transmission carries its own timing in the form of start and stop bits. In synchronous transmission, a timing source which is usually a clock built into a modem furnishes the timing signal to enable the devices to identify the characters as they are being transmitted or received.

1.10 In serial transmission, the bits that comprise the character to be transmitted are sent in sequence over one line. In parallel transmission, the characters are transmitted serially but the bits that represent the character are transmitted in parallel.

Most communications systems are serial transmission since only one communications line is required.

1.11 A terminal on a point-to-point line does not have to be addressable as it is the only device on the line when a connection is made to a computer system. In comparison, a multidrop line can allow many terminals to be connected to a common computer and the computer must then be able to direct its responses to specific terminals. This can only be accomplished if the terminals are addressable.

Since many terminals share the use of a common line in a multidrop environment, terminals must have buffers to store data while the computer is servicing a different device. In comparison, terminals used on a point-to-point line may not require buffers.

1.12 Morse code is unsuitable for transmission by terminal devices due to the variable number of dots and dashes that represent characters in this code and the lack of a method to distinguish when one character starts and when it is completed.

1.13 In the Baudot code, 2 of the 32 normally unique characters are used to define a 'letters shift' and a 'figures shift' character. The reception of either character informs the receiver as to which set of characters (letters or figures) it should use to interpret succeeding characters and extends the number of characters that can be uniquely represented.

1.14 ASCII A is 1000001. ASCII a is 1100001.

1.15

Character	ASCII code	Odd parity	Even parity
A	1000001	1	0
E	1000101	0	1
I	1001001	0	1
O	1001111	0	1
U	1010101	1	0

1.16 The major limitation of parity checking is its inability to detect the occurrence of multiple bit errors.

1.17

If the bit errors occur randomly and are singular in occurrence per transmitted character, we can expect 22 characters to be received in error.

3000	lines of data
×60	characters/line
180,000	characters of data

$$\begin{array}{rl} 180{,}000 & \text{characters of data} \\ \times 8 & \text{bits/character} \\ \hline 1{,}440{,}000 & \text{bits} \end{array}$$

$$\frac{1{,}440{,}000}{100{,}000} \text{ bits} \times 1.5 \text{ errors}/10^5 = 21.6 \text{ bit errors}$$

1.18 The checksum is computed by adding the ASCII values of the 128 characters in the XMODEM block, dividing by 255 and using the remainder as the checksum. Thus,

$X = 1011000_2 = 88_{10}$

$88 \times 128 = 11264$

$$\frac{11264}{255} = Q44 + R44$$

The remainder (44) then becomes the checksum.

1.19 When the transmitted and the internally generated cyclic redundancy check characters match, no transmission error is considered to have occurred and the receiver then issues a positive acknowledgement to the transmitting device. When no match occurs, a transmission error is considered to have occurred. This results in the receiver issuing a negative acknowledgement to the transmitter which is interpreted as a request to retransmit the previously transmitted block.

1.20 Alternating DLE 0 and DLE 1 characters are transmitted instead of ACK characters as a positive acknowledgement in bisynchronous transmission to prevent a single message block from becoming lost or garbled.

1.21 Every time a sequence of 5 one bits is encountered, an HDLC transmitter will insert a zero bit into the data stream. This process is known as zero insertion and is required to prevent naturally occurring data from being misinterpreted as an HDLC flag.

1.22 Bit-oriented protocols are naturally transparent to data and have the capability to support full-duplex transmission. In comparison, most character oriented protocols require a special sequence of control characters to place them into a transparent mode of operation and may not be capable of full-duplex operations.

1.23 When a secondary station responds to the poll of a primary station by setting N(R) equal to five in its response, this indicates that frames 0 through 4 were received correctly and the secondary station is expecting frame number 5.

1.24 As the rate of data transmission increases, the width of the transmitted pulses becomes narrower. Since narrower pulses are more susceptible to a fixed amount of noise than a wider pulse, a lower data rate has more immunity to a fixed amount of noise than a higher data rate.

Both RS-232-C and CCITT V.24 interface standards specify a maximum cabling distance of 50 feet at all data rates from zero to 19.2 kbps. Since these standards do not take into consideration the fact that low data rates have wide pulses that are less susceptible to a transmission impairment than narrower pulses, we can normally exceed the 50-foot cable limit of these standards at low data rates.

Circuit	Function
Request-to-send	This circuit becomes active when the terminal has data to send to the modem
Clear-to-send	This circuit becomes active when a modem is ready to transmit
Data set ready	This circuit informs the terminal of the status of the connected modem
Data terminal ready	This circuit is used to prepare a modem attached to a terminal to connect to the telephone line

1.26 db (decibels) is a measurement of power gain or power loss where:

number of decibels $= 10 \times \log_{10} P1/P2$

with $P1$ the larger power, and $P2$ the smaller power.

dbm is also a measurement of power gain or power loss, however, a 1 milliwatt power level is used by the telephone company as a reference level for comparing gains or losses in a circuit.

1.27 The losses and gains are added algebraically, resulting in an overall loss of -4 dbm.

1.28 A null modem cable reverses the transmit and receive conductors to enable two DTE devices to communicate with one another. To obtain compatible control signals at the interface, Request-to-send and Clear-to-send are tied together and connected to Data carrier detect at the opposite end of the cable. In addition, Data terminal ready will be connected to Data set ready at each end of the cable.

Due to the omission of transmit and receive clocks, a null modem cannot be used for synchronous transmission.

1.29 When the 'telephone set controls the line' option is selected, calls are originated or answered with the telephone by lifting the handset off-hook. When the 'data set controls the line' option is selected, calls can be automatically originated or answered by the modem without lifting the telephone handset.

1.30 The goal of the ISO Open System Interconnection Reference Model is to provide a framework for standardizing communication systems.

1.31 On an ISDN facility, data is transmitted in a digital format, with repeaters used to regenerate pulses. This insures that the distortion of digital pulses does not become unacceptable. In comparison, analog lines use amplifiers to boost the level of the transmitted signal, however, they also boost any distortion in the signal.

1.32 Transmission on the switched telephone network is normally limited to 9600 bps. In comparison, a Basic Access ISDN circuit permits a data rate of 144 kbps, which exceeds the analog circuit rate by a factor of 15.

CHAPTER 2: DATA TRANSMISSION EQUIPMENT

2.1 An acoustic coupler is acoustically connected to a telephone line by the placement of a telephone handset into its built-in cradles or fittings. In comparison, a modem is directly connected to a telephone line via a plug-jack connection.
2.2 The term 'Bell System' compatibility means that a modem is constructed to operate according to a specific set of characteristics that makes it operationally compatible with a specific Bell System modem.
2.3

Originate mode		Receive mode
Transmit mark f_1	→	Receive mark
Transmit space f_2	→	Receive space
Receive mark f_3	←	Transmit mark
Receive space f_4	←	Transmit space

2.4 Originate mode modems and couplers are normally connected to terminal devices that originate data calls. Answer mode modems and couplers are normally connected to computer ports that receive calls.
2.5 If an American used a portable personal computer in Europe that was built in North America, its internal modem would probably be Bell System compatible. If so, the modem would not be compatible with low-speed CCITT modems which are primarily used throughout Europe.
2.6 A carrier signal is a tone that will be varied to impress information onto the signal. By itself it does not convey any information because it continuously repeats itself which does not provide knowledge of information. When the carrier signal is modulated, it then changes its state and this change in the carrier represents information impressed onto the carrier.
2.7 A carrier signal can be altered by changing its amplitude, frequency, phase or a combination of two or more of the preceding characteristics of the signal.
2.8 A bit per second is a measurement of the speed of transmission while baud is a measurement of the rate of signal change. If each bit entering a modem results in one signal change, then the terms bps and baud are equivalent. When two or more bits are combined to represent one signal change, the terms are not equivalent.
2.9 The Nyquist relationship states that there is a maximum signaling rate on a circuit which, if exceeded, will result in one signal interfering with another and causing intersymbol interference. Thus, instead of increasing the baud rate, modem designers must pack more bits into each signal change to enable modems to operate at high data rates.
2.10 The signal constellation pattern of a modem represents a plot of all possible signal points that represent all of the data samples possible based upon the modulation method used by the modem. The denser the signal pattern the more susceptible the modem is to a transmission impairment, since a small shift in the phase of a signal becomes more susceptible to misinterpretation.

2.11 In Trellis coded modulation, redundant bits are encoded into each symbol interval that results in only certain sequences of signal points in the constellation pattern being valid. Thus, although the density of a Trellis coded modulation pattern exceeds the density of a conventional quadrature amplitude modulation modem, it is less susceptible to transmission impairments since a shift in phase may result in the shift to an invalid constellation point.

2.12 A reverse channel can be used to acknowledge data blocks transmitted using a half-duplex synchronous protocol. Doing so alleviates the necessity of reversing the transmission direction after each data block is transmitted.

A reverse channel supports transmission in one direction only. In comparison, a secondary channel supports bidirectional transmission.

2.13 A multiport modem is a modem that contains a limited functioning, synchronous operating multiplexer. This modem should be considered for use when two or more synchronous data sources within a geographical area require routing to a common computer at a distant location.

2.14 Although the V.22 modem's 2400 Hz tone varies significantly from the Bell 212A modem's 2225 Hz standard, when the V.22 transmits data, its primary frequency of 2250 Hz is usually close enough to the Bell standard to illicit a response.

2.15 Bell System type-202 modems are incompatible with CCITT V.23 modems due to the different frequencies used to encode marks and spaces by each modem.

2.16 The phase change patterns used by a pair of V.26 modems must be identical. Otherwise, as an example, the dibit 00 would be encoded as a zero phase change by a V.26 modem employing the pattern A set of phase changes. Since a modem employing the V.26 pattern B set of phase changes does not recognize a zero phase shift, the dibit could never be decoded.

2.17 The V.29 signal constellation pattern forms a mirror image of points around the *X* and *Y* axes based upon the assignment of amplitude and phase change to each distinct quadbit. This assignment ensures that the points in the constellation pattern are rotated equidistant among 8 axes, with each axis at a 45-degree separation from the preceding axis to enable the points in the pattern to form a mirror image of one another.

2.18 In the United States, any modem built to comply with the Federal Communications Commission Equipment Registration Program can be directly connected to the switched telephone network without requiring the use of a Data Access Arrangement.

2.19 You could first run a local test on each modem. If each test was successful, this would indicate that the modems are correctly modulating and demodulating the test pattern generated by the local test. Next, you could attempt a digital loop-back self-test from either modem. If this test fails, it would then indicate that although the modems are correctly modulating and demodulating data generated locally, they cannot do so when the data flows onto the circuit. This would indicate that a transmission impairment on the circuit is the most likely cause of your inability to transmit data.

2.20 A wraparound unit is a device that is physically cabled to both ends of one vendor's modem to enable it to be used with a different vendor's network control systems (NCS). The wraparound unit responds to commands issued by the NCS to perform such operations as placing the attached modem in a test mode or issuing a status report.

2.21 Since a multiport modem combines the functions of modulation and demodulation as well as multiplexing within one device, only one power supply, logic unit, and housing is required. This usually makes the device more economical than obtaining separate devices to perform both functions built into the multiport modem.

2.22 A DCE option on a communications device permits one DCE device to be connected to another device. Since a DCE device receives data on pin 2 and transmits data on pin 3, using a conventional straight-through conductor cable to connect two DCE devices would be impossible. The DCE option therefore reverses the pin 2 and pin 3 conductors within the device. The DCE option could be obtained through the use of a cable by reversing pins 2 and 3 on the cable as illustrated below.

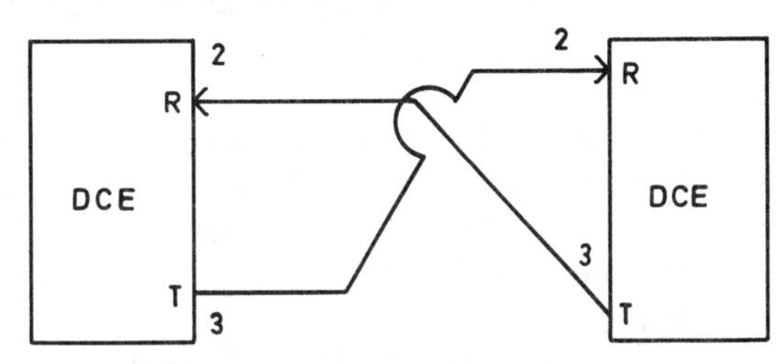

2.23 The key difference between a multipoint modem and a conventional modem is the design goal used for the former device, which is oriented toward reducing its internal and RTS/CTS delay times.

Increasing the average size of transmitted data blocks will increase the efficiency of transmission up to a point. In general, as the block size increases more data is transmitted per fixed unit of modem internal and RTS/CTS delay times as well as other delay times to include propagation and processing delay. Since the probability of a transmission error occurring increases as the block size increases, at a certain block size the inefficiency resulting from the retransmission of data blocks due to transmission errors will negate the efficiency of transmitting more data per fixed unit of delay. Based upon the preceding, the efficiency of transmission initially increases as the size of the transmitted block increases. This increase in efficiency peaks at a certain block size and then decreases as the block continues to increase in size.

2.24 Multipoint modems have a minimal effect upon the use of satellite circuits due to the insignificant reduction in the internal and RTS/CTS delay times of the modem in comparison to the lengthy propagation delay times associated with the transmission of each block of data via satellite.

2.25 As the data rate increases, digital pulses become narrower and more susceptible to noise. As the wire gauge decreases, the diameter of the conductor increases, providing less resistance to the flow of data. Thus, installing a smaller gauge wire can extend the transmission distance between devices at a fixed data rate while increasing the data rate on a wire of fixed gauge will decrease the transmission distance.

2.26 A line driver samples the line for the rise of a pulse and then regenerates the pulse. Thus, the input and output of the device are digital signals. A modem accepts digital signals and modulates those signals into an analog signal, in effect converting a digital signal into an analog signal.

2.27 (a) *Using direct cabling*

The terminals cannot be directly cabled since digital signals distort at 400 feet and a 600-foot cable length is required to connect each terminal to the computer.

(b) *Using line drivers*

8 cables × 600 ft/cable × \$0.50/ft	=	\$2400.00
8 line drivers at \$100	=	800.00
		\$3200.00

(c) *Using multipoint limited distance modems*

2 cables × 600 ft/cable × \$0.50/ft	=	\$ 600.00
4 limited-distance modems at \$250	=	1000.00
		\$1600.00

2.28 On a digital transmission facility, repeaters regenerate digital pulses to their original condition. On an analog transmission facility, amplifiers are used to increase the transmitted signal. Unfortunately, an amplifier increases not only the signal but, in addition, any distortion that may have previously occurred to the signal.

2.29 In bipolar signaling, alternating polarity pulses are used to represent marks while a zero pulse is used to represent a space. This type of signaling inhibits the buildup of dc voltages on the line, permitting repeaters to be placed further apart in comparison to the use of other signaling techniques.

2.30 A parallel interface extender does not require the use of teleprocessing software. Not only is the device locally connected to the computer but devices attached remotely to the interface extender also appear to the computer as locally attached devices. This means that an application program operating on the computer can transmit a report to a specific remotely located printer by having the application direct its output to the logical unit of the computer assigned to the printer. Additional printers can receive the output of the program by rerunning the application and directing its output of the program by rerunning the application and directing its output to different logical unit addresses. Since this requires multiple program executions, a better method would be to first direct the program's output into a general System Output (SYSOUT) queue and then copy the SYSOUT data into specific queues assigned to one or more logical devices. Then, each time one of the logical devices becomes active it can retrieve the data placed into its SYSOUT queue.

CHAPTER 3: DATA CONCENTRATION EQUIIPMENT

3.1 In frequency division multiplexing, the available line bandwidth is subdivided into two or more data bands or derived channels. This subdivision permits two or more simultaneous transmissions to occur on a common line.

In time division multiplexing, the aggregate capacity of the line is subdivided by time and each data source connected to the multiplexer is assigned to a specific time slot. This subdivision by time only permits one data source to have exclusive use of the line capacity at a particular point in time.

3.2 The FDM channel spacing required at 300 bps is 480 Hz. Since a voice-grade line has a bandwidth of 3000 Hz, six 300 bps data sources can be multiplexed with FDM equipment. The number of 300 bps data sources that can be multiplexed by a traditional time division multiplexer is dependent upon the operating rate of the high-speed line connected to the multiplexer. If, as an example, the line is operated at 9600 bps, thirty-two 300-bps (9600/300) data sources could be multiplexed.

3.3 Since frequency division multiplexing subdivides a line by frequency into derived channels, channel sets can be connected to the line that are tuned to operate at the specific frequencies used for each channel. This enables terminals to be connected to an FDM line at varying locations and share the use of the line by frequency without requiring terminals to be addressable or the use of poll and select software.

3.4

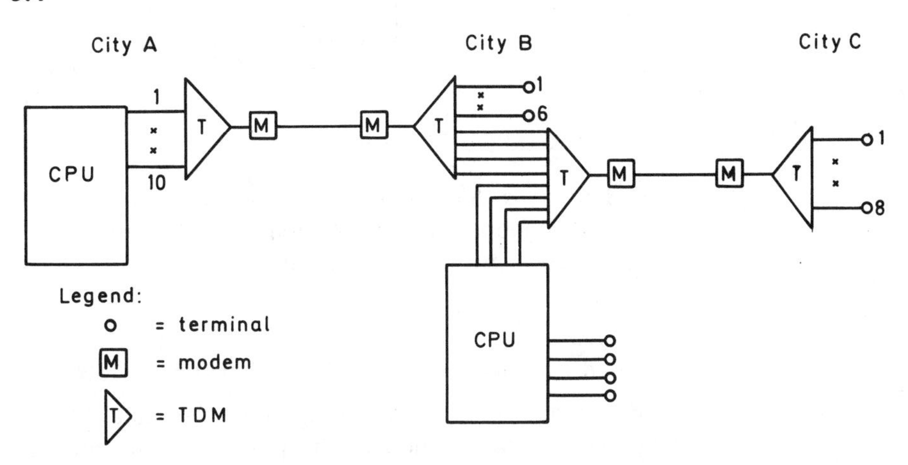

3.5 Front-end substitution is not common due to the requirement to program the computer to perform demultiplexing. Since many multiplexer vendors have proprietary multiplexing techniques, there is no standard software available and software development costs can exceed the hardware savings associated with this technique unless a large number of multiplexers can be replaced by the result of the software development effort.

3.6

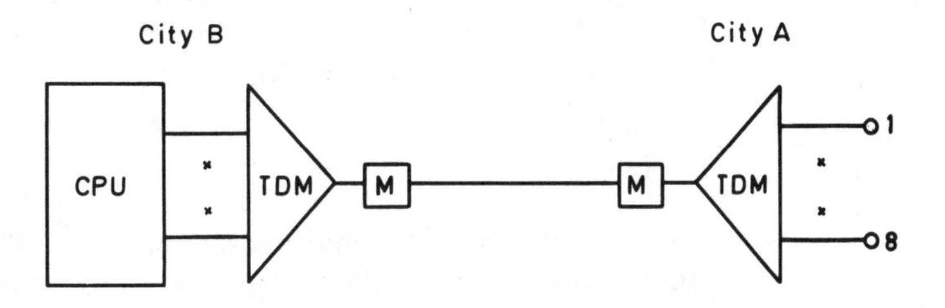

One-time costs:	
two 8-channel TDMs at $2000	$4000
two 9600-bps modems at $3000	$6000
	$10,000
Recurring costs:	
one line at $1000/month × 12	$12,000
Total cost	$22,000

3.7 The multiplexer and terminals located at city A would appear as follows:

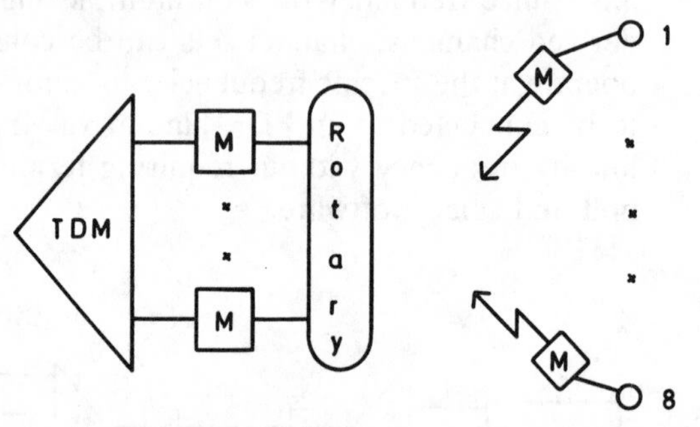

The additional cost of the network is:

1 rotary at $50/month × 12 months	$ 600
16 telephone lines at $30/month × 12 months	$5760
16 1200-bps modems at $200	$3200
	$9560

Total network cost is $22,000 + $9560 = $31,560.

3.8 In a conventional multiplexer, the data in the multiplexer's frame corresponds by position to the port where it entered the multiplexer. This relationship enables the data to be easily demultiplexed and output to its destination port. When data is multiplexed statistically, the relationship between the position of the data in the frame and its port of origin is destroyed. Thus, an address must be added to the data when multiplexed to enable it to be routed to the correct port after it is demultiplexed.

3.9 The throughput through a statistical multiplexer depends upon the activity of all data sources routed through the device. If multidrop circuits are multiplexed, the response times on such circuits can be severely affected by the activity of other data sources routed through the statistical multiplexer.

3.10 For a conventional multiplexer:

$$\frac{9600}{1200} = 8$$

Since the statistical multiplexer has an efficiency 2.5 times that of a conventional TDM it can multiplex 8 × 2.5 = twenty 1200-bps data sources.

3.11 If the multiplexer discussed in question 3.10 uses a 4800-bps bandpass channel, only 4800 bps (9600−4800) is available for the statistical multiplexing of asynchronous data.

Since only 4800 bps is available for statistical multiplexing with an efficiency 2.5 times that of a conventional multiplexer, the statistical multiplexer can support:

$4800 \times 2.5 = 12{,}000\text{-bps}$

Then,

$$\frac{12{,}000}{1200} = 10 \text{ terminals}$$

3.12

Synchronous	Asynchronous
Terminal speed 2400	1200
Times number of terminals × *4*	× *18*
Aggregate throughput 9600	21,600
Divide by efficiency ratio ÷ *1.5*	÷ *3*
Statistical rate 6400	7200

The total statistical rate of 7200 plus 6400 bps exceeds the line operating rate of 9600 bps. Based upon this, the multiplexer could not support the problem as stated.

3.13 By creating several test files on a computer and having terminal operators list these files by accessing the computer through the statistical multiplexer, one would know the data flow into the multiplexer. By examining the statistical loading and comparing it with the aggregate data routed through the multiplexer, one can determine if the vendor's cited efficiency is a reasonable figure.

3.14

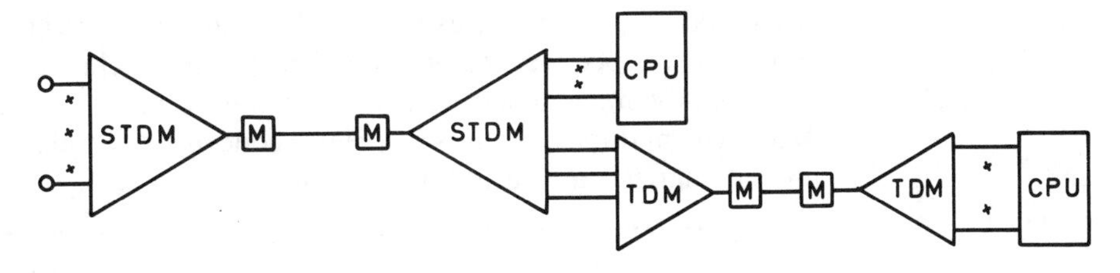

Legend:

STDM = statistical time division multiplexer with port contention option

TDM = time division multiplexer

M = modem

3.15 T1 carriers operate at different data rates in the United States and Europe due to the different numbers of voice channels that are combined onto this carrier in each area of the world.

3.16 Current methods used to digitize voice require a digital signal rate ranging from 16 to 64 kbps. Assuming the resulting voice quality is the same for each method, the method with the lower signal rate enables more voice circuits to be multiplexed onto a T1 carrier.

3.17 Inverse multiplexing cost:

2 inverse multiplexers at $500	$1000
4 9600-bps modems at $300	$1200
2 voice-grade lines at $1250	$2500
	$4700

Wideband cost:

1 wideband line at $9000	$ 9000
2 19,200-bps modems at $700	$ 1400
	$10,400

Based upon the preceding, it would be more economical to use inverse multiplexers.

3.18 The primary advantages of inverse multiplexing in comparison to wideband transmission are in the areas of line availability, backup and circuit compensation.

Wideband circuits are not as readily available as leased voice-grade lines. Since an inverse multiplexer supports two or more voice-grade lines, it permits one voice-grade circuit to backup the other circuit. In addition, if one of the voice-grade circuits becomes inoperative, the switched telephone network can be used to compensate for the circuit outage. In comparison, a wideband data rate could not be compensated for by the use of the switched telephone network.

3.19 Most inverse multiplexers are limited to two channels since each additional channel requires two modems and a voice-grade circuit. From an economic perspective, the additional cost of more channels may rapidly negate the cost advantage associated with the use of an inverse multiplexer.

3.20 The major differences between a concentrator and a statistical multiplexer are summarized in the following table.

	Concentrator	Statistical multiplexer
Processor	Minicomputer	Microprocessor
Attached peripheral devices	Can include tape, disk, card reader, line printer	Usually limited to RS-232 port for monitoring
Programming	Loaded into RAM	ROM based

Specialized communications requirements requiring customized software or the use of high capacity peripheral devices would probably justify the use of a concentrator instead of a statistical multiplexer. One example of this would be a store and forward message switching system which is normally implemented by the use of concentrators.

3.21 Read-only memory in concentrators is designed to facilitate the down-line loading of the concentrator. In comparison, ROM in a statistical and intelligent multiplexer contains the coded instructions that govern the operation of the multiplexer.
3.22 The number of processor cycles that can occur during the 300 ms required for a failing system to inform the remaining operational system to take control is:

$$\frac{300 \times 10^{-6}}{500 \times 10^{-9}} = 600 \text{ cycles}$$

With 3 processor cycles per instruction, 600/3 or 200 instructions can be coded to initiate cutover.
3.23 Remote network processing should be considered when both remote batch processing and terminal concentration requirements exist at a remote location.
3.24 When the aggregate data input exceeds the data transmission rate of a high-speed line connected to a statistical multiplexer, the multiplexer will attempt to inhibit the input data stream. This is normally accomplished by the multiplexer transmitting an XOFF character, dropping the CTS signal or lowering the clocking rate. In comparison, a concentrator will buffer the excess data to a disk for retrieval at a later time when the aggregate data input falls below the data transmission rate of the high-speed line connected to the concentrator.
3.25

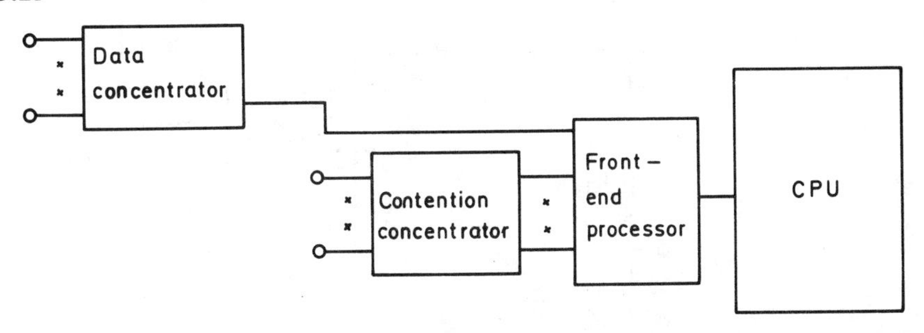

3.26 A data concentrator only requires a connection to one port on a front-end processor since the FEP, in effect, demultiplexes the data traffic. In comparison, one front-end processor port is required for each data source multiplexed by a statistical multiplexer.
3.27 A channel-attached communications controller is directly cabled to a host computer system and is located in close proximity to that computer. In comparison, a link-attached communications controller is connected to the host computer system via a communications line and can be located at a significant distance from the computer.
3.28 Modem and line-sharing units are used on point-to-point circuits. The major difference between devices is the internal timing source included in the line-sharing unit which is omitted from the modem-sharing unit since it obtains its timing from an attached modem.
3.29 The primary constraint one should consider when employing port-sharing units in conjunction with modem-sharing or line-sharing units is the total

number of terminals to be polled and selected through a common port as well as the anticipated activity of each terminal.

3.30 Sharing units are passive devices that are transparent to the flow of data. In comparison, control units are more sophisticated devices that direct the operation of attached displays and printers.

3.31 Since the IBM 3270 Information Display System is a coaxial cable based system, a converter is required to attach RS232/CCITT V.24 devices into this type of network. Common products marketed to provide this conversion feature include terminal emulators, control unit emulators and protocol converters.

3.32 The major benefit of a port selector is its ability to make more efficient use of front-end processor ports. This is accomplished by making a large number of data sources contend for access to a lesser number of computer ports.

3.33 If a port selector has queuing and queuing position display capability, users that cannot obtain an immediate cross-connect can see their position in the queue. Then, based upon that position, users can decide whether to hang up and try again at a later time or to wait for a cross connection.

3.34 When a large number of terminals colocated in a building must contend for access to a communications device such as a multiplexer, a port selector may be economically justified.

3.35

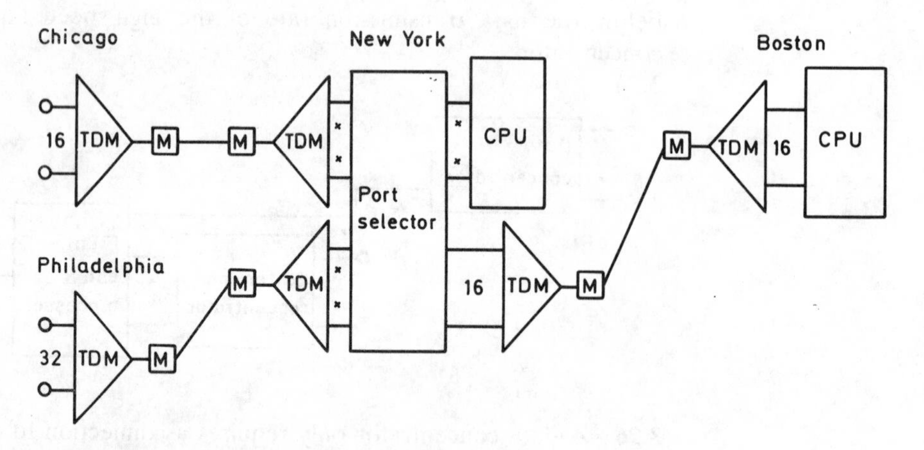

CHAPTER 4: REDUNDANCY AND RELIABILITY AIDS

4.1 Two types of switches used in networks are EIA and telco switches. The former switch transfers 24 leads of the RS232 or CCITT V.24 interface since ground is not switched. Telco switches transfer 4-wire leased or 2-wire dial-up telephone lines.

4.2 The following schematic illustrates how the public switched telephone network (PSTN) can be easily used as backup for a leased line without requiring the recabling of any network device.

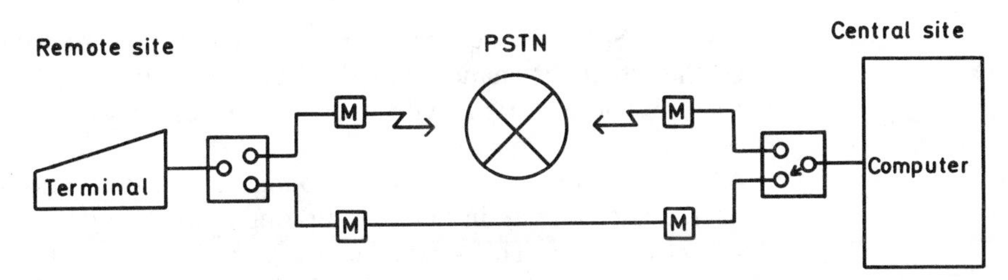

4.3 The bypass switch permits any input on one side of the switch to be cross-connected to a line on the other side of the switch. This line is usually interfaced to a space device. One use of this switch at a computer center would be to connect n leased lines to $n + 1$ modems, with the extra modem serving as a spare device. Then, if a modem becomes inoperative, the bypass switch can be used to transfer a line to the spare modem, bypassing the failed modem.
4.4 A bypass switch permits the transfer of any input to a specific output. A fallback switch permits the transfer of one input to one of two outputs. In comparison, a matrix switch permits any input to be transferred to any output, increasing routing flexibility.
4.5 Suppose two terminals located at a remote site must access two colocated computers at the computer center, however, they must access one computer in the morning and the other computer in the afternoon. Through the use of a cross-over switch at the remote site, this requirement could be satisfied.
4.6 Four fallback switches could be used to obtain a bypass switch capability linking m terminals to $m + 1$ modems as illustrated in the following diagram.

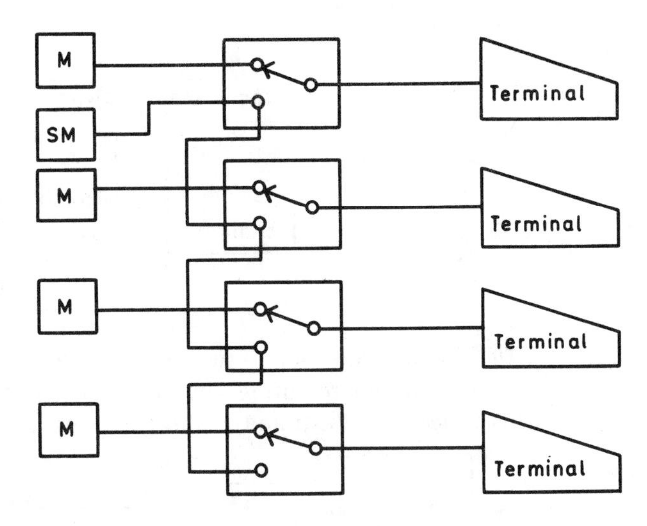

Legend:

M = modem

SM = spare modem

4.7 The major differences between manual and automatic switches are in the areas of cutover time and cost. The manual switch requires operator intervention to switch the device while a code can be sent to the automatic switch to effect cutover. Since the automatic switch must be built to recognize this code, it is costlier than the manual switch. The automatic switch should be considered for use where instant switchover is required or where unattended operations occur.

4.8 In a hot-start concentrator configuration, two concentrators are interconnected and operate in tandem, although only one device is actually connected to active lines. The second concentrator monitors the status of the active concentrator and if the primary device fails the second device takes over and the lines are automatically switched to that device. In the cold-start method, the concentrators are not interconnected and when the lines are switched upon the failure of one concentrator the backup concentrator must be initialized. This initialization requires some time, causing data to be lost at the time the primary device failed. Since switchover between concentrators requires a period of time, the switches transferring the lines are usually manual devices.

4.9 Assuming fallback and matrix switches are used to connect terminals to two multiplexers, the cabling might occur as indicated by the following diagrams.

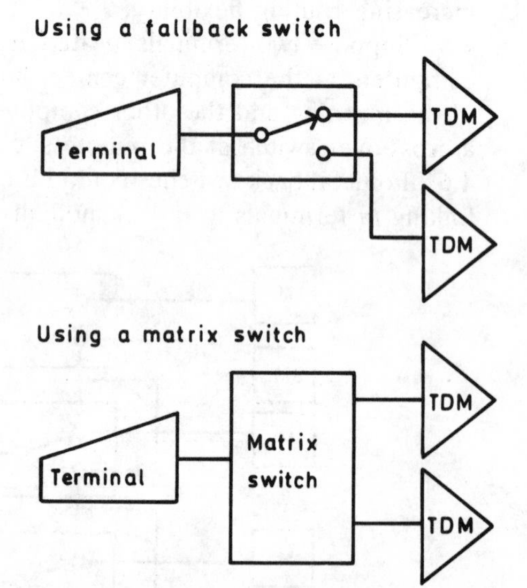

4.10 By the utilization of two matrix switches, a common backup circuit could be used without requiring any additional ports at the central computer site. This concept is illustrated in the following diagram.

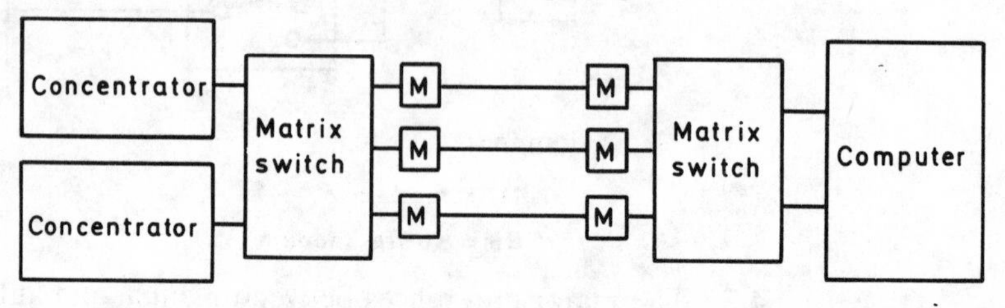

4.11 Line restoral units are paired devices because one device must dial the other over the PSTN to re-establish communications when the circuit quality on a leased line inhibits transmission.

4.12 The line restoral unit monitors the signal strength on a leased line since this parameter is the decision criteria for initiating a backup call over the PSTN.
4.13 The delay circuits in a line restoral unit are designed to prevent the constant switching between the PSTN and the leased line when the signal strength is marginal and varies between an acceptable and an unacceptable level.
4.14 A line restoral unit must make two calls over the switched telephone network to backup a full-duplex high-speed transmission that previously occurred on a leased line. Since the PSTN is a 2-wire circuit, two calls are necessary to obtain a 4-wire circuit.
4.15 It is more advisable to use an intelligent or statistical multiplexer with a line restoral unit instead of a conventional TDM because the former devices have the capability to inhibit data flow. This permits the restoral unit to initiate backup at a lower data rate without having to obtain a restoral unit that can inhibit one or more channels on the multiplexer, since this would be required if a conventional TDM was used.

CHAPTER 5: AUTOMATIC ASSISTANCE DEVICES

5.1 The ring indicator signal is normally used to activate an automatic answering unit.
5.2 On an automatic answering unit the timer abort option should normally be set to ON to prevent misdialed telephone numbers from seizing the automatic answering unit.
5.3 On an automatic answering unit the loss of carrier disconnect option should normally be set to ON to enable the device to receive other calls if the current user simply hangs up or if the line connection is broken.
5.4 An 801A-type calling unit is designed for use on rotary dial telephone systems while the 801C-type device is designed for use on touch-tone dialing systems.
5.5 A chain store that has point-of-sale terminals could use an automatic calling unit at its computer site to dial each store after normal business hours. The computer could transfer recorded sales information and use this data for inventory control and reordering purposes.
5.6 RS-232 transfers data serially while RS-366 transfers data in parallel.
5.7

	Digit values			
Number	D_4	D_3	D_2	D_1
B	0	0	1	0
O	0	1	1	0
Y	1	0	0	1
G	0	1	0	0
I	0	1	0	0
R	0	1	1	1
L	0	1	0	1

5.8 The main advantage in using an RS-232 interfaced automatic calling unit is the ability to locate the device at a distance from the controlling computer. This would enable a calling unit to be located in New York City, as an example, although the computer controlling the calling unit might be located in Chicago.

5.9 For the situation where many calls must be made to a similar but remotely located area with respect to one's computer center it may be more economical to locate the calling unit in a distant location. Even though a leased line between the computer center and the remotely located calling unit will be required, the cost of the leased line may be exceeded by the savings obtained by making local calls instead of long-distance calls.

5.10 A calling line selector can be used to establish many concurrent calls originated by a common calling unit. The calling line selector can be effectively used when the terminals to be polled are located within the same calling area.

CHAPTER 6: SPECIALIZED DEVICES

6.1 Limitations of security access implemented through passwords include the potential for unauthorized persons viewing the entry of the password, its impact upon a terminal ribbon which might be read and the fact that many users post passwords on terminals to facilitate remembering the code.

6.2 The number of password combinations depends upon the number of positions in the password and the number of characters that can be used in each position. As characters are reserved for communications or system functions they are no longer available for use in the password, reducing the number of available password combinations.

6.3 A personal computer can use a software program to encode a file prior to its transmission or an encoder could be connected to the computer to encode data during transmission.

6.4 If both the header and the text of a message to be transmitted on a Telex or TWX network are encoded, the message will never reach its destination. This is caused by the inability of the network to read the encoded header information that contains the destination of the message.

6.5 Data to be transmitted 10110010
Key 01101010
Modulo 2 addition
(encoded) text 11011000

6.6 Received data 11001001
Key 01010110
Modulo 2 subtraction
(clear) text 10011111

6.7 If a rotary is used to access a multiplexer, an encoder is difficult to use unless all terminals accessing the rotary require transmission security. This is because an encoder connected to a specific port on the multiplexer cannot be easily accessed since dialing the main rotary number will result in a connection to a specific port based upon the number of calls currently active through the rotary.

6.8 Common features incorporated into most security modems are password verification and the dialback over the switched telephone network to the originator of the call. The potential economic problem from the use of this type of modem is the cost of the second call required during the dialback operation.

6.9 A security switch can be used to provide a common point of access to two or more devices. As an example, a security switch could provide access via the PSTN to two different computers after the user enters the required password followed by an appropriate routing code.

6.10 The primary use of speed and code converters is to obtain compatibility between the operating characteristics of a terminal and the operating characteristics of a communications line.

6.11 The different levels of operational conversion performed by protocol converters include device functionality, device operation, protocol, data code/speed and physical and electrical interface. Speed and code converters perform data code/speed conversion and, if necessary, perform physical/electrical conversion.

6.12 Since a multiport modem only accepts synchronous input, an asynchronous to synchronous converter must be used to connect an asynchronous terminal to a multiport modem. This connection is illustrated in the following diagram.

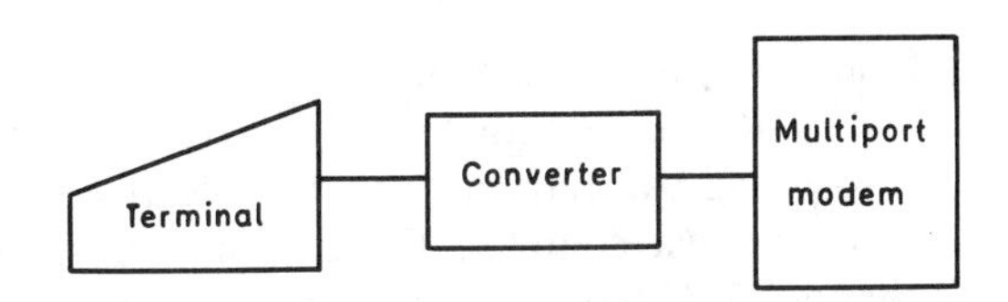

6.13 The market for data access arrangements is contracting because such devices are only required for use with devices built prior to the Federal Communications Commission Equipment Registration Program. Thus, all current devices are built to comply with the Equipment Registration Program and as older devices become obsolete and are removed from service, the requirement for data access arrangements also becomes obsolete.

6.14 A voice adapter is designed to operate in conjunction with a specific modem and permits alternate voice or data or simultaneous voice and data to be transmitted on a leased line. When simultaneous voice and data occurs, a portion of the bandwidth of the line is used for the voice conversation while the remaining bandwidth is used by the modem for data transmission.

A voice digitizer converts an analog conversion into a digital data stream that can be integrated into a data network. This permits the output of a voice digitizer to be multiplexed and enables many voice conversions to share the use of a common line. In comparison, the output of the voice adapter is an analog signal which cannot be integrated into the digital portion of a network.

6.15 The following diagram illustrates how two switchboards could be interconnected using voice digitizers and multiport modems to obtain three simultaneous voice channels on one leased line.

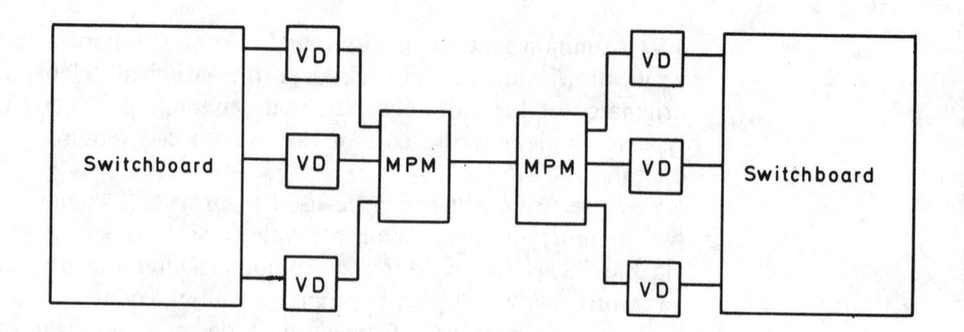

Legend:
VD – voice digitizer
MPM – multiport modem

If the voice digitizers operate at 2400 bps, a 2400 bps channel on the 9600 bps multiport modem is available for data transmission.

CHAPTER 7: INTEGRATING COMPONENTS

Network integration I

Since the pulse width of the terminals becomes distorted at a distance of 600 feet and the terminals are located 800 feet from the computer, directly connecting the terminals to the computer via a cable is not possible. Due to this we must compare the use of line drivers, multiport modems and eight-channel TDMs to link the terminals to the computer.

Multiport modems

Two pairs of multiport modems operating at 9600 bps could be employed, with each pair used as illustrated in the following.

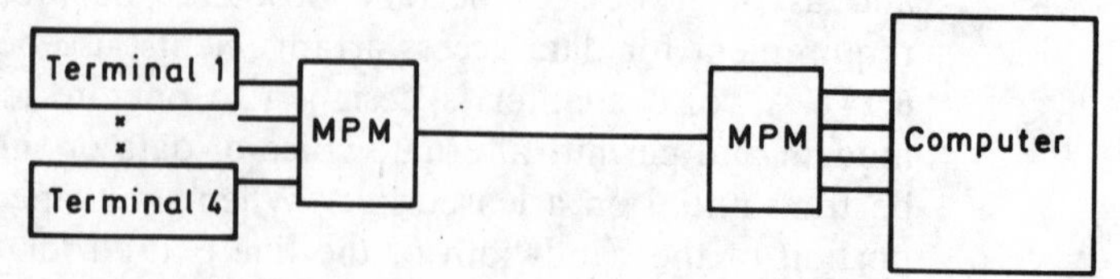

The cost of the multiport modem configuration would be:

4 modems at \$500	\$2000
2 cables × 800 ft × \$1/ft	1600
	\$3600

Line drivers

Each terminal could be cabled to the computer with the line driver inserted into the cable to boost the transmission distance as illustrated below.

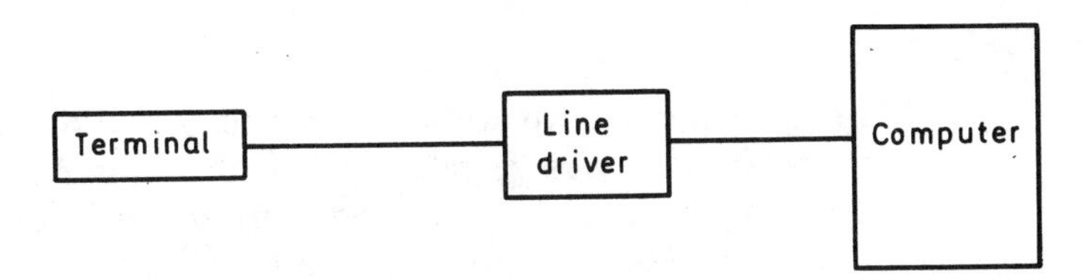

The cost of this configuration would be:

8 line drivers at $100	$ 800
8 cables × 800 ft × $1/ft	6400
	$7200

TDMs

Since the high-speed output of the TDM distorts after 450 feet, we must also use a line driver between TDMs as illustrated in the following.

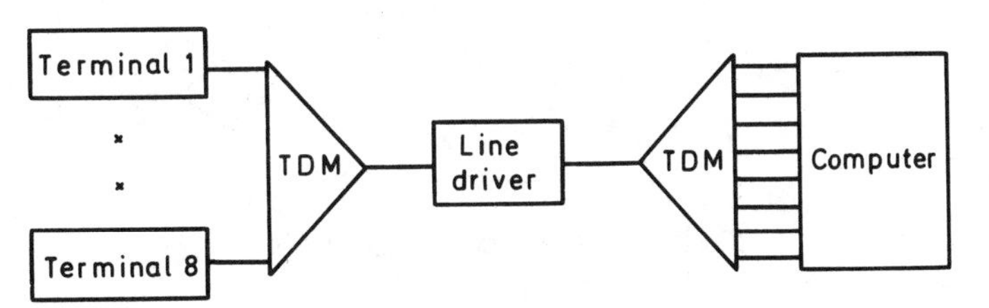

The cost of this configuration would be:

2 eight-channel TDMs at $1000	$2000
1 line driver at $100	100
1 cable × 800 ft × $1/ft	800
	$2900

From an economic standpoint, using TDMs is the best approach. Unfortunately, there are three critical communications components in this approach whose failure would cause all terminals to become inoperative.

At the other end of the spectrum, the highest cost solution of eight lines using line drivers only results in one terminal becoming inoperative if a critical communications component should fail. From this problem, one should understand that economics and the ability to communicate may at times be mutually exclusive.

Network integration II

A variety of network configurations can be designed to satisfy the requirement outlined in the problem. Potential solutions to the problem can be based upon the use of a conventional 4-channel TDM as well as the use of statistical multiplexers.

4-channel TDM

In this solution, we will assume that a 4-position rotary is installed at the multiplexer to receive calls originated by the terminal operators. Since each operator must use one telephone to accept incoming orders, ten additional telephones must be installed, one for each terminal. At the rotary, four dial-in lines and auto answer modems will be installed to service up to four simultaneous calls from the operator. The output of the 4-channel TDM will then be routed into one port of a 9600-bps modem while the second port of the modem will service the RBT. At the central computer site, five computer ports will be required, one for each TDM port and one for the RBT. The network schematic illustrating the 4-channel TDM solution follows.

REMOTE SITE

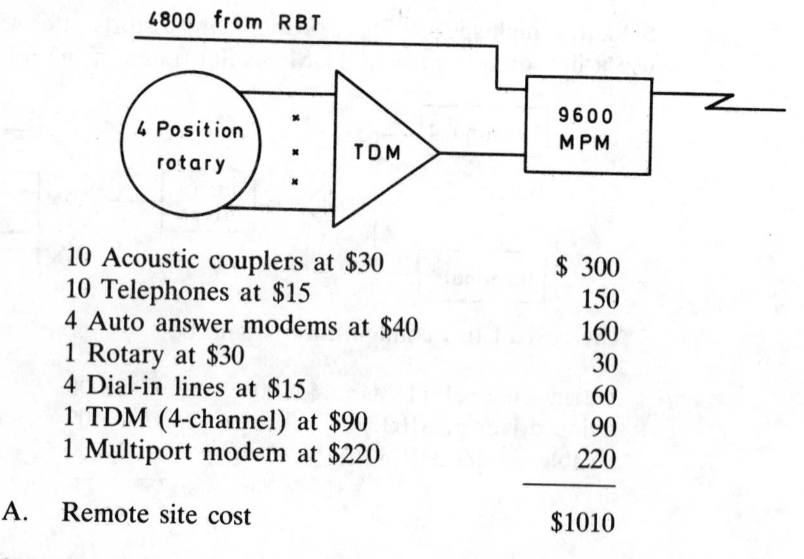

10 Acoustic couplers at $30	$ 300
10 Telephones at $15	150
4 Auto answer modems at $40	160
1 Rotary at $30	30
4 Dial-in lines at $15	60
1 TDM (4-channel) at $90	90
1 Multiport modem at $220	220
A. Remote site cost	$1010

CENTRAL SITE

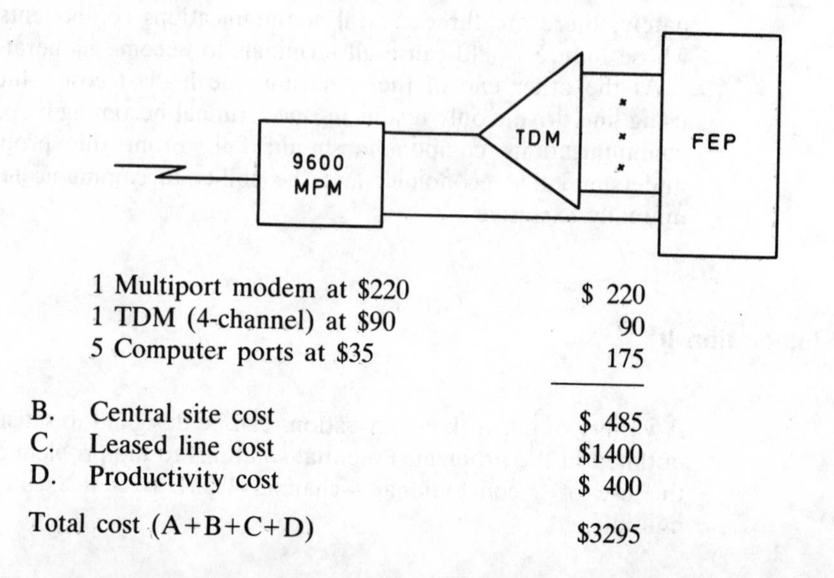

1 Multiport modem at $220	$ 220
1 TDM (4-channel) at $90	90
5 Computer ports at $35	175
B. Central site cost	$ 485
C. Leased line cost	$1400
D. Productivity cost	$ 400
Total cost (A+B+C+D)	$3295

8-channel TDM

Since an 8-channel TDM would operate at 9600 bps, an additional line would be required to service the RBT. Immediately we can recognize that this would add $1000 per month to our network cost. Since the cost of two statistical multiplexers does not exceed the cost of two TDMs by anywhere near $1000, we should examine the use of statistical multiplexers to ascertain if they will enable us to continue to use just one line. If so, we can exclude the use of 8-channel TDMs.

6-channel statistical multiplexer

We will examine the cost of using a 6-channel statistical multiplexer based upon the requirement of the problem where servicing six terminal users at one time does not result in any loss of productivity.

The use of a 6-channel statistical multiplexer by terminal operators is similar to the 4-channel TDM solution, with each terminal operator requiring a telephone and acoustic coupler. At the multiplexer, six dial-in lines and auto answer modems are now required. The only other configuration change is the ability of some statistical multiplexers to multiplex synchronous data, which we have assumed to be true here. This enables the RBT data to be multiplexed and eliminates the requirement for multiport modems. The network configuration and cost of this solution follows.

REMOTE SITE

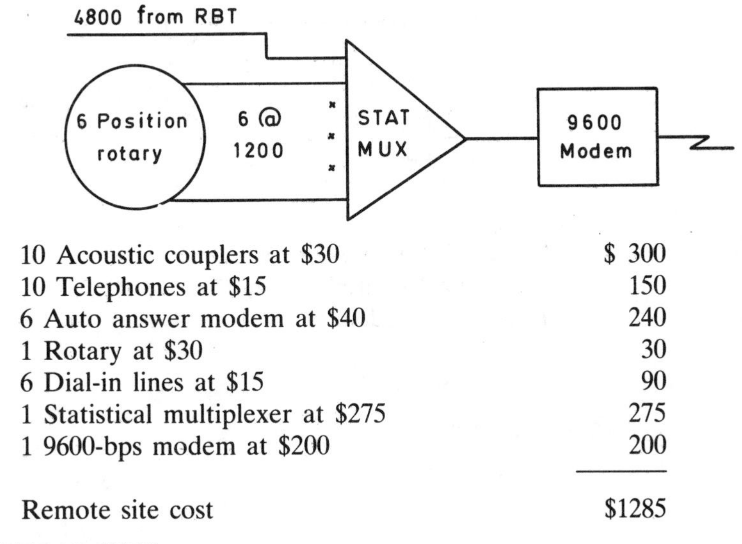

10 Acoustic couplers at $30	$ 300
10 Telephones at $15	150
6 Auto answer modem at $40	240
1 Rotary at $30	30
6 Dial-in lines at $15	90
1 Statistical multiplexer at $275	275
1 9600-bps modem at $200	200
A. Remote site cost	$1285

CENTRAL SITE

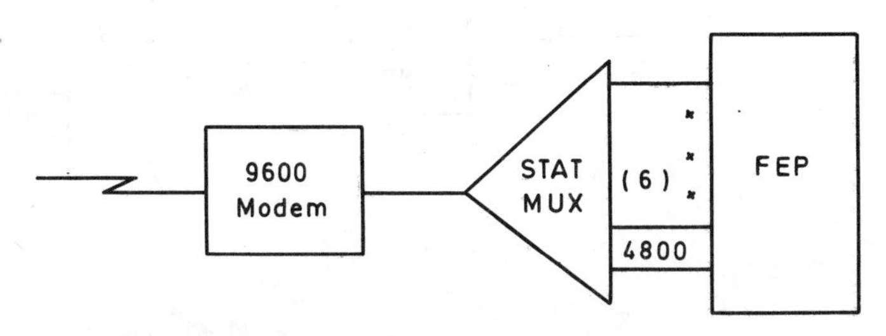

1 9600-bps modem at $200	$ 200
2 Statistical multiplexers at $275	275
7 Computer ports at $35	245
	$ 720
B. Central site cost	$ 720
C. Leased line cost	$1400
D. Productivity loss	$ 0
Total cost (A+B+C+D)	$3405

Line drivers

The last solution we will examine illustrates the effect of bypassing the telephone company whenever possible. In this solution, each terminal is directly cabled to a port on the statistical multiplexer, eliminating the requirement for acoustic couplers, dial-in lines, auto answer modems and a rotary. The network configuration and cost associated with the use of line drivers follows.

REMOTE SITE

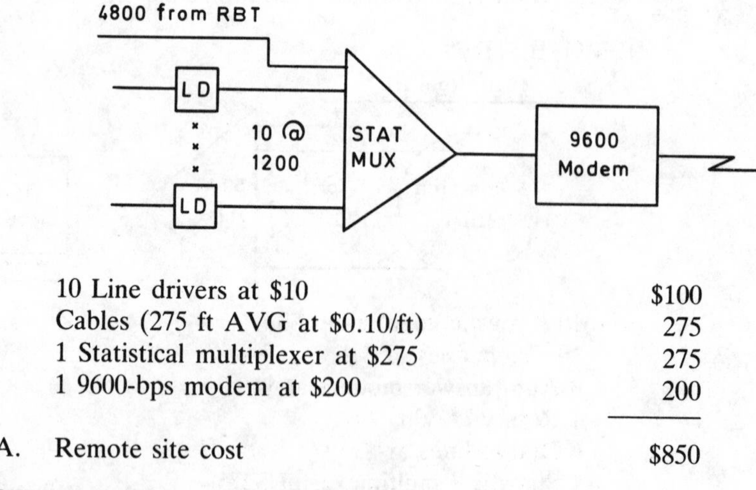

10 Line drivers at $10	$100
Cables (275 ft AVG at $0.10/ft)	275
1 Statistical multiplexer at $275	275
1 9600-bps modem at $200	200
A. Remote site cost	$850

CENTRAL SITE

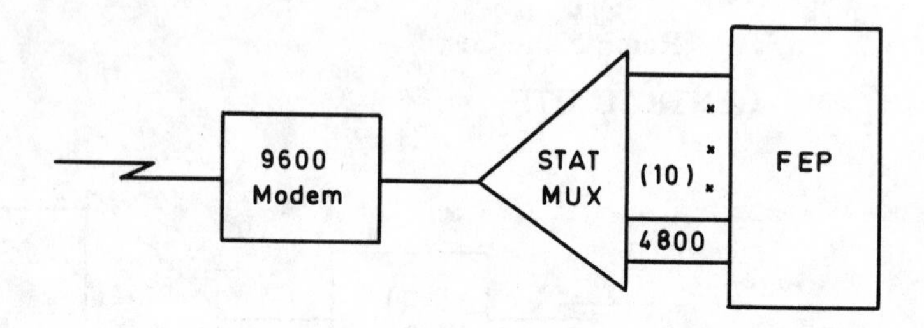

	1 9600-bps modem at $200	$ 200
	1 Statistical multiplexer at $275	275
	11 Computer ports at $35	385
B.	Central site cost	$ 860
C.	Leased line cost	$ 1400
D.	Productivity loss	$−100
Total cost (A+B+C+D)		$ 3010

Network integration III

The two more common methods that can be used to satisfy the requirements of the problem both involve the use of a tail circuit between cities B and C. The first solution involving the use of TDMs will require the installation of a new front-end processor since TDMs do not save computer ports. The second solution which involves the use of a modem sharing unit assumes poll and select software is available for operation of the existing front-end processor. If so, this solution alleviates the necessity of obtaining a second front-end processor.

Using TDMs

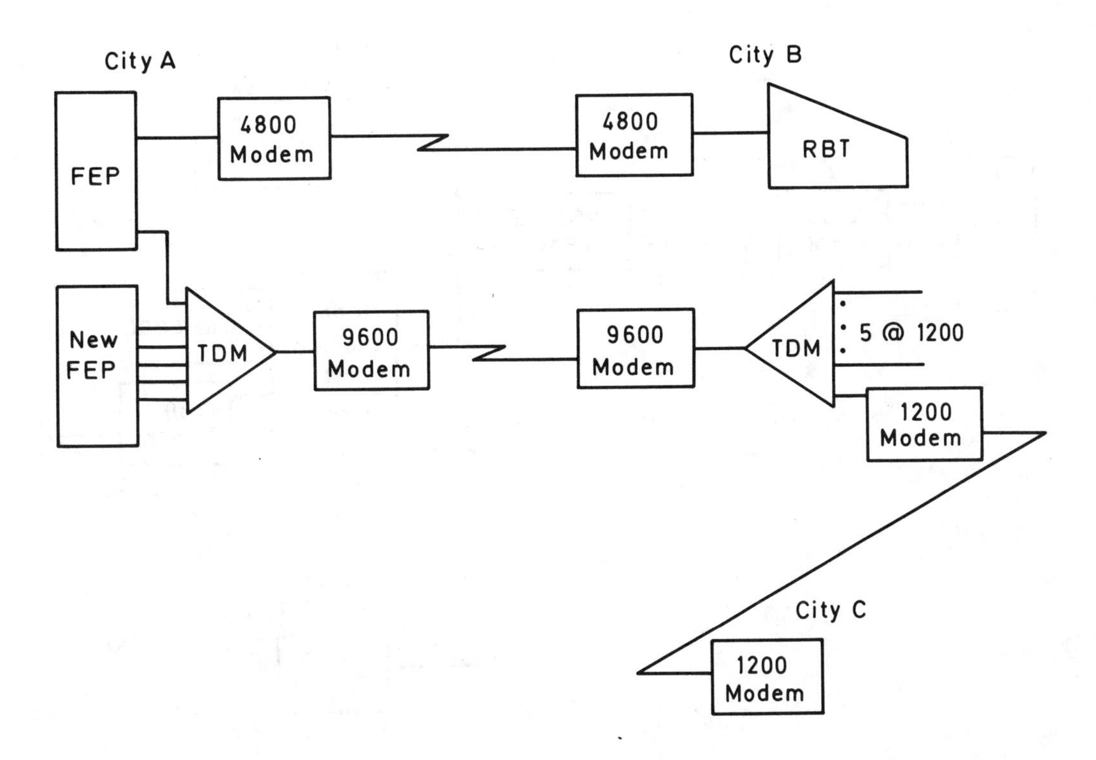

Monthly cost using TDMs:

Line cost	
2 between city B–city A at $200	$ 400
1 between city C–City B at $250	250
Modem cost	
2 1200 Modems at $40	80
2 4800 Modems at $120	240
2 9600 Modems at $200	400
Front-end processor (FEP)	
7 Channel (port) costs at $35	245
Additional FEP	1500
Terminal cost	
6 1200 bps at $100	600
1 Remote batch terminal at $500	500
Multiplexers	
2 8-channel TDM at $160	320
Total monthly cost	$4535

Clustering the terminals

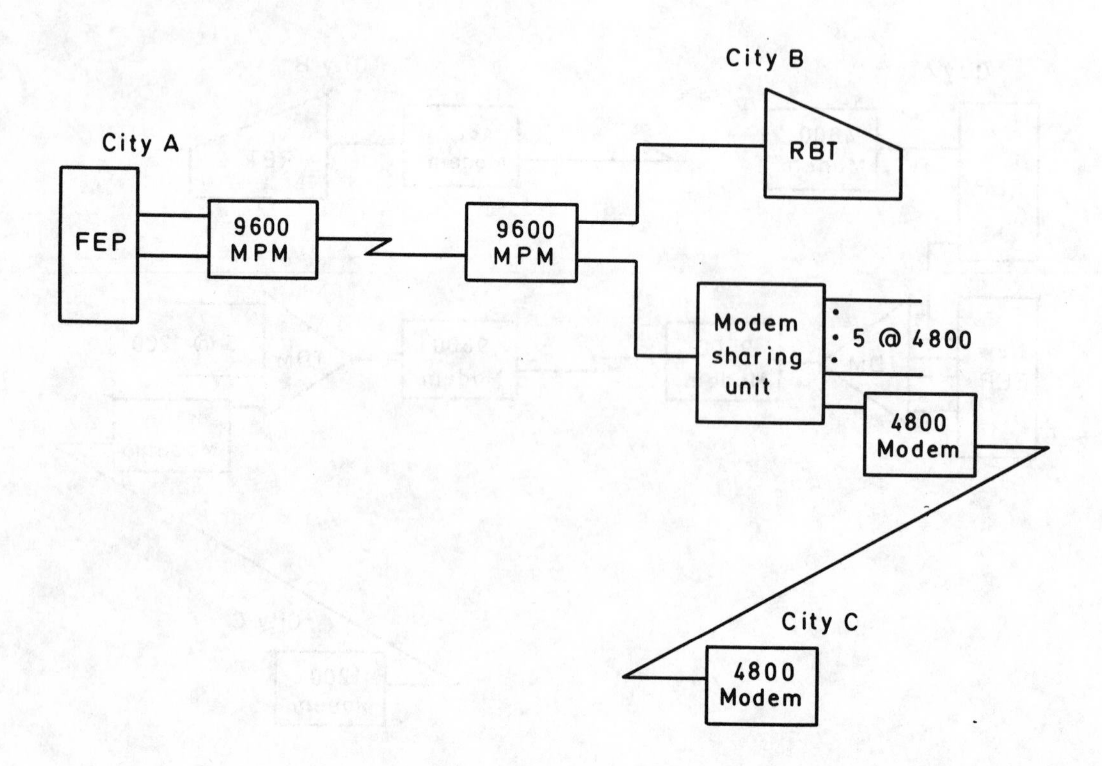

Monthly cost of clustering:

Line cost	
1 between city B–city A at $200	$ 200
1 between city C–city B at $250	250
Modem cost	
2 4800 modems at $120	240
2 9600 multiport modems at $220	440
FEP cost	
2 Channel (port) costs at $35	70
1 Modem-sharing unit at $35	35
Terminal cost	
6 4800mbps at $125	750
1 Remote batch terminal at $500	500
Total monthly cost	$2485

INDEX